· A ·
Mathematics
Sampler
Topics for
the Liberal Arts

FOURTH EDITION

William P. Berlinghoff
COLBY COLLEGE

Kerry E. Grant
SOUTHERN CONNECTICUT STATE UNIVERSITY

Dale Skrien
COLBY COLLEGE

ARDSLEY HOUSE, PUBLISHERS, INC., NEW YORK

Address orders and editorial
correspondence to:
Ardsley House, Publishers, Inc.
320 Central Park West
New York, NY 10025

ISBN: 1–880157–23–3

Printed in the United States of America

10 9 8 7 6 5 4 3 2 1

To Phyllis, Sandy, and Pam

Contents

PROBLEMS AND SOLUTIONS 1

MATHEMATICS OF PATTERNS: NUMBER THEORY 19

MATHEMATICS OF AXIOM SYSTEMS: GEOMETRIES   67

MATHEMATICS OF CHANCE: PROBABILITY AND STATISTICS   125

MATHEMATICS OF INFINITY: CANTOR'S THEORY OF SETS   228

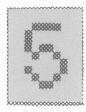

MATHEMATICS OF SYMMETRY: FINITE GROUPS   280

APPENDICES

Preface

This book grew out of a liberal-arts course at Southern Connecticut State University taken by nonscience majors, usually in their first year. Our experiments with the form and content of this course for more than a decade seemed to be most successful when several different mathematical areas were introduced to students and covered in some depth. From these experiments emerged text units, which ultimately became chapters in this book. Thus, *A Mathematics Sampler* is a collection of independent chapters covering a broad spectrum of mathematics, old and new, pure and applied, traditional and unusual. The course on which it is based was cited as an innovative approach to liberal-arts mathematics in Lynne Cheney's report, "50 HOURS: A Core Curriculum for College Students," published by the National Endowment for the Humanities.

Previous editions of *Sampler* have been used at more than 100 colleges and universities. Our revisions for this 4th Edition are based on the suggestions of many colleagues around the country who have been kind enough to share their thoughts with us, as well as on our own teaching experiences. Major modifications and additions are as follows:

■ Chapter 5 of the 3rd Edition, on computers and programming, has been replaced by a new chapter on computer science (Chapter 9 of this 4th Edition), featuring the Traveling Salesman Problem and the analysis of algorithms.

▨ Approximately 400 additional exercises, including 20 new writing exercises, have been added to the other eight chapters.

▨ The statistics part of Chapter 4 has been expanded to include much more material on various forms of data representation (bar graphs, stem-and-leaf plots, boxplots, etc.) and more examples of measures of center and spread.

▨ Chapter 3, retitled "Mathematics of Axiom Systems: Geometries" to reflect its axiomatic emphasis, includes an expanded discussion of Euclidean geometry.

▨ James O. Bullock's article, "Literacy in the Language of Mathematics," originally published in the *American Mathematical Monthly*, has been reprinted as Appendix C and is used as the basis for new writing exercises in several chapters.

Of course, we have also made many minor adjustments and corrections throughout the text, including the acknowledgement of significant mathematical events of the past few years, such as Andrew Wiles' proof of Fermat's Last Theorem.

WRITING ASSIGNMENTS

In recent years there has been widespread national interest in incorporating writing as a tool for learning science and mathematics. Writing assignments have always been an important part of the course as we have taught it. This text includes numerous writing exercises, many of them already class-tested by us. These exercises come in two forms:

1. At the end of virtually every section there is a short list of **Writing Exercises**, most of which call for answers of more than a sentence and less than a page. These questions are suitable for use as homework or as in-class exercises to help focus on the ideas at hand, or even for incorporation into course journals.

2. At the end of each chapter there is a section entitled **Topics for Papers**. These longer assignments require formal papers of 4–6 pages in length, or perhaps even more at times. They are intended to encourage the student to draw together the main ideas of the chapter and/or to extend those ideas to some independent exploration.

PREREQUISITES

The wide variety of student backgrounds in a first-year course for nonmajors makes it unwise to require significant skill or sophistication as a prerequisite. We assume only that the student has had a first course in high-school algebra and is familiar with some elementary geometry. This book will not remedy a student's lack of basic skills. On the other hand, it *will* reinforce those skills by drawing on them in solving interesting problems relevant to the understanding of real mathematics and a real world.

LINK SECTIONS

A primary function of an introductory mathematics course in a student's liberal education is to exhibit the interrelationships between mathematics and the various other fields of human knowledge. Too often, we mathematicians tend to think of those interrelationships merely as applications of mathematical methods and results to other subject areas. Although this is one kind of LINK between mathematics and other disciplines, there are many others—LINKS of analogy, of common historical influence, of personal achievement. These LINKS provide a sense of the underlying unity of all knowledge and shed some light on both the power and the limitations of mathematics as a tool for understanding the world of our common experience and as an area of individual aesthetic achievement. The LINK sections of *Sampler* connect specific mathematical ideas with many other subject areas. The connections examined are as varied as the fields explored—art, chemistry, coding, demographics, fiction, genetics, management, marketing, music, philosophy, politics, psychology, and social planning. These LINKS are intended to kindle a spark of interest, rather than to teach a technique, so they are brief, informal, and open-ended. They may be used for outside reading, as a basis for class discussions, or as leads for independent student projects.

SYNOPSES OF THE CHAPTERS

The book begins with an introduction to problem solving in mathematics, followed by an exploration of eight different mathematical topics. Each chapter after the first starts "from scratch" and proceeds to develop a significant mathematical idea, illustrating what mathematicians do in that area. *These eight chapters may be covered in any combination and in any order.* Some chapters have several natural stopping places to allow for either introductory coverage or in-depth exploration of the topic. Although the book contains enough material for a longer course, it is most frequently used as a one-semester text, thereby affording considerable flexibility in the choice of topics to be covered. Selections from three or four chapters, used in conjunction with Chapter 1 and the history appendix, provide ample material for a 3-credit course. Here are some suggestions for using the various chapters:

▪ **Chapter 1**, on problem solving, sets the tone for the entire book. We recommend it as an appropriate first chapter in any choice of topics.

▪ **Chapter 2** investigates the topic of perfect numbers as a way of exhibiting the power and elegance of number theory. The students see that analytical observation, educated guesswork, and trial-and-error investigation are good sources of mathematical conjecture, but that patterns can be deceiving. To be most effective, this chapter should be covered in its entirety, although some of the proofs along the way are easily omitted.

▪ **Chapter 3** uses geometry as a vehicle for examining the axiomatic structure of formal mathematics. After a brief overview of the principles of Euclidean geometry, the historical controversy over Euclid's Parallel Postulate is used as a steppingstone to consistency and independence results in formal axiom systems, culminating in a discussion of the modern view of the relationship between pure mathematical systems and real-world models.

▪ **Chapter 4** briefly treats set language and notation, which are then utilized in discussing probability questions. (Section 4.2 may be used independently as a review of or an introduction to basic set-theoretic ideas.) The next two sections present the fundamental idea of probability and some combinatoric principles. The chapter may be terminated after Section 4.5 if those ideas are sufficient for the needs of the course. Sections 4.6 and 4.7 provide a further exploration of probability, and LINK Section 4.8 is another natural intermediate stopping place. Sections 4.9 through 4.12 examine some fundamental ideas of statistics, including elementary data analysis, normal distributions, and the Central Limit Theorem.

▪ **Chapter 5** begins with the elementary notions of *set* and *1-1 correspondence*, and progresses to an explanation and proof of Cantor's Theorem. Discussions of paradoxes, the Continuum Hypothesis, and philosophies of mathematics are provided. The proof of Cantor's Theorem in Section 5.7 may be omitted without seriously affecting the integrity of the chapter. A brief treatment might bypass the computational techniques of Section 5.4 and end before the proof in Section 5.7, omitting Section 5.8 entirely.

▪ **Chapter 6** exhibits the power and utility of abstraction, as it pursues its goal of explaining Lagrange's Theorem. Tables for finite groups are used to examine properties of operations. Section 6.8, the proof of Lagrange's Theorem, may be omitted without disturbing the continuity of the chapter. A brief treatment could end with Section 6.5 or 6.6.

- **Chapter 7** introduces four-dimensional geometry by way of analogy with lower dimensions. The earlier sections provide practice and review of the concepts of *coordinate*, *interval*, *path*, and *distance* in 1-, 2-, and 3-space, which are generalized so that visual intuition may be replaced naturally by coordinate-algebraic techniques in making the transition to 4-space. Sections 7.6 and 7.7 extend these ideas to several different types of figures; these sections may be omitted, if desired. The LINK section on 4-space in fiction and art should be covered in any event.

- **Chapter 8** investigates the twin existence problems of an Euler path and a Hamilton circuit in a graph. The problems are easily understood and appear to be quite similar in form. The radical dissimilarity of results, by contrast, provides an instructive insight into the nature of mathematics. The LINK section illustrates how graph theory is applied to project management.

- **Chapter 9** assumes no prior computer knowledge. It begins with a historical summary of computing machines, and then introduces students to computational complexity. The Traveling Salesman Problem is used throughout as a vehicle for the discussion of complexity theory. By comparing two simple sorting techniques, students are led into more general issues, including the difference between polynomial-time and exponential-time algorithms. The chapter culminates in a discussion of NP-Completeness. A brief treatment could end after Section 9.6.

- **Appendix A** is a self-contained, concise reference unit on the basic principles of mathematical logic. It can be taught as a short unit in its own right, it can be used as an independent study assignment, or it can just be consulted, as needed, for definitions or examples during the course.

- **Appendix B** is a compact history of mathematics as it relates to the major events that shaped the development of Western civilization, from the earliest-known evidence of mathematical achievements through the latter part of the twentieth century. The authors recommend it as an early outside-reading assignment. This material can then be used throughout the course as a historical context in which to place the ideas covered in the other chapters.

- **Appendix C** contains an essay on the role of mathematics as an integral part of a liberal education. Its author, James O. Bullock, is a research physiologist at the Cancer Research Center in Columbia, Missouri. Beginning with the assertion that mathematics is a language, he argues that much of the power of mathematics stems from the metaphors it presents. Bullock's thought-provoking opinions in this essay provide fertile soil for class discussions and/or writing projects.

- Answers and/or hints to most of the odd-numbered exercises are provided at the back of the book.

INSTRUCTOR'S MANUAL

An Instructor's Manual (*ISBN*: 1–880157–50–0), which is provided to adopters of this text, is divided into four major sections:

- A brief overview of each chapter, including suggestions, based on the authors' experiences in teaching this material, on ways to use the chapter in a course.

- Answers to all the exercises, section by section.

- Two sample tests for each chapter (except Chapter 1), each designed for fifty minutes of class time. (The suggested percentage values for the questions are based mainly on the time required to complete the answers.)

- Answers to the sample tests.

ACKNOWLEDGEMENTS

As preparation of this text progressed through various stages, many people provided help and encouragement. We are grateful to Ross Gingrich, Michael Meck, Dorothy Schrader, Michael Shea, J. Philip Smith, and other colleagues at Southern Connecticut State University for their expert advice, to Martin Zuckerman of The City College of the City University of New York for a wealth of helpful suggestions, and to Joseph Moser and his colleagues at West Chester University for their comments. We offer special thanks to H. T. (Pete) Hayslett, Jr. and Dexter Whittinghill of Colby College for their valuable assistance during the expansion and reworking of the statistics sections of Chapter 4, to Krishna Kumar of Colby for his constructive commentary during the development of Chapter 9, and to Claudia Henrion of Middlebury College for steering us to relevant material on women in mathematical history.

Finally, we wish to express our appreciation to the many students in MA 103 at Southern Connecticut and MA 111 at Colby, whose struggles, insights, questions, and criticisms have made this book much more user-friendly for the students still to come.

We gratefully acknowledge the many helpful suggestions and criticisms of each of the following users of the 3rd Edition:

Jules Albertini, Ulster Community College
Peter Anspach, Lake Forest College
Thomas L. Bartlow, Villanova University
Joe Broderick, National Security Agency
Rotraut G. Cahill, University of Wisconsin Center—Washington County
Garnet Crites, West Virginia University, Parkersburg
Karl Dilcher, Dalhousie University
Mary Jefferson, Rockford College
Leo Andrew Lusk, West Virginia University
Jerald McCloy, Eastern Nazarene College
Chris L. Morin, Blackburn College
Lesley O'Connor, Queens College, Charlotte
Jean Probst, Macalester College
Randy Pruim, Calvin College
Carla Schultes, Salisbury State University
Joseph Straight, State University of New York, College at Fredonia
Carolyn Sur, Mount Mary College
Norman Sweet, State University of New York, College at Oneonta
Daniel Ullman, George Washington University
Ian Walton, Mission College
Herbert Wilf, University of Pennsylvania

To the Student

As you begin this topical tour of mathematics, we who seek to guide you have a few words of advice and perspective to offer. The mathematics in this book is not highly technical, but the concepts are often challenging. Mastery of them will at times test your patience and perseverance. We hope you will find our writing style comfortable and our explanations clear. But do not expect to read this book—or any mathematics book—like a mystery novel. Expect to read and reread thoughtfully. Build a habit of having paper and pencil at hand to answer questions raised in the text, to work through exercises, and to create examples of your own. Examine your own understanding frequently. Do not go on to new material until you understand the old, or at least until you know exactly what it is that you do not understand. And do not hesitate to ask questions of your instructor!

A word of warning: *Just doing the assigned exercises is not your main job in this course!* If you regard any text material that does not explain "how to" as superfluous, you have fallen into a dangerous trap. The *primary* purpose of this book is not to refresh old mathematical techniques or to teach you new ones, even though techniques are used in the course. Rather, we seek to show you some mathematical ideas you may not even regard as part of mathematics. The text discussions you will read and the exercises you will do are important details in these larger pictures, like pieces of a jigsaw puzzle that make little sense until you begin to see the outlines of the picture as a whole. As you work your way along, then, spend some time thinking about how these details fit into the larger picture of the topic you are studying. When you recognize and understand the broad view of the landscape, you will have arrived at your destination. We hope you enjoy the trip!

CHAPTER

1

PROBLEMS AND SOLUTIONS

WHAT IS MATHEMATICS?

This book is written by three mathematicians. Because we are mathematicians and because mathematicians are fond of precise definitions and careful logic, we would like to begin by defining "mathematics" for you. Then we could proceed to unfold that definition to exhibit in logical sequence the various topics covered in the rest of the book. We would like to do that—but we can't. Mathematics, a subject utterly dependent on definitions and logic, cannot itself be defined in a clear, comprehensive sentence or two, or even in a short paragraph. In fact, mathematicians themselves often delight in giving examples to show that any proposed definition of their subject is deficient in one way or another. The many diverse ideas, methods and results of modern mathematics defy simple description. Faced with this ironic situation, how, then, are we to make any sense at all of the label "mathematics"?

Let us first look at what mathematics has been. At different times and different places in the early history of mankind, civilization brought with it the need for counting and measuring. Traders needed to tally their goods and profits; builders needed to cut their stone and wood to exact sizes; landowners needed to mark precisely the edges and corners of their fields; navigators needed to fix their positions by using the stars to compute distances and angles; the list goes on and on. These specialized tasks gave rise to skilled craftsmen of various

1

Accounting by the Passamaquoddy Division of the Abnaki Tribe.

Picture Collection, The Branch Libraries, The New York Public Library

sorts—counting-house clerks, draftsmen, surveyors, astronomers, etc. Their crafts, though different, had some common features: They all used numbers and/or basic shapes, and they all sought precise quantitative descriptions of some part of their world. This array of specialized counting and measuring techniques was all of mathematics for the first several thousand years of recorded history.

Greek philosophical speculation during the five or six centuries before Christ brought about a major change in mathematics. As the Greeks tried to solve the mysteries of truth, beauty, goodness, and life itself, they turned human reason into formal logic and then applied that logical system to their observations of the world. Their belief that logic was the key to understanding reality led them to apply it to all areas of thought, especially to mathematics. As they systematically recast the techniques of counting and measuring into logical systems based on "self-evident" assumptions about the real world, mathematics changed from craft to science. Complex mathematical statements were deduced from simpler ones, and they were all applied to solve the problems of the physical world. Architecture, surveying, astronomy, mechanics, optics—all these fields grew out of mathematics applied to solve real-world problems, and those solutions provided observable confirmation that the physical world behaved as mathematics predicted it would. Thus, for more than two thousand years mathematics was regarded as the "queen of sciences," the only reliable means for finding scientific truth.

Then came the nineteenth century. The development of the non-Euclidean geometries in the early 1800s provided mankind with three separate geometric systems, each one logically correct by itself, each one providing a perfectly good way of describing physical space, and

❡ Ein ander Exempel.

Ein Fraw oder Haußmutter gehet auff
den marckt / kaufft vberhaupt ein Körblin
mit Rebnerbyrn / darumb gibt sie achtzehen
pfenning / so sie heim kompt / findet sie im
körblin hundert vnd achtzig byrn / Ist die
frag / wie vil byren sie vmb ein pfenning ha=
be? Thu / als obgelert / so kompt dir zehen /
Also vil byren hat sie vmb einen pfen=
ning / Vnd ist wolfeyl
drumb.

The Problem of the Market Woman. From Köbel's
Rechenbüchlein of 1514 (1564 edition). Traders needed
to tally their goods and profits.

each one contradicting the other two! If mathematics were truly a science—that is, if mathematical truths were experimentally verifiable
facts about the real world—then this paradoxical situation could never
occur. Mathematics, therefore, must be a study that is essentially independent of the real world, and hence its pursuit need not be bound
by human experience. Mathematicians suddenly found themselves
free to explore any questions their minds could ask, unhampered by
worries about whether their results had anything to do with reality.
They found themselves in "a world of pure abstraction;. . . 'the wild-
ness of logic' where reason is the handmaiden and not the master."[1]
Mathematics became a subject essentially different from the (other)
sciences: "In other sciences the essential problems are forced upon the
subject from external sources, and the scientist has no control over the
ultimate end. The mathematician, however, is free to prescribe not
only the means of realizing the end, but also the end itself."[2]

Mathematics responded to this newfound freedom with a mush-
roomlike expansion in all directions. Results began to appear at such a
rate that by the 1990s, mathematicians were publishing more than
200,000 *new* theorems every year! Some of these theorems solve old
problems, but most of them solve new ones, problems suggested by results published only a short time earlier or even suggested for the first
time in that same paper. Some of these theorems are profoundly signi-

---

1. Marston Morse, "Mathematics and the Arts." *The Yale Review*, 40 (4): 1951, p. 612.
2. Raymond G. Ayoub, in a book review of *Mathematics: The Loss of Certainty. MAA
   Monthly* 89 (9): 1982, p. 716.

ficant, some are utterly useless, most are somewhere in between—and, with few exceptions, no one yet knows how to tell the difference. Modern mathematics has become an art, the art of posing and solving problems by logical reasoning. To the "pure" mathematician, the significance of these problems outside of mathematics is irrelevant; it is enough to seek them out and put them to rest.

Thus, the bewildering variety of topics currently labeled as "mathematics" is unified by a common theme—problem solving. Of course, solving problems is not limited to mathematics; it is part of almost every human activity. Nevertheless, the problem-solving process is most clearly defined in the simple, abstract world of mathematics. It is at the heart of all mathematical activity—from primitive tallying and measuring, through its many scientific applications, to its most modern abstract speculations. All the mathematical topics you will see in the later chapters of this book are united by this one common bond. There are many differences among them, differences not only in the topics themselves, but also in the ways that they are approached. Different mathematicians study different topics in different ways, and if you are to learn something about mathematics as a whole, then you must come to terms with this basic fact. But you should also see that all these different things are appropriately labeled "mathematics" because they are all held together by a common bond. That bond is a way of thinking, a way of using reason to approach and solve problems of any sort. In the rest of this chapter we examine that way of thinking—the problem-solving process itself.

1.1   EXERCISES

WRITING EXERCISES

1. Write a brief definition of "mathematics" that is broad enough to include all the mathematics you have seen in elementary school and high school, but is specific enough to exclude biology, music, and social studies. Keep this definition until the end of the course; at that time, check to see if it includes all the mathematics you have covered.

2. Look up the definition of "mathematics" in a dictionary, and comment on whether or not it covers all the mathematics you have seen in elementary school and high school.

3. Does the term "quantitative thinking" encompass all of mathematics? Why or why not?

4. Appendix C contains the essay, "Literacy in the Language of Mathematics," by James O. Bullock. In the first part of this essay, the author asserts that mathematics is essentially a language. Does his view conflict with this section's claim that mathematics is characterized by problem solving, or do the two viewpoints complement each other? Read the first part of the essay; then write a paragraph commenting on this question.

PROBLEM SOLVING

The word *heuristic* is often used to describe any method of discovering or inventing something, or any approach to solving a problem or finding a proof. Writings on heuristics date back to early Christian times. Pappus, a third-century Greek geometer, devoted quite a bit of attention to problem-solving methods. In the seventeenth century, Descartes and Leibniz both tried to formulate systematic treatments of heuristic processes. However, in later years, intellectual activity became more specialized, and this odd mixture of logic, philosophy, and psychology was largely neglected. Recently there has been renewed interest in problem solving, especially among mathematicians and mathematics educators. Much of this interest is due to the remarkable work of George Pólya (1888–1985), a Stanford University mathematician who spent much of his long professional life teaching and writing about heuristic methods. The first of his several books on the subject appeared in 1945, a small, insightful volume entitled *How to Solve It*. Much of the material in the rest of this section is based on that book.

Pólya begins with a four-part overview of how to solve any problem:

- ■ Understand the problem.
- ■ Devise a plan.
- ■ Carry out the plan.
- ■ Look back at the completed solution.

Taken all by itself, this list is of little use, except perhaps to say, "Slow down! A worthwhile problem is not likely to be solved at first glance." But if we examine each of the four parts, expanding on them a bit, a truly valuable scheme to help any problem solver in any situation emerges. Let us look at each part in turn.

Early Navigation. The Roman Navy, from Livy, Venice, 1493. Navigators needed to fix their position by using the stars to compute distances and angles.

Early Methods of Measuring Distance. From Belli's *Libro del Misvar con la vista*, Venice, 1569. This was essentially the method used by the ancient Greeks.

Understand the Problem

Exactly what are we trying to find or prove? What facts are at our disposal? What parts of the context of the problem might be relevant to what we are trying to find or prove? In order to answer these questions you will have to *check the definitions* of all the words used in the problem to make sure you understand them thoroughly. A useful way to check your understanding is to try to *restate the problem* in as many different forms as you can think of. Sometimes it is helpful to *draw a diagram* to give your visual intuition an image of the problem as a whole. You might also want to *introduce appropriate notation* to symbolize essential pieces of the data or the question for ease in remembering and working with them.

Devise a Plan

This is the hard part, but usually also the part that's the most fun. The main trouble here is having to cope with the uncertainty of not knowing what to do first, or how long the job will take, or even if you will be able to solve the problem at all. In this aspect of problem solving, the main difference between college students and professional mathematicians is not that the mathematicians are brighter or quicker, but rather that they have developed more patience. Pólya provides some comforting advice for all of us:

> The first rule of discovery is to have brains and good luck. The second rule of discovery is to sit tight and wait till you get a bright idea.[3]

Of course, while you "sit tight," there are many things you can do to help you get a bright idea. You can *look for a pattern* in the data you

---

3. George Pólya, *How to Solve It*, 2nd ed. Princeton: Princeton University Press, 1957, p. 172.

have. If you don't have much data, you might *construct examples* to generate some. You can find a similar problem that has already been solved and try to *argue by analogy.* You might try to solve a *simpler problem* by assuming additional useful information, then try to generalize that solution. If the problem asks you to find something, you might *approximate the answer,* then check to see how far off you are and why; some approaches of this kind are called "trial-and-error" methods. If the problem asks you to prove something, it is often useful to *reason backwards from the desired conclusion.* Finally, you might check to *see if you used all the data.* Remember: none of these techniques is guaranteed to produce a plan for solving the problem, but some combination of them usually works *if* you have the patience and perseverance to work at them until you get the right "bright idea."

Carry Out the Plan

Once you have devised a plan to solve the problem, you must carry it out step by step, checking each step carefully to see that you have not made a reasoning error or introduced some unjustified assumption. This part of the problem-solving process is not nearly as much fun as the excitement that comes when you first see your plan, but it must be done to insure that you actually have a correct solution. (It's like having to split the firewood after cutting down the tree.) Sometimes there are surprises in store for you at this stage; sometimes the plan turns out to be less complete or accurate than you first thought. Good plans often yield dividends when you carry them out; the step-by-step logical development may turn up unexpected connections with other ideas, other problems. And in the event that your plan is wrong in

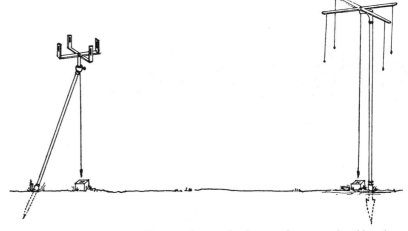

Surveyor's Tools. The Nineteenth Century Surveyor's Cross (on the left) and the Roman First Century A.D. Groma. Landowners needed to make the precise boundaries of their fields.

Picture Collection, The Branch Libraries, The New York Public Library some fundamental way, learning to cope with the disappointment and frustration as you force yourself to search for another plan is itself a valuable lesson.

Look Back at the Completed Solution

This is a critical part of the process for any researcher or serious problem solver. Once a problem is solved, it is sometimes possible to find an easier way to comprehend the whole situation. Often you can check your solution for hidden errors that way; you might also check it by finding a simple approximation of the answer. Looking back also gives you a chance to see if you can *generalize the solution* to get a stronger result, and to look for any new questions suggested by your solution.

For convenient reference, we provide here a summary list of the dozen specific problem-solving techniques discussed in this section. You might find it useful to refer back to these suggestions as you wrestle with the questions and exercises in the other chapters.

- Check the definitions.
- Restate the problem.
- Draw a diagram.
- Introduce appropriate notation.
- Look for a pattern.
- Construct examples.
- Argue by analogy.
- Solve a simpler problem.
- Approximate the answer.
- Reason backwards from the desired conclusion.
- See if you used all the data.
- Generalize the solution.

1.2   EXERCISES

Exercises 1–9 refer to Problem A, which follows.

   Problem A: What is the sum of the measures of the angles of a decagon?

1. List and define all the important terms in Problem A.

2. Draw a diagram for Problem A, labelling important features.

3. Restate Problem A using the diagram and labels of Exercise 2.

4. Find two simpler problems analogous to Problem A. Explain how they are analogous.

5. Solve the two problems you found in Exercise 4.

6. Devise two different plans for solving Problem A.

7. Solve Problem A.

8. Generalize Problem A.

9. Solve the generalized problem you stated in Exercise 8.

Exercises 10–16 refer to Problem B, which follows.

Problem B: Prove that the product of any three consecutive integers is a multiple of 6.

10. List and define all the important terms in Problem B.

11. Restate Problem B in a different form.

12. Construct four different examples that satisfy the conditions of Problem B.

13. Reason backward from the conclusion in Problem B: What might you prove about the product of the integers that would allow you to conclude that this product is a multiple of 6?

14. Devise a plan for solving Problem B.

15. Solve Problem B.

16. Formulate two problems analogous to Problem B. Explain how they are analogous.

Exercises 17–21 refer to Problem C, which follows.

Problem C: A piece of music lasting $22\frac{1}{2}$ minutes is recorded in a single band on a $33\frac{1}{3}$ r.p.m. phonograph record. When the record is played, the music starts when the needle is 30 centimeters from the center of the record and stops when it is 5 centimeters from the center. If the groove containing the recorded music were in a straight line, how long would the line be?

17. List and define all the important terms in Problem C.

18. Draw a diagram for Problem C. (A diagram need not be a work of art.)

19. Restate Problem C in a different form.

20. Estimate the answer to Problem C.

21. Determine whether your estimate in Problem 20 is likely to be greater than or less than the correct answer. Explain.

Exercises 22–26 refer to Problem D, which follows.

Problem D: Assuming that the Earth is a sphere with radius 6370 kilometers, what is the average width of a time zone at 60° North Latitude?

22. List and define all the important terms in Problem D.

23. Draw a diagram for Problem D. (You may find that several related diagrams are desirable.)

24. Restate Problem D in a different form.

25. Reason backwards: What intermediate problem or problems could you solve along the way to solving Problem D?

26. Devise a plan for solving Problem D.

WRITING EXERCISES

1. What does *analogy* mean? Specifically:

    (a) Give an example of an analogy (drawn from anywhere, not necessarily from mathematics) and use it to help you explain *analogy* as if you were talking to a high-school junior.

    (b) What is the difference between an analogy and a metaphor? (They are not exactly the same.)

2. Explain the meaning of the word *pattern* as it is used in reference to problem solving. Go beyond a mere dictionary definition to clarify what is meant by talking about a pattern of ideas. Provide an example to illustrate your explanation.

3. Do you think an approximate answer to a quantitative question might ever actually be *better* than an exact one? Why or why not? Illustrate your answer with an example, if appropriate.

IT ALL ADDS UP

1.3

We have spoken in general terms about problem solving in the first two sections. Let us now see how some of the suggested techniques can be used to solve a problem. Here is a problem that is based on a very elementary arithmetic idea—adding whole numbers. The problem is easy to state, but it is not so easy to solve.

(1.1)   Find the sum of all positive integers less than 1,000,000.

First, let's make sure that we *understand the problem*. Although there is nothing complicated or tricky about this problem, it would be a mistake to ignore this first phase of problem solving. Probably, the greatest source of error or failure in problem solving is failure to understand the problem.

In this case we need to *check the definitions* of the terms: *sum*, *positive integers*, and *less than*. "Sum," of course, means the result of addition; so we know we are dealing with an addition problem. We recall that the "positive integers" are another name for the natural numbers, 1, 2, 3, 4, ... It is not easy (and not necessary) to give a precise definition of "less than" here; it is important that we know what it means. In particular, we know that the largest number we will include in our sum is 999,999 (not 1,000,000).

We can *restate the problem*, then, as, "What is the sum of all positive integers from 1 to 999,999, inclusive?" Or we might state it using the notation of arithmetic:

Find $1 + 2 + 3 + \ldots + 999,999.$

The suggestion of *drawing a diagram* might seem inappropriate for this problem because it deals only with numbers and their sum. Frequently, though, numerical problems translate into geometric ones, or vice versa. In this problem, picture a side view of a staircase with 999,999 steps, with each step 1 unit by 1 unit. (See *Figure 1.1*.) Because 1 square unit is under the first step, 2 square units must be under the second step, and so on, we can restate Problem (1.1) geometrically by asking

(1.2)   What is the area of the figure pictured in *Figure 1.1*?

We now turn to the second phase, *devising a plan*. You might think that the plan is predetermined in this problem; there's only one way to add, isn't there? Not at all. We might plan to write down all 999,999 integers on paper and add them by hand, or we might plan to punch the numbers into a hand calculator, or we might plan to program a

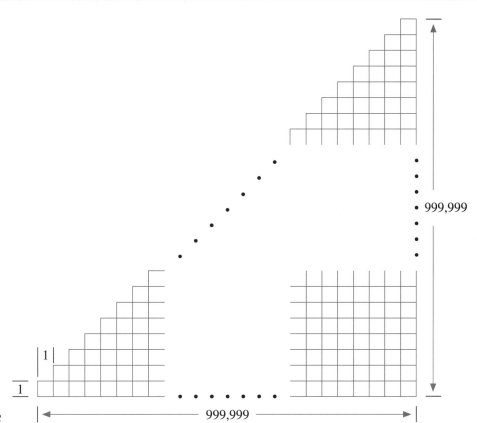

Figure 1.1 Problem: How many 1-by-1 squares are represented by this figure?

computer to do the problem. The first two plans would take an enormous amount of time, and the third, though practical, relies on machinery and skills we will not presume to have at our disposal.

Several of the suggestions in the preceding section apply here. We could *solve a simpler problem* by changing the upper limit from 1,000,000 to something more manageable. Maybe we can learn from some trivial analogous problems, like the following:

(1.3)

$$1 + 2 = 3$$
$$1 + 2 + 3 = 6$$
$$1 + 2 + 3 + 4 = 10$$
$$1 + 2 + 3 + 4 + 5 = 15$$

We can *look for a pattern* in these problems and their answers. Notice in Display (1.3) that 6 is 3 greater than 3, 10 is 4 greater than 6, and 15 is 5 greater than 10. This is a pattern that we can predict to continue; it simply reflects the fact that the next successive (positive) integer is added in each step. However, that doesn't appear to be much help in devising a plan (other than the ones we have already identified).

We might *approximate the answer* by using the fact that we are adding, roughly, 1,000,000 integers whose average magnitude is about 500,000 (half of 1,000,000). This would suggest an approximate answer of 1,000,000 times 500,000, or 500,000,000,000. This is an interesting piece of information; but it seems unlikely that we can count on it to be an exact answer, and we have no way to determine the size of the error in the approximation.

Let us examine our reasoning in finding this approximation, though, and see if we can develop a precise answer. Our approximation is the product of two quantities—the number of integers being added and the average of these numbers. If we had precise values for these two quantities, our answer would be precise. Here is a new plan, then!

We know the number of integers in the sum; it is exactly 999,999. But what is the *exact average* of all the numbers being added? The usual way to find an average of a set of numbers is to add them up and divide by the number of numbers in the set. Thus, the average of 2, 5, and 11 is 6 because $(2 + 5 + 11) \div 3 = 6$. But we don't know the sum of these numbers; if we did, we would already have our answer, and we wouldn't need the average. On the other hand, we are not dealing here with some arbitrary collection of numbers, but with a set that has a *very special pattern*.

Let us turn to *a simpler problem* and look at the problems in Display (1.3), where we do know the sums and can easily figure out the averages. These are recorded in *Table 1.1*.

Here we see a striking pattern, but one that is hardly surprising when we think about it. As the number of integers increases by one, the average increases by one half. Another way to state the relationship, in a general form, is to say that the average of the integers 1, 2, 3, . . . , n is simply the average of the first and the last numbers, that is,

$$\frac{1 + n}{2}$$

Now we have a plan that can be carried out in a reasonable amount of time, without a computer. We *carry out the plan* by computing the average of the integers 1, 2, 3, 4, . . . , 999,999, which is

$$\frac{1 + 999,999}{2}, \text{ or } 500,000$$

Then we multiply by the number of integers, 999,999, and obtain the desired answer, 499,999,500,000.

If we *look back at the completed solution*, we can easily *generalize this solution* to the sum of the set of all positive integers from 1 to n,

Numbers	Average
1, 2	$1\frac{1}{2}$
1, 2, 3	2
1, 2, 3, 4	$2\frac{1}{2}$
1, 2, 3, 4, 5	3

Table 1.1 Some simple averages.

inclusive, where n can represent any positive integer. In arithmetic notation,

(1.4)

xxxxxxxxxxxxxxxxxxxxxxxxxxxxxxxxxxxxxxxxxxxxxxxxxxxxxxxxx

$$1 + 2 + 3 + 4 + \ldots + n = \frac{n \cdot (n + 1)}{2}$$

xxxxxxxxxxxxxxxxxxxxxxxxxxxxxxxxxxxxxxxxxxxxxxxxxxxxxxxxx

As you reflect back on this problem and its solution, notice that the "obvious" plan of solution is not at all the simplest. And notice that one of the suggested techniques (solving a simpler problem) proved to be helpful, *not* the first time we used it, but, rather, the second time. Perseverance pays off; be open to trial and error *and retrial*.

Problem: *How many squares are determined by the lines on a standard chessboard?* Illustration from Jacobus de Cessolis, *Game of Chess*, Westminster, about 1483.

1.3   EXERCISES

1. Devise a plan to solve the problem of this section, as restated in its geometric form (1.2). (*Hint*: Try working with two copies of *Figure 1.1*.) Carry out the plan and solve the problem.

2. What is the sum of all positive integers from 1000 to 100,000, inclusive?

3. Generalize the problem of Exercise 2 and solve your generalized problem.

4. How many squares are determined by the lines on a standard chessboard?

5. Generalize the problem of Exercise 4 to a square analogous to a chessboard. Restate this in arithmetic terminology.

6. Generalize the problem of Exercise 5 to a three-dimensional analogue. Restate the generalization in arithmetic terminology.

7. How many rectangles are determined by the lines on a standard chessboard?

8. Give two different generalizations of the problem of Exercise 7.

9. Fifty points are marked on a circle, and all possible chords are drawn with these as endpoints. How many chords are there?

10. Generalize the problem of Exercise 9 and solve this generalized problem.

11. (The Towers of Hanoi Problem) In a hidden monastery near Hanoi, embedded in the floor of a room are three silver spikes; on one of them there are sixty-four gold disks, shaped like large washers, which are graduated in diameter so as to form a conical tower. The monks work tirelessly at the task of transferring the disks to re-form the tower on a different spike, according to the following rules: Only one disk at a time may be moved; each disk must be on one of the spikes except when being moved; and a larger disk may not be placed on top of a smaller one. What is the minimum number of moves that the monks must make to complete their task? Assuming that they make no wasted moves and move one disk every second, how long will it take them to finish?

12. Generalize the Towers of Hanoi Problem (Exercise 11) in two different ways.

13. Explore an alternative way of arriving at Formula (1.4) for the sum of the first n positive integers, starting with a simpler problem, as follows:

(a) Observe that the sum
$$1 + 2 + 3 + \ldots + 10$$
can be found by "peeling off" the first and last numbers in the string and adding them, "peeling off" and adding the first and last numbers of the remaining string, and so on, until all the numbers have been used up. That is, we can write the desired sum in the form
$$(1 + 10) + (2 + 9) + \ldots + (5 + 6)$$
Now, each summand in parentheses is 11, and there are 5 of them. How is 11 related to 10? How is 5 related to 10?

(b) What is the analogous form for the sum of the first 20 numbers? What is the value of each two-number summand? How many are there? How are these two answers related to 20?

(c) What are the analogous answers to Part (b) for the sum of the first 100 numbers?

(d) What pattern seems to be emerging from Parts (a)–(c)? Is this pattern disturbed if the largest number in your sum is odd?

(e) Can you relate the pattern you see to Formula (1.4)? If so, how?

WRITING EXERCISES

1. Write a dialogue in which you, as questioner, lead a hypothetical friend (roommate, companion marooned with you on a desert island, or whomever) to discover the pattern that occurs when you add up successive even positive integers, starting with 2. Get your friend to describe the general pattern *in words* as clearly as he/she can.

2. Try to devise a problem for which drawing a diagram would *not* be a helpful problem-solving technique. If you can think of one, explain it and explain why a diagram would not be helpful. If you can't think of one, justify the claim that diagrams are helpful for *all* problems.

THE MATHEMATICAL WAY OF THINKING

Each of the other chapters in this book focuses on a particular mathematical problem. Despite differences among the areas they explore, the methods they develop, and the results they obtain, all these chapters have a fundamental similarity in the way their problems are approached. This general methodology is sometimes called the *mathematical way of thinking*. It is based on the assumption that the key to mathematical truth is logical reasoning. The correctness of mathematical statements and solutions to problems can only be proven true by checking, step by step, the logical validity of the arguments used. But constructing these proofs is only half the story, usually the second half. The first (and more interesting) half of the mathematical way of thinking is the problem solving itself, which was discussed in general in Section 1.2. Its main features appear in every chapter, emphasizing the underlying unity of this vast, sprawling field called mathematics.

As you proceed to sample the various topics in the chapters to follow, you will see some of the diversity of mathematics. Be forewarned, though, that the common methods of problem solving are not bound to appear exactly as you saw them in this chapter. For instance, restating the problem will not always be a shift from algebra to geometry. It may be a shift from a formula to a graph, from words to symbols, or from space to time.

In the preceding section we saw a problem for which an appropriate diagram was a geometric figure whose area represented the solution to the problem. At the beginning of Chapter 4 the appropriate diagram to help understand sets is a *Venn diagram*, for which the important feature is overlapping circles. Later in the same chapter, counting problems are solved by using *tree diagrams*, in which the most important feature is the number of branches. In Chapters 3 and 7 the diagrams are geometric figures that are literally the topic of discussion, whereas in Chapter 5 an appropriate diagram is an array of fractions, organized so as to communicate visually a property of those fractions not otherwise obvious. In many chapters, "diagrams" may be charts, graphs, tables, or some other visually structured aid.

Patterns and examples are used to solve the problems in virtually every chapter, but, of course, the structure of the patterns and the content of the examples depend on the problem. For example, Chapter 2 looks at patterns in the prime factors of certain numbers. Chapter 6 draws conclusions from patterns of operations such as addition and multiplication. Chapter 7 finds a way of "seeing" figures and distances in four-dimensional space by looking at the patterns for those ideas in two- and three-dimensional space.

The technique of solving a simpler problem appears in every chapter. Sometimes "simpler" means *smaller* numbers (as in Chapter 2) and sometimes it means *fewer* numbers (as in Chapter 4). Sometimes it means a *lower* dimension (as in Chapter 7), sometimes it means a *shorter* task (as in Chapter 9), and sometimes it means *finite* rather than *infinite* (as in Chapter 5). Be prepared to shift your ideas and labels on what is the "simpler" problem and what is the "general" one. Often you will find the "general solution" of one section in a chapter becomes the springboard "simpler problem" of the next section.

Your understanding of mathematics as a liberal art with a methodology all its own will be greatly enhanced if, as you work through the details of each topic, you watch for the many instances where heuristics are used. And remember to use these problem-solving techniques yourself as you attempt the exercises. Mathematics is not a spectator sport!

TOPICS FOR PAPERS

1. Write a detailed explanation of the Tower of Hanoi Problem (Exercise 11 of Section 1.3) that illustrates how to use as many as possible of the dozen problem-solving techniques listed at the end of Section 1.2. (*Note*: Section 1.3 itself is a somewhat analogous treatment of the problem of summing successive positive integers; it might be a helpful guide for you.)

2. George Pólya is one of the most interesting personalities in twentieth-century mathematics. Write a human-interest biography of Pólya that highlights his personal qualities, as if you were writing an article for a popular magazine. Be sure to provide a bibliography that lists any sources you used in writing the paper, and be careful that you do not simply copy or paraphrase the material from your sources. Any material not explicitly quoted should have been mentally digested by you, so that your own words describe what are really your own ideas when you write them down.

FOR FURTHER READING

1. Averbach, Bonnie, Patricia Brewer, and Orin Chein. *Mathematics: Problem Solving through Recreational Mathematics.* San Francisco: W. H. Freeman and Co., 1980.

2. Gardiner, A. *Discovering Mathematics: The Art of Investigation.* New York: Oxford University Press, 1987.

3. Hayes, John R. *The Complete Problem Solver.* Hillsdale, NJ: L. Erlbaum Associates, 1989.

4. Hughes, Barnabas. *Thinking through Problems: A Manual of Mathematical Heuristic*. Palo Alto: Creative Publications, 1976.

5. Pólya, George. *How to Solve It*, 2nd ed. Princeton: Princeton University Press, 1957.

6. _____. *Mathematical Discovery* (2 vols.). New York: John Wiley and Sons, Inc., 1962, 1965.

7. Schoenfeld, Alan H. *Mathematical Problem Solving*. Orlando, FL: Academic Press, 1985.

8. Whimbey, Arthur. *Problem Solving and Comprehension*. Philadelphia: Franklin Institute Press, 1982.

9. Wickelgren, Wayne A. *How to Solve Problems*. San Francisco: W. H. Freeman and Co., 1974.

Calculo. subduco, ungo fine fine refoluens
Sum numeros numeris. conforiare potens

Arithmetica Carrying a Weight-driven Clock. From a sixteenth-century engraving
by Martin de Vos.

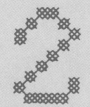

MATHEMATICS OF PATTERNS: NUMBER THEORY

WHAT IS NUMBER THEORY?

2.1

One of the most intriguing fields of mathematics concerns the **natural numbers**, that is, the numbers 1, 2, 3, 4, 5, ... Throughout the Greek era (600 B.C. to A.D. 400), this field embraced all of theoretical mathematics, including much of what we would now call geometry and algebra. Number theory as a specialized discipline first developed in the seventeenth century with the leisure-time activities of Pierre de Fermat; it became firmly established by Karl Friedrich Gauss in 1801. Despite the length of its history and the ingenuity of its many brilliant devotees, number theory remains "the last great uncivilized continent of mathematics."[1] Its jungles of primes, multiples, and factorizations draw nourishment from countless streams of results whose common headwaters lie in unexplored territory that always seems to be just beyond the next problem.

The purpose of this chapter is to provide a taste of number theory and with that taste some flavor of "pure" mathematics. The main topic will be *perfect numbers*, together with some necessary preliminaries and several related areas. In number theory the familiar operations of arithmetic (addition, subtraction, multiplication, and division) are used more as the starting point of intriguing investigations than as topics of

1. Eric Temple Bell, "The Queen of Mathematics." In *The World of Mathematics* (James R. Newman, ed.). Redmond, WA: Tempus Books, 1988, p. 490.

19

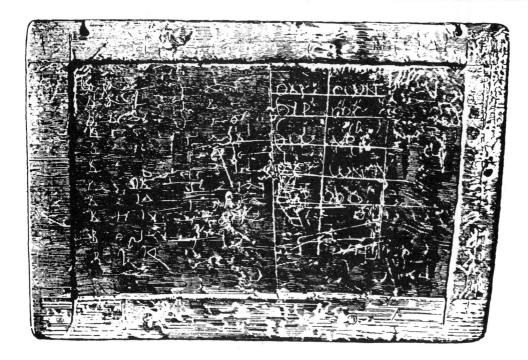

Greek Multiplication on a Wax Tablet.

primary interest. The field is concerned with finding relationships and patterns as well as with investigating the structure of the natural numbers. For example, consider this "parlor trick" with numbers:

Step 1. Pick any three-digit number (for example, 239).

Step 2. Form a six-digit number by repeating the three-digit number chosen (for example, 239,239).

Step 3. Divide this number by 13, obtaining a quotient and a remainder.

Step 4. Divide the quotient obtained in *Step 3* by 7, obtaining a second quotient and remainder.

Step 5. Divide the quotient obtained in *Step 4* by 11.

This last quotient, found in *Step 5*, will equal the original number!

If you repeat this process several times using several different starting numbers, you should be persuaded that it always works. [Of course, unless you try all 900 possible numbers, or discover some general proof, you don't know for sure that it *always* works.] Clearly, this success is not an accident. But what is the explanation?

At this point, instead of producing the answer, let us consider some related questions. Is there a different set of divisors (other than 13, 7, and 11) that would produce the same result? Is there a similar procedure for repeated four-digit numbers (for example, for the eight-digit

number 48,764,876, which is the four-digit number 4876 repeated)? for
five-digit numbers? etc. Is there an analogous procedure for numbers
repeated a third time (for example, for 479,479,479)? a fourth time?
etc. The answers to all these questions are well within your grasp, and
they will be addressed in the exercises for Section 2.3. For now, note
that these are the kinds of generalizations that mark number theory.
Indeed, the hallmark of mathematicians, in general, is not primarily
whether they can answer these or similar questions, but the fact that
they pose such questions in the first place.

Turning to another example, consider the following:

$$1 + 3 = ?$$
$$1 + 3 + 5 = ?$$
$$1 + 3 + 5 + 7 = ?$$

If there were such persons as "arithmeticians," they would, presumably, be interested only in the answers to these addition problems: 4,
9, and 16, respectively. However, number theorists would note the
similarity in the three problems (sums of odd numbers) and in the form
of their answers: 4, 9, and 16, which are, respectively, 2×2, 3×3, and
4×4. Struck by this curiosity, they would go on to ask whether these
problems and their solutions are isolated instances or are part of some
general pattern.

You may already have noticed a specific pattern; namely, that the
sum of consecutive odd numbers, starting with 1, is a square. If you
test this conjecture for several more cases, you will find it verified.
You might even make a more precise observation about this pattern,
which could be formulated as:

(2.1)   The sum of consecutive odd numbers, starting with 1, is the square of
the number of odd numbers added.

Here we begin to have problems with language. Statement (2.1) is
wordy and so may tend to confuse, rather than to clarify, the pattern
we have been studying. Fortunately, we can use letters to represent
numbers. You have encountered this process in algebra, where letters
are used to label unknown numbers. ("Let Mary's age equal x.") The
letters are then used like numbers in equations, the equations are
manipulated in accordance with the rules of arithmetic, and the number (solution) is determined. Letter representation can also convey
information about a large, sometimes infinite, set of numbers. For example, we can abridge the verbal statement:

> The sum of any two numbers is unchanged when we
> reverse the order in which the two are added

to

$$x + y = y + x$$

To be precise, the latter formulation should include a verbal or symbolic phrase indicating the set of numbers whose elements can replace x and y in the statement. *In this chapter we shall make an agreement that, unless otherwise stated, any letter used to represent a number may be replaced by any natural number.* This means that any general statement or formula without explicit restriction is understood to be true for all natural numbers, but is not necessarily true, or even meaningful, for other numbers.

Returning to our example, we can replace the verbal Statement (2.1) by:

(2.2)   If the first n odd numbers are added, the sum is n^2.

We can symbolize even further by representing "the first n odd numbers added" as

$$1 + 3 + 5 + \ldots$$

This would be ambiguous, though, for it doesn't indicate where we intend the sequence of numbers to stop. To correct this, we need an expression to represent the nth odd number. You can figure that out yourself, if you study the following three sequences:

$$
\begin{array}{cccccccc}
1, & 2, & 3, & 4, & 5, & 6, & 7, & \ldots \\
2, & 4, & 6, & 8, & 10, & 12, & 14, & \ldots \\
1, & 3, & 5, & 7, & 9, & 11, & 13, & \ldots
\end{array}
$$

The first sequence is merely the sequence of natural numbers, which we use as a reference sequence. The second, of course, is the sequence of even numbers, and the third is the sequence of odd numbers. The even numbers form a "bridge" to clarify the connection between the first and third sequences. If we continued listing the terms in each sequence, which numbers would line up below 50? By generalizing the pattern begun, you should see that the 50th even number is 100 and the 50th odd number is 1 less, or 99. Similarly, the 123rd even number must be 246 and the 123rd odd number must be 245.

In general, the nth even number is $2n$ and the nth odd number is $2n - 1$; we can use this to rewrite Statement (2.2) as:

xxxxxxxxxxxxxxxxxxxxxxxxxxxxxxxxxxxxxxxxxxxxxxxxxxxxxxxxxxxxx

(2.3)
$$1 + 3 + 5 + 7 + \ldots + (2n - 1) = n^2$$

xxxxxxxxxxxxxxxxxxxxxxxxxxxxxxxxxxxxxxxxxxxxxxxxxxxxxxxxxxxxx

The advantage of this form, Formula (2.3), over that of Statement (2.2) is that it is an equation, and thus can be worked with using the formal rules of arithmetic or algebra.

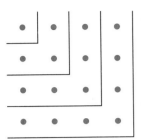

Figure 2.1 Sums of successive odd numbers.

In this example we have a generalization that is plausible because it is verified in all the instances we have checked. But to assert it as a *factual property* of all natural numbers, we must *prove* it. And our proof cannot consist of checking *every* number; if it did, we would be faced with an unending task. We must have a general means of proof to match the generality of what we claim to be true.

There are a variety of arguments available to demonstrate the universal truth of our statement about adding odd numbers, varying from pictorial and intuitive arguments to those of considerable formality. For example, a proof known to the Greeks no later than the second century B.C. uses a pictorial approach. (See *Figure 2.1*.) This figure shows a square array consisting of 25 dots, 5 on a side. The separating lines emphasize the fact that this array can be thought of as "growing" from one dot by the addition of 3 dots in a "bent row," 5 dots in the next "bent row," 7 dots in the next, and 9 in the last. We could add the next "bent row," which would contain 11 dots, and have a 6-by-6 array. In general, it should be clear that each successive "bent row" of dots will add on the next odd number, and the result will always be a square. In other words:

The sum of the first n odd numbers is the square of n.

An alternative proof that relies on arithmetic and algebraic notation is also available. Suppose we did not already know the sum of the first n odd numbers and chose to represent this sum by some letter, say s. Then

$$s = 1 + 3 + 5 + 7 + \ldots + (2n - 1)$$

It follows that

$$s = (2n - 1) + (2n - 3) + (2n - 5) + \ldots + 5 + 3 + 1$$

Don't let this step throw you off. We have merely written the same sum in reverse order. Of course, we are writing in a few of the numbers that are implied, but not written, in the first step [found by decreasing $2n - 1$ by 2, by 4, etc., following the pattern of odd numbers]; but nothing is changed. If we now add the two equations, combining corresponding terms of each on the right, we obtain

$$2s = 2n + 2n + 2n + 2n + \ldots + 2n + 2n$$

On the right, there are n numbers [why?]. Thus, we can write

$$2s = n \times 2n$$

or

$$s = n \times n = n^2$$

and we are finished.

2.1   **EXERCISES**

1. What is the 100th even number?

2. What is the 235th even number?

3. What is the 994th even number?

4. What is the 235th odd number?

5. What is the 994th odd number?

6. What is the 1996th odd number?

In Exercises 7–10, the *ordinal position* is the numerical position of a number in the sequence; for example, 12 is the 6th even number; so it is in ordinal position 6 in the sequence 2, 4, 6, 8, 10, 12, 14, ...

7. What is the ordinal position of 578 in the sequence of even numbers?

8. What is the ordinal position of the even number 994?

9. What is the ordinal position of the odd number 995 in the sequence of odd numbers?

10. What is the ordinal position of the odd number 4321?

11. What is the sum of the first 100 odd numbers?

12. What is the sum of the first 995 odd numbers?

13. What is the sum of all odd numbers from 1 to 995, inclusive?

14. What is the sum of all odd numbers from 1 to 1987, inclusive?

15. What is the sum of all odd numbers from 1 to 4321, inclusive?

16. What is the sum of all odd numbers from 1989 to 4321, inclusive? (*Hint*: Look at the two preceding exercises. How do those two exercises relate to this one? How does the answer to this exercise relate to those two answers?)

17. Devise an alternative method for finding the answer to Exercise 16 by using the fact

that 1989 = 1988 + 1, 1991 = 1988 + 3, 1993 = 1988 + 5, etc.

18. What is the sum of all odd numbers from 135 to 579, inclusive?

Exercises 19–22 refer to the sequence of numbers that starts:

$$3, 6, 9, 12, 15, 18, . . .$$

19. What is the next number in this sequence?

20. What is the 10th number in this sequence?

21. What is the 200th number in this sequence?

22. Determine an expression for the *n*th number in this sequence.

Exercises 23–26 refer to the sequence of numbers that starts:

$$1, 4, 7, 10, 13, 16, . . .$$

23. What is the next number in this sequence?

24. What is the 10th number in this sequence?

25. What is the 200th number in this sequence?

26. Determine an expression for the *n*th number in this sequence.

Exercises 27–32 refer to the sequence of numbers that starts:

$$10, 17, 24, 31, 38, 45, . . .$$

27. What is the next number in this sequence?

28. What is the 10th number in this sequence?

29. What is the 200th number in this sequence?

30. Determine an expression for the *n*th number in this sequence.

31. What is the sum of the first 200 numbers in this sequence? (*Hint*: Follow the procedure used in the second proof of Statement (2.3).)

32. What is the sum of the first *n* numbers in this sequence? (See the *hint* for Exercise 31.)

The sequence of odd numbers, the sequence of even numbers, and those exemplified above are examples of *arithmetic sequences*; that is, sequences in which successive terms increase by the addition of a constant amount. We can represent a typical such sequence by:

$$a, a + d, a + 2d, a + 3d, \ldots$$

Exercises 33–36 refer to this generalized arithmetic sequence.

33. What is the next term in this sequence?

34. What is the 200th term in this sequence?

35. Determine an expression for the nth term in this sequence.

36. Determine an expression for the sum of the first n numbers in this sequence. (See the *hint* for Exercise 31.)

Exercises 37–43 refer to the sequence of numbers that starts:

$$1, 3, 6, 10, 15, 21, \ldots$$

These numbers were called *triangle numbers* by the Greeks. Thus, the first triangle number is 1, the second is 3, the third is 6, and so on.

37. Why are these numbers called triangle numbers? (*Hint*: Look at the arrangement of the 10 pins in a bowling alley or of the 15 balls on a pool table, as shown in *Figure 2.2*.)

38. What is the 7th triangle number?

39. What is the 10th triangle number?

40. Determine an expression for the nth triangle number. (*Hint*: Note that the nth triangle number is equal to

$$1 + 2 + 3 + \ldots + (n - 1) + n$$

Follow the procedure used in the second proof of Statement (2.3).)

41. What is the 100th triangle number?

42. What is the result when a pair of consecutive triangle numbers is added? Formulate a general statement. Prove your generalization, if possible.

43. On the hour, a clock chimes the correct number of times to indicate that hour, and it chimes once on the half hour. How many times does it chime in a twelve-hour period?

WRITING EXERCISES

1. There are many situations in which a sequence of numbers occurs. In some cases the sequence is natural or intrinsic, whereas in others it is arbitrary or contrived, no more appropriate to the situation than many other possible sequences. Describe one particular instance of *each* of these two types of sequences as clearly as you can. For the first kind, explain how and why the sequence is natural; for the second, explain how and why it is arbitrary.

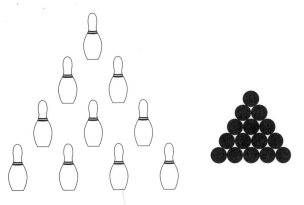

Figure 2.2 Triangular numbers in bowling and in pool.

2. Exercises 37–43 deal with *triangle numbers;* you are already familiar with *square numbers.* Both take their names from plane geometric figures; the (equilateral) triangle and the square are the simplest examples of regular polygons,[2] but such figures can be generalized to any number of sides. How would you generalize from these two kinds of numbers to define **figurate numbers**? Which regular polygons of plane geometry correspond naturally to number patterns and which do not? Identify the first few numbers of each type you consider, and explain why they correspond to the polygon with which you identify them.

3. Research and report on what the Greeks considered to be figurate numbers. Give specific examples besides triangle and square numbers.

4. How might you generalize the idea of *square number* to *rectangle number*? Explain the pros and cons of doing this.

5. How might you generalize the idea of *triangle number* to

   (a) *right triangle number*

   (b) *isosceles triangle number*

   (c) *scalene triangle number*

   Explain the pros and cons of doing this.

6. The 3-dimensional analogue of *square number* is *cube* or *cubic number.* What is a 3-dimensional analogue of *triangle number*? Explain how you would define such a number and why it is an appropriate analogue. Include in your description the first five such numbers.

DIVISIBILITY

2.2

If we consider the four basic arithmetic operations as applied to the set of natural numbers, we find some simple, but significant, differences. Thus, the *sum* or *product* of any two natural numbers is always a natural number, but this is not the case, in general, for the *difference* or *quotient* of two natural numbers. For some choices of x and y, y − x is a natural number; for other choices, it is not. For example, 5 − 3 equals the natural number 2, but 3 − 5 is not a natural number. If, for a particular x and y, y − x happens to be a natural number, then that indicates some special relationship must hold between x and y. This relationship, of course, is that x **is less than** y (or equivalently, that y **is greater than** x), which is denoted by the familiar symbolism

$$x < y \qquad (\text{or} \qquad y > x)$$

Thus, because 5 − 3 is a natural number, we write 3 < 5, and say, "3 is less than 5."

---

2. A *regular* polygon is one with all sides equal and all angles equal.

A completely analogous situation holds for division. Sometimes $y \div x$ is a natural number; sometimes it is not. When it is, there is a special relationship that holds between x and y.

DEFINITION

If $y \div x$ is a natural number, then we say that x **divides** y (or x is a **divisor** of y, or x is a **factor** of y, or y is a **multiple** of x, or y is **divisible by** x).

NOTATION

The statement "x divides y" (or any of the equivalent forms) is denoted by

$$x \mid y$$

Just as we denote "does *not* equal," the negation of "equals" by putting a slash mark through the equality symbol (as in, $5 \neq 8$ or $2 + 2 \neq 3$), we denote the negation of "divides" with a slash mark through the divisibility symbol:

$$x \nmid y$$

Thus, all of the following are true statements:

$$2 \mid 6, \quad 8 \mid 8, \quad 2 \nmid 9, \quad 12 \nmid 9$$

Of course, with some pairs of numbers we may have to perform some arithmetic operation to determine whether one divides the other.[3] For instance, it is not immediately obvious whether 37 divides 703. This is settled by dividing 703 by 37, obtaining the quotient 19. Note that the statement "37 | 703" does not contain the information that 19 is the quotient of 703 divided by 37. It only states that *some* (natural number) quotient exists.

It is worth noting that

$$x \mid y$$

is equivalent to

For some natural number z, $x \cdot z = y$

This follows, of course, from the familiar connection between multiplication and division. This "multiplication form" of the definition, however, exhibits the symmetric role of x and z, and verifies that z is also a factor of y. These two general principles are restated for emphasis in the following list of statements, along with several others that should be equally clear after a bit of reflection.

---

3. The same is true of the relation "less than." For example, it is not immediately obvious whether 5^7 is less than 7^5, or, looking beyond the set of natural numbers, whether $\frac{12}{19}$ is less than $\frac{103}{163}$.

XX

(2.4) $x \mid y$ if and only if there is some natural number z such that $x \cdot z = y$.[4]

(2.5)   If $x \mid y$, then $\frac{y}{x}$ is a natural number, and $\frac{y}{x} \mid y$.

(2.6) $1 \mid y$.

(2.7) $y \mid y$.

(2.8)   If $x \mid y$ and $y \mid z$, then $x \mid z$.

(2.9)   If $x \mid y$ and $x \mid z$, then $x \mid (y + z)$.

(2.10)   If $x \mid y$, then $x \leq y$.

XXXXXXXXXXXXXXXXXXXXXXXXXXXXXXXXXXXXXXXXXXXXXXXXXXXXXX

2.2   EXERCISES

For Exercises 1–23, determine whether the given statement is true or false.

1. $9 \mid 99$
2. $7 \nmid 59$
3. $12 \mid 6$
4. $8 \mid 300$
5. $17 \nmid 102$
6. $37 \nmid 444$
7. $5 \mid 24{,}680$
8. $12 \mid 153{,}624$
9. $2 \nmid 123{,}456{,}789$
10. $3 \mid 123{,}456{,}789$
11. $3 \mid 525$
12. $5 \mid 525$
13. $15 \mid 525$
14. $4 \mid 228$
15. $6 \mid 228$
16. $24 \mid 228$
17. $2^2 \mid 2^3$
18. $38 \nmid 3^{11}$
19. $5^7 \mid 10^7$
20. $5^7 \mid 7^5$
21. If $x \leq y$, then $x \mid y$.
22. If $x \mid y$ and $y \mid x$, then $x = y$.
23. If $x \mid y$ and $x \mid (y + z)$, then $x \mid z$.

24. Suppose that b and c are two numbers with the property that $b \mid c$ and $b \mid (c + 1)$. What are the possible values for b?

25. Suppose that b and c are two numbers with the property that $b \mid c$ and $b \mid (c + 6)$. What are the possible values for b?

26. Fill in the blanks to create an accurate definition: "If _____ divides _____, then n is an *even number.*"

27. Use Statement (2.9) to explain why the sum of two even numbers is even. (See Exercise 26.)

28. Use Statement (2.8) to explain why the product of any number and an even number is even. (See Exercise 26.)

29. Explain why the product of any two consecutive numbers is even. (See Exercise 28.)

30. If $5 \mid k$, what is the smallest number larger than k that is a multiple of 5?

31. Construct a table with six columns, as follows: in the first column, list the natural numbers from 1 to 30; head the column "n." In the second column, next to each number in the first, enter all the divisors of

---

4. If you are unfamiliar with the phrase "if and only if," you can find an explanation of it in Section A.3 of the Appendix.

that number; head this column "divisors." In the third column, tally the number of divisors entered in the second column; head this column "D(n)." (For example, the entries corresponding to 18 in the second and third columns would be "1, 2, 3, 6, 9, 18" and "6," respectively.) Leave the last three columns blank. They will be used for Exercises 1 and 2 of Section 2.5. (This table will be used as a source of data for the Topics for Papers at the end of the chapter.)

WRITING EXERCISES

1. The use of "|" to stand for divisibility and of letters to stand for arbitrary numbers of a system typifies the mathematical use of symbols as a convenient shorthand.

   (a) Write each of the Divisibility Statements (2.4) through (2.10) completely in words, without using *any* symbolic notation, not even letters to represent numbers. (Be careful to *quantify* the statements properly. That is, if a statement is intended to be a property of *all* numbers, make sure that your translation reflects that fact; if only *some* numbers have this property, indicate which numbers.)

   (b) Discuss the pros and cons of using symbols in presenting mathematical ideas.

2. The text identifies an analogy between *subtraction* and *less than*, on the one hand, and *division* and *divisibility*, on the other. Elaborate on this analogy by showing how any four of the Divisibility Statements (2.4) through (2.10) translate into analogous statements about *less than*. Are there any of these seven statements that *cannot* be translated by means of this analogy? Justify your answer.

3. The text identifies an analogy between *subtraction* and *less than*, on the one hand, and *division* and *divisibility*, on the other; the latter might just as well have been compared with *subtraction* and *less than or equal to*. Discuss the distinction between these two analogies. In which particulars is the analogy suggested by the text better? In which is the alternative better? Overall, is one choice preferable to the other?

4. Write a short paper discussing and comparing the two words *divisor* and *factor*. What is the etymology of these words? Do the non-mathematical meanings of these words correspond to their mathematical meanings? Which do you think is more appropriate? Why? Can you suggest other words or phrases that you think would be better? On what basis?

COUNTING DIVISORS

In Exercise 31 of Section 2.2, you were asked to tabulate the number of divisors of each natural number from 1 to 30. In this section we pursue the generalization of that notion. In other words, we attempt to discover a formula or process that will allow us to calculate, in as simple and efficient a manner as possible, the number of divisors of any natural number. For brevity in dealing with this notion, we adopt the following notation:

NOTATION The number of divisors of *n* will be denoted by $D(n)$. (This is read "*D* of *n*.")

For example, there are 4 divisors of 10—namely, 1, 2, 5, and 10—thus, we write $D(10) = 4$. Similarly, because there are 9 divisors of 36—namely, 1, 2, 3, 4, 6, 9, 12, 18, and 36—we write $D(36) = 9$.

So in terms of this notation, our goal in this section is to determine $D(n)$ for any number *n*.[5] There is, of course, the "brute force" method whereby we test every number less than or equal to *n*, recording (or at least counting) those that divide *n*. [Statement (2.10) points out that it would be fruitless to test numbers larger than *n* because none can possibly be divisors of *n*.] This is precisely the method suggested in constructing the table for Exercise 31 of Section 2.2. Unfortunately, this method is not very practical for large numbers. Imagine, for instance, testing 10,000 numbers! Of course, with the aid of a computer this could be done in a reasonable amount of time; but even with such assistance, testing every number less than or equal to *n* is highly inefficient.

A great deal of time and effort can be saved by observing the import of Statement (2.5). This says, for example, that because 40 | 10,000, then $\frac{10,000}{40}$ (which equals 250) is also a divisor of 10,000. Put another way, when we test 40 as a divisor of 10,000 and obtain 250 as a quotient, not only do we establish that 40 is a divisor, but also that 250 is a divisor. In other words, divisors occur in pairs. Thus, in testing for divisors of 10,000, we find pairs: 1 and 10,000, 2 and 5000, 4 and 2500, and so on. The real advantage of this pairing process is seen in the observation of which numbers are *not* divisors. For example, when we test 3 and find that it is not a divisor of 10,000, we also establish that there are no divisors of 10,000 between 2500 and 5000.

Divisibility by 5. From Calandri's *Arithmetica*, Florence, 1491.

---

5. Recall that throughout this chapter, unless otherwise specified, the numbers referred to are natural numbers.

For if there were, there would be a "companion" divisor between 2 and 4—which is clearly false.

If we identify the divisors of 10,000 in this way, namely in pairs, we only have to test part way. A casual reflection might suggest then that we need to test half the numbers, from 1 to 5000; that would be appropriate if we were identifying pairs of numbers whose *sum* is 10,000. But for pairs of numbers whose *product* is 10,000, the appropriate "halfway" point is 100—the *square root* of 10,000. Each divisor less than 100 must pair up with a divisor greater than 100, and vice versa, so that when we have found all divisors from 1 to 100, with their paired divisors from 100 to 10,000, we have thereby found all the divisors of 10,000. In general:

> To find all the divisors of a number, say n, test all the numbers up to $\sqrt{n}$, identifying pairs of divisors.

For most values of n, $\sqrt{n}$ is not a natural number, of course; but the reasoning is just the same in either case. In practice, if we test for divisibility by successive integers, starting with 1, we merely keep going until we obtain a quotient less than or equal to the divisor being tested. At that point we know we have passed $\sqrt{n}$ and can stop testing. It is not necessary to know or compute $\sqrt{n}$.

EXAMPLE 2.1

Problem: What is $D(200)$?

Solution: To find all the divisors of 200, we divide by 1, 2, 3, and so on, obtaining a quotient and remainder in each case. The results of this arithmetic are recorded in *Table 2.1*. Each time we

TEST DIVISOR	QUOTIENT	REMAINDER
1	200	0
2	100	0
3	66	2
4	50	0
5	40	0
6	33	2
7	28	4
8	25	0
9	22	2
10	20	0
11	18	2
12	16	8
13	15	5
14	14	4

Table 2.1 Testing for divisors of 200.

obtain a remainder of 0, we have found *two* divisors of 200. There is no need to test numbers larger than 14 because when 200 is divided by 14, the quotient is 14 (with a nonzero remainder). Testing numbers larger than 14 would produce quotients less than 14, and we already know which of these are divisors. Thus, the divisors of 200 are: 1 and 200, 2 and 100, 4 and 50, 5 and 40, 8 and 25, 10 and 20. Therefore, $D(200) = 12$. (Incidentally, we have also found that $\sqrt{200}$ is between 14 and 15.)

This method is significantly better than our first idea of testing *all* numbers less than or equal to n. And it can be marginally improved by noting that we need not test for divisibility by any number that is a multiple of one known not to be a divisor. (In Example 2.1, testing 6, 9, and 12 was unnecessary, given that 3 is not a divisor.) But it is not the best we can do. Look at the table that you were asked to construct in Exercise 31 of Section 2.2. Reorganize that information by asking which numbers have exactly one divisor, which have exactly two divisors, and so on. The resulting information is found in *Table 2.2*.

We do not have a lot to go on, but some observations can be made. Clearly, 1 is unique in having only one divisor. The numbers with exactly two divisors are probably familiar to you. They play a very important role in number theory.

DEFINITION

A **prime number** is a number that has exactly two divisors.

An equivalent statement is:

xxxxxxxxxxxxxxxxxxxxxxxxxxxxxxxxxxxxxxxxxxxxxxxxxxxxxxxx

(2.11)

A number is prime if and only if it is greater than 1 and cannot be written as the product of two smaller numbers.

xxxxxxxxxxxxxxxxxxxxxxxxxxxxxxxxxxxxxxxxxxxxxxxxxxxxxxxx

k	Numbers with Exactly k Divisors
1	1
2	2, 3, 5, 7, 11, 13, 17, 19, 23, 29, . . .
3	4, 9, 25, . . .
4	6, 8, 10, 14, 15, 21, 22, 26, 27, . . .
5	16, . . .
6	12, 18, 20, 28, . . .
7	64, . . .
8	24, 30, . . .

Table 2.2 Numbers classified by the number of their divisors.

The first ten prime numbers are identified in *Table 2.2;* the next five are 31, 37, 41, 43, and 47. We will see examples of much larger prime numbers later in the chapter; in fact, there is no limit on the magnitude of prime numbers. Observe that 1 is not a prime number.

Note that we have not explained why certain numbers have exactly two divisors. We have only established a label for such numbers. There are many fascinating theorems and conjectures involving prime numbers. We will see some of these as we proceed, but we will not digress at this point from our pursuit of a formula or process for finding $D(n)$.

In *Table 2.2*, looking now at numbers with exactly three divisors, we find 4, 9, 25 (and others, if we were to look beyond 30). A bit of reflection shows that these numbers are just the squares of the prime numbers. If we test this idea a bit, we find, for example, that

the only divisors of 7^2 (or 49) are 1, 7, and 49;

the only divisors of 11^2 (or 121) are 1, 11, and 121

In general, it should be clear that if a number n is the square of a prime p (that is, if $n = p^2$), it will have the three distinct divisors: 1, p, and p^2. But does that guarantee that $D(n) = 3$? For example, 227 is a prime number, so 227^2, which equals 51,529, has *at least* the three divisors 1, 227, and 51,529; but does it have any others? In general, for the squares of some primes, might there not be more than the three divisors 1, p, and p^2? To answer this question, we need a bit more terminology and one more basic fact about numbers.

DEFINITION

Any representation of the number n as a product of two or more numbers, or the number n (by itself), is called a **factorization** of n. If all factors in a factorization are prime, it is called a **prime factorization**.

EXAMPLE 2.2

The following are all factorizations of 36:

4×9	$3 \times 4 \times 3$	6^2
$2 \times 3 \times 2 \times 3$	$1 \times 2 \times 3 \times 6$	$2^2 \times 3^2$
$1 \times 2 \times 2 \times 1 \times 3 \times 3$	1×36	36

Of these, $2 \times 3 \times 2 \times 3$ and $2^2 \times 3^2$ are prime factorizations of 36. Note that $1 \times 2 \times 2 \times 1 \times 3 \times 3$ is *not* a prime factorization because 1 is not a prime number. Also, among the factorizations of 36 is the number 36, itself.[6]

---

6. Some texts define "factorization" in a way that excludes a single number from being called a factorization of itself.

Observe that every number has factorizations, and every number except 1 has prime factorizations. The latter fact is true because we can omit 1 from a factorization of any number other than 1, and we can write nonprime numbers as products of smaller numbers, which in turn are either prime or are products of smaller numbers. Eventually, the process of breaking numbers into products must terminate in a product of numbers that cannot be written as a product of smaller numbers, that is, prime numbers. This is the easy part of the proof of a basic fact about numbers:

xxxxxxxxxxxxxxxxxxxxxxxxxxxxxxxxxxxxxxxxxxxxxxxxxxxxx

The Fundamental Theorem of Arithmetic: Every number greater than 1 has a prime factorization that is unique aside from the order in which the primes are written.

xxxxxxxxxxxxxxxxxxxxxxxxxxxxxxxxxxxxxxxxxxxxxxxxxxxxx

The proof of the "uniqueness" part of this theorem is somewhat involved and will not be given here. (See the references at the end of the chapter.) Consider the import of the theorem for, say, 1711, which equals 29×59; because 29 and 59 are both prime, this product (29×59) is *the* prime factorization of 1711. No other prime numbers can be multiplied to produce 1711. Or, as was noted in Example 2.2, $2 \times 3 \times 2 \times 3$ and $2^2 \times 3^2$ are prime factorizations of 36; they differ only in the order in which the primes are written (and grouped, via exponential notation. Given the variety of possible arrangements of the prime factorization, we usually give preference to the latter arrangement, with the pimes listed in increasing order and with repeated primes expressed in exponential form.) Again, this is *the* prime factorization of 36, and no other primes, or, more precisely, no other powers of primes, can be multiplied to produce 36.

The Fundamental Theorem effectively states, then, that every number greater than 1 has a unique set of *powers of primes* associated with it. (Of course, there may be only one prime, and some or all of the primes may only be raised to the first power; we normally do not write 1 as an exponent.) The product of these powers of primes is the number, and no other set of powers of primes multilied together will produce that number.

Thus, every number other than 1 is either prime or is uniquely expressible as the product of two or more (possibly repeated) primes. This prompts another definition:

DEFINITION

A number is a **composite number** if it is the product of two or more (possibly repeated) primes.

For example, 6 (= 2 × 3) and 9 (= 3 × 3) are composite numbers. Notice that 1 does not fit this definition, and thus it is neither prime nor composite. In conjunction with the definition of a prime number and Statement (2.11), we can characterize composite numbers as follows:

×××

(2.12)   A number is composite if and only if it has three or more divisors.

×××

Simply put, a prime number is so called because it is "primal" relative to factorization; that is, it cannot be factored as the product of smaller numbers. A composite number is composed (by multiplication) of smaller numbers, and ultimately, of a unique collection of (powers of) prime numbers. An illuminating, if imperfect, analogy can be drawn between numbers and chemical compounds, with primes corresponding to atoms and composites to compounds.

The Fundamental Theorem guarantees a unique prime factorization of any number n (greater than 1); so any method that repeatedly factors n (and avoids the trivial factor of 1) will ultimately produce the prime factorization. When attempting to find a unique prime factorization, arithmetic skill, a hand calculator, and common sense can all be used. One way to accomplish this goal is to factor a composite number into two smaller numbers, and to repeat this for any composite factors. You probably learned this method, using what is usually called a "factor tree." For example,

$$144 = 12 \times 12 = 3 \times 4 \times 3 \times 4 = 3 \times 2 \times 2 \times 3 \times 2 \times 2 = 2^4 \times 3^2$$

Another process is to factor all the 2s, then all the 3s, then all the 5s, and so on.

EXAMPLE 2.3

Problem: Determine the prime factorization of 5280.

Solution:

$$
\begin{array}{r|r}
2 & 5280 \\
2 & 2640 \\
2 & 1320 \\
2 & 660 \\
2 & 330 \\
3 & 165 \\
5 & 55 \\
 & 11
\end{array}
$$

Thus, $5280 = 2^5 \times 3 \times 5 \times 11$.

| EXAMPLE 2.4 | **Problem:** Determine the prime factorization of 126,000. |

Solution: Here, we can save some time by recognizing that each terminal zero represents a factor of 10, or 2×5. Thus,

$$2^3 \times 5^3 \underline{| \; 126000}$$
$$2 \underline{| \; 126}$$
$$3 \underline{| \; 63}$$
$$3 \underline{| \; 21}$$
$$7$$

Consequently, $126{,}000 = 2^4 \times 3^2 \times 5^3 \times 7$.

| EXAMPLE 2.5 | **Problem:** Determine the prime factorization of 20,889. |

Solution: 20,889 is not divisible by 2.

$$3 \underline{| \; 20889}$$
$$3 \underline{| \; 6963}$$
$$2321$$

2321 is not divisible by 3, 5, or 7, but

$$11 \underline{| \; 2321}$$
$$211$$

211 cannot be divisible by 2, 3, 5, or 7. (Otherwise, 2321 would be; see Statement (2.8)). It is not divisible by 11, and thus we try the next prime, 13. It is not divisible by 13, so we try 17. Now, 17 is not a factor, but $2321 \div 17$ equals 12, with remainder 7; the fact that the quotient obtained is less than 17 tells us that we need go no further. (Why not?) Therefore, 211 is a prime number. So

$$20{,}889 = 3^2 \times 11 \times 211$$

The Fundamental Theorem provides the key to the question posed earlier about divisors of the square of a prime, and indeed about $D(n)$ in general, by yielding the following characterization of divisibility:

xxxxxxxxxxxxxxxxxxxxxxxxxxxxxxxxxxxxxxxxxxxxxxxxxxxxxxxxxxxxxxx

(2.13)   For numbers m and n, $m \mid n$ if and only if every power of a prime that divides m also divides n.

xxxxxxxxxxxxxxxxxxxxxxxxxxxxxxxxxxxxxxxxxxxxxxxxxxxxxxxxxxxxxxx

The proof of this fact is fairly easy, given the Fundamental Theorem. (This theorem is trivially satisfied if $m = 1$, in which case there are no prime divisors of m.) If a power of a prime divides m, and $m \mid n$, then that power of the prime divides n, by Statement (2.8). Conversely, if every power of a prime that divides m also divides n, then the prime factorization of m is part (possibly all) of the prime factorization of n. The product of the "excess" factors, if any, in the prime factorization of n forms a number k such that $m \cdot k = n$. Therefore, by Statement (2.4), $m \mid n$. (If there are no "excess" primes, then $m = n$ and $k = 1$.)

For example, the prime factorizations of 15 and 180 are

$$3 \times 5 \quad \text{and} \quad 2 \times 2 \times 3 \times 3 \times 5 = 2^2 \times 3^2 \times 5$$

respectively, demonstrating that $15 \mid 180$. The "excess" primes in 180, relative to 15, are 2, 2, and 3, whose product equals 12, and

$$12 \times 15 = 180$$

As another example, note that 8, which equals $2 \times 2 \times 2$, has one more factor of 2 than does 180; thus, Statement (2.13) tells us that $8 \nmid 180$.

Statement (2.13), applied to the square of any prime, $n = p^2$, shows that the only divisors other than 1 are p and p^2. There are only three divisors, so $D(n) = 3$. Similarly, if n is the cube of any prime, p, its only divisors are 1, p, p^2, and p^3, and hence $D(n) = 4$. Generalizing, we observe:

xxxxxxxxxxxxxxxxxxxxxxxxxxxxxxxxxxxxxxxxxxxxxxxxxxxx

If p is any prime, then $D(p^k) = k + 1$.

xxxxxxxxxxxxxxxxxxxxxxxxxxxxxxxxxxxxxxxxxxxxxxxxxxxx

In general, Statement (2.13) tells us that the number of divisors of a number n is the number of distinct products of powers of primes that can be formed from the set of powers of primes in the prime factorization of n. We can even fit the divisor 1 into this description by observing that 1, the only number with no prime factors, corresponds to the "product" of no primes.

EXAMPLE 2.6

Problem: Find all divisors of 144. Also, find $D(144)$.

Solution: We begin by factoring 144; as we saw above,

$$144 = 2^4 \times 3^2$$

We then list all the distinct products of powers of primes that can be formed from these four factors of 2 and two factors of 3. This is done in *Table 2.3(a)*, where the divisor 1 is also listed. Without multiplying out each of these products, it is easy to count that there are 15 of them; so $D(144) = 15$. In *Table 2.3(b)*, the corresponding collections of powers of primes are listed.

A closer inspection of the pattern in *Table 2.3(a)* shows that we actually did more than necessary to compute $D(144)$. There are 15 numbers in the table because it is composed of 3 rows of 5 columns each. The first row contains the divisors of 144 with no factors of 3; the second row contains divisors with exactly one factor of 3; the third row contains divisors with two factors of 3. The columns are similarly described in terms of the number of factors of 2 (none, 1, 2, 3, and 4, respectively). Thus, the numbers of rows and columns in the table necessarily correspond to the number of choices of factors of 3 and the number of choices of factors of 2, respectively. But these numbers can be determined *without forming the table*!

EXAMPLE 2.7

Problem: What is $D(25,000)$?

Solution: The prime factorization of 25,000 is

$$2^3 \times 5^5$$

If we formed a table listing all the divisors, it would be 4–by–6 in size because there are 4 divisors of 2^3:

$$1, 2, 2^2, \text{ and } 2^3$$

and there are 6 divisors of 5^5:

$$1, 5, 5^2, 5^3, 5^4, \text{ and } 5^5$$

Therefore, there are 24 divisors of 25,000. Symbolically,

$$D(25,000) = D(2^3 \times 5^5)$$
$$= D(2^3) \times D(5^5)$$
$$= (3 + 1) \times (5 + 1)$$
$$= 4 \times 6$$
$$= 24$$

1	2	2^2	2^3	2^4
3	2×3	$2^2 \times 3$	$2^3 \times 3$	$2^4 \times 3$
3^2	2×3^2	$2^2 \times 3^2$	$2^3 \times 3^2$	$2^4 \times 3^2$

(a)

$\{1\}$	$\{2\}$	$\{2^2\}$	$\{2^3\}$	$\{2^4\}$
$\{3\}$	$\{2, 3\}$	$\{2^2, 3\}$	$\{2^3, 3\}$	$\{2^4, 3\}$
$\{3^2\}$	$\{2, 3^2\}$	$\{2^2, 3^2\}$	$\{2^3, 3^2\}$	$\{2^4, 3^2\}$

(b)

Table 2.3 *All the divisors of 144 (a) as products of primes, (b) as collections.*

Early Illustration of Calculations on a Blackboard. From Böschensteyn's *Rechenbüchlin*, Ausburg, 1514.

The procedure developed in the two preceding examples generalizes to numbers whose prime factorization involves three or more distinct primes. We should note that the visual device of representing the divisors of a given number in a table does not carry over conveniently to numbers with three or more distinct prime factors. We could generalize by relying on a series of two or more such tables, the first representing the first two distinct powers of primes in the prime factorization, and each subsequent table representing the results of the preceding table together with a subsequent power of a prime. But, as noted just before Example 2.7, a table is not really necessary, nor is its visualization. We only need to calculate the number of distinct combinations of powers of primes, a simple matter of multiplying the appropriate numbers.

EXAMPLE 2.8

Problem: What is $D(7056)$?

Solution: The prime factorization of 7056 is

$$2^4 \times 3^2 \times 7^2$$

As was seen in Example 2.6, there are 15 divisors of $2^4 \times 3^2$. We can multiply each of these 15 divisors by 1 or 7 or 7^2 (the 3 divisors of 7^2) to produce 45 distinct divisors of 7056. Thus,

$$\begin{aligned}
D(7056) &= D(2^4 \times 3^2 \times 7^2) \\
&= D(2^4) \times D(3^2) \times D(7^2) \\
&= (4 + 1) \times (2 + 1) \times (2 + 1) \\
&= 5 \times 3 \times 3 \\
&= 45
\end{aligned}$$

Our process for determining $D(n)$, then, for $n > 1$, can be summarized as follows:

Step 1. Determine the prime factorization of n.

Step 2. Collect repeated primes and write the prime factorization in exponential form.

Step 3. Compute $D(p^k)$ for every power of each distinct prime by adding 1 to each exponent k.

Step 4. Multiply the resulting numbers obtained in *Step 3*.

The simple calculation in *Step 3* determines the number of choices of factors of each power of a prime; adding 1 to the exponent accounts for the choice of using no factors of that prime. *Step 4* computes the total number of combinations.

EXAMPLE 2.9	**Problem:** What is $D(9828)$?

Solution: The prime factorization of 9828 is $2^2 \times 3^3 \times 7 \times 13$. Adding 1 to each exponent, we obtain 3, 4, 2, and 2, respectively. (Don't forget that 7 and 13 have unwritten exponents of 1.) Thus, $D(9828) = 3 \times 4 \times 2 \times 2 = 48$.

The problem of determining $D(n)$ is thus ultimately reduced to finding the prime factorization of n and doing some trivial arithmetic.

2.3   EXERCISES

Determine the prime factorization for each of the following sixteen numbers. Express your answer in exponential form. (Your answers will be useful for the exercises for Sections 2.4 and 2.5.)

1. 91	2. 120	3. 299	4. 313
5. 347	6. 496	7. 836	8. 1878
9. 3817		10. 4173	
11. 21,168		12. 30,030	
13. 64,372		14. 100,100	
15. 299,299		16. 1,000,000	

17–32. Determine $D(n)$ for each of the numbers in Exercises 1–16, and identify which of these numbers n are prime and which are composite.

33. **(a)** If p and q are distinct primes, what is $D(pq)$?

   (b) If p, q, and r are distinct primes, what is $D(pqr)$?

   (c) If n is the product of k distinct primes, what is $D(n)$?

34. Show that if $n > 1$ and if n has no prime divisor less than or equal to $\sqrt{n}$, then n must be prime. (See Examples 2.1 and 2.5.)

35. *Prove or disprove*: $D(n)$ is odd if and only if n is the square of some number.

36. *Prove or disprove*: If p is prime, then $p^2 - 2$ is prime.

37. *Prove or disprove*: If p is an odd prime, then $D(3 + p) = D(3) + D(p)$.

38. **(a)** Find five examples of pairs of numbers m and n for which
$$D(m \times n) = D(m) \times D(n)$$

   (b) Find five examples of pairs of numbers m and n for which
$$D(m \times n) \neq D(m) \times D(n)$$

   (c) Formulate a conjecture, consistent with the evidence of Parts **(a)** and **(b)**, about what properties, in general, distinguish pairs of numbers m and n for which
$$D(m \times n) = D(m) \times D(n)$$

In Exercises 39–44, you are asked to reason backward from information about $D(n)$ to infer something about n. For example, if $D(n) = 6$, then the form of the prime factorization of n is either p^5 or $p^2 \times q$, where p and q are distinct primes.

39. What are the possible forms of the prime factorization of n if $D(n) = 10$?

40. What are the possible forms of the prime factorization of n if $D(n) = 11$?

41. What are the possible forms of the prime factorization of n if $D(n) = 12$?

42. What is the smallest number that has exactly 12 divisors?

43. What is the smallest number that has exactly 18 divisors?

44. Determine all numbers less than 200 with exactly 12 divisors.

45. What is the prime factorization of 1001? Use your answer to explain the "parlor trick" of Section 2.1. (See page 20.)

46. The prime factorization of 10,001 is

$$73 \times 137$$

Design a "parlor trick" similar to that on page 20 that uses this fact.

47. (a) What is the prime factorization of 10,101?

    (b) Use your answer to Part (a) to describe a "parlor trick" like that on page 20. (See Exercise 45.)

48. According to Statement (2.12), composite numbers have three or more divisors. Characterize the composites that have exactly three divisors.

WRITING EXERCISES

1. The text mentions an analogy between prime and composite numbers in mathematics, and atoms and compounds in chemistry.

   (a) Explore this analogy in some detail, discussing how it is apt and how it is inappropriate. (In particular, what law in chemistry is analogous to the Fundamental Theorem of Arithmetic in mathematics?)

   (b) Identify and discuss an analogy in some other subject area that mirrors, at least partially, the relationship between prime numbers and composite numbers.

2. Each of the statements given at the end of this question purports to be a definition of *prime number*. However, all but one of them are inaccurate (gramatically, syntactically, or logically). Identify the correct statement, and justify how the one you picked describes exactly the same thing as the text definition. Then choose any three others and explain why they are incorrect.

   (a) A prime number is a number divisible by itself and one.

   (b) A prime number is a number that cannot be written as the product of two numbers without using itself as a factor.

   (c) A prime number is a number that is divisible only by itself and one.

   (d) A prime number is a number that is divisible by just one number that is smaller than itself.

   (e) A prime number is one like 2, 3, 5, 7, etc.

   (f) A prime number is a number with two different factors.

   (g) A prime number is a number that has no factor that is larger than one and smaller than itself.

   (h) A prime number is when you try to divide it by something besides one or itself but you can't.

3. In an introductory text on number theory (see *For Further Reading* at the end of the chapter), study and report on the proof that there are an infinite number of prime numbers. Who, if anyone, is credited with discovering this proof? Does the proof depend on the Fundamental Theorem of Arithmetic?

4. Explain in your own words how the Fundamental Theorem of Arithmetic justifies the correctness of the four-step process for determining $D(n)$. How do you know

   ■ that every factor of n is counted?

   ■ that no numbers are counted which are not factors?

   ■ that no factor is mistakenly counted twice?

SUMMING DIVISORS

2.4

We now turn our attention to the problem of *adding* all the divisors of a given number. As with the problem of determining the number of divisors, our practical challenge is not how to get an answer, but how to do so efficiently. As in the preceding section, we introduce a notational definition for the sake of brevity.

NOTATION

We denote the sum of all the divisors of n by $S(n)$. For instance, the divisors of 10 are 1, 2, 5, and 10; so

$$S(10) = 1 + 2 + 5 + 10 = 18$$

Similarly,

$$S(24) = 1 + 2 + 3 + 4 + 6 + 8 + 12 + 24 = 60$$

Much of the work necessary to determine $S(n)$ is already behind us. In fact, the development of our process of determining $D(n)$ identified that each combination of powers of primes which is a subcollection of those in the prime factorization of n produces a divisor of n. Thus, we already have an efficient method of listing all the divisors of a number n.

EXAMPLE 2.10

Problem: List all the divisors of 675.

Solution: The prime factorization of 675 is

$$3^3 \times 5^2$$

Thus, its divisors are:

1	3	3^2	3^3
5	3×5	$3^2 \times 5$	$3^3 \times 5$
5^2	3×5^2	$3^2 \times 5^2$	$3^3 \times 5^2$

We could now go on to compute each of the divisors listed in Example 2.10, add them up, and obtain $S(675)$. Let us do this in several carefully chosen steps. The twelve divisors of 675 are listed in four columns, each consisting of three numbers. Let us add each column, and then add up the subtotals. The sum of the numbers in the first column is $1 + 5 + 25$, which equals 31. We can now use the internal pattern of the list of divisors to help us. Because each number in the second column is exactly 3 times the corresponding number in the first column, the sum of these three numbers must be 3×31. Similarly, the sum of the numbers in the third column must be $3^2 \times 31$ and the sum of those in the fourth column must be $3^3 \times 31$. Thus, the grand total of all twelve numbers is

$$31 + (3 \times 31) + (3^2 \times 31) + (3^3 \times 31)$$

or

$$(1 \times 31) + (3 \times 31) + (9 \times 31) + (27 \times 31)$$

which, in turn, equals

(2.14)
$$(1 + 3 + 9 + 27) \times 31 = 40 \times 31 = 1240$$

Notice that 31, the sum of the first-column numbers, is the sum of the divisors of 5^2. Also 40, the sum of the multipliers of 31 in Formula (2.14), is the sum of the divisors of 3^3, which can be obtained by adding the numbers in the first row. In summary, since

$$675 = 3^3 \times 5^2$$

$$S(675) = S(3^3 \times 5^2) = 1240$$

and

$$1240 = (1 + 3 + 9 + 27) \times 31 = S(3^3) \times S(5^2)$$

we see that

$$S(3^3 \times 5^2) = S(3^3) \times S(5^2)$$

EXAMPLE 2.11

Problem: Determine $S(288)$.

Solution: The prime factorization of 288 is

$$2^5 \times 3^2$$

If we were to list the eighteen divisors in six columns of three rows in an analogous way to the listing in the preceding example, the first column would contain 1, 3, and 3^2, whose sum is 13. And the six multipliers of 13 would be 1, 2, 2^2, 2^3, 2^4, and 2^5, whose sum is 63. Thus,

$$S(288) = S(2^5) \times S(3^3) = 63 \times 13 = 819$$

Notice that in Example 2.11 it is not necessary to determine or list most of the divisors of 288. The problem reduces to finding $S(2^5)$ and

$S(3^2)$ and multiplying the results. This pattern holds true for any number. Thus, to compute $S(n)$:

Step 1. Determine the prime factorization of n.

Step 2. Collect repeated primes and write the prime factorization in exponential form.

Step 3. For each power of a prime, p^k, in this factorization, compute $S(p^k)$.

Step 4. Multiply the resulting numbers.

(Notice the strong parallel between this process and the one discussed in Section 2.3.)

EXAMPLE 2.12

Problem: Determine $S(1800)$.

Solution: The prime factorization of 1800 is

$$2^3 \times 3^2 \times 5^2$$

Here,

$$S(2^3) = 1 + 2 + 4 + 8 = 15$$
$$S(3^2) = 1 + 3 + 9 = 13$$
$$S(5^2) = 1 + 5 + 25 = 31$$

Therefore,

$$S(1800) = 15 \times 13 \times 31 = 6045$$

Thus, the problem of computing $S(n)$ is reduced to factoring n and computing $S(p^k)$ for each prime factor p. The former has been treated earlier. For the latter, as noted in the preceding section, the divisors of p^k are:

$$1, p, p^2, p^3, \ldots, p^{k-1}, p^k$$

If k is small, the simplest computation is obtained by adding the few numbers. For large k, the following formula can be used:

××

(2.15)   If $n > 1$, then $1 + n + n^2 + \ldots + n^{k-1} + n^k = \dfrac{n^{k+1} - 1}{n - 1}$

××

EXAMPLE 2.13

Problem: Determine $S(11^2)$.

Solution: $S(11^2) = 1 + 11 + 11^2 = 1 + 11 + 121 = 133$

EXAMPLE 2.14	**Problem:** Determine $S(3^{10})$.

Solution: $S(3^{10}) = 1 + 3 + 3^2 + \ldots + 3^{10}$. By Formula (2.15), the expression on the right side equals

$$(3^{11} - 1)/(3 - 1) = (3^{11} - 1)/2 = 88{,}573$$

Note that the arithmetic involved in the last step is not much easier than that involved in adding the eleven numbers. (It is considerably easier with a hand calculator that has an exponent, or "y^x," key.) For some of our purposes, though, it will be useful to be able to represent the desired sum as a fraction, even if we don't do the implied arithmetic.

2.4   EXERCISES

For Exercises 1–6, list, in factored form, all the divisors of the given numbers.

1. 91
2. 120
3. 288
4. 405
5. 496
6. 756

For each of the numbers n in Exercises 7–22, determine $S(n)$. (You have already determined the prime factorizations in Exercises 1–16 of Section 2.3.)

7. 91
8. 120
9. 299
10. 313
11. 347
12. 496
13. 836
14. 1878
15. 3817
16. 4173
17. 21,168
18. 30,030
19. 64,372
20. 100,100
21. 299,299
22. 1,000,000

23. What is $S(2^5)$?

24. What is $S(3^5)$?

25. What is $S(2^5 \times 3^5)$?

26. What is $1 + 6 + 6^2 + 6^3 + 6^4 + 6^5$? Is this $S(6^5)$? (See the preceding exercise.)

27. Prove Formula (2.15). (*Hint:* Let

$$1 + n + n^2 + \ldots + n^{k-1} + n^k = z$$

Determine a comparable equation for $n \times z$. Combine these two equations by subtracting corresponding sides, and then solve for z.)

28. Generalize Formula (2.15) to produce a formula for

$$n^a + n^{a+1} + n^{a+2} + \ldots + n^k$$

where $a < k$. (*Hint:* Use Formula (2.15) twice and subtract.)

29. Suppose you had a job that paid 1 cent the first day and that doubled in pay each successive day. How much money would you earn in 30 days at this job?

WRITING EXERCISES

1. Describe how at least three of the dozen problem-solving tactics of Section 1.2 can be used to help in the solution of Exercise 27. (Even if you can't complete the proof for Exercise 27, you should be able to do this writing exercise.)

2. In what way does the four-step process for computing $S(n)$ depend on the Fundamental Theorem of Arithmetic? Explain.

PROPER DIVISORS

2.5

Numbers can be classified in many ways. They are either even or odd; they are prime or composite, or the number 1 itself; they may have exactly 12 divisors. A classification introduced by the ancient Greeks, which will be of interest to us, depends on the sum of the divisors of a number. Greek mathematicians thought of divisors slightly differently from the way presented here; they considered them to be "parts" of that number, and understandably, did not consider a number to be part of itself. To match up with this classification, we introduce the following definition.

DEFINITION

A number x is a **proper divisor** of y provided that $x \mid y$ and $x < y$.

Note that y is then the only "improper" divisor of y; all other divisors are proper. We can now approach the Greek classification determined by sums of proper divisors.

NOTATION

We denote the sum of all proper divisors of a number n by $P(n)$.

EXAMPLE 2.15

Problem: Determine $P(15)$.

Solution: The proper divisors of 15 are 1, 3, and 5; so it follows that

$$P(15) = 1 + 3 + 5 = 9$$

EXAMPLE 2.16

Problem: Determine $P(1800)$.

Solution: In Example 2.12 we noted that $1800 = 2^3 \times 3^2 \times 5^2$; so $D(1800) = 36$. Thus, 1800 has 35 proper divisors. It would be tedious to determine these 35 numbers and then add them; fortunately, there is no need to do so. In that same example, we found that $S(1800) = (15)(13)(31) = 6045$. Because this is the sum of all 36 divisors and because 1800 is the lone "improper" divisor included in the sum,

$$P(1800) = 6045 - 1800 = 4245$$

Example 2.16 illustrates that in general:

xxxxxxxxxxxxxxxxxxxxxxxxxxxxxxxxxxxxxxxxxxxxxxxxxxxxxxxxxxxxxx

(2.16)
$$P(n) = S(n) - n$$

xxxxxxxxxxxxxxxxxxxxxxxxxxxxxxxxxxxxxxxxxxxxxxxxxxxxxxxxxxxxxx

We can thus use the techniques of Section 2.4 to compute $S(n)$, and then, by Formula (2.16), subtract n to determine $P(n)$. Example 2.15 indicates that it is not always necessary to compute $S(n)$ first, however. It would be a waste of time when there are only a few divisors. Use common sense to select your method.

Finally, we can observe an interesting distinction in the results of Examples 2.15 and 2.16. It is not surprising that $P(1800)$ is much larger than $P(15)$. What is noteworthy is that $P(1800)$ is larger than 1800, but $P(15)$ is smaller than 15. Thus, we can categorize the two numbers differently, as did the Greeks.

DEFINITION

If $P(n) > n$, then n is **abundant**.

If $P(n) < n$, then n is **deficient**.

If $P(n) = n$, then n is **perfect**.

The simple connection between $P(n)$ and $S(n)$ suggests the following simple characterizations:

xxxxxxxxxxxxxxxxxxxxxxxxxxxxxxxxxxxxxxxxxxxxxxxxxxxxxxx

(2.17)   The number n is deficient if and only if $S(n) < 2n$.

(2.18)   The number n is abundant if and only if $S(n) > 2n$.

(2.19)   The number n is perfect if and only if $S(n) = 2n$.

xxxxxxxxxxxxxxxxxxxxxxxxxxxxxxxxxxxxxxxxxxxxxxxxxxxxxxx

Thus, using Statements (2.17) and (2.18), we see that 15 is deficient and 1800 is abundant. As the exercises show, there are infinitely many numbers that are abundant and infinitely many that are deficient. Questions about perfect numbers will occupy us for the next two sections.

2.5   EXERCISES

1. Determine $S(n)$ and $P(n)$ for each number from 1 to 30. Using the table you constructed for Exercise 31 of Section 2.2, record these values in the fourth and fifth columns, under the headings "$S(n)$" and "$P(n)$," respectively.

2. Classify each number from 1 to 30 as *deficient*, *abundant*, or *perfect*. Use the last column of the table you constructed for Exercise 31 of Section 2.2 to record these answers, under the heading "type."

For each of the numbers in Exercises 3–18, determine $P(n)$. (You have already determined $S(n)$ for these numbers in Exercises 7–22 of Section 2.4.)

3. 91

4. 120

5. 299

6. 313

7. 347

8. 496

9. 836

10. 1878

11. 3817

12. 4173

13. 21,168

14. 30,030

15. 64,372

16. 100,100

17. 299,299

18. 1,000,000

19–34. Classify each number in Exercises 3–18 as *deficient*, *abundant*, or *perfect*.

WRITING EXERCISES

1. Until the latter half of this century, number theory texts tended to use the term "aliquot part" for the concept now more commonly identified as "proper divisor." Write a short paper discussing and comparing these two phrases. What is the etymology of the words? Which do you think is more appropriate? Why? Can you suggest another word or phrase that you think would be better? On what basis?

2. In what ways are the words "abundant," "deficient," and "perfect" appropriate labels for the concepts they represent? In what ways are they inappropriate? Can you suggest an alternative set of labels for these types of numbers that is more descriptive or more appropriate? If so, give an argument to support your suggestion.

35. Show that any prime number is deficient. (Notice that this implies that there are infinitely many deficient numbers. Why?)

36. Show that any multiple of 12 is abundant. (*Hint:* Some, but not necessarily all, of the proper divisors of $12x$ are x, $2x$, $3x$, $4x$, and $6x$.) (This implies that there are infinitely many abundant numbers. Why?)

37. Show that every multiple of an abundant number is abundant. (See the preceding exercise.)

38. Show that 2^k is deficient for every value of k.

39. Give an example of an odd abundant number, if possible; if this is not possible, explain why not.

3. Describe how at least three of the dozen problem-solving tactics of Section 1.2 can be used to help in the solution of Exercise 36. (Even if you can't complete Exercise 36, you should be able to do this writing exercise.)

EVEN PERFECT NUMBERS

2.6

In working through the exercises of Section 2.5, you will have encountered several perfect numbers:

$$6 \qquad 28 \qquad 496$$

This is not a lot; but we should expect perfect numbers to be rare. After all, there are many more ways for $P(n)$ to be unequal to n than to be equal to it. Our goal is to identify the characteristic properties of perfect numbers. Are the three numbers cited here the only perfect numbers? If there are more, how many more? Is there a formula that will determine every perfect number, just as there is a formula that expresses the nth triangle number as

$$\frac{n(n+1)}{2}$$

(See Exercise 40 of Section 2.1.)

All that we have to go on in our pursuit of answers to these questions are the three perfect numbers identified so far. What do they have in common? Certainly, they are all even numbers; thus, we might restrict our attention to even numbers for the present. Among the numbers from 1 to 30, we know there are only two perfect numbers, 6 and 28, and that they are relatively close together. The only other example we know of at this point is 496, a long way from 28. The distribution of these three perfect numbers strongly suggests that there may be some additional ones between 28 and 496. We could test every number in this interval; however, before launching into such a sizable task, we should attempt some intelligent guesswork, perhaps along the following lines. Because 6 and 28 differ by 22, we might guess that another perfect number could occur at about 22 more than 28. Following this idea, let us test the number 50, as well as a few even numbers in the neighborhood of 50.

EXAMPLE 2.17

Problem: Determine whether any of the numbers 46, 48, 50, 52, or 54 is perfect.

Solution: $P(46) = 26,$ $P(48) = 76,$ $P(50) = 43,$
$P(52) = 46,$ $P(54) = 66$

Thus, 46, 50, and 52 are deficient, whereas 48 and 54 are abundant; none of these numbers are perfect.

The idea of an additive pattern from one perfect number to the next doesn't seem to work, so let's explore the possiblity of a multiplicative pattern. Note that 28 is between 4×6 and 5×6, but that 496 is between 17×28 and 18×28. It seems plausible to look for a potential "missing" perfect number that fills this gap by being approximately 4×28 and at the same time approximately $\frac{1}{4}$ of 496. Because $4 \times 28 = 112$ and $\frac{1}{4}$ of 496 equals 124, we might reasonably look at even numbers in the range from 112 to 124.

| EXAMPLE 2.18 | **Problem:** Determine whether any even number from 112 to 124 is perfect. |

Solution: $P(112) = 136,$ $P(114) = 126,$ $P(116) = 94,$
$P(118) = 62,$ $P(120) = 240,$ $P(122) = 64,$
$P(124) = 100$

Thus, 116, 118, 122, and 124 are deficient; 112, 114, and 120 are abundant; none of these numbers are perfect.

n	PRIME FACTORIZATION
6	6×3
28	$2^2 \times 7$
496	$2^4 \times 31$

Table 2.4 The prime factorizations of three perfect numbers.

Neither of these ideas has worked. Perhaps we did not look far enough. Perhaps we have been looking in the wrong place. If we reflect on the things we have seen in the earlier sections, we must note that determining $D(n)$, $S(n)$, and $P(n)$ depends on first finding the prime factorization of n. In fact, the Fundamental Theorem of Arithmetic provides the proof that the prime factorization of a number determines all the properties of the number that are definable in terms of its divisors. This suggests that we should search for the key to perfect numbers in their prime factorizations.

Table 2.4 shows a very strong pattern—in fact, several patterns. We have a perfect number with the factor 2 appearing once, one with the factor 2 appearing twice, and one with the factor 2 appearing four times. And each perfect number has a single odd prime factor, with the magnitude of that prime increasing as the number of 2s increases. Perhaps we should look for a perfect number with the factor 2 appearing three times (one may be missing from the list) and then one with 2 appearing five times, six times, and so on. Or we might look for a perfect number with five or more 2s in its prime factorization, as part of some number pattern that starts with 1, 2, 4, the three powers of 2 observed so far.

To pursue the first idea, we should examine numbers of the form $2^3 \times p$, where p is some odd prime between 7 and 31. The information about these numbers is given in *Table 2.5*. Although we do not find another perfect number among these numbers, we do see some interesting information there. When multiplied by 2^3, smaller odd primes

Table 2.5 Testing the primes between 7 and 31.

p	$n = 2^3 \times p$	$P(n)$	CLASSIFICATION	$n - P(n)$
11	88	92	Abundant	−4
13	104	106	Abundant	−2
17	136	134	Deficient	2
19	152	148	Deficient	4
23	184	176	Deficient	8
29	232	218	Deficient	14

produce abundant numbers whereas larger primes produce deficient numbers. If we consider the numerical difference n minus $P(n)$, we find it closest to zero (the difference for a perfect number) when the value of the prime is 13 or 17, and we see that this difference is farther away from zero as we pick odd primes farther away in either direction.

The differences recorded in the last column of *Table 2.5* suggest that we should try 15, the number halfway between 13 and 17, and hope for a difference of $n - P(n) = 0$, which would mean that n is a perfect number. However, as we saw in Example 2.18,

$$P(2^3 \times 15) = P(120) = 240$$

and thus 120 is "very" abundant. This is to be expected, of course, because 15 is composite, which makes 120 quite unlike the numbers tested in *Table 2.5*. Each of those numbers has exactly 8 divisors, whereas 120 has 16 divisors, an abundance of them. What is called for is a prime number halfway between 13 and 17. Of course, no such number exists; but if it did, we would have the perfect makings of another perfect number. For if 15 were prime, there would be only eight divisors of 120, namely 1, 2, 4, 8, 15, 30, 60, and 120, and the sum of the proper divisors would be 120.

Another way to put all this is as follows: Each perfect number in *Table 2.4* consists of a power of 2 times an odd prime number that happens to occur in "the right place"; but there happens to be no prime number in the right place to match up with 2^3. Observe that 3, 7, and 31 are primes, and are also factors of perfect numbers. Further note that 15 is an odd number that fits neatly into the gap between 7 and 31; however, it is composite, and thus fails to yield a perfect number.

We have not yet found another perfect number, but we have made excellent progress in that direction. We now know "good" odd numbers to be 3, 7, 15, and 31. It is not too difficult to see a pattern in these numbers. Each is twice the preceding number, plus 1. Another way to describe the pattern is to note that each number is, respectively, 1 less than 4, 8, 16, and 32. Thus,

$$3 = 2^2 - 1, \qquad 7 = 2^3 - 1, \qquad 15 = 2^4 - 1, \qquad 31 = 2^5 - 1$$

The former observation is useful for successively computing terms of the sequence; the latter description is preferable for identifying a formula for a typical term—namely, $2^k - 1$. The next number in this sequence of odd numbers is 63, which should be multiplied by 2^5 in order to continue the pattern of *Table 2.5*. Because 63 is composite, we might expect a situation similar to that encountered with $2^3 \times 15$. This is precisely the case, as you are asked to verify in the exercises.

If we go to the next step in this progression, we find

$$(2 \times 63) + 1 = 127$$

which is a prime number. And when we examine

$$2^6 \times 127 = 8128$$

we find that

$$\begin{aligned} S(8128) &= S(2^6) \times S(127) \\ &= (2^7 - 1) \times 128 \\ &= 127 \times 128 \\ &= 16{,}256 \end{aligned}$$

Therefore,

$$P(8128) = 16{,}256 - 8128 = 8128$$

Thus, 8128 is a perfect number!

We should now have enough information to formulate a precise conjecture and, hopefully, to prove it.

×××

(2.20)   If $2^k - 1$ is prime, then $(2^{k-1})(2^k - 1)$ is perfect.

×××

The proof of the conjecture is merely a generalization of the computation done above for the number 8128. We will calculate $S(n)$, where $n = (2^{k-1})(2^k - 1)$:

$$S(n) = S(2^{k-1}) \times S(2^k - 1)$$

as we saw in Section 2.4. But

$$S(2^{k-1}) = 2^k - 1$$

by Formula (2.15), and

$$S(2^{k-1}) = (2^k - 1) + 1 = 2^k$$

because $2^k - 1$ is prime. Therefore,

$$S(n) = (2^k - 1)(2^k)$$

This is almost the prime factorization of n (with the order of the factors interchanged). But $S(n)$ contains a factor of 2^k, rather than of 2^{k-1}. Consequently,

$$S(n) = (2^k - 1)(2^k) = (2^k - 1)(2^{k-1})(2) = 2n$$

and, by the characterization in Statement (2.19), n is perfect. We have proved Statement (2.20), and can label it a theorem.

This theorem was known to the ancient Greeks and was recorded in Euclid's *Elements*. Perfect numbers of this form are thus known as "Euclidean (perfect) numbers."

DEFINITION   A number of the form $(2^{k-1})(2^k - 1)$ is called a **Euclidean number**, and is denoted E_k.

Thus, 6 is a Euclidean number because

$$6 = 2 \times 3 = 2^1 \times (2^2 - 1) = (2^{2-1})(2^2 - 1)$$

Specifically, $6 = E_2$. Similarly,

$$E_3 = 4 \times 7 = 28, \qquad E_4 = 8 \times 15 = 120$$

and so on. Using this notation, our theorem could be reworded:

Counting Board.
Medieval woodcut.

Credit: Smithsonian Institution

XX

(2.21)   If $2^k - 1$ is prime, the Euclidean number E_k is perfect.

XXXXXXXXXXXXXXXXXXXXXXXXXXXXXXXXXXXXXXXXXXXXXXXXXXXXXX

So, our investigation of the initial three even perfect numbers has led us to identify them as examples of Euclidean perfect numbers—numbers that consist of a power of 2 times a single odd prime. But might there not be other even perfect numbers, with prime factors not of the form $2^k - 1$, or with two or more odd prime factors? Surprisingly, the answer is *no*, as the following statement indicates.

XXXXXXXXXXXXXXXXXXXXXXXXXXXXXXXXXXXXXXXXXXXXXXXXXXXXXX

(2.22)   If n is an even perfect number, it is of the form $(2^{k-1})(2^k - 1)$, where $2^k - 1$ is prime.

XXXXXXXXXXXXXXXXXXXXXXXXXXXXXXXXXXXXXXXXXXXXXXXXXXXXXX

In other words, if n is an even perfect number, it has a single odd prime factor which is related to the power of 2 in precisely the manner that makes it a Euclidean number. The proof of this statement is considerably more involved than the others presented here, and thus it is omitted.

2.6   EXERCISES

1. What is E_5?

2. What is E_1?

3. Verify that $2^5 \times 63$ is abundant.

4. Show that $2^5 \times p$ is abundant if p is any odd prime less than 63, and is deficient if p is any prime greater than 63. (See *Table 2.5*.)

5. *Prove or disprove*: If an even number has two or more odd prime factors, it is abundant.

6. *Prove or disprove*: The sum of the reciprocals of the divisors of a perfect number is 2. $\left(\text{The reciprocal of } x \text{ is } \frac{1}{x}.\right)$

7. *Prove or disprove*: Every Euclidean number is a triangle number. (See Exercises 37 and 40 of Section 2.1.)

8. Show that any multiple kn of a perfect number n is abundant, provided that $k > 1$. (*Hint*: See Exercise 37 of Section 2.5.)

9. Show that the final digit of 2^k must be 2, 4, 6, or 8 (and hence the final digit of $2^k - 1$ must be 1, 3, 5, or 7.) (*Hint*: Look at the pattern for small values of k.)

10. Show that the final digit of E_k, for $k > 1$, must be 0, 6, or 8. (See the preceding exercise.)

WRITING EXERCISES

1. Discuss how at least four of the problem-solving techniques of Section 1.2 are used in developing the ideas of this section. Identify specific passages in the text to illustrate your discussion.

2. **(a)** Write Statement (2.20) completely in words, without using *any* symbolic notation, not even letters to represent numbers.

   (b) Write the proof of Statement (2.20) completely in words, without using *any* symbolic notation, not even letters to represent numbers.

   (c) Which parts (if any) of Statement (2.20) and its proof are easier without symbols? Which parts are harder? Which parts are impossible? Justify your answers, briefly.

   (d) Use Parts **(a)** and/or **(b)** to illustrate a general discussion of the pros and cons of using symbols in presenting mathematical ideas.

MERSENNE PRIMES

2.7

In Section 2.6 we found that every odd prime number of the form $2^k - 1$ is a factor of a Euclidean perfect number and also that every even perfect number is Euclidean and has exactly one odd prime factor, which is of the form $2^k - 1$. The search for additional even perfect numbers, then, boils down to an examination of numbers of the form $2^k - 1$, to determine which are prime.

DEFINITION   A number of the form $2^k - 1$ is called a **Mersenne number** and is denoted by M_k.[7]

In terms of this notation, we can rephrase Statement (2.21) as follows:

xxxxxxxxxxxxxxxxxxxxxxxxxxxxxxxxxxxxxxxxxxxxxxxxxxxxxxx

(2.23)   If M_k is prime, then E_k is perfect.

xxxxxxxxxxxxxxxxxxxxxxxxxxxxxxxxxxxxxxxxxxxxxxxxxxxxxxx

---

7. Marin Mersenne (1588–1648) was a Franciscan friar, an important music theorist, and an amateur mathematician. He spent a great deal of time corresponding about a variety of matters, including mathematics, with prominent intellectuals of his day, fostering communication among them. His interest in number theory is honored by the numbers that bear his name.

Note

Statement (2.22) implies that the converse of Statement (2.23) is also true.[8]

k	M_k	CLASSIFICA-TION
1	1	
2	3	Prime
3	7	Prime
4	15	Composite
5	31	Prime
6	63	Composite
7	127	Prime

Table 2.6 The first seven Mersenne numbers.

To determine whether or not M_k is prime for a given k, we could always resort to the definition of prime numbers. Given sufficient time, M_k can be tested (for a specific value of k) to determine if it is prime. But, as we have said so often before, there ought to be a better way. Let us examine what we already know in an organized fashion, and look for some pattern that may guide us to a general truth; what we have already observed about M_k is displayed in *Table 2.6*.

The next Mersenne number, M_8 (= 255), is clearly composite, which suggests the possibility that the Mersenne numbers are alternately prime and composite, after an initial anomaly. Unfortunately, this hypothesis is disproved by the fact that $M_9 = 511 = 7 \times 73$, a composite number.

A different conjecture, which corresponds to all the data so far, is suggested by comparing the subscript k with the classification of M_k. We note that M_k is prime when k equals 2, 3, 5, or 7 and is composite when k equals 4, 6, 8, or 9. The pattern is striking in its simplicity, and we formulate it as follows:

xxxxxxxxxxxxxxxxxxxxxxxxxxxxxxxxxxxxxxxxxxxxxxxxxxxxxxxxxx

(2.24)

Conjecture: M_k is prime if k is prime, and composite if k is composite.

xxxxxxxxxxxxxxxxxxxxxxxxxxxxxxxxxxxxxxxxxxxxxxxxxxxxxxxxxx

Note

The only number that is neither prime nor composite is 1. In the special case of 1 as the subscript, we even have $M_1 = 1$.

Conjecture (2.24) suggests that M_{10} should be composite, M_{11} prime, M_{12} composite, and so on. Furthermore, note that there are two parts to this conjecture; let us separate them, taking the "composite" part first. We can obtain a lead to a more precise connection between k and M_k by observing some of the divisors of the composite occurrences of M_k; to this end, we list the prime factorizations of these Mersenne numbers in *Table 2.7*.

Table 2.7 suggests that a Mersenne prime is always among the prime factors of a Mersenne number. Specifically,

$$3 \ (= M_2) \text{ is a factor of } M_4, M_6, M_8, \text{ and } M_{10};$$
$$7 \ (= M_3) \text{ is a factor of } M_6 \text{ and } M_9;$$
$$31 \ (= M_5) \text{ is a factor of } M_{10}.$$

---

8. The term *converse* is explained in Section A.3 of the Appendix.

Again, there is a striking correlation among the Mersenne numbers and their subscripts, suggesting the following:

xxxxxxxxxxxxxxxxxxxxxxxxxxxxxxxxxxxxxxxxxxxxxxxxxxxxxxxx

(2.25)   If $c \mid k$, then $M_c \mid M_k$.

xxxxxxxxxxxxxxxxxxxxxxxxxxxxxxxxxxxxxxxxxxxxxxxxxxxxxxxx

Statement (2.25) turns out to be fairly easy to prove, using Formula (2.15) and some laws of exponents. We consider the fraction

$$\frac{M_k}{M_c} = \frac{2^k - 1}{2^c - 1}$$

Because $c \mid k$, we know there is some integer b such that $k = b \cdot c$; so our fraction becomes

$$\frac{M_k}{M_c} = \frac{2^{bc} - 1}{2^c - 1} = \frac{(2^c)^b - 1}{2^c - 1}$$

This fraction has the same form as the right side of the Formula (2.15), with $n = 2^c$ and $k + 1 = b$; hence, substituting 2^c for n and b for $k + 1$ in the left side of that formula yields

$$1 + 2^c + (2^c)^2 + (2^c)^3 + \ldots + (2^c)^{b-1}$$

Thus, $\frac{M_k}{M_c}$ equals the sum of a series of whole numbers, and hence must itself be a whole number, implying that $M_c \mid M_k$.

As a simple corollary, if k is composite, it has a divisor c such that $1 < c < k$. Then

$$1 < M_c < M_k \qquad \text{and} \qquad M_c \mid M_k$$

establishing that M_k is composite. This is the proof of

xxxxxxxxxxxxxxxxxxxxxxxxxxxxxxxxxxxxxxxxxxxxxxxxxxxxxxxx

(2.26)   If k is composite, then M_k is composite.

xxxxxxxxxxxxxxxxxxxxxxxxxxxxxxxxxxxxxxxxxxxxxxxxxxxxxxxx

k	M_k	Prime Factorization of M_k
4	15	3×5
6	63	$3^2 \times 7$
8	255	$3 \times 5 \times 17$
9	511	7×73
10	1023	$3 \times 11 \times 31$

Table 2.7 Factorizations of Mersenne numbers.

Thus, we have proved *half* of Conjecture (2.24). The other half, asserting that M_k is prime when k is prime, turns out to be false! It is not true for $k = 11$ and for many other prime values of k. Although 11 is prime, M_{11} is composite, as you are asked to verify in the exercises.

This is one of the classic examples in mathematics of a persuasive pattern turning out to be misleading. Many mathematicians believed in the truth of the conjecture about Mersenne primes, and as recently as the beginning of this century printed works appeared with the erroneous assertion that M_{11} is prime.[9] The falsehood of such a persuasive conjecture that was based on several verified examples is a paramount example of why mathematics insists on rigorous proofs for every assertion, even the seemingly obvious ones. Simplicity of form, successful prediction of examples, and majority belief can all be wrong.

So, where do we stand in our search for Mersenne primes, which are literally the key factors of even perfect numbers? Statement (2.26) narrows the search to Mersenne numbers with prime subscripts; but each such number must be tested to determine whether it is prime or composite. Although some additional winnowing can be effected by considering the form of the prime, it is basically true that, to date, no pattern has been found to characterize Mersenne primes. Special algorithms for testing the primeness of large numbers and high-speed computers have extended the list of known Mersenne primes (see *Table 2.8*), but no general characterization of Mersenne primes has been proved.

Table 2.8 identifies all known Mersenne primes (as of the printing of this edition). As you can see, the smaller primes often generate corresponding Mersenne numbers that are prime. For example, the first four primes (2, 3, 5, and 7) generate Mersenne primes, as do three of the next four (13, 17, and 19). But this frequency is deceptive as a general guide, as the overall occurrence of Mersenne primes is very rare and highly sporadic. The magnitude of these numbers, after the first few, is truly monumental. And, of course, each corresponding perfect number has a factor of the form 2^{k-1}, which is approximately the same size as the Mersenne prime factor. Thus, the thirty-third (and largest known, to date) perfect number contains 517,430 digits. This number, if printed in the type size and page style of this book, with the requisite 172,476 commas, would fill 180 pages; if printed on a single line, it would stretch 3600 feet, or nearly $\frac{7}{10}$ of a mile!

To round out our discussion of perfect numbers, we consider now the question of *odd* perfect numbers. Briefly, nothing definitive is known about odd perfect numbers. More precisely, no examples of such numbers are known, and theoretical methods have shown that

9. This is so despite the fact that M_{11} was shown to be composite by Cataldi in 1603.

any possible odd perfect numbers must be extremely large. (It can be proved, for example, that an odd perfect number, if one exists, must have at least six distinct prime factors; and it has been shown that there are no odd perfect numbers less than 100,000,000,000,000,000,000.) So the evidence strongly suggests, and many mathematicians believe, that no odd perfect numbers exist; so far, however, this conjecture has defied proof.

k	M_k	Number of Digits	Discoverer and Date
2	3	1	Unknown (before 300 B.C.)
3	7	1	Unknown (before 300 B.C.)
5	31	2	Unknown (before 300 B.C.)
7	127	3	Unknown (before 300 B.C.)
13	8191	4	Regius (1536)
17	131,071	6	Cataldi (1603)
19	524,287	6	Euler (1772)
31	2,147,483,647	10	Euler (1772)
61		19	Seelhoff and Pervusin (1886)
89		27	Powers and Cunningham (1914)
107		33	Uhler (1952)
127		39	Uhler (1952)
521		157	Robinson (1952)
607		183	Robinson (1952)
1279		386	Robinson (1952)
2203		664	Robinson (1952)
2281		687	Robinson (1952)
3217		969	Lucas and Lehmer (1962)
4253		1281	Lucas and Lehmer (1962)
4423		1332	Lucas and Lehmer (1962)
9689		2917	Lucas and Lehmer (1962)
9941		2993	Lucas and Lehmer (1962)
11,213		3376	Gillies (1964)
19,937		6002	Tuckerman (1971)
21,701		6533	Nickel and Noll (1978)
23,209		6987	Noll (1979)
44,497		13,395	Slowinski (1979)
86,243		25,962	Slowinski (1982)
110,503		33,265	Colquitt and Welsh (1988)
132,049		39,751	Slowinski (1983)
216,091		65,050	Slowinski (1985)
756,839		227,832	Slowinski (1992)
859,433		258,716	Slowinski (1993)

Table 2.8 The known Mersenne primes.

2.7   EXERCISES

In Exercises 1–8, state whether the following are prime or composite.

1. M_{11} 2. M_{12} 3. M_{19} 4. M_{25}
5. M_{29} 6. M_{39} 7. M_{89} 8. M_{121}

In Exercises 9–16, state whether or not the following are perfect.

9. E_{11} 10. E_{12} 11. E_{19} 12. E_{25}
13. E_{29} 14. E_{39} 15. E_{89} 16. E_{121}

17. What number can be multiplied by M_{89} to produce a perfect number?

18. What number can be multiplied by $(2^{2203} - 1)$ to produce a perfect number?

19. What odd number can be multiplied by 2^{12} to produce a perfect number?

20. What number can be multiplied by 2^{10} to produce a perfect number?

In Exercises 21–34, determine whether the given statement is *true* (in all instances), or *false* (*not* true in *all* instances), or whether its truth or falsehood is unknown.

21. If k is prime, then M_k is prime.

22. If M_k is prime, then k is prime.

23. If k is composite, then M_k is composite.

24. If M_k is composite, then k is composite.

25. If M_k is prime, then E_k is prime.

26. If M_k is prime, then E_k is perfect.

27. If E_k is perfect, then k is prime.

28. All perfect numbers are Euclidean.

29. All Euclidean numbers are perfect.

30. All Euclidean numbers are even.

31. All Mersenne numbers are odd.

32. All perfect numbers are even.

33. There are an infinite number of Mersenne numbers.

34. There are an infinite number of perfect numbers.

In Exercises 35–40, find one prime divisor of the given number.

35. M_{11} 36. M_{12} 37. M_{19} 38. M_{25}
39. M_{39} 40. M_{121}

41. Determine two distinct prime factors of M_{11}.

42. Determine three distinct prime factors of M_{20}.

43. Determine three distinct prime factors of M_{21}.

44. What is the prime factorization of M_{15} (which equals 32,767)?

45. Show that the final digit of an even perfect number must be either 6 or 8. (See Exercise 10 of Section 2.6.)

WRITING EXERCISES

1. Look up Marin Mersenne in a book on the history of mathematics and also in a book on the history of music; then write a short biographical sketch of him. (Be sure to list the sources you used in writing the paper, and be careful not to simply copy or paraphrase the material from your sources.)

2. Look up one (or one of the pairs) of the twentieth century discoverers of Mersenne primes listed in *Table 2.8*. Consult appropriate library sources and write a paragraph or two summarizing what you find.

3. If you were setting out to search for an odd perfect number, where and how would you start? Look back at Section 1.2 for help in answering this question.

LINK: NUMBER THEORY AND CRYPTOGRAPHY

Cryptography: The Design and Use of Codes

2.8 We have strongly suggested that number theory is an example of "pure" mathematics, and this is so. But there are numerous applications of this field within and outside of mathematics. From the pencil-and-paper algorithms of arithmetic to the design of computer software, one finds uses of number theory. A rather surprising use of some basic properties of numbers is found in the area of cryptography—design and use of codes—and cryptanalysis—the science of code breaking.

The desire for secrecy in communication and the countereffort to discover others' secrets are probably as old as man himself. And from the earliest stages of those secret communications, number patterns have been used to systematize the coding and to aid in the code breaking. For example, Julius Caesar used a simple code in his military messages, substituting for each letter the one occurring three letters earlier in the alphabet. Of course, this simplest of coding methods is also simple to figure out.

As societies grew in complexity, so did their cryptography and crypt-analysis. By the Renaissance, cryptography had reached the level of a science. In the sixteenth century Blaise de Vigenere formalized the so-called "polyalphabetic" system of coding. In this system, one uses several of the 26 cyclic re-arrangements of the alphabet, such as

U V W X Y Z A B C ... R S T

in a substitution code. The choice and pattern of substitute alphabets is determined by a key word or sequence of numbers, hopefully known only to the sender and receiver of the message. Later generations have used noncyclic rearrangements of the alphabet to tap the millions of such variations, again maintaining utility of these by use of longer and longer key sequences of letters or numbers.

Each escalation of complexity in cryptography has called forth an equal complexity in cryptanalysis, drawing not only on number theory, but on probability, statistics, abstract algebra, and graph theory. And until recently, virtually every code and key used by the cryptographer has been unravelled and discovered by the cryptanalyst.

In today's world of high-speed communication, banks, corporations, law-enforcement agencies, and other institutions need to transmit confidential information over public phone lines, airwaves, or computer networks. This would be feasible with coded messages if each different pair of correspondents shared a different code key for their messages. Of course, the keys themselves could not be transmitted openly, and would require personal meetings or safe couriers; the maintenance of accurate and secure records of which correspondent matched which key would be a bookkeeper's nightmare and a security agent's undoing.

To meet these communication problems, cryptographers have developed "public-key" systems. In such a system both the method of coding and a "one-way" key for each potential receiver are public information. In one such system, the RSA (developed by Ronald Rivest, Adi Shamir, and Leonard Adleman of M.I.T.), a user selects a pair of large (say 100-digit) prime numbers. After calculating the product of these two primes (a number we shall call a) and the number of natural numbers less than a that have no prime factors in common with a, the user (arbitrarily) chooses another number that has no prime factors in common with a. This number and the number a are made public. Anyone wishing to send this individual a coded message uses this public key, following a known encoding process. In order to decode the message, it is essential to be able to compute the original prime factors of the product key number. Without this factorization, even someone who knows both the coded message and the encoding process would not be able to break the code. To date, no one has found a practical method for factoring such large composite numbers. Although, as we saw in this chapter, any number can be factored uniquely into primes, this is a theoretical fact, not a practical method for handling large (200-digit) numbers. Any procedures currently known, even with the largest and fastest computers, would require many thousands of years to factor such a number!

In 1982, a ripple of concern ran through the National Security Agency, the nation's code-watching authority, when two Europeans, Hendrik Lenstra and Henri Cohen, devised a practical computer method for identifying large prime numbers of arbitrary type (as opposed to primes of special types, such as Mersenne primes.) The fear was that similar methods might solve the factorization problem for large composite numbers. So far, this has not proved to be the case, and some feel that the RSA public-key crypto system will continue to be unbreakable.

Whatever the ultimate fate of factorization and the current coding system, cryptography and its use of both simple and increasingly sophisticated mathematics seems destined to be with us for the foreseeable future.

TOPICS FOR PAPERS

1. Examine the six-column table you constructed for Exercises 31 of Section 2.2 and 1 and 2 of Section 2.5. Look for some interesting property that a few, but not too many, of the numbers between 1 and 30 have. For example, looking at the data for 18, you might note that $D(18) = 6$, $S(18) = 39$, and $P(18) = 21$; all three of these numbers are divisible by 3, and no other numbers on the list share this property! In other words, 18 is the only number between 1 and 30 for which $D(n)$, $S(n)$, and $P(n)$ are all multiples of 3. There must be others, one would think. What would they look like? How can they be found? Is there a formula that describes all of them? or some of them? Is there some property that would exclude certain kinds of numbers from consideration?

   Once you have found one or two properties like this, get your instructor's approval to use one of them as a topic. (If you can't find any such property, ask your instructor to help you.) The type of number described by the topic you choose will be yours to name and investigate, *using as many of the problem-solving techniques of Chapter 1 as you find helpful*. Your paper is to be a description of this investigation, reporting the specific questions you asked yourself, describing in detail the particular problem-solving tactics you pursued, showing where they led, etc. (In this type of investigation, even a blind alley is something to report, provided you can describe the alley.) The paper should begin with a clear definition of your kind of number, and it should end with some questions that seem to be promising avenues for further investigation. If you generate more numerical data, explain how it was done. Be as creative as you like in looking for patterns and making conjectures, but try to prove your conjectures or at least supply some evidence or argument to support them.

Remember that the main significance of the paper is *how you go about investigating your kind of number.* The results of your investigation, while interesting to report, are less important than the procedures you used to find them.

2. *For Further Reading,* which follows, lists several introductory texts on number theory. Research one or more of these or a text on the history of mathematics for some of the following topics:

   - semiperfect numbers;
   - pseudo-perfect numbers;
   - multiply perfect numbers (doubly perfect, triply perfect, etc.);
   - weird numbers;
   - amicable pairs.

Write a descriptive paper that explains in detail one or two of these topics. For each topic you discuss, your first paragraph should include a definition of the type of number in question, and the last paragraph should include some questions that seem to be promising avenues for further investigation of the topic.

FOR FURTHER READING

1. Armendariz, Efraim P., and Stephen J. McAdam. *Elementary Number Theory.* New York: Macmillan Publishing Co., Inc., 1980.

2. Beck, A., M. Bleicher, and D. Crowe. *Excursions into Mathematics.* New York: Worth Publishers, Inc., 1969.

3. Burton, David M. *Elementary Number Theory,* 2nd Ed. Dubuque, Iowa: Wm. C. Brown Publishers, 1989.

4. Devlin, Keith. *Microchip Mathematics: Number Theory for Computer Users.* Cheshire, England: Shiva Publishing Limited, 1984.

5. Dudley, Underwood. *Elementary Number Theory,* 2nd Ed. New York: W. H. Freeman and Company, 1978.

6. Dunham, William. *Journey Through Genius: The Great Theorems of Mathematics.* New York: John Wiley and Sons, Inc., 1990, Chapter 3.

7. Edgar, Hugh M. *A First Course in Number Theory.* Belmont, CA: Wadsworth Publishing Company, 1988.

8. Gardner, Martin. "Mathematical Games: A short treatise on the useless elegance of perfect numbers and amicable pairs." *Scientific American,* 218 (3): 121–26, 1968.

9. Hellman, Martin E. "The Mathematics of Public-Key Cryptography." *Scientific American,* 241 (2): 146–57, 1979.

10. Kahn, David. *The Code Breakers*. New York: The Macmillan Co., Inc., 1967.

11. Niven, Ivan, and H. S. Zuckerman. *An Introduction to the Theory of Numbers*, 4th Ed. New York: John Wiley & Sons, 1980.

12. Riebenboim, Paulo. "Prime Number Records." *The College Mathematics Journal*, 25 (4): 280, 1994.

13. Slowinski, D. "Searching for the 27th Mersenne Prime." *Journal of Recreational Mathematics*, 11 (4): 258–67, 1979.

Euclid Greeting Students at the Outer Gate of the Circle of Knowledge.

MATHEMATICS OF AXIOM SYSTEMS: GEOMETRIES

WHAT IS GEOMETRY?

3.1

The key to understanding geometry lies in its history. As you will see in this chapter, the meaning of geometry has evolved greatly over the last three thousand years or so. The word, "geometry," meaning "earth measurement," describes its own beginnings. The mathematical ancestors of modern geometers were the land surveyors of ancient Egypt, whose job was to reestablish boundaries that had been washed away by the periodic flooding of the Nile. They, along with the builders of Egypt and Babylon and the navigators of the coastal-trading civilizations along the Mediterranean, were among the first people to use the mathematical properties of *lines*, *angles*, and *circles* in systematic ways to produce useful results. However, just as we can use a power saw without knowing the electrical principles that make it work, so those ancient craftsmen used their geometric tools without probing for any unifying theory to explain their reliability.

Geometry as a mathematical discipline began with Thales, a wealthy Greek merchant of the sixth century B.C. He is considered to have been the first Greek philosopher, as well as the father of geometry as a deductive study. The significance of Thales's work is that he began the search for unifying rational explanations of reality, thus moving away from the previous reliance on religion and mythology to explain the phenomena of nature. His search for some underlying unity in geometric ideas led him to investigate how some geometric statements

Ancient Egyptian Surveyors. Geometric methods were developed to reestablish property boundaries obliterated by the Nile's annual flooding.

could be derived logically from others. The statements themselves were well-known, but the process of linking them by logic was new.

The Pythagoreans and other Greek thinkers continued the logical development of geometric principles, which culminated in Euclid's *Elements.* Building on some three centuries of work, Euclid organized and extended all that was known about mathematics in the Greece of 300 B.C. Although the thirteen books of the *Elements* contain considerable material on arithmetic and number theory, Euclid paid special attention to geometry. His goal was to systematize the various observable relationships among spatial figures, which he, like Plato, Aristotle, and the other Greek philosophers, regarded as ideal representations of physical entities.

Euclid listed a small number of basic statements that appeared to capture the essential properties of *points*, *lines*, *angles*, etc.; he then proceeded to derive the rest of geometry from these basic statements by means of a series of carefully proved propositions. Euclid's work in geometry was so comprehensive and clear that his *Elements* became the universally accepted source for the study of plane geometry from his time on. Even the geometry studied in high school today is essentially an adaptation of Euclid's *Elements.*

Euclid's work was so good, in fact, that it took over two thousand years to find any flaw in his system. Not until the nineteenth century did mathematicians begin to suspect that there might be other ways of looking at geometry. They gradually realized that the "space" we live in need not conform to Euclid's assumptions, and that some spatial properties can actually be explained more easily from a different viewpoint, that is, in another geometric system. The question that flows immediately from this realization is the motivating theme of the present chapter:

If Euclidean geometry is not *the* geometry, then what *is* geometry?

This question is almost as difficult as "What is mathematics?," and replies to it seem nearly as numerous as geometers themselves. Rather than giving you a glib, shallow definition, in this chapter we try to capture a sense of the historical evolution of the subject, employing Euclid's answer to the question as the basis for discussing more recent viewpoints. In the course of doing this, we briefly describe several different kinds of geometry. After you have seen these various types of geometry "in operation," we shall be able to draw some conclusions about geometry in general, and also about the relationship between modern mathematics and the world we live in.

| 3.1 | **EXERCISES** |

WRITING EXERCISES

1. **(a)** Find three English words besides *geometry* that begin with *geo*, and indicate how their meanings relate to "earth."

   (b) Find three English words besides *geometry* that contain *metr*, and indicate how their meanings relate to "measure."

2. Suppose you want to measure and mark off a 100-by-200-foot rectangular shore lot at the edge of a lake in such a way that one short side of the rectangle runs along the (straight) lake shore. The only tools you have are a supply of marking stakes and a 50-foot cloth tape measure. Describe how you would do it and what principles of plane geometry your method would depend on. (Remember that you must produce straight sides and right angles.)

EUCLIDEAN GEOMETRY

Euclidean geometry is essentially the same as the geometry you studied in high school; so our discussion of it will presuppose some familiarity with its basic language. However, several difficulties that often arise in the treatment of this material need to be pointed out, especially insofar as they are due in large measure to Euclid himself.

Recall that Euclid's goal was to derive as many geometric statements as possible by strictly logical argument from a small number of initial statements, assumed as a foundation because they were so "obviously" true. The derived statements he called **theorems**; the initial statements were divided into two categories, some he called **axioms**, others **postulates**.[1] In addition, Euclid started off most of the thirteen books of the *Elements* with a list of **definitions**, describing the technical terms that would appear in that book. For example, Book I has twenty-three definitions, covering such terms as *point, line, angle, circle, triangle, quadrilateral,* and *parallel*. For later reference, let us state just two of these:

> A **circle** is a plane figure consisting of all points at a given distance from one point within the figure, called its **center**.

> Two lines are **parallel** if they are contained in the same plane, but do not meet, no matter how far they are extended in either direction.[2]

The *axioms*, also known as *common notions*, were general statements about quantity and logic; Euclid asserted the following five:

1. Things which are equal to the same thing are equal to each other.
2. If equals are added to equals, the results are equal.
3. If equals are subtracted from equals, the results are equal.
4. Things which coincide with each other are equal.
5. The whole is greater than any of its parts.

The *postulates* were specific geometric ideas that were assumed to be true. Euclid based all of his *plane geometry* on the following five postulates:

E1. A line segment can be drawn from any point to any other point.

E2. A line segment can be extended continuously in a straight line.

E3. A circle can be drawn with any center and distance (that is, *radius*).

E4. All right angles are equal to one another.

E5. Through a given point not on a given line exactly one line can be drawn parallel to a given line.

We should note that Postulate **E5**, as stated here, is an alternate and logically equivalent form of **Euclid's Fifth Postulate**; it is called

---

1. Modern mathematics no longer distinguishes between the terms *axiom* and *postulate*; they are considered synonyms that refer to any statements assumed to be true without proof.
2. The language in these definitions, as in the statement of Euclid's axioms and postulates, is not a literal translation of Euclid's language, but uses terminology that is more modern, and hopefully more familiar.

Playfair's Postulate, named after the British mathematician John Playfair (1748–1819) who formulated it.[3]

Euclid's conception of points, lines, triangles and circles was that they were objects that could be drawn; this notion is apparent in his postulates and is corroborated in many of his proofs. Of course, the drawing of figures was not to be thought of as the physical drawing on a tablet or papyrus, but rather, as the construction of idealized objects in an "ideal" space. But this concept of geometric figures, real or ideal, allowed "proof by picture" to creep unnoticed into many of Euclid's deductive arguments. For instance, there is nothing in Euclid's definitions or postulates to guarantee that a straight line passing through the center of a circle must actually have a point in common with the circle, and yet many of his proofs depend on the location of such an intersection point.

"But," you say, "it is obvious that a line cannot pass from the inside to the outside of a circle without crossing it at some point, because there are no gaps in a circle; it is continuous."

A moment's reflection should convince you that you are deriving this property of circles from the way we visualize them, not from what Euclid explicitly defined or postulated. Euclid tacitly assumed many similar ideas, such as the following:

(3.1)   A line that intersects one side of a triangle but does not pass through any vertex of the triangle must intersect one of the other sides.

(3.2)   Given any three distinct points on the same line, one of them is between the other two.

These statements are "obvious" only if we think of geometry in terms of pictures; they cannot be deduced from Euclid's axioms and postulates. It is not that these and similar properties are incorrect or undesirable; a review of Euclid's work makes it obvious that he accepted them, used them, and fully intended that they be part of his system. But Euclid had set out on a revolutionary path in that he intended all properties of geometric figures *either* to be proved from more fundamental properties *or* to be identified as axiomatic assumptions. Assertions (3.1) and (3.2), though, fall into neither category. Just as they are "obvious" to us, they were so obvious to Euclid that he failed to notice that he was *using* these properties of geometric figures (and others) as axiomatic fundamentals in the deductive development of his theorems. Consequently, he neglected to identify them as such.

---

3. A complete list of Euclid's original definitions, axioms, and postulates can be found on pages 38–39 of Item 4 in the list *For Further Reading* at the end of this chapter.

In order to circumvent such logical difficulties, David Hilbert, in the early part of this century, renovated the foundations of Euclidean geometry by selecting a full set of assumptions that accurately describes geometry as Euclid envisioned it. In Hilbert's system all the things that "ought to work" in Euclidean geometry are established from his set of explicit axioms. From this point on, we will ignore the historical distinction between Euclid's geometry and Hilbert's, treating this study as a single type called **Euclidean geometry** to distinguish it from the other geometric systems of modern mathematics.

Despite what has just been said, a large part of the legitimate study of Euclidean geometry throughout its history has consisted in drawing pictures, or, more precisely, in determining which figures *can be drawn* using only the tools corresponding to Euclid's first three postulates, **E1**, **E2**, and **E3**. The first two of these allow the construction of a (straight) line segment joining two points or the extension of such a segment to a longer one, and thus they correspond to drawing a line physically guided by a straightedge. The third postulate provides for the construction of a circle with given center and radius, and hence corresponds to the physical drawing accomplished with a compass. Collectively, these are known as the **Euclidean tools**, and any figure that can be constructed only using them is called a **compass-and-straightedge construction**. The challenge of such a construction is to draw a specified figure in the Euclidean plane (an "ideal" flat surface), using only the Euclidean tools. This is one of the great classical "games" of mathematics, perhaps the greatest in terms of the depth of investigation and the number of players.

Note that the axioms do not provide for any measurement of length, so neither tool is allowed to "have a memory"; the straightedge cannot be marked in any way, like a standard ruler, nor can the compass preserve a fixed opening after its points have been lifted from the plane. This latter restriction may surprise you because the compasses used in high school geometry classes can preserve length, and that use may seem to be "cheating," in the sense of using a tool not allowed in the "rules of the game." The dishonesty is only an illusion, however; it can be shown that the distance-preserving compass is no more powerful a tool than the collapsing one. (See Exercise 4.)

Each compass-and-straightedge construction involves a specific geometric object with one or more special properties, and the goal is to prove that this object can be constructed using only the Euclidean tools. To be more precise, the *construction* should depend only on the Euclidean tools, but the other axioms and the common notions are often needed to complete the argument that the figure thus constructed actually has all the desired properties. Notice that the process is deductive by nature; *the construction need not be carried out, but merely proved possible.* Thus, the entire construction "game" could be played without actually drawing a picture, although this is seldom done because visual representations are very helpful to the imagination and memory of the player. Euclid's first two theorems involve compass-and-straightedge constructions.

×××

(3.3)   Given a line segment $\overline{AB}$, an equilateral triangle with one side $\overline{AB}$ can be constructed.

×××

Proof

Construct circle #1 with center A and radius AB; construct circle #2 with center B and radius AB, as shown in *Figure 3.1*. Let C be one of the points of intersection of these two circles. (The existence of such a point is one of those pictorially obvious items that slips through the logical cracks in Euclid's postulates.) Connect A and C and also B and C. Clearly, $\overline{AC} = \overline{AB}$ because both segments are radii of circle #1; similarly, $\overline{BC} = \overline{AB}$ because both segments are radii of circle #2; finally, $\overline{AC} = \overline{BC}$ because things equal to the same thing are equal to each other. Thus, ABC is an equilateral triangle.

Note   Strictly speaking, the statement about line segments being *equal* should be rephrased in terms of *congruence* in the sense that the two figures have the same size and shape but they do not contain the same points. However, it has been traditional to apply the word *equal* to congruent line segments and congruent angles. We follow this practice, relying on common sense to maintain the distinction.

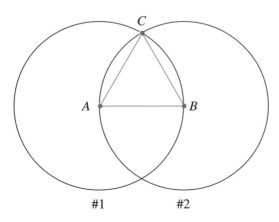

Figure 3.1 An equilateral triangle.

"The Measure of Any Height in a High Place." Woodcut from *Tanner's Manual*, London, 1587.

✕✕✕

(3.4)   Given a line segment $\overline{AB}$, and a point P not on this segment, a segment $\overline{PQ}$ equal to $\overline{AB}$ can be constructed.

✕✕✕

Proof

Starting with the given line segment $\overline{AB}$ and point P, draw segment $\overline{BP}$, as shown in *Figure 3.2*. Then construct the equilateral triangle BPC with $\overline{BP}$ as one side. (The preceding example justifies this construction.) Next, construct a circle with center B and radius $\overline{BA}$, and extend $\overline{CB}$ beyond B to intersect this circle at point F. Now construct a circle with center C and radius $\overline{CF}$, and extend $\overline{CP}$ beyond P to intersect this circle at point Q. Then $\overline{CF} = \overline{CQ}$ (why?) and $\overline{CB} = \overline{CP}$ (why?) Subtracting $\overline{CB}$ from $\overline{CF}$ and $\overline{CP}$ from $\overline{CQ}$, we conclude that $\overline{BF} = \overline{PQ}$. But $\overline{BF} = \overline{BA}$ because both segments are radii of the same circle; so $\overline{PQ} = \overline{BA}$, as desired.

In both of the preceding theorems, the construction phase and the proof phase are separated. This is faithful to the style used by Euclid, but is not essential. In more complicated constructions, it can become artificial and confusing to adhere blindly to such a stylistic device. What *is* essential is that each step be justified in terms of the axioms, postulates, and previously proved theorems. Note that the construction of Theorem (3.3), once proved, becomes a useful tool for the construction of Theorem (3.4). Both of these constructions, in turn, are theorems and consequently can be used as tools in subsequent constructions. Each successful construction not only solves a problem posed, but also becomes a new device in our Euclidean "toolbox," available as called for in such constructions as copying or bisecting a given line segment or angle, constructing a line perpendicular to a given line, or constructing triangles and rectangles of given dimensions. From such humble beginnings, the game proceeds to more complex constructions, such as a line tangent to a circle, a regular 15-sided figure inscribed in a circle, or a square equal in area to a given parabolic segment.

We have emphasized the construction theorems in Euclid's treatment of plane geometry; but they are only a part of the *Elements*. Interspersed among the construction theorems, Euclid also presented theorems that logically established various properties of given figures. For example, one of the early theorems proved by Euclid is the following:

XXXXXXXXXXXXXXXXXXXXXXXXXXXXXXXXXXXXXXXXXXXXXXXXXXXX

Vertical angles are equal.

XXXXXXXXXXXXXXXXXXXXXXXXXXXXXXXXXXXXXXXXXXXXXXXXXXXX

(**Vertical angles** are any pair of opposite angles formed when two lines intersect; for example, in *Figure 3.3*, $\angle PQR$ and $\angle SQT$ are vertical angles, as are the pair $\angle PQS$ and $\angle RQT$.)

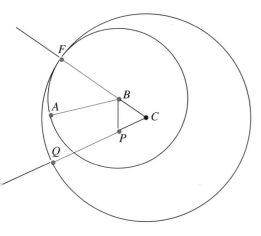

Figure 3.2 $\overline{BF} = \overline{AB} = \overline{PQ}$

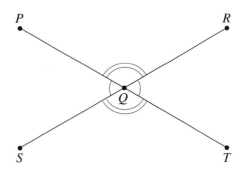

Figure 3.3 Vertical angles:
$\angle PQS = \angle RQT;$
$\angle PQR = \angle SQT.$

Probably most familiar among the theorems about properties of plane figures are the various triangle-congruence theorems; for example:

xxxxxxxxxxxxxxxxxxxxxxxxxxxxxxxxxxxxxxxxxxxxxxxxxxxxx

If two sides and the included angle of a triangle are equal, respectively, to the corresponding parts of a second triangle, then the two triangles are equal (congruent).

xxxxxxxxxxxxxxxxxxxxxxxxxxxxxxxxxxxxxxxxxxxxxxxxxxxxx

This theorem is commonly abbreviated as the "side-angle-side" theorem, or simply "SAS." In the same genre are the congruence theorems ASA, SSS, and AAS, and the "hypotenuse-leg" congruence theorem for right triangles.

Perhaps less familiar, but no less important, is a theorem relating the *interior* and *exterior* angles of a triangle. An **exterior angle** of a triangle is an angle formed *outside* the triangle by extending one side. Thus, for triangle ABC of *Figure 3.4*, $\angle ACD$ is an exterior angle obtained by extending side BC beyond vertex C. The **remote interior angles** relative to this exterior angle are those with vertices other than C—in other words $\angle BAC$ and $\angle ABC$.

xxxxxxxxxxxxxxxxxxxxxxxxxxxxxxxxxxxxxxxxxxxxxxxxxxxxx

The Exterior Angle Theorem: An exterior angle of a triangle is greater than either remote interior angle.

xxxxxxxxxxxxxxxxxxxxxxxxxxxxxxxxxxxxxxxxxxxxxxxxxxxxx

Proof

Given a triangle, ABC, we must show that an exterior angle, say $\angle ACD$, is greater than a remote interior angle, say, $\angle BAC$. (See *Figure 3.4*.) Letting E be the midpoint of $\overline{AC}$, we connect B and E and extend $\overline{BE}$ its own length to F; then we connect C and F. Observe that $\overline{AE} = \overline{CE}$ because E is the midpoint; $\overline{EB} = \overline{EF}$, by construction; and $\angle AEB = \angle CEF$

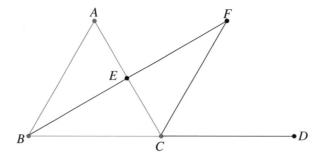

Figure 3.4 The Exterior Angle Theorem. $\angle ACD$ is greater than $\angle BAC$.

because they are vertical angles. Hence, $\triangle AEB$ is congruent to $\triangle CEF$, by SAS. So $\angle BAE = \angle FCE$ because they are corresponding angles of congruent triangles; but $\angle FCE$ is part of, and hence less than, $\angle ACD$, by Axiom 5. Therefore, $\angle BAE$ is less than $\angle ACD$, which was to be proved.

The Exterior Angle Theorem is useful in proving several subsequent results, particularly those in which inequalities are involved. Perhaps the most noteworthy, however, the Alternate Interior Angle Theorem, is one that does not involve inequalities, at least not in an obvious way. We present the theorem and its proof as an example of proof by contradiction.[4] This theorem is the first of several in Euclid's work that examine various relationships in a configuration with a **transversal**, a line intersecting each of two other lines. Among the eight angles formed in this configuration, the four between the pair of lines are called **interior angles** and the remaining four, **exterior angles**. Any two of the four (interior or exterior angles) on *opposite* sides of the transversal and not sharing a vertex are called **alternate angles**. Thus, in *Figure 3.5*, t is a transversal to lines l and m, forming, among others, alternate interior angles $\angle PTU$ and $\angle SUT$.

×××

The Alternate Interior Angle Theorem: If two lines are cut by a transversal so as to form equal alternate interior angles, then the lines are parallel.

×××

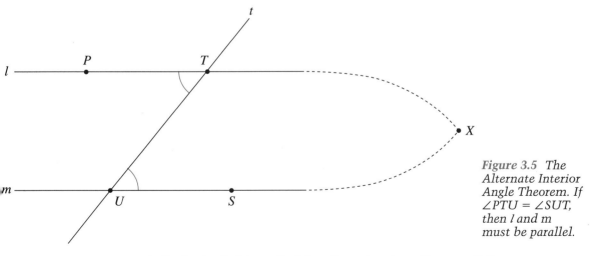

Figure 3.5 The Alternate Interior Angle Theorem. If $\angle PTU = \angle SUT$, then l and m must be parallel.

4. See Section 3 of Appendix A for a discussion of proof by contradiction.

Proof

The two lines, *l* and *m*, are cut by the transversal *t*, at *T* and *U*, respectively, forming equal alternate interior angles ∠*PTU* and ∠*SUT*, as shown in *Figure 3.5*. Suppose, contrary to the desired conclusion, that *l* and *m* are *not* parallel, and hence, extended sufficiently far, they meet at some point *X*, which is therefore on both lines. Then Δ*TUX* would be formed, and would have an exterior angle, ∠*PTU*, because *X* is on line *l*. Then ∠*PTU* would be equal to a remote interior angle, ∠*TUX*, because *X* is also on line *m*. This contradicts the Exterior Angle Theorem. Hence the supposition must be false, and *l* and *m* must be parallel.

This theorem and its nearly identical successor in Euclid's *Elements*, theorems 27 and 28 of Book I, are followed there by the converse.

xxxxxxxxxxxxxxxxxxxxxxxxxxxxxxxxxxxxxxxxxxxxxxxxxxx

If two parallel lines are cut by a transversal, the alternate interior angles formed are equal.

xxxxxxxxxxxxxxxxxxxxxxxxxxxxxxxxxxxxxxxxxxxxxxxxxxx

Proof

The parallel lines *l* and *m* are cut by transversal *t* at *T* and *U*, respectively, as shown in *Figure 3.6*. Suppose, contrary to the desired

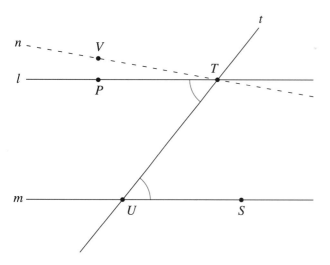

Figure 3.6 *The converse of the Alternate Interior Angle Theorem. If l is parallel to m, then* ∠*PTU* = ∠*SUT*.

Figure 3.7 *A double cone.*

conclusion, that $\angle PTU \neq \angle SUT$; then copy $\angle SUT$ at T as $\angle VTU$. Because $\angle VTU \neq \angle PTU$, $\overline{VT}$ is not the same line segment as $\overline{PT}$, and consequently n is not the same line as l. But, by the Alternate Interior Angle Theorem, n is parallel to m, and we have two distinct lines through T, l and n, both parallel to m, a contradiction of Postulate **E5**. Hence, the supposition must be false, and $\angle PTU = \angle SUT$.

Again, the form of the proof is by contradiction; but in this case the contradiction hinges on Euclid's Fifth Postulate, the focus of several millennia of attention within mathematics. We shall return to this major thread in Section 3.6.

Euclidean geometry is by no means confined to examining circles and straight lines. The ancient Greeks made extensive studies of various types of curves. Notable examples of such figures are the **conic sections**, first thoroughly treated by Apollonius about 225 B.C. These curves represent the intersection of a plane with a (double) cone. The cone can be thought of as obtained by rotating two intersecting lines about an axis that bisects their angle of intersection. The cone itself is the collection of all points in space that the rotating lines "pass through," as shown in *Figure 3.7*. There are four essentially different ways in which a plane that does not contain the intersection point of the cone's generating lines can intersect that cone. They are shown in *Figure 3.8*. The four kinds of figures that result are called the **circle**, **ellipse**, **parabola**, and **hyperbola**. One of the truly remarkable facts of mathematics is that, once René Descartes (in 1637) had connected geometry with algebra by means of his coordinate system and mathematicians had begun to study the geometric representation of algebraic

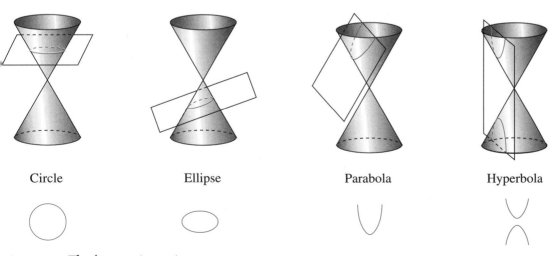

Circle   Ellipse   Parabola   Hyperbola

Figure 3.8 The four conic sections.

equations, it was discovered that every conic section can be repre-sented by a quadratic (2nd-degree) equation in two variables![5]

Geometry can be regarded in one sense as the theory of compari-son of shapes, with different norms of comparison yielding different geometries. In Euclidean geometry, shapes or figures are compared in terms of either *congruence* or *similarity*. It is often said that two fig-ures are congruent if one can be made to coincide with the other by a "rigid motion." This is a bit misleading, since there really is no move-ment involved in the notion of congruence; but the general idea is cor-rect. The formal mathematical definition of **congruence** captures the idea of superimposing one figure on the other, thus checking to see that both *shape* and *size* are preserved. Other kinds of geometry also deal with congruence of figures, but only Euclidean geometry studies similarity as a concept distinct from congruence.

The formal definition of similarity provides for a distinction be-tween size and shape, and figures are called **similar** if they have the same shape but not necessarily the same size. For example, any two circles are similar; but to be congruent they must have the same radius. Any two equilateral triangles are similar as are any two squares; but to be congruent two equilateral triangles or two squares must have sides of the same length. By contrast, two rectangles are not similar unless their corresponding sides have proportional lengths.

We continue this sketchy sampling of Euclidean geometry with a proof of one of the most familiar and important statements in all of mathematics—the *Pythagorean Theorem*. Since it is both a numerical statement and a geometric one, this theorem applies to a remarkably wide variety of mathematical topics. From among the almost incred-ible variety of proofs of this theorem, the one attributed to the nine-teenth-century German geometer Jacob Steiner stands out because of its simplicity and elegance. It depends solely on classical constructive methods, using no numerical or algebraic arguments at all. Thus, it is a fitting example of classical Euclidean geometry "in action."

xxxxxxxxxxxxxxxxxxxxxxxxxxxxxxxxxxxxxxxxxxxxxxxxxxx

The Pythagorean Theorem: The square of the hypotenuse of a right triangle is equal to the sum of the squares of its legs.[6]

xxxxxxxxxxxxxxxxxxxxxxxxxxxxxxxxxxxxxxxxxxxxxxxxxxx

---

5. For example, the equation $y = x^2$ represents a parabola, and the equation $x^2 + y^2 = 1$ represents a circle. (See Chapter 7 for a review of the Cartesian coordinate system.)

6. Recall that the hypotenuse of a right triangle is the side opposite the right angle, and the legs are the sides adjacent to the right angle.

(All terms of this theorem can and should be interpreted in a strictly geometric sense. Thus, the square of a line segment is a square whose side is equal in length to the line segment,[7] and the sum of two squares will be considered equal to another square if the first two squares can be "cut into pieces and rearranged" to form the third. The latter "cut and paste" sense of figure equality implies that the areas are equal, but is logically more fundamental than relying on area formulas. These intuitive ideas can be made rigorous; but a pictorial approach enhances the elegance of the argument and best suits our purpose. The justifications for most steps of the construction should be clear, so they have been omitted here for the sake of brevity. In Exercise 19 you are asked to justify each step.)

Proof

Starting with right triangle ABC:

    (a) Construct the square $ABDE$ as shown in *Figure 3.9*.

    (b) Extend line segment $\overline{AC}$, and

    (c) locate a point F on that line such that $\overline{AF} = \overline{BC}$. Draw $\overline{EF}$. Now

    (d) $\angle ABC$ is the complement of $\angle BAC$, and

    (e) $\angle EAF$ is the complement of $\angle BAC$; so

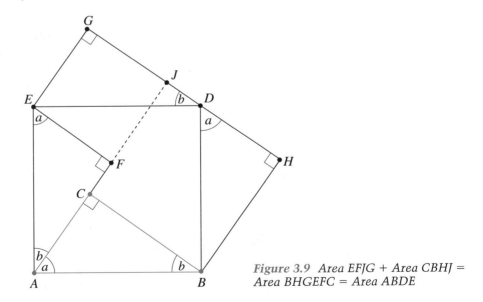

Figure 3.9 *Area EFJG + Area CBHJ = Area BHGEFC = Area ABDE*

---

7. In Euclid's statement and proof of Pythagoras's famous theorem, the language used is "the square *on* the hypotenuse" and the squares "*on* the legs," identifying that each square referred to has one of its sides identical to a side of the right triangle.

(f) $\angle ABC = \angle EAF$. In addition,

(g) $\overline{EA} = \overline{AB}$, so

(h) $\triangle ABC$ is congruent to $\triangle EAF$.

(i) Construct two more congruent copies of triangle ABC, on sides $\overline{BD}$ and $\overline{DE}$, as shown. Because

(j) $\angle EDG = \angle ABC$ and $\angle BDH = \angle BAC$,

(k) $\angle EDG$ and $\angle BDH$ are complementary; so

(l) G, D, and H lie on a straight line. Observe that

(m) $BCJH$ is a square with $\overline{BC}$, one leg of the original triangle ABC, as a side; and

(n) $EFJG$ is a square with $\overline{EF}$ as a side. Furthermore,

(o) $\overline{EF}$ is equal to $\overline{AC}$, the other leg of triangle ABC.

Together, the two squares in Steps (m) and (n) form the irregular hexagon (6-sided figure) $BHGEFC$. But this hexagon can also be decomposed into triangle DBH plus triangle EDG plus the irregular pentagon (5-sided figure) $BDEFC$. But

(p) this combined area is equal to that of triangle EAF plus triangle ABC plus pentagon $BDEFC$, which constitutes the square $ABDE$.

The hexagonal sum of the two smaller squares has been "rearranged" to form the original large square. This completes the proof of the theorem.

3.2   EXERCISES

In the following exercises you may use any of the familiar Euclidean theorems about congruent triangles as well as the visually "obvious" properties of figures mentioned in the text. You may also use the result of any exercise preceding the one you are considering.

1. Follow the steps described in Theorem (3.4), starting with a configuration of the "given" line segment and point that is *different* from the one pictured in *Figure 3.2*. Does the figure look the same? Does the logical argument still work?

2. Suppose, in Theorem (3.4), that the point P is *on* the given segment $\overline{AB}$, between A and B. The first step of the construction is now superfluous; but do the remaining steps still work? If not, which steps do not work, and why not?

3. Suppose, in Theorem (3.4), that the point P *equals* the the point A.

   (a) Do the steps of the proof still work? If not, which steps do not work, and why not?

(b) Explain why the construction and proof are superfluous in this case.

In Exercises 4–12, show how a (collapsing) compass and (unmarked) straightedge can be used to produce each of the constructions called for.

4. Given a segment $\overline{AB}$, a line l that does not contain $\overline{AB}$, and a point P on l, construct a point Q on l such that $\overline{PQ} = \overline{AB}$. (This construction shows that a collapsing compass and an unmarked straightedge can be used to transfer length; this adds a *marked* straightedge to our "toolbox.")

5. Given an angle, construct its bisector. (*Note*: An **angle bisector** is a line segment that has one end at the vertex of the angle and that divides the angle into two equal adjacent angles.)

6. Given a line segment, find its midpoint.

7. Given a line segment and a point on it, construct a perpendicular segment with that point as one endpoint.

8. Given a line and a point not on it, construct a perpendicular line segment with that point as one endpoint.

9. Given triangle ABC and line l, construct a triangle congruent to ABC with one side on l.

10. Given angle ABC, line l, and a point Q on l, construct angle PQR, equal to angle ABC with one side on l.

11. Given a line segment, construct a square with that segment as one side.

12. Given two line segments, construct a rectangle with adjacent sides equal to the two segments.

13. The text refers to triangle congruence theorems abbreviated as SAS, ASA, SSS, and AAS. Show by example why there is no theorem corresponding to SSA.

14. Prove that the base angles of an isosceles triangle are equal. (*Hint*: No construction is necessary; just use SAS judiciously.)

15. Prove that if two angles of a triangle are equal, then the triangle is isosceles. (*Hint*: Use proof by contradiction.)

16. Prove that the sum of any two angles of a triangle is less than 180°. (*Hint*: Study *Figure 3.4*.)

17. Prove that if a triangle has any pair of sides unequal, then the angle opposite the greater side is larger than the angle opposite the smaller side. (*Hint*: Copy the smaller side onto the larger, forming an isosceles triangle, and use the Exterior Angle Theorem.)

18. Prove that if a triangle has any pair of angles unequal, then the side opposite the greater angle is longer than the side opposite the smaller angle. (*Hint*: Suppose the contrary, and use Exercises 14 and 17.)

19. Give a reason to justify each of the Steps **(a)**–**(p)** in the proof of the Pythagorean Theorem.

WRITING EXERCISES

1. A careful study of how Euclid used his third axiom would suggest the following somewhat physical image for that axiom:

    Any line segment can be held fixed at one end while the other end rotates; but the segment as a whole cannot be moved.

    Explain in your own words the difference between a collapsing compass and a noncollapsing compass, and explain how the image described here corresponds to the former.

2. Criticize the following "proof" of the statement that there are two perpendiculars from a point to a line:

Proof:

Consider two intersecting circles with centers O and O', as in *Figure 3.10*; call one of their intersection points A. Draw the diameter of each circle from A, and let B and C be the other endpoints of these diameters. Draw BC, intersecting the two circles at points D and E. Draw AD and AE. Both $\angle ADB$ and $\angle AEC$ are right angles because each is inscribed in a semicircle. Therefore, there are two perpendiculars from the point A to the line segment BC.

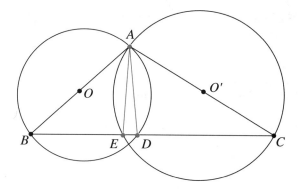

Figure 3.10 *Are there two perpendiculars from point A?*

AXIOM SYSTEMS

3.3

The organization of information into axiomatic systems can be traced back to the Greeks, who as early as 600 B.C. began to study the logical connections among mathematical facts. In approximately 300 B.C., Euclid organized most of the known mathematics of his time so that virtually all statements were proved from a small collection of definitions and axioms, and thus the **axiomatic method** was born. Today, the axiomatic method is the distinctive structure of mathematics (and much of science). A staggering amount of material has been added to mathematical knowledge in the more than twenty-two centuries since Euclid. Nevertheless, no mathematical claim is accepted unless it can be proved from basic axioms; and whenever a new mathematical theory emerges, mathematicians follow Euclid's pattern in working to identify an axiomatic basis for it. Such is the influence of the axiomatic method.

Simply put, an **axiom system** has four essential components: *defined terms*, *undefined terms*, *axioms*, and *theorems*. As we describe each of these in turn, it should also become clear how they fit together.

When we define terms in mathematics, we are merely assigning name tags to ideas. A **definition** is the statement of a single, unambiguous idea that the word, phrase, or symbol being defined represents from that point on—no more and no less. The single, unambiguous idea is called the **characteristic property** of the definition; it is a condition such that, given any object whatsoever:

1. We can determine whether or not that object satisfies the condition.

2. The term being defined is used to label everything that satisfies the condition and is not used to label anything else.

EXAMPLE 3.1

The statement "A square is a four-sided figure" is accurate, but it is not a good definition of a *square* because there are four-sided figures that we do not want to call squares. We might use "A square is a rectangle with four equal sides" to define *square*, provided we have already defined *rectangle*.

"Defining" a word simply by giving a synonym (another word that means the same thing), a method common in some dictionaries, is either meaningless or useless from a mathematical viewpoint. If we do not know what the synonym means, we have no way of applying it as a characteristic property; if we do know what the synonym means, then that word is a perfectly good name for the idea and there is no need to confuse the issue by supplying another. Moreover, that practice can lead to a problem of "circularity." For instance, if we define *number* as *quantity*, define *quantity* as *amount*, and define *amount* as *number*, we have stated that there are three different labels for the same idea; but we haven't said what the idea is! In general, then, a good definition must not be **circular**; that is, its defining condition must neither use the term itself nor use terms that are themselves defined using the term being defined.

EXAMPLE 3.2

The assertions, "An *odd number* is a whole number that is not even," and "An *even number* is a whole number that is not odd," taken together, amount to a circular pair of definitions because the characteristic property of each one depends on the other. Either definition taken by itself is fine, provided that the other term has been defined independently.

The requirement that definitions cannot be circular means that some terms must be **undefined**. We cannot have an infinite regression of terms, each defined, in turn, by more basic terms; so there must be some level of "most basic" terms that are not themselves defined at all. The fact that a term is undefined does not mean that it is meaningless. For example, the terms *point* and *line* are normally taken as undefined in a formal treatment of geometry. But these words are not meaningless; rather, their meaning is determined by the way they are used within the underlying assumptions or assertions of the system, the axioms of geometry. In other words, the undefined terms of a mathematical system acquire their meanings from the context.

Just as we cannot expect every term to be defined, so we cannot expect every mathematical statement to be proved from previously developed statements. We must have some statements to start with, statements that are assumed without proof. These statements are called **axioms**. In an ideal mathematical system, all other statements are derived from the axioms by strict logical proof. Statements that are proved from the axioms are called **theorems**.

Euclid recognized the need for undefined terms and axioms in his treatment of geometry. He regarded geometry as a description of the physical world, and he tried to systematize that description by putting it on a deductive foundation. He defined all the technical terms he used, including *point, line, plane, angle, circle*; but in so doing, he used other words such as *part, length, equal*, etc., without defining them.

As we saw in Section 3.2, Euclid distinguished between axioms and postulates. His axioms were statements of common ideas that he regarded as obvious, such as

The whole is greater than any of its parts.

His postulates were statements dealing explicitly with geometry, such as

A straight line can be drawn from any point to any point.

These statements (the postulates) he regarded as true observations about the physical world, rather than as arbitrary assumptions. In that sense, Euclid's geometry typifies what is now known as "material axiomatics." A **material axiomatic system** considers its undefined terms to have meanings derived from reality and its underlying axioms to be statements that are "obviously" true in the real world.

In contrast to material axiomatics is the modern deductive approach to mathematical systems called "formal axiomatics." A **formal axiomatic system** begins by assuming that its undefined terms have no meaning at all and behave much like algebraic symbols. A small set of statements are presented using these terms and, although these statements have no necessary connection with truth or reality outside the system, they are assumed to be "true" within the system itself. These assumed statements are the *axioms* or *postulates*. (These two words are interchangeable in formal axiomatics.) From there on, all new terms used in the system are defined from the undefined terms and all other statements (theorems) are proved from the axioms. The mathematics constructed in this way is called **pure mathematics**.

EXAMPLE 3.3

Undefined terms: squirrel, tree, climb

Axioms:

1. There are exactly three squirrels.
2. Every squirrel climbs at least two trees.
3. No tree is climbed by more than two squirrels.

If we consider this to be a formal axiom system, its terms have no meanings other than what they acquire from their use in the axioms. The assertions being made are not to be understood as referring to "real" squirrels or trees, or the familiar physical meaning of "climb." The meaning of the system is completely derived from the *form* of the axioms. Hence, this system is formally the same as the following one:

Undefined terms: hinkle, pinkle, babble

Axioms:

1. There are exactly three hinkles.
2. Every hinkle babbles at least two pinkles.
3. No pinkle is babbled by more than two hinkles.

Starting with the undefined terms and axioms of a formal system, we can define other technical terms and prove theorems phrased in terms of them. For example, we might note that the axioms do not require that every tree is climbed by some squirrel; so we might choose to distinguish such entities by defining a "free tree" as one that is climbed by no squirrels. In the second version of this system, we might choose "parsimonious pinkle" as the analogous terminology. The different choices of adjectives inevitably carry different connotations; but we must ignore any such connotations within the formal system. Mathematics is only interested in the denotative content of a definition. The exercises explore briefly some of the theorems that are logical consequences of these axioms.

In a highly formal axiomatic development, not only are there undefined and defined terms, axioms, and theorems, but there are also explicit rules of logic, usually quite restrictive. Such systems are often highly symbolic, rather than verbal, because symbols are less likely to invite errors caused by unjustified assumptions about the "real meanings" of terms or axioms. Words, even in mathematics, can be deceiving.

In a less formal development, rules of logic are not specified; instead, we allow ourselves a sort of "undefined" logic—the formalized (and sometimes symbolized) mathematical logic that is merely a sharpened form of common sense. This is the logic used in everyday discourse and practiced in arithmetic, algebra, science, (occasionally, politics), and, especially, geometry. Even in developments of this sort, a considerable spread of precision and formalism is found. In some treatments, nothing is accepted as "understood" about the system except what is contained in the definitions and axioms. Even a slight variation, such as in the tense or mood of a verb, is not allowed unless formally introduced by a definition or language rule. In other treatments, familiar variations of grammar are accepted without question, and results of other familiar parts of mathematics are used without formal justification.

We shall adopt this last, most informal mode for the rest of the chapter. It is likely the style you saw in your high school geometry course, and it is the one used in the preceding two sections. There is an obvious comfort in allowing ourselves to use logical and grammatical "common sense," but we must be aware of the danger of this approach. We can grow so casual that we believe we have proved a theorem, when, in fact, we have been duped by a persuasive, but flawed, argument. To help you avoid such difficulties, you might find it useful to peruse the basic principles of logic summarized in Appendix A.

3.3   EXERCISES

Exercises 1–25 concern definitions. If some of these terms are not familiar to you, look them up in an unabridged dictionary.

(a) Determine whether the given statement accurately defines the given word or phrase (in its usual sense);

(b) if not, identify whether the statement is *circular* or is *not characteristic*;

(c) If the statement is not characteristic, give an example (verbally or by means of a figure) that fits the given description but not the word or phrase described, or vice versa.

1. Two angles are supplementary if each is the supplement of the other.

2. An angle bisector of $\angle ABC$ is a line segment with B as its endpoint.

3. An angle bisector of $\angle ABC$ is a line segment $\overline{BD}$ such that $\angle ABD = \angle CBD$.

4. An acute triangle is a triangle with no obtuse angles.

5. An acute triangle is a triangle with all of its angles equal.

6. A right triangle is a triangle containing a right angle.

7. A right triangle is a triangle containing two acute angles.

8. A right triangle is a triangle containing two angles whose sum equals the third angle of the triangle.

9. A right triangle is a triangle whose sides have length 3, 4, and 5, respectively.

10. A right triangle is a triangle with one pair of sides perpendicular to each other.

11. An obtuse triangle is a triangle that is obtuse.

12. An obtuse triangle is a triangle with one side longer than each of the other two.

13. An isosceles triangle is a triangle in which one side is not equal to either of the other two.

14. An isosceles triangle is a triangle in which one side is either smaller than the other two or greater than the other two.

15. An isosceles triangle is a triangle in which two sides are equal.

16. An isosceles triangle is a triangle in which two angles are equal.

17. A parallelogram is a figure with opposite sides parallel.

18. A parallelogram is a quadrilateral with opposite sides parallel.

19. A parallelogram is a quadrilateral with opposite sides equal.

20. A parallelogram is a quadrilateral with opposite angles equal.

21. A rectangle is a quadrilateral that is rectangular.

22. A rectangle is a quadrilateral with four right angles.

23. A rectangle is a quadrilateral with three right angles.

24. A rectangle is a quadrilateral with two right angles.

25. A rectangle is a figure with all angles equal.

Exercises 26–29 refer to the (first) axiom system of Example 3.3. These statements are theorems of that axiom system; in each instance, prove the given assertion; that is, give an argument that shows that the given assertion follows logically from the axioms. You may find it useful to use the results of earlier exercises in proving some of the later ones.

26. There are at least three trees. (*Hint*: Suppose that there were only two trees and show that this would logically conflict with what the axioms assert.)

27. If there are only three trees, then every squirrel climbs exactly two trees. (*Hint*: Suppose that some squirrel climbs all three trees and show that this would logically conflict with what the axioms assert.)

28. If there is some tree that is climbed by only one squirrel, then there are at least four trees. (*Hint*: Use Exercises 26 and 27.)

29. If every tree is climbed by only one squirrel, then there are at least six trees.

WRITING EXERCISES

1. Identify several "most basic" words of the English language, words that you learned by example and context as a young child. Are these words "undefined" in the same sense as the undefined words of a mathematical axiom system? Justify your position.

2. Is it possible to have *no* defined terms in an axiom system? For example, in geometry we could avoid using the term *quadrilateral* by using the phrase "four-sided rectilinear figure" whenever we needed to refer to one. Can this be done for all the defined terms of geometry? If so, what are the advantages and disadvantages?

3. Relying on what you know about real squirrels and real trees, which of the assertions about them in Example 3.3 are true? In light of your response, write a short paper on the pros and cons of choosing abstract or nonsense terms, as in the second version of that axiom system, versus terms with some meaning in everyday language.

MODELS

All the emphasis on formality and on the absence of meaning in the preceding section might lead you to believe either that pure mathematics is a useless game or that mathematicians are somehow psychic or phenomenally lucky in what they choose to play with. That is not the case at all. Mathematicians do not pick arbitrary symbols and statements at random and just happen to come across useful systems most of the time; they usually have in mind a concrete system they are trying to describe, just as Euclid did. The essential difference is that, unlike Euclid, modern mathematicians recognize that the concrete system they are describing does not *predetermine* the abstract system they are formulating. The axioms are not statements that are true by nature; they are formal sentences that are considered true in the system merely by agreement. The abstract system is, in general, subject to many interpretations, of which its prototype is only one.

DEFINITION   An **interpretation** of an axiom system is any assignment of specific meanings to the undefined terms of that system. If an axiom becomes a true statement when its undefined terms are interpreted in a specific way, then we say that the interpretation **satisfies** the axiom. A **model** for an axiom system is an interpretation that satisfies all the axioms of the system.

Thus, the word *model* is used in mathematics in much the same way as it is used in ordinary English. Consider, for example, an architect's model of a proposed building. In this instance, the model builder is often working from the blueprints of the proposed building. Different model builders might construct different, but equally accurate, models from the same set of blueprints; the choice of materials, scale of the model, landscaping, etc., would not be prescribed by the blueprints, so considerable variance might appear in the models. This kind of modeling is mirrored in the mathematical use of the word. Our blueprints are the axioms of a system. A model for an axiom system makes its ideas more realistic, just as an architect's model makes the design ideas more concrete and visible. And, because there are usually many things *not* specified by the axioms, a considerable variety of models may be possible. **Applied mathematics** can be described as the study of models of abstract mathematical systems in fields other than mathematics (physics, biology, sociology, etc.).

Let us look at an example of how an axiom system is taken from its original setting, formalized, and then modeled. In this instance, the artistic problem of picturing objects realistically led to an axiom system for a special kind of geometry, thereby providing a fine example of the interplay between reality and abstraction. The story begins more than five hundred years ago.

As Europe passed from Middle Age night to Renaissance morning, man's awareness of the world around him yawned, stretched, and began to search for its slippers. While scientists and philosophers explored the physical world, artists searched for ways to mirror this reality on paper and canvas. One of their problems was that of dimension—how to portray depth on a flat surface. The artists of the fifteenth century realized that this problem was primarily geometric, so they began a mathematical investigation of the properties of spatial figures as the eye sees them. The problem was approached by considering the surface of a picture to be a window through which the artist views the object to be painted or drawn. As the lines of vision from the object converge to the point where the eye is viewing the scene, the picture captures a cross section of the figure formed by them. (See *Figure 3.11*.) This simple principle became the basis for the artistic theory of perspective, pioneered by Leonardo da Vinci and Albrecht Dürer.

Abstracting this approach from the artistic complications of color, shading, motion, and the like, we obtain a relatively simple geometric process called *projection*—a way of making the points of an object correspond to the points of its representation on a planar surface.

One well-known illustration of perspective is the image of railroad tracks stretching into the distance. Although we know they are the same distance apart both near and far away, they appear to be meeting at a point just beyond the horizon. Indeed, if they are painted, the artist must treat the track lines on the canvas plane as lines that meet somewhere "just beyond" the picture. Thus, for the artist, a realistic theory of representing parallel lines requires that those lines be regarded as intersecting lines. This requirement, along with a few other basic assumptions about points and lines, became the basis for an important type of geometric system with applications far beyond the artistic questions that spawned it. The system is called **projective geometry**. An axiom system for the projective geometry of a plane is as follows:

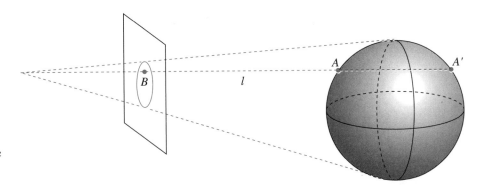

Figure 3.11 The lines of vision from the sphere to the eye form a cross section on the plane.

Institutiones Geometricae, by Dürer.

P1. There exist at least one point and one line.

P2. If A and B are distinct points, then exactly one line passes through both of them.

P3. If l and m are distinct lines, then exactly one point is common to both of them.

P4. There are at least three points on any line.

P5. Not all points are on the same line.

Now, if we consider *point*, *line*, and the various ways of expressing *on* (including "pass through") as undefined terms, this axiom system can be restated as follows:

Undefined terms: X, Y, related to

Axioms:

P1. There exist at least one X and one Y.

P2. If a and b are distinct Xs, then exactly one Y is related to both of them.

P3. If c and d are distinct Ys, then exactly one X is related to both of them.

P4. At least three Xs are related to any Y.

P5. Not all Xs are related to the same Y.

In this form, not only is it unclear that we are talking about geometry, but even the relationships among the various terms of the system become obscure. The only information that has been retained by the terms of this formal system comes from the syntax (grammatical form) of the original statements. Specifically, X and Y are understood as nouns of some sort, and "related (to)" is understood as some sort of connective verb form that can be used with these nouns.

Let us build a model for this axiom system using brass rings for the Xs, wires for the Ys, and "attached" for *related to*. Take seven brass rings (labeled A, B, C, D, E, F, G) and seven wires (labeled 1, 2, 3, 4, 5, 6, 7), and attach them as shown in *Figure 3.12*.

It is not hard to see that this interpretation is a model for Axiom System **P1–P5**; the five abstract axioms, when interpreted according to the given prescription, become:

P1. There exist at least one brass ring and one wire.

P2. Any two distinct brass rings are attached to exactly one wire.

P3. Any two distinct wires are attached to exactly one brass ring.

P4. Three brass rings are attached to each wire.

P5. Not all brass rings are attached to the same wire.

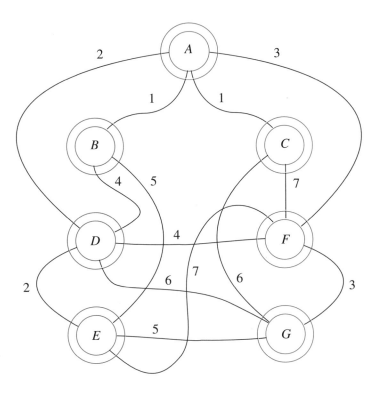

Figure 3.12 A ring-and-wire model for axiom system P1–P5.

Consider, in particular, what Axiom **P2** asserts, and note that it corresponds to what is pictured in *Figure 3.12*. For example, note that rings *B* and *F* are both attached to wire 4, and to no other; and rings *E* and *C* are both attached to wire 7, and to no other. If we continue, we find that each pair of rings is attached to one and only one wire—that is, Axiom **P2** is satisfied. Similarly, as Exercise 1 asks you to check, each of the interpreted statements is a true assertion about *Figure 3.12*. In other words, this figure is a model for the axiom system.

Of course, this little collection of rings and wires is not the only model of our axiom system. Many other models exist, including the standard projective plane used by artists. You will be asked to find yet another model in Exercise 6.

It is also worth observing that, like many models, the ring-and-wire contraption shown in *Figure 3.12* can be used as a model for other axiom systems, as well. Consider, for example, the following axiom system, which describes some of the basic properties of the Euclidean plane:

Undefined terms: point, line, on

Axioms:

    A1. There is at least one point.

    A2. Every point is on at least two lines.

    A3. Every line is on at least two points.

    A4. Given any two points, there is exactly one line on both of them.

The phrasing of Axioms **A3** and **A4** may seem a bit odd, in that it is common to think of points being on lines, but not the reverse. But "on" is an undefined term, as are "point" and "line"; so this axiom system is free to use them in any syntactically meaningful way. When formulating a formal axiom system, we are not constrained by ordinary usage. If we interpret *point* as "brass ring," *line* as "wire," and *on* as "attached," then *Figure 3.12* is a model for this axiom system.

| EXAMPLE 3.4 | Another (different) model for Axiom System **A1–A4** can be constructed as follows:[8] |

Let *point* mean "one of the numbers 1, 2, or 3"; let *line* mean "one of the sets {1, 2}, {1, 3}, or {2, 3}"; let *on* mean "contains" or "is contained in," whichever is appropriate.

It is easy to check that all four axioms are true in this case. Notice that this model differs from the ring-and-wire model in some essential ways. Not only are the objects involved apparently different, but the

8. See Section 4.2 for an introduction to the notation and terminology of set theory.

models actually have some different properties. For instance, this model only has 3 *points* and 3 *lines*, whereas the ring-and-wire model has 7 *points* and 7 *lines*. Nevertheless, both models satisfy all axioms of the system.

The axiomatic approach to mathematics has one vitally important benefit:

Any statement that can be proved from the axioms of a system *must be true in every model* of the system.

That is, a single proof of a theorem in an abstract axiom system guarantees the truth of that statement as it is interpreted in all models of that axiom system, regardless of how different the various interpretations appear to be. Besides being a valuable observation about applied mathematics, this fact will be useful in understanding the ideas of the next section.

ᐟ

EXAMPLE 3.5

Prove the following theorem for Axiom System **A1–A4**:

Theorem: Given a line, there is a point not on that line.

Proof:

Let L be the given line.

There is a point on L (by **A3**); label it p.

There is (at least) one other line, M, on p (by **A2**).

Then there is (at least) one other point, q, on M (by **A3**).

This point q cannot be on L; for if it were, M and L would be the same (by **A4**).

Thus, no matter how odd or unusual the meanings of *point*, *line*, and *on* in any model of Axiom System **A1–A4**, we are guaranteed that, for each *line* of that model, there is a *point* not *on* it.

3.4   **EXERCISES**

1. Show that the ring-and-wire contraption shown in *Figure 3.12* is a model for Axiom System **P1–P5** by verifying that each of the axioms, interpreted as described in the text, is true. (Check all possibilities.)

2. Show that the ring-and-wire contraption shown in *Figure 3.12* is a model for Axiom System **A1–A4** by verifying that each of the axioms, interpreted as described in the text, is true. (Check all possibilities.)

3. Verify that the model of Example 3.4 is a model for Axiom System **A1–A4**, with the interpretations for point and line interchanged.

4. Verify that the model of Example 3.4 is also a model for the "squirrels-and-trees" axiom system of Example 3.3, with the interpretations of *point, line* and, *on* applied to *squirrel, tree,* and *climb,* respectively.

5. Give another model for the "squirrels-and-trees" axiom system of Example 3.3, different from that of Example 3.4. (See the preceding exercise.) Describe how it differs.

6. Find a model for Axiom System **P1–P5** that differs from the ring-and-wire model given in this section and also from the standard projective-plane model. Verify that your model actually satisfies the axioms; then describe how it differs from the other two models.

7. Construct an axiom system containing three axioms; then find two different models for it.

For Exercises 8–13, find a model for the given axiom system:

8. **Undefined terms**: pencil, word, write

   Axioms:

   (a) There are exactly three pencils.

   (b) All pencils write the same word.

   (c) There is exactly one word that no pencil writes.

9. **Undefined terms**: cat, mouse, catch

   Axioms:

   (a) All cats catch mice.

   (b) Some mice do not catch cats.

   (c) There are at least two cats.

10. **Undefined terms**: point, line, on

    Axioms:

    (a) There are at least four points.

(b) Every point is on two lines.

(c) Some line is not on any point.

11. **Undefined terms**: spade, heart, trump

    Axioms:

    (a) There is at least one heart.

    (b) For any two distinct spades, there is exactly one heart that trumps them.

    (c) Each heart trumps at least two distinct spades.

    (d) Given any heart, there is at least one spade it does not trump.

12. **Undefined terms**: X, Y, related to

    Axioms:

    (a) There are exactly five Xs.

    (b) At least one Y is related to each X.

    (c) No X is related to every Y.

    (d) Every X is related to at least two Ys.

13. **Undefined terms**: gump, lump, bump

    Axioms:

    (a) Every gump bumps at least one lump.

    (b) Not every lump bumps a gump.

    (c) There are at least three gumps.

    (d) No lump bumps exactly one gump.

14. Verify that the theorem proved in Example 3.5 is satisfied in both models for Axiom System **A1–A4** given in this section.

15. Prove the following theorem for Axiom System **A1–A4**:

    There are at least three points.

16. Prove the following theorem for Axiom System **A1–A4**:

    There are at least three lines.

17. Prove the following theorem for Axiom System **A1–A4**:

    Given a point, there is a line not on that point.

8. Consider the following statement in Axiom System **A1–A4**:

   Every line is on at least three points.

   (a) Is this statement verified in the model of *Figure 3.12*?

   (b) Is this statement verified in the model of Example 3.4?

(c) Is this statement a theorem of the axiom system?

19. In general, how can you show that a given statement is not a theorem of a given axiom system? (*Hint*: See Exercise 18.)

VRITING EXERCISES

. Discuss how the mathematical usage of the word "model" is analogous to its nonmathematical usage.

. Does changing the undefined terms in an axiom system—as was done in this section in the axioms for projective geometry—have any effect on the existence or variety of models? Justify your answer by appropriate argument or examples.

. If there is a model that faithfully represents each of two different axiom systems, does

this indicate that the two abstract systems are equivalent, aside from purely linguistic differences? Justify your answer by appropriate argument or examples.

4. Create your own example of a formal axiom system using "snark" as one of your undefined terms. Design your system so that no axiom explicitly states that there are an infinite number of snarks, but any model of the system must actually interpret "snarks" as an infinite set. Explain why this is so.

CONSISTENCY AND INDEPENDENCE

Consider the following axiom system:

Undefined terms: *A*, *B*, match

Axioms:

  1. *A* matches *B*.
  2. *A* does not match *B*.

This clearly is an example of an abstract mathematical system because it is a collection of statements about some undefined terms; yet our common sense tells us that something is wrong. Axiom **2** is the logical opposite of Axiom **1**; therefore, regardless of the interpretations we assign to *A* and *B*, one of the axioms must be true and the other false. However, because they are *axioms* of our system, Axioms **1** and **2** must be true within the system. In other words, this axiom system is useless because there can be no interpretation that will preserve the truth of the axioms; so there are no models. In general, any system containing contradictory axioms is of no practical value at all.

Let us pursue this matter a little further. Suppose the axioms of a system do not appear to be contradictory; but at some point in the development of the system, two contradictory statements can be proved by correct deductive reasoning. What can be said about the axiom system? The assumption that the axioms are all true and that the deductive arguments are all valid leads inescapably to the fact that two contradictory statements are both true. Thus, the contradiction was at least latently present in the original system, even though it may have been disguised by the phrasing of the axioms. As in the trivial case just given, this means that there is no possible model of such an axiom system. Such a system is said to be *inconsistent*. In slightly more formal terms:

DEFINITION

The **negation** of a statement is another statement that is true when the original statement is false, and is false when the original statement is true.[9]

DEFINITION

An axiom system is **inconsistent** if it is possible to prove both a statement and its negation from the axioms. A system is **consistent** if it is not inconsistent.

As we have already remarked, any system that has a model cannot involve a contradiction within itself. This fact provides us with a convenient way of testing the consistency of an axiom system—all we have to do is find a model for it. For example, the existence of the ring-and-wire model for the projective geometry Axiom System **P1–P5**, described in Section 3.4, guarantees that the geometry built on this axiom system will be free of contradiction, no matter how complicated its theorems become! This test of consistency is based mainly on physical models, but all or part of an abstract system that has already been proven consistent may also serve as a model to verify the consistency of another system. For instance, since Euclidean plane geometry has been proven consistent, any plane geometric figure can serve as a legitimate model for another axiom system.

EXAMPLE 3.6

The squirrels-and-trees axiom system given in Example 3.3 can be proved consistent by using a triangle as a model. If we interpret the *squirrels* as the three vertices, the *trees* as the three sides, and *climb* as "on," then it is clear that all three axioms are satisfied:

---

9. See Appendix A for examples and further discussion of negation.

1. There are exactly three vertices.
2. Every vertex is on (exactly) two sides.
3. No side is on more than two vertices.

Sometimes it is helpful to construct a model of a new system in terms of an older, more familiar, system, even though the latter has not itself been proven consistent. In this case, the consistency of the new system depends on that of the older one; thus, we call the new system **relatively consistent** with respect to the old one. Any model of a system that uses our usual arithmetic is an example of a relative consistency proof because our arithmetic is a simple, workable, but *not* provably consistent system.

Although consistency is all that is really necessary to insure the logical reliability of an axiom system, the natural desire for simple efficiency leads us to another consideration. In specifying axioms for a system, we would surely avoid stating the same axiom twice because repeating a statement adds no new information to the system. Similarly, we would wish to avoid giving as an axiom any statement that could be proved from the other axioms of the system because such a statement also adds no new information to the system. If the statement were dropped as an axiom, there would be no change in the theorems that could be proved or in the models that are possible for the system; in this sense, the inclusion of such a statement as an axiom is superfluous. The logical dependence of such a statement on the other axioms can be characterized by the fact that, if it were replaced as an axiom by its negation, the resulting system would contain a contradiction. This observation leads to a useful definition.

DEFINITION

An axiom *A* in a consistent axiom system is said to be **independent** if the axiom system formed by replacing *A* with its negation is also consistent. An axiom is **dependent** if it is not independent. An axiom system is **independent** if each of its axioms is independent.

The notion of classifying a statement as independent or dependent is not restricted to just the axioms of a system. If either a statement or its negation could be added to a set of axioms to produce a consistent axiom system, then that statement is independent. If a statement or its negation is inconsistent with a set of axioms, it is dependent. Note that this latter case is equivalent to saying that either the given statement or its negation (whichever is consistent with the axioms) must be a theorem. To show that a statement is dependent, then, we give some logical argument, either one showing that it is a theorem derived from the axioms, or one showing it is inconsistent with the axioms.

To prove that an axiom (or any other statement) in a consistent system is independent, we can use models. Recalling that the negation of a statement is true if and only if the original statement is false,[10] what is required to show an axiom independent is a model for which that axiom is *false* while all the other axioms are true. This model insures the consistency of the "new" axiom system, and it necessarily will have to be different from the model used to prove the consistency of the original system. Similarly, for a statement other than an axiom, we can show independence by a pair of models of the axiom system, one in which the statement is true, the other in which the negation of the statement is true.

| EXAMPLE 3.7 | Consider the abstract form of the projective geometry Axiom System, **P1–P5**, as it was given in Section 3.4. To show that Axiom **P4** is independent, we must construct a model in which there are fewer than three Xs related to some Y, but all the rest of the axioms are true statements. A triangle can be used for such a model, with its vertices as Xs, its sides as Ys, and *is related to* interpreted as "is on" (or "connects," when grammatically more comfortable). In this model Axioms **P1**, **P2**, **P3**, and **P5** are all true: |

P1. There exist at least one vertex and one side.

P2. If a and b are distinct vertices, then exactly one side connects both of them.

P3. If c and d are distinct sides, then exactly one vertex is on both of them.

P5. Not all vertices are on the same side.

Moreover, Axiom **P4** is false because no side has three vertices on it. Thus, we have shown that Axiom **P4** is independent of the other axioms.

To prove an entire axiom *system* independent, we need to show that each axiom is independent. This requires a different model for each axiom, so that each one, in turn, can be modeled as false while the others are modeled as true. These models, together with a model for which all the axioms are true, form a proof of the consistency and independence of an axiom system. Thus, to prove the independence of Axiom System **P1–P5**, we would need four more models. Some of these are addressed in the exercises.

10. See Section 3 of Appendix A for a discussion of "if and only if."

There is a logical "trick" that sometimes comes in handy for constructing models. It involves axioms that are "conditional statements" —that is, statements that can be put in the form

If p, then q

In order for such a statement to be false, its hypothesis p must be true *and* its conclusion q must be false. (See Section A.3 of the Appendix.) Thus, for example, if you state

If I get home in time, then I'll call.

you would be judged as having spoken falsely only if you *do* get home in time and you *don't* call; thus, your statement should be considered true if you *don't* get home in time, whether or not you call. The "trick" in making models related to this is summed up in the following statement:

If you construct a model in which the hypothesis of an axiom is *false*, then that axiom is *true* in this model.

Thus, we can satisfy a conditional axiom with an interpretation *either* by making both its hypothesis and its conclusion true or by making its hypothesis false (regardless of the truth or falsity of its conclusion). In the latter case we say the statement is **vacuously satisfied**. As the next example shows, an extremely trivial interpretation, appropriately chosen, can make the work of verifying the desired properties of a model easy by vacuously satisfying many axioms.

EXAMPLE **3.8**

We can prove that Axiom **P1** of the projective geometry system of Section 3.4 is independent as follows: Interpret the symbol "#" to be the only X, let there be no Ys, and let *related to* mean "connected to." Axiom **P1** is false in this case because there are no Ys. The other axioms are vacuously satisfied for the following reasons:

P2: Because there are not two distinct Xs, the hypothesis of this axiom, as interpreted, is false; so the statement itself is true.

P3: Because there are no Ys at all, the hypothesis of this axiom is false.

P4: This axiom can be rephrased as "If there is a Y, then at least three Xs must be related to it." Because there are no Ys, the hypothesis of this statement is false.

P5: This axiom can be rephrased as "If there is a Y, then there is some X not related to it." As in the previous case, because there are no Ys, the hypothesis is false.

3.5   EXERCISES

Exercises 1–8 refer to the following axiom system:

  Undefined terms: letter, envelope, contain
  Axioms:
  (a) There are at least two envelopes.
  (b) Each envelope contains exactly three letters.
  (c) No letter is contained in all the envelopes.

1. Prove that the system is consistent.

2. Write the negation of each axiom.

3. Prove that Axiom **(b)** is independent.

4. Which of the following assertions are logically equivalent to Axiom **(c)**?
  (a) There is a letter contained in two different envelopes.
  (b) There is an envelope that does not contain any letter.
  (c) There is an envelope that does not contain all the letters.
  (d) If *l* is a letter, then it is not contained in every envelope.
  (e) If *l* is a letter, then it is contained in some envelopes, but not in all envelopes.
  (f) If *l* is a letter, then there is some envelope that does not contain *l*.

5. Prove that Axiom **(c)** is independent.

6. Is the statement
      There are at least three envelopes.
  dependent or independent? Justify your answer.

7. Is the statement
      There are at least four letters.
  dependent or independent? Justify your answer.

8. Is the statement
      There are at least six letters.
  dependent or independent? Justify your answer.

9. Prove that the following axiom system is inconsistent:
  Undefined terms: point, line, on
  Axioms:
  (a) Any two points are on exactly one line.
  (b) There are exactly four points.
  (c) Every line is on exactly two points.
  (d) There are exactly five lines.

Exercises 10–14 refer to the following axiom system:

  Undefined terms: X, Y, related
  Axioms:
  (a) There are at least two Xs and two Ys.
  (b) Each pair of Xs is related to exactly one Y.
  (c) Each Y is related to at least one X.

10. Is the system consistent? Prove your answer.

11. Write the negation of each axiom.

12. Is Axiom **(b)** independent? Prove your answer.

13. Is Axiom **(c)** independent? Prove your answer.

14. Is the statement
      There are at least three Xs.
  dependent or independent? Justify your answer.

Exercises 15–19 refer to the following axiom system:

  Undefined terms: box, crate, in
  Axioms:
  (a) There are exactly four boxes.
  (b) There is at least one crate.
  (c) Every box is in at least two crates.
  (d) Not all boxes are in the same crate.

5. Prove that the system is consistent.

6. Prove that one of the axioms is dependent upon the others.

7. Prove that three of the four axioms are independent.

8. Supply a fifth axiom that makes the system inconsistent. Prove your answer.

9. Supply a fifth axiom that is independent of the other four. Prove your answer.

xercises 20–25 refer to the following axiom ystem:

Undefined terms: point, line, contain

Axioms:

(a) **There are at least two lines that contain the same number of points.**

(b) **There are at least two lines that contain no points in common.**

(c) **Each line contains at least two points.**

(d) **Not all lines contain the same number of points.**

(e) **There are at least four points.**

0. Prove that the system is consistent.

1. Prove that Axiom (a) is independent.

2. Prove that Axiom (d) is independent.

3. One of the axioms is dependent upon the others. Which one is it, and why?

4. Is the statement

There are only two lines.

dependent or independent? Justify your answer.

5. Is the statement

Two distinct lines contain at most one common point.

WRITING EXERCISES

. Explain why the following definition of *dependent axiom* is equivalent to the definition given in the text:

An axiom *A* in an axiom system *S* is dependent if and only if it is a theorem

dependent or independent? Justify your answer.

Exercises 26–31 refer to the following axiom system:

Undefined terms: student, book, read

Axioms:

(a) **There is at least one student.**

(b) **Some students read at least two books.**

(c) **At least one student does not read any books.**

(d) **Every book is read by at least two students.**

26. Prove that the system is consistent.

27. Is Axiom (a) dependent or independent? Justify your answer.

28. Is Axiom (d) dependent or independent? Justify your answer.

29. Is the statement

There are at least three students.

dependent or independent? Justify your answer.

30. Is the statement

There are exactly three students.

dependent or independent? Justify your answer.

31. Is the statement

If there is only one student, then there are at least ten books.

true or false in this system? Why?

32. Complete the proof that the projective geometry Axiom System **P1–P5** of Section 3.4 is independent by finding three more models to establish the independence of Axioms **P2**, **P3**, and **P5**.

in the axiom system consisting of all the axioms of *S* other than *A*.

2. (a) Explain why the deletion of a dependent axiom from an axiom system has no

effect on which theorems can be proved or on which models the system has.

(b) If an axiom system has two dependent axioms, does it follow that *both* could be deleted with no substantive effect on the system? Explain your answer.

3. Consider the following interpretation of the axiom system in Exercises 26–31: let *student* be "Moe," "Larry," or "Curly"; let *book* be "War and Peace" or "Moby Dick"; let *read* be described by "Moe reads no books, Larry reads both *War and Peace* and *Moby Dick*, and Curly reads both *War and Peace* and *Moby Dick*." Is this a model for the axiom system? Explain your answer. As part of your answer, address what distinguishes Larry from Curly?

4. Consider the following interpretation of the axiom system in Exercises 26–31: let *student* be one of the numbers 1, 2, or 3; let *book* be either 2 or 3; let *read* be described by "*x* reads *y* if and only if *x* times *y* is greater than three." Verify that this is a model for the axiom system, and discuss what this example suggests about the differences among the undefined terms of an axiom system.

NON-EUCLIDEAN GEOMETRIES

3.6

From its earliest days, the axiomatic approach to mathematics has been motivated by a desire to base the topics it treats upon the simplest possible foundation. Long before formal axiomatics raised such issues and provided the methodology of models to test them, mathematicians' concern for the independence of axiom systems led them to ferret out unnecessary assumptions wherever they could be found. The problem of finding such redundancies often involved great difficulties. Mathematicians had to rely on their "logical instinct" for some intuitive hint that an axiom or postulate might be dependent, and only then would the task of proving that statement from the other assumptions begin.

Almost from the very moment Euclid proposed his five postulates for geometry, that logical instinct began to stir doubts about the fifth of these statements, often called the **Parallel Postulate**:

> Through a given point not on a line, there is exactly one line parallel to the given line.

(In reading this section, you might find it helpful to review all five of Euclid's postulates, stated as Postulates **E1–E5** at the beginning of Section 3.2.)

Even Euclid himself seemed to be bothered by the possible dependence of this postulate; he avoided using it until the proof of his twenty-ninth theorem, and subsequently did not use it again, though many later

Picture Collection, The Branch Libraries, The New York Public Library

Railroad Tracks Stretching into the Distance.

proofs depend upon it via that twenty-ninth theorem. (The essence of the twenty-ninth theorem is the converse of the Alternate Interior Angle Theorem, as stated and proved on pages 77–78. As was noted there, the final step of the proof relies on Euclid's Fifth Postulate.) The belief that the Parallel Postulate could be proved from the other four spread rapidly, and over the centuries many scholars established their mathematical reputations by constructing "proofs" of it in which the flaws were so subtle as to escape notice during their lifetimes. However, even these flaws were uncovered eventually, and the independence of the Parallel Postulate remained an open and tantalizing question through the end of the eighteenth century.

In the early 1700s, an Italian teacher and scholar named Girolamo Saccheri made the first noteworthy attempt to attack the problem indirectly. He proposed to negate the parallel postulate and then to probe the resulting geometric tangle until he found a contradiction. The existence of that contradiction would establish the dependence of the postulate. (This is the approach discussed in the preceding section.) Now, the Parallel Postulate specifies that there is *exactly one* line through a given point parallel to a given line; so its negation takes the form of two alternatives:

There is a line and a point not on it with the property that

i. there are *no* lines parallel to this line through the given point, or

ii. there are *at least two* lines parallel to this line through the given point.

Using the assumption that Euclid's Second Postulate requires straight lines to be infinitely long, Saccheri found a contradiction resulting from the first alternative. The second case was much more stubborn, though; convinced that his thread of logic must ultimately snag, Saccheri actually knotted it himself. After skillfully proving many valuable results, he ended by forcing a weak and vague conclusion about lines that merge at infinity, which he partially persuaded himself to be a logical contradiction. It apparently convinced almost no one, and even Saccheri himself was sufficiently skeptical to attempt another solution. However, his second effort was no more successful than the first.

Saccheri's work had little influence on the mathematical thought of his day. Apparently, everyone was so convinced of the dependence of the Parallel Postulate that no attention was paid to the alternative possibility. This alternative would imply that the *negation* of the Parallel Postulate combined with Euclid's Postulates **E1–E4** do not form an inconsistent system, but rather are the basis for a new, logically consistent geometry that is essentially different from the system Euclid described. Such a situation was not, it seems, within the realm of speculation for the scholars of the eighteenth century.

Almost a century later, Saccheri's indirect approach was revived independently and almost simultaneously by four men. Shortly after 1800, the great German mathematician Carl Friedrich Gauss was the first to recognize that a new geometry is created if the Parallel Postulate is replaced by the assertion that through a given point there are at least two lines parallel to a given line; but he did not publish his findings. It has been speculated that his reluctance stemmed from the dominance of Immanuel Kant's philosophy within the European intellectual community of that day. Kant's theory of metaphysics was based in part on the assertion that the human perception of space is necessarily Euclidean, and hence the assertion of an equally valid geometry with different properties would bring Gauss into direct conflict with Kant. Perhaps even Gauss did not want to pit his reputation as Germany's greatest mathematician against that of its greatest philosopher.

The first publication of this revolutionary possibility appeared in 1829, written by a Russian mathematician, Nicolai Lobachevsky, who devoted much of his life to developing this type of geometry. Part of his work was anticipated by Janos Bolyai, a young Hungarian army officer; but Bolyai did not publish his results until 1832. Bolyai was also interested in developing what he called "absolute" geometry, a system based on Euclid's first four postulates alone. A geometry based on these four postulates and the assertion that there are no lines parallel to a given line appeared in 1854, when Bernhard Riemann (also of Germany) proved that Saccheri's contradiction in this case could be avoided. Riemann's insight was based on shifting from the commonly accepted understanding of Euclid's Second Postulate, which calls for the ability to extend a line segment continuously; if one understands this to imply that straight lines must be unlimited in extension, a very natural inference, then Saccheri's argument is correct. But if one suspends the usual *visualization* of a line, then the requirement of continuous extendibility does *not* inescapably lead to infinitely long lines, and

Nicolai Lobachevsky.
He postulated that more than one line can be constructed parallel to a given line.

a consistent geometry with no parallels is possible.[11] (Think of a circle instead of a line, for example.)

The geometries developed by Lobachevsky and Riemann are called **non-Euclidean geometries**. They differ from Euclidean geometry only in their rejection of the Parallel Postulate. There are two types of non-Euclidean geometry, corresponding to the two parts of the negation of Postulate **E5**:

■ The type that postulates no line through a given point parallel to a given line is called **Riemannian geometry**.

■ The type that postulates more than one line through a given point parallel to a given line is called **Lobachevskian geometry**.

We will not attempt to pursue either of these rich axiomatic systems in any detail. (See the references in the list *For Further Reading* at the end of the chapter.) But we can get some appreciation for the consequences of changing the Parallel Postulate by a consideration of a model for each alternative. Many different models are possible for each of these axiom systems, but we shall confine our study to an informal discussion of one planar model for each geometry.

To visualize a model of Riemannian geometry, consider the *plane* to be the surface of a sphere. *Points* are then the points on the sphere and *lines* are great circles (that is, circles that divide the sphere into two equal parts, like the equator or the lines of longitude on the Earth). Because the shortest path between any two points on a sphere is an arc of a great circle through those points, it is natural for such circles on the sphere to be analogous to straight lines in the Euclidean plane. This model represents Riemannian **double elliptic geometry**.[12] It is a fairly good interpretation of Euclid's first four postulates (with just a little refinement needed), but the Parallel Postulate does not hold because any two great circles intersect. Thus, given a point P not on a line l, any line through P must intersect l, (in exactly two diametrically opposite points, Q_1 and Q_2, as in *Figure 3.13*).

The *plane* for our model of Lobachevskian geometry is a little harder to visualize. Imagine a little child walking along, pulling a toy attached to a string. If the child makes an abrupt left turn, the toy will

Figure 3.13 Any two distinct great circles must intersect at two points.

11. Subsequently, Riemann described an alternative version of his geometry-without-parallels in which infinitely extendable lines were possible; but the trade-off necessary to accomplish this was a denial of one of Euclid's unstated (and hence even more fundamental) assumptions—that a line in a plane separates that plane into two distinguishable parts.

12. The label corresponds to the property of the model that distinct "lines" meet in two distinct points. There is an alternative model in which distinct "lines" meet in only one point, known as Riemannian single elliptic geometry.

Figure 3.14 The forma-
tion of a pseudosphere.

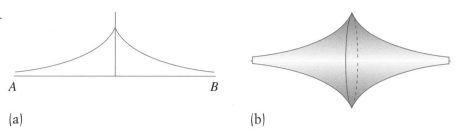

A   B

(a)   (b)

trail behind, not making a sharp corner, but slowly curving around until it is almost directly behind the child once more. The curved path that the toy travels is called a **tractrix**, and it is used to generate the surface we want for our model. We take two opposite copies of a tractrix, as shown in *Figure 3.14(a)*, and rotate this double curve about the line through A and B to produce the object in *Figure 3.14(b)*. The resulting surface is called a **pseudosphere**; it looks somewhat like two trumpet bells joined together, with the narrow part of each one tapering gradually as it becomes infinitely long. This is the *plane* of the model;[13] *points* are the points on this pseudosphere, and *lines* are "straight" paths on it, with straightness characterized by the minimum distance between points, analogous with great-circle arcs as the minimum distance between two points on a sphere.

It is challenging to draw or visualize pictures on the pseudosphere, but this surface has the property that through any point not on a line there are many parallels to the given line. *Figure 3.15* illustrates how this would look on a flat surface; if that picture were "wrapped around" the pseudosphere, all the lines in it would be straight. (Notice that parallel lines in this geometry may not be everywhere equidistant from one another.)

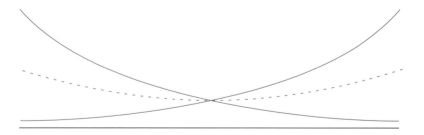

Figure 3.15 Several parallels
to a line through a point on a
pseudosphere.

13. Actually, the pseudosphere is a model for only a limited portion of the Lobachevskian plane; there are anomalous features at the portion of the pseudosphere where the two "trumpets" join, so that Lobachevskian figures that would extend across that boundary are not faithfully represented.

To see how these two young geometries compare with their Euclidean ancestor, we return to the work of Saccheri. His approach to the problem of parallels provides a result that typifies the differences among the three systems. For simplicity, we shall use pictures taken from our models of these geometries, with the plane, sphere, and pseudosphere as the basic surfaces. Nevertheless, it should be noted that the results described do not depend on the models, but can be proved directly from the axiomatic foundations of their respective systems.

Saccheri began his investigation by constructing an *isosceles birectangular quadrilateral*; that is, he formed a four-sided figure with two opposite sides of equal length and two right angles. This figure, also called a **Saccheri quadrilateral**, serves as our starting point.

In any of these geometries we may begin with a line segment $\overline{AB}$; at each of its endpoints we construct perpendicular segments $\overline{AD}$ and $\overline{BC}$ of equal length, and we join them by the line segment $\overline{CD}$. The figures thus formed on the three surfaces share some features, but are quite different in other respects, as shown in *Figure 3.16*. In all three cases, it can be shown that the line containing $\overline{CD}$ is parallel to the line containing $\overline{AB}$ and that the angles formed at C and D are equal to each other. The planar figure is a rectangle with $\overline{CD} = \overline{AB}$ and right angles at C and D. On the sphere, however, $\overline{CD}$ is shorter than $\overline{AB}$, and the angles at C and D are obtuse. (The size of an angle may be visualized by considering tangents to its sides at the vertex as shown by the dotted lines in Figure *3.16(b) and (c)*.) On the pseudosphere, $\overline{CD}$ is longer than $\overline{AB}$, and the angles at C and D are acute.

We can use this information to prove a theorem comparing a property of triangles in the three geometries:

✕✕

(3.5)   The sum of the angles of a triangle is equal to two right angles in Euclidean geometry, is greater than two right angles in Riemannian geometry, and is less than two right angles in Lobachevskian geometry.

✕✕

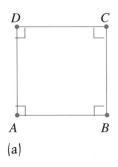 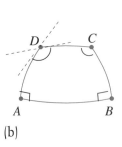 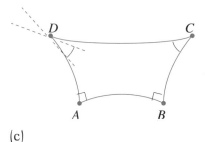

Figure 3.16 Saccheri quadrilaterals on (a) the plane, (b) the sphere, and (c) the pseudosphere.

(a)   (b)   (c)

The steps of the following proof are independent of the Parallel Postulate, so they apply to all three geometries. *Figure 3.17* illustrates the construction. This figure, which is *not necessarily "flat,"* will look approximately the same if it is drawn on a Euclidean plane, on a small portion of a large sphere, or on a small portion of a large pseudosphere.

Proof

Let ABC be a given triangle, and let D and E be the respective mid-points of sides $\overline{AC}$ and $\overline{AB}$. Draw the line through D and E. Drop perpendiculars $\overline{AF}$, $\overline{BG}$, and $\overline{CH}$ from the three vertices of the original triangle to line $\overline{DE}$. In each geometry it can be shown that triangles ADF and CDH are congruent and that triangles AFE and BGE are congruent. It follows that

$$\angle 1 = \angle 3 \quad \text{and} \quad \angle 2 = \angle 6$$

Thus, the sum of the angles of triangle ABC, which is

$$\angle 1 + \angle 2 + \angle 4 + \angle 5$$

is equal to

$$\angle 3 + \angle 6 + \angle 4 + \angle 5$$

which can be regrouped in the form

$$(\angle 3 + \angle 4) + (\angle 5 + \angle 6)$$

But, because $\overline{BG} = \overline{AF} = \overline{CH}$, by the congruence of the triangles identified, $HGBC$ is a Saccheri quadrilateral. This implies that

$$\angle 3 + \angle 4 \quad \text{as well as} \quad \angle 5 + \angle 6$$

are right angles, obtuse angles, or acute angles, depending upon the geometry they are in, and hence Theorem (3.5) is proved.

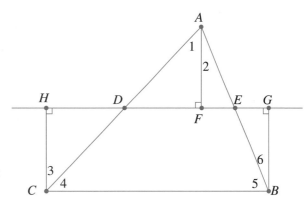

Figure 3.17 The sum of the angles of a triangle.

3.6   EXERCISES

Exercises 1–8 refer to Riemannian geometry. We define the *excess of a triangle ABC* as the sum of the interior angles of the triangle minus two right angles; we denote this quantity by exc(*ABC*). Similarly, we define the *excess of a quadrilateral ABCD* as the sum of the interior angles of the quadrilateral minus four right angles; we denote this by exc(*ABCD*).

1. Show that congruent triangles have the same excess.

2. Let *PQRS* be a quadrilateral with diagonal *PR*. Prove that
$$\text{exc}(PQRS) = \text{exc}(PQR) + \text{exc}(PSR)$$

3. Referring to *Figure 3.17*, show that
$$\text{exc}(ABC) = \text{exc}(BCHG)$$

4. Let *D* be a point on side *BC* of triangle *ABC*. Show that
$$\text{exc}(ABC) = \text{exc}(ABD) + \text{exc}(ACD)$$

5. For any triangle, show that there must exist another triangle whose excess is at most half that of the given triangle. (See Exercise 4.)

6. Show that there must exist a triangle whose excess is as small as we like. (See Exercise 5.) What is the sum of the angles of such a triangle?

7. What is the "upper limit" for the excess of a triangle? (The *upper limit* of the sequence of numbers
$$\frac{1}{2}, \frac{3}{4}, \frac{7}{8}, \frac{15}{16}, \cdots$$
for example, is 1). What is the sum of the angles of such a triangle?

8. What is the upper limit for the excess of a quadrilateral? (See Exercise 7.) What is the sum of the angles of such a quadrilateral?

Exercises 9–16 refer to Lobachevskian geometry. We define the *defect of a triangle ABC* as two right angles minus the sum of the interior angles of the triangle; we denote this quantity by def(*ABC*). Similarly, we define the *defect of a quadrilateral ABCD* as four right angles minus the sum of the interior angles of the quadrilateral; we denote this by def(*ABCD*).

9. Show that congruent triangles have the same defect.

10. Let *PQRS* be a quadrilateral with diagonal *PR*. Prove that
$$\text{def}(PQRS) = \text{def}(PQR) + \text{def}(PSR)$$

11. Referring to *Figure 3.17*, show that
$$\text{def}(ABC) = \text{def}(BCHG)$$

12. Let *D* be a point on side *BC* of a triangle *ABC*. Show that
$$\text{def}(ABC) = \text{def}(ABD) + \text{def}(ACD)$$

13. For any triangle show that there must exist another triangle whose defect is at most half that of the given triangle. (See Exercise 12.)

14. Show that there must exist a triangle whose defect is as small as we like. (See Exercise 13.) What is the sum of the angles of such a triangle?

15. What is the "upper limit" for the defect of a triangle? (See Exercise 7.) What is the sum of the angles of such a triangle?

16. What is the upper limit for the defect of a quadrilateral? (See Exercise 7.) What is the sum of the angles of such a quadrilateral?

Exercises 17 and 18 refer to Riemannian geometry. Refer also to the spherical model described in the text. (See *Figure 3.13*.)

17. In the model:
    (a) What kind of figure represents a line?
    (b) What kind of figure represents a circle?
    (c) What is the distinction between these two figures?

18. Consider a circle in the Riemannian plane (or its representation in the model). Let *R* be the ratio of the circumference to the diameter.

(a) Is R less than, equal to, or greater than π?

(b) Is the value of R the same for all circles in the Riemannian plane?

(c) How small could R be?

(d) How large could R be?

19. Answer the questions of Exercise 18 for the Lobachevskian plane. (You will probably find it easier to reason by analogy, rather than to use the model of the pseudosphere.)

WRITING EXERCISES

1. Could there be other non-Euclidean geometries besides the Riemannian and Lobachevskian geometries? Explain your answer.

2. Defend the thesis:

   Non-Euclidean geometry could not have been developed before Euclidean geometry.

3. Defend the thesis:

   Non-Euclidean geometry could have been developed before Euclidean geometry.

AXIOMATIC GEOMETRY AND THE REAL WORLD

3.7

The emergence of the non-Euclidean geometries provoked a major shift in Western scientific thought. What started as an attempt to show the dependence of Euclid's Fifth Postulate, presumably an exercise internal to mathematics, ended in a dramatic change in the concept of what mathematics is, what "the truth" about the real world is, and how the two relate to each other.

To begin to appreciate this shift, let us follow up on the concepts introduced in Section 3.6. We saw there that it is logically consistent to deny Euclid's Parallel Postulate and to assert in its place that there may be a line and a point not on that line through which there are either many parallels (Lobachevskian geometry) or none at all (Riemannian geometry). Certainly, both of these alternatives seem contrary to our intuition, to our experience of the world around us. This intuition and experience, bolstered by our study of high school geometry, might easily persuade us that the geometries of Lobachevsky and of Riemann are merely intellectual curiosities. "After all," we might say, "the three geometries—Euclidean, Riemannian, and Lobachevskian—are mutually contradictory, so one of them must be true and the other two false when applied to reality. And, *obviously*, the real world follows the properties described by Euclid."

But on what basis do we assert that the geometry of the real world is Euclidean? As we saw earlier, a geometry is a formal system whose basic terms have no meaning. In dealing with geometries, it is the form and internal consistency of the systems that concerns mathematicians, not the way the systems relate to the real world. Statements about points and lines could just as well be about rings and wires, or sets and numbers, or *X*s and *Y*s. In the language of this chapter, then, the question of the correct geometric description of the real world is a question of which of the geometries is satisfied when we use the real world as a model; so far, at least, we have little more than Euclid's word to justify calling reality a model for his geometry, and his alone.

"This is just a semantic issue," you say. "In the terminology of models, can we not assert that when the terms *point*, *line*, and *on* are interpreted in the real world, we find that Euclidean geometry is satisfied and not the non-Euclidean geometries?"

It's not that simple. First, we must be precise about what, in the real world, is the interpretation of a *point* and of a *line* and what we mean by *on*; only then can we test the truth of Euclid's axioms with that interpretation. Suppose, for example, we interpret *point* as a speck of graphite made by pressing a pencil to paper, and *line* as a streak of the same stuff laid down by drawing along the edge of a ruler. Then how do we test the parallel property? Given a streak and a speck, how can we assert that there is one and only one streak through that speck which does not meet the given streak, no matter how far the streaks are extended? Surely, on any piece of paper we can make two or more streaks through the given speck, neither of which intersect the original streak. Does that mean the geometry of the real world is Lobachevskian?

Perhaps we were a bit hasty in our choice of interpretations. You might opt for some description such as "a location in space" for *point*. But now we are speaking of an abstract concept created in our minds, not something that exists in the physical universe. Thus, whatever we conclude about such "points" may well describe the geometry we are *thinking* about, but it has no necessary connection with the geometry of the universe around us.

There is another, more serious, problem with making a physical interpretation. No matter what we choose for it, we are also doomed to failure in trying to distinguish among the three geometries by looking directly at parallelism. The distinction, at least between Euclidean and Lobachevskian geometry, *requires* that we be able to extend lines indefinitely in order to decide whether or not certain pairs of lines really *never* meet. However, in the real world we only have access to a finite portion of the universe, so we would never be able to know with certainty whether or not (any) lines were parallel.

This problem suggests that we seek a test that avoids the direct confrontation with the differing parallel axioms, and instead, attempts to distinguish among them on the basis of some derived property. An obvious and promising candidate is the sum of the angles of a triangle. Theorem (3.5) at the end of Section 3.6 says that the sum of the angles of a triangle varies with the type of geometry we are in. Therefore, if we construct a physical model of a triangle and measure its angles carefully, then the sum of those angles should tell us which geometry the physical triangle is in. That is, the physical world will be Lobachevskian, Euclidean, or Riemannian, depending on whether the angle sum is less than, equal to, or greater than two right angles, respectively.

We again face the problem of interpretation; but, bypassing that for the moment, suppose we carefully draw a triangle using a pencil with a very sharp point and measure the angles with a protractor. As you know, particularly if you have done any lab work in physics or chemistry, when two different people measure something, or even the same person at two different times measures it, the observed results will vary slightly, no matter how much care is taken. That is why measurements in physical experiments are normally done several times, and the average value is used, often with some expression of the range of possible error. Thus, any careful triangle construction and angle measurement can be expected to produce a result like

179.95 degrees ± .15 degrees

Unfortunately, this measurement says that the "true" angle sum might be less than, equal to, or greater than two right angles (180°); so we have not distinguished among the geometries.

Perhaps, having carefully studied the exercises for Section 3.6, you have some hint of why this problem arises and how it might be fixed. In looking at the concept of "excess" in Riemannian geometry and of "defect" in Lobachevskian geometry, you may have noted that "small" triangles in either geometry have angle sums that are very close to two right angles. In fact, it turns out that all the properties of these two geometries are approximately Euclidean for "small" portions of space. Thus, even if we have reasonable physical interpretations for point and line, as long as we are dealing with a "small" part of the physical universe, the shortcomings of physical measurement will render the distinction among geometries impossible. The obvious solution, then, is to work with a "big" triangle. We might, for example, use thin poles in a large field, measuring angles with a surveyor's transit. Or we might try three mountain tops. Or we might think really big and use two widely separated locations on the earth and a reflector on the moon. (Note that in all these situations we are interpreting *line* as the path taken by a ray of light.) Gauss himself tried this experiment in the early part of the nineteenth century. Using three mountain peaks in Germany (Brocken, Hohehagen, and Inselsberg), he built a signal fire and set up mirrors to form a large triangle of reflected light rays. The sides of the triangle were 69, 85, and 197 km (about 43, 53, and 123 miles), and his angle-sum measurement was

180 degrees + 14.85 seconds

However, this excess, corresponding to less than a quarter of a thousandth of a percent, was far smaller than the margin of error for measurement in that experiment, so the results proved nothing.

Could we argue that the mere fact that all these results are very close to 180° (that is, within the limits of error of measurement) demonstrates the Euclidean nature of the real world? Unfortunately, we cannot! After all, compared to the size of the physical universe, even just the part of it we know, distances between mountains—or between planets in the solar system—are incredibly small. We are still working with little triangles. It is not surprising that, until relatively recently, Euclidean geometry was the only one recognized, and no wonder that we have built up such a strong bias toward it. But we should take care not to mistake that bias for proof that Euclidean geometry is "true." As long as there is physical measurement involved, some error must be expected and must be taken into account.

Gauss' Experiment. Using three mountain peaks, he built a signal fire and set of mirrors to form a triangle of reflected light rays.

And as long as there is some error, it is impossible to conclude such an experiment in favor of Euclidean geometry.

Interestingly enough, it is conceivable that careful measurement *might* support one or the other of the non-Euclidean options. For instance, a triangle sum of $181° \pm .5°$ could only be true for a Riemannian triangle. Some experiments have shown just such a discrepancy from Euclidean predictions, though the experiment was not on the sum of the angles of a triangle. Specifically, an astronomical experiment performed by Sir Arthur Eddington in 1919, confirming a prediction made by Albert Einstein, showed that light from a distant star followed a path that was *not* a Euclidean line. One way to explain the phenomenon is to say that the geometry of space really is Euclidean, but the light was following a curved path. Another explanation is that the light went in a straight line, but the geometry is Riemannian. The "truth" is that *either explanation is acceptable*; you may pick either, depending on whether your bias is stronger for Euclidean geometry or for the straightness of light paths. But it *is* a matter of bias or of convenience, not really one of "truth."

In other words, asking about the "true" geometry of the physical universe is pursuing a presumably unanswerable and misleading question. The universe is whatever way it is, and geometry is a mathematical concept, created by human mathematicians. A more reasonable question might concern which geometry is most useful to apply in some particular physical situation, with certain appropriate interpretations of the geometric terms. Clearly, for most measuring, building, travelling, etc., where we are dealing with a tiny portion of physical space, Euclidean geometry is the simplest to use. It works for such purposes. For large-scale measurements (galactic size and up), many scientists argue for the use of Riemannian geometry as the simplest way to explain and predict phenomena. It works well in these instances. In still other situations, Lobachevskian geometry is the type that works, or works best. The type of geometry used in a scientific investigation, then, is determined simply by which one best fits the observable data, not by a claim to truth *a priori*.

The so-called "big bang" theory of the origin of the universe provides a striking modern example of this approach. Building on Einstein's general theory of relativity, Alexander Friedmann, a Russian physicist-mathematician, concluded in 1922 that the universe must be expanding. His theoretical findings were verified in the late 1920s by the observations of American astronomer Edwin Hubble and others. Subsequent scientific investigations over the next four decades pointed to the likelihood that the universe as we know it began with a mighty cosmic explosion of matter and energy and that the universe has been expanding "outward" from that initial burst. Assuming that this is the case, we are faced with three distinct models of the expansion of the universe, all of which fit the big-bang theory:

- The expansion of the universe is slowed by the gravitational attraction between galaxies to the extent that it will eventually stop. Then the universe will begin to contract, pulling back into a single infinitely dense point at some distant future time (the "big crunch").

- The expansion is proceeding so rapidly that gravity can never stop it. The universe will continue to expand indefinitely, getting bigger and bigger, at a rate that will eventually become nearly constant (due to the retarding effect of gravity).

- The gravitational attraction between galaxies will eventually counterbalance the expansive force of the "big bang" almost exactly. That is, the expansion of the universe will slow almost to a halt, continuing just enough to avoid reversal, gradually approaching (and never exceeding) some maximum finite size.

The most convenient geometry for describing the universe depends on which of these models you accept. The first one is most easily thought of as a Riemannian universe, the second one as Lobachevskian, and the last as Euclidean. Which one is the "real" universe? We don't know. To date, there is no conclusive scientific evidence to eliminate any of these three models. Thus, we might be in a Riemannian or a Lobachevskian universe, rather than a Euclidean one. But, even if we are, we can still *choose* to use the Euclidean axioms for our geometry because they provide a convenient approximation of the behavior of our small, known part of the large, unknown universe.[14]

Now perhaps you can begin to see the far-reaching effects of the development of the non-Euclidean geometries. At the beginning of the nineteenth century, it was generally accepted that truth about reality was to be found through logical analysis of observed data. The so-called "scientific method" typifies this approach to truth, as it proceeds from observation to hypothesis to experimental verification to "law." The origins of Euclidean geometry are firmly rooted in such observation of the real world, and no area of human knowledge has received a more thorough logical analysis. Yet, paradoxically, the final result of all that analysis is the realization that "the truth" about the geometry of the real world does not really exist.

---

14. For a fuller discussion of this topic, see Chapter 3 of Item 5 in the list *For Further Reading* at the end of this chapter.

EXERCISES

WRITING EXERCISES

1. When you *say* that something is true, what do you mean? Has anything you have seen in this chapter led you to rethink your own use of the word "true"? If so, what? If not, defend your use of the word as being consistent with the ideas presented in this section.

2. **(a)** In light of this section, and recalling that the literal meaning of geometry is "earth measurement," comment on the propriety of using the term *geometry* to describe the Euclidean and non-Euclidean systems.

   (b) Is it more appropriate to apply the term geometry to the Euclidean system than to the Riemannian or Lobachevskian systems? Why or why not?

LINK: Axiom Systems and Society

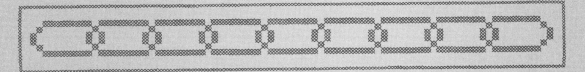

3.8 In Section 3.7, we saw a truly ironic twist in the history of mathematics. Geometry, one of man's oldest formal studies, a study whose very name—"earth measurement"—describes its concrete, practical roots, has now evolved to the point where technically it is no longer a study of the earth or of anything physical at all, but is rather an abstract study of consistency and internal logical relationships.

Actually, this deliberately overstates the case a bit. Even today, Euclidean geometry is constantly being applied to the real world, with dependable results. Moreover, in recent years the non-Euclidean geometries have been put to use in studying the realities of our universe—Riemannian geometry has been employed in theoretical physics, and other non-Euclidean geometries have been developed and used in examining the cosmos. Nevertheless, the work of the past 150 years has established a fundamental separation between the subject itself and the reality that spawned it. Today it is understood that geometries are

The Supreme Court Building, Washington, D.C.

created and studied in the abstract world of pure mathematics, and then applied to real-world problems "from the outside" as a tool to be used, just as a hammer used to build a house is not part of the house itself. The recognition of that separation has thereby given this abstract tool a much broader applicability to human affairs because geometry is no longer confined to the spatial limitations of the physical universe.

Even before the nineteenth century's recognition of geometry as an abstract study of form, the "axiomatic method" was being used to bring the order and persuasion of logic to ideas outside the realm of traditional mathematics and physical science. We close this chapter with two examples of nonmathematical works that clearly trace their form to the axiomatic method crystallized so many centuries ago by Euclid.

One of the most striking examples of the axiomatic form in a nonmathematical setting appears in a major work of the seventeenth-century Dutch philosopher Benedict de Spinoza. Although Jewish by birth, Spinoza's independence of thought resulted in his excommunication from Judaism at the age of twenty-three. He devoted most of his life to the study of philosophy, supporting himself (modestly) as a lens maker. Spinoza's *Ethics*, published in 1677 (the year of his death), is constructed in five parts, each consisting of a list of definitions, a set of axioms, and a number of propositions derived from the axioms. For example, "Part One, Of God," defines eight technical terms, including *cause of itself*, *substance*, *attribute*, and *eternity*. The text then gives seven axioms, such as

> A true idea must agree with that of which it is the idea.

Thirty-six propositions are then deduced from that beginning, including the necessary existence of God.

It is not our purpose here to examine Spinoza's philosophy, but rather, to point out a classic example of the pervasive nature of the axiomatic form. The fact that the form of the *Ethics* is obviously borrowed from Euclid's *Elements* suggests more than just the use of a good literary or expository style. It was Spinoza's intention to do in philosophy what Euclid had done in geometry—to express a set of basic statements (the axioms) that the reader should find so compelling as to be undeniable, and to follow them with a step-by-step logical development of inescapable consequences.

Spinoza's use of the axiomatic form is obvious. Other uses outside of mathematics are less obvious, but no less exemplary. For example, in the late eighteenth century, a group of men gathered together to fashion a brief political document. Although they did not use the word *axiom* in stating their premises, they surely thought of them that way.

> We hold these truths to be self-evident, that all men are created equal; that they are endowed by their Creator with certain unalienable rights; that among these are life, liberty, and the pursuit of happiness. That, to secure these rights, governments are instituted among men, deriving their just powers from the consent of the governed.

Euclid may or may not have agreed with these assertions, but he would certainly have understood their form. These are the axioms from which one can logically derive necessary propositions. Notice that several distinct models of government, such as a confederation of states, a republic, or a democracy, could fit these axioms. But a monarchical form of government is *not* a model allowed by these axioms. The only logical conclusion for the framers of this axiom system was that their society, the United States of America, must become free and independent states.

One might argue that any human society—family, club, nation, world—determines its operational rules, however formal or informal, on the basis of fundamental principles, stated or understood. These are the axioms and theorems, and the resulting society is merely a model (perhaps one of many possible) of that axiom system. In some instances the principles are written down, as in the Declaration of Independence or the Constitution of the United States; in others they are not. Sometimes those fundamentals seem to change and evolve (and perhaps they do) with the agreement of the society. Some social change results from the discovery that certain rules are not axiomatic, but are only accidental properties of one model. In this political society called the United States we have a group of individuals whose explicit charge is to safeguard our axiom system, the Constitution, by confirming modifications of our model that conform to the axioms and discarding those that do not. This group is called the Supreme Court.

We could give other, less lofty, examples, or we could study some of the logical consequences of these examples. Either of those paths leads to interesting places, but we need not travel any further. Our point has already been made: Whatever role geometry has played or will play in mathematics and science, its mark has been indelibly left on human society in the form of the axiomatic method.

TOPICS FOR PAPERS

1. The purpose of this project is to help you understand the role of axiom systems as a bridge between the "real world" and the world of abstract science. Write a paper about axiom systems, constructed according to the following outline:

   (a) Describe informally a real-life situation you know, involving sports, a game, an organization, a political structure, your dorm, a class, or anything else, that involves rules or procedures.

   (b) Describe *some aspect* of that situation formally by means of an axiom system consisting of three or four basic terms and three or four axioms. (Do yourself a favor—keep them simple!)

   (c) Rework your axiom system, if necessary, to insure that
      i. the system is consistent and
      ii. all the axioms are independent.

      Prove these two assertions.

   (d) State at least one theorem and prove it from the axioms.

   (e) Treating the basic terms of your axiom system as undefined, find an interpretation for the system that is different than the one you started with. Confirm the difference by stating a property of the basic terms in the original situation that is not shared by the basic terms in your new interpretation.

2. Research one of the following historical figures and write a detailed report on his attempt to prove Euclid's Fifth Postulate:

> Poseidonios (c. 135–51 B.C.)
> Proclus (410–85)
> John Wallis (1616–1703)
> Johann Heinrich Lambert (1728–77)
> John Playfair (1748–1819)
> Adrien-Marie Legendre (1752–1833)

Be sure to provide a bibliography that lists any sources you used in writing the paper, and be careful that you do not simply copy or paraphrase the material from your sources. Any material not explicitly quoted should have been mentally digested by you, so that your own words describe what really are your own ideas when you write them down.

3. Write a fictional short story wherein the central theme is the non-Euclidean geometry of the universe in which the characters live. The story may or may not be directly about geometry or mathematics, and may or may not touch upon the possible existence of other types of geometry, at your discretion.

FOR FURTHER READING

1. Bonola, Roberto. *Non-Euclidean Geometry: A Critical and Historical Study of Its Development*, transl. by H. S. Carslaw, 1911. Reprint. New York: Dover Publications, 1955.

2. Davis, Philip J., and Reuben Hersh. *The Mathematical Experience*. Boston: Birkhäuser, 1981.

3. Dunham, William. *Journey Through Genius: The Great Theorems of Mathematics*. New York: John Wiley and Sons, Inc., 1990, Chapters 1 and 2.

4. Eves, Howard, and Carroll V. Newsom. *An Introduction to the Foundations and Fundamental Concepts of Mathematics*, Rev. Ed. New York: Holt, Rinehart and Winston, 1965.

5. Hawking, Stephen. *A Brief History of Time*. New York: Bantam Books, 1988.

6. Kline, Morris. *Mathematical Thought from Ancient to Modern Times*. New York: Oxford University Press, 1972.

7. Lieber, H. G., and L. R. Lieber. *Non-Euclidean Geometry*. Lancaster, PA: The Science Press Printing Co., 1931.

8. Newman, James R., ed. *The World of Mathematics*, Vols. 1–4. New York: Simon & Schuster, Inc., 1956.

9. Prenowitz, Walter, and Meyer Jordan. *Basic Concepts of Geometry*, 1965. Reprint. New York: Ardsley House, Publishers, Inc., 1989.

10. Trudeau, Richard. *The Non-Euclidean Revolution*. Boston: Birkhäuser, 1987.

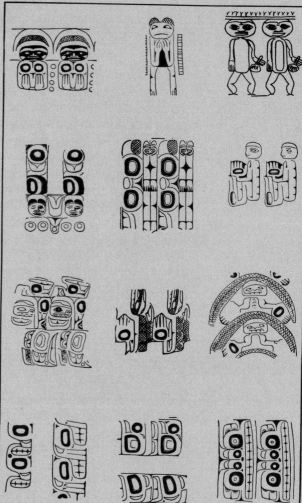

Games of Chance. The study of probability originated from problems concerned with gambling. Pictured are games of chance from various parts of the world in different eras. *From upper left and proceeding downward:* Card Party, Early Sixteenth Century, Western Europe. Gambling with a Revolving Pointer, China. Three-handed Game of Monte, Mexico. *At the right:* Gambling Sticks of the Haida Tribe, Queen Charlotte Islands.

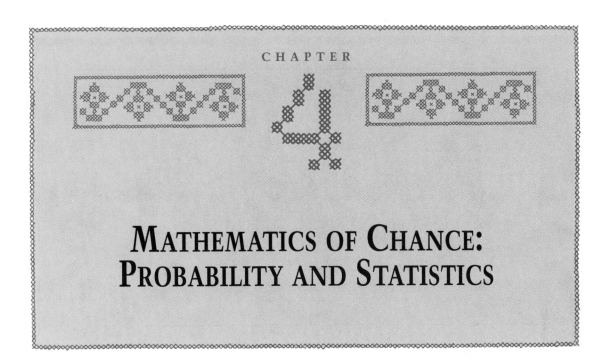

MATHEMATICS OF CHANCE: PROBABILITY AND STATISTICS

THE GAMBLERS

Much of mathematics is concerned primarily with deducing valid conclusions from premises that are assumed to be true. This process is logically satisfying, but any attempt to apply it to the world we live in is hampered by a major obstacle: There is no way of verifying with certainty most assumptions that deal with reality. The twin branches of mathematics called *probability* and *statistics* face this problem squarely and try to bridge the gap between hypothesis and fact. They consider uncertain situations, examine the conclusions that can be drawn from the possible alternatives, and make quantitative judgments about the reliability of those conclusions.

The mathematical theory of probability dates back to 1654, when the Chevalier de Méré, a wealthy French nobleman with a taste for gambling, proposed to the mathematician Blaise Pascal a problem involving the distribution of stakes in an unfinished game of chance. This was the famous "problem of points," in which a simple gambling game between two players is ended before either wins. The question is how to divide the prize money if the partial scores of the players are known, and its answer is based on the likelihood of each player winning the game. Pascal communicated the question to Pierre de Fermat, another prominent French mathematician, and from their correspondence a new field of mathematics emerged. Both Pascal and Fermat arrived at

Are YOU ÆTNA-IZED?

Over 100,000 Accident Claims Paid by the Ætna Life Insurance Company

Railroad, Street car and Steamboat	31,634	$2,613,145.98
Falls on pavement, stairs, etc.	20,624	1,475,359.86
Horse and carriage	8,705	751,565.66
Automobile	2,362	519,949.76
Fire Arms	628	392,751.23
Bathing or drowning	737	288,932.39
Burns or scalds	3,257	273,456.69
Bicycle	2,835	192,489.50
Septic wounds (blood poisoning)	2,021	177,811.91
Athletic sports	3,201	154,482.82
Falling of heavy weights	2,957	149,407.97
Cuts with edged tools or glass	5,030	139,035.36
Fingers crushed	4,503	120,144.73
Eye injuries	2,281	108,541.76
Elevator	320	89,420.29
Machinery	1,653	70,903.84
Assaults	434	52,998.86
Toes crushed	1,577	50,418.01
Stepping on nails or glass	1,668	45,647.52
House accidents (contact with furniture)	483	45,285.66
Hands lacerated on hooks, nails, etc.	1,465	41,162.50
Bites by dogs or insects	783	31,973.20
Tripped over mats or rugs	237	19,514.54
Splinters in hands or feet	508	18,284.70
Motor boats	147	10,505.47
Fingers caught in electric fans	147	6,892.06
Miscellaneous accidents	7,803	476,365.40
Total	**108,000**	**$8,316,447.67**

Picture Collection, The Branch Libraries, The New York Public Library

Insurance. The investigation of death records for insurance purposes was another early impetus to the study of probability and statistics. Pictured here is an insurance ad that appeared in *Life* magazine in 1911.

Credit: Smithsonian Institution

Blaise Pascal. One of the founders of the theory of probability.

the same answer to the problem, but their methods of solution were different. They generalized the problem and its solution, then extended their investigations to other games of chance. Their discussions aroused considerable interest in the European scientific community, and soon other scholars took up the challenge of analyzing gambling games.[1]

As Pascal, Fermat, and others in Europe were investigating games of chance, researchers in England were hard at work studying a slightly more respectable, but equally costly, form of gambling. The many hazards of life in the sixteenth and seventeenth centuries had established an eager market for a remarkable new service called "life insurance." The groups of people who backed these first life-insurance policies must be ranked among the most daring of gamblers, for they often risked large sums of money on a single turn of fate. In order to understand this life-and-money game of chance better, people began to investigate death records. The first important mortality tables were compiled by Edmund Halley in 1693 as a basis for his study of annuities. This second type of probabilistic investigation is known as *statistics*.

The first book devoted entirely to probability and statistics was Jakob Bernoulli's *Ars Conjectandi*, published in 1713. In this book Bernoulli anticipated by more than 200 years the practical importance of the subjects by suggesting their application to questions of government, law, economics, and morality. As the study of statistics has developed since that time, it has provided the means by which probability theory is applied not only to the fields suggested by Bernoulli, but to education, the social sciences, and many other areas as well.

Probability and **statistics** are two distinct, but closely related, fields of the mathematics of "uncertainty"; in fact, they are devoted to investigating opposite sides of the same fundamental problem. The question asked of probability is:

Given a *known collection* of objects, what can be said about the characteristics of an *unknown sample* of that collection?

The "problem of points" is such a question; the set of all possible outcomes of the gambling game is known, but the particular outcome of an unfinished game is not. Statistics attempts to answer the converse question:

1. For further information about Pascal and Fermat, see Section B.6 of the Appendix.

Given a *known sample* of an *unknown collection* of objects, what can be said about the characteristics of the collection as a whole?

For instance, the statistical question of death rates may be phrased: "Knowing the life span of each person in a relatively small group, what can be said about the life spans of people in general?"

The principles of elementary probability are easy extensions of a "common-sense" approach to chance situations. A little experience with flipping a coin or playing a simple card game should provide the insight needed for the fundamentals of the theory presented in this chapter. Statistical investigations rely heavily on the principles of probability theory, and elementary statistical concepts also flow easily from common-sense notions. However, the powerful advanced techniques in both probability and statistics require an understanding of calculus and are beyond the scope of this book; we present only a general discussion of these areas.

4.1   EXERCISES

WRITING EXERCISES

1. What is *gambling*? Is the insurance business really gambling? If so, who are the gamblers? If not, why not?

2. Is investing in the stock market really gambling? If so, who are the gamblers? If not, why not?

3. Describe an aspect of or situation in one of the following areas that you think involves probabilistic and/or statistical thinking (beyond mere data gathering) in some significant way:

   (a) government   **(b)** law   **(c)** morality
   (d) psychology   **(e)** sociology

THE LANGUAGE OF SETS

4.2

The brief discussion of probability and statistics in Section 4.1 shows that both areas are fundamentally concerned with collections of objects. In mathematics (of any sort), collections of objects are usually called *sets*. The language of sets is basic to virtually all of modern mathematics. The importance of set-theoretic language comes from the fact that it applies to so many different areas and thereby acts as a unifying force within the subject. A significant application occurs in this chapter. In our study of probability, we restate the problems in set-theoretic language. This restatement makes it easier to explain new ideas by using precise terminology and notation. As a bonus, the use of set language here may help you to notice analogies between parts of probability theory and other mathematical areas.

If you are familiar with the language of sets, then this section may be skimmed quickly in order that you see the notation we use in this book. However, if you haven't seen much set-theoretic language before or if what you have seen is a little rusty in your memory, we invite you to study this section with some care. It contains a succinct summary of all the basic set theory needed for understanding the rest of this chapter.

Any collection of objects is called a **set**, and the objects themselves are called **elements** of that set. We use the term **collection** as a synonym for "set." (Notice that we do *not* have a formal definition of set here, because of the obvious circularity between the terms *set* and *collection*. It is presumed that at least one of these two words has a familiar, common-sense meaning. A similar observation can be made about the terms *element* and *object*.) The standard notation used to write sets and elements includes the following conventions:

- Sets are usually denoted by capital English letters, such as A or B or S, and elements by small letters, such as a, b, c, etc. We assume, unless otherwise specified, that different small letters denote different elements.
- The symbol "$\in$" stands for the phrase *is an element of* (or *is in*). Thus, "$a \in S$" is read "a is an element of the set S."
- Negation is denoted by a slash mark through the symbol being negated. For example, we write "b is not an element of S" as "$b \notin S$." (This usage is similar to the custom of writing "1 does not equal 2" as "$1 \neq 2$.")
- If a set is described by listing its elements, that list is enclosed in braces. For example, the set consisting of the first four letters of the alphabet is written {a, b, c, d}.

Sometimes it is either inconvenient or impossible to list all the elements in a set. In that event there are two alternatives. One of them only makes sense if there is an obvious sequential pattern to the set. Then an ellipsis (. . .) may be used to indicate that the pattern is continued, either to a specified stopping point or indefinitely. Thus, the set of all letters of our alphabet is {a, b, c, . . . , z}, and the set of all even whole numbers is {0, 2, 4, 6, . . .}.

The other, more general, alternative is to specify some property or properties that characterize the elements included in the set. (By "characterize" we mean that they describe all the elements in the set and nothing else.) In this way it becomes possible to "build" the set from its description. In such cases we use a standard form of notation, called (not surprisingly) **set-builder notation**:

$$\{x \mid x \text{ has a certain property}\}$$

This is read "the set of all x such that x has a certain property." The vertical bar stands for "such that" in this context, and the variable x (or whatever expression is used in its place) represents the general form of a single typical element of the set.

EXAMPLE 4.1	The set of all digits can be denoted by

$$\{0, 1, 2, 3, 4, 5, 6, 7, 8, 9\}$$

or by $\{x \mid x \text{ is a digit}\}$, which is read "the set of all x such that x is a digit."

EXAMPLE 4.2	The set

$$\left\{1, \frac{1}{2}, \frac{1}{3}, \frac{1}{4}, \dots\right\}$$

is the same set as

$$\left\{\frac{1}{n}, \mid n \text{ is a positive integer}\right\}$$

Of course, the description must be good enough to tell us exactly what belongs in the set and what does not. If that is the case, we say the set is **well-defined**. For example, the set $\{a, b, c\}$ is well-defined because it is clear exactly what is in the set (the first three letters of our alphabet) and what is not (everything else). The set $\{1, 2, 3, \dots, 100\}$ is well-defined, provided that we assume that the ellipsis indicates the obvious pattern; thus, this set includes all natural numbers between 3 and 100.

EXAMPLE 4.3	"The set of all tall American citizens" and "the set of all lovable puppies" are not well-defined sets because their descriptions do not allow us to decide unambiguously what is in each set and what is not. On the other hand, "the set of all American citizens who are more than six feet tall" and "the set of all beagles less than six months old" are well-defined sets.

The set $\{x \mid x \text{ has a certain property}\}$ is well-defined if and only if the statement "x has a certain property" is always unambiguously true or false, no matter what x is. If the defining condition for a set is *always* false whenever anything is substituted for the variable, then the set has no elements in it; but it is considered to be a set, just the same. This convention simplifies the language of set theory.

DEFINITION   The set containing no elements is called the **empty set** (or **null set**) and is denoted by ∅.

EXAMPLE 4.4

The empty set has many possible descriptions, such as:
- The set of all three-headed people,
- $\{x \mid x$ is a planet within 5 miles of the sun$\}$,
- $\{n \mid n$ is a whole number whose square is 3$\}$.

All of these describe the same set, ∅.

DEFINITION   A set A is a **subset** of a set B if every element of A is also an element of B. We write this as "A $\subseteq$ B."

Of course, it may happen that every element of A is in B, but B contains nothing else (that is, every element of B is also in A). In this case, the sets A and B are indistinguishable as sets because these two collections contain precisely the same things.

DEFINITION   Two sets A and B are **equal** if A is a subset of B and B is a subset of A. Here, write "$A = B$."

DEFINITION   A set A is a **proper subset** of a set B if A is a subset of B, but A does not equal B. Here, write "$A \subset B$."

In Examples 4.5 and 4.6, the given sets consist of letters of our alphabet.

EXAMPLE 4.5

The relationship
$$\{a, b, c\} \subseteq \{a, b, c, d, e\}$$
holds true. In fact,
$$\{a, b, c\} \subset \{a, b, c, d, e\}$$
because the two sets are not equal.

EXAMPLE 4.6

The equality
$$\{a, b, c\} = \{c, b, a\}$$
holds true because every element of each set is an element of the

other. Notice that the order in which the elements are written does not matter; the two sets are collections of exactly the same elements, and hence the sets are equal.

EXAMPLE 4.7	Let

$$A = \{1, 2, 3, 4, 5, \ldots\}, \qquad B = \{x \mid x \text{ is a natural number}\}$$

and

$$C = \{2, 4, 6, 8, \ldots\}$$

Then the following statements all are true:

$$A \subseteq B, \qquad C \subseteq B, \qquad B \subseteq A, \qquad A = B, \qquad C \subset B, \qquad A \nsubseteq B$$
$$B \nsubseteq A, \qquad B \nsubseteq C, \qquad A \neq C, \qquad A \subseteq A, \qquad A \nsubseteq A, \qquad A = A$$

Often it is important to know what is *not* in a set. For instance, if we ask for things that are not in the set $\{1, 2, 3, 4\}$, many different kinds of answers are possible: 5, 100, $\frac{1}{2}$, you, this book, the lions in the Bronx Zoo, and so forth. Some of these answers are more appropriate in one context, some in another. The appropriate kind of response can be inferred from the general setting of the question. In this case, for example, it would be reasonable to regard the answers "you, this book, the lions in the Bronx Zoo" as beyond the intent of the question; but which of the other suggested answers might fit the intent is not clear. Hence, sometimes it is necessary to specify the set of all elements being considered in a particular discussion. This set is called the **universal set** for that discussion; it is denoted by $\mathcal{U}$. With this in mind we can determine unambiguously what is *not* in a set.

DEFINITION

The set of all elements in the universal set that are not in a particular set A is called the **complement** of A, and is denoted by A'. In symbols,

$$A' = \{x \mid x \in \mathcal{U} \text{ and } x \notin A\}$$

EXAMPLE 4.8	Consider the set $A = \{0, 1, 2, 3, 4, 5\}$. If the universal set consists of the ten digits, then $A' = \{6, 7, 8, 9\}$. If the universal set is the set **W** of all whole numbers, then $A' = \{x \mid x \in W \text{ and } x > 5\}$; that is,

$$A' = \{6, 7, 8, 9, 10, 11, \ldots\}$$

It is useful to have a way of picturing relationships among sets. The most common way of doing this is by drawing the sets as regions enclosed by circles or ovals inside a rectangular box. The box repre-

Figure 4.1 Venn diagrams for (a) a proper subset (B ⊂ A) and (b) a complement, A′ (shaded in color).

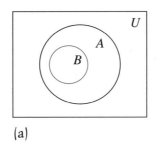 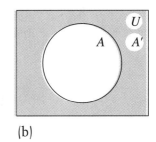

(a)   (b)

sents the universal set. If a particular region is to be emphasized, that region is shaded. Figures drawn this way are called **Venn diagrams**.[2] (See, for example, *Figure 4.1*.)

DEFINITION   The set of all elements that are in both a set A and a set B is called the **intersection** of A and B, and is written $A \cap B$. In symbols,

$$A \cap B = \{x \mid x \in A \text{ and } x \in B\}$$

DEFINITION   If A and B are sets such that $A \cap B = \varnothing$, we say that A and B are **disjoint**. (That is, A and B are disjoint if they have no elements in common.)

EXAMPLE 4.9   Let $A = \{1, 2, 3, 4\}$, $B = \{3, 4, 5, 6\}$, and $C = \{5, 6, 7, 8\}$. Then $A \cap B = \{3, 4\}$ and $B \cap C = \{5, 6\}$, but $A \cap C = \varnothing$.

The intersection of two sets is usually pictured by the Venn diagrams of *Figure 4.2*.

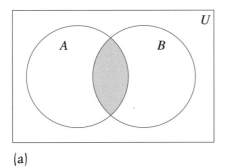 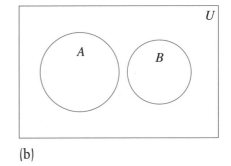

(a)   (b)

Figure 4.2 Venn diagrams for intersection. (a) $A \cap B \neq \varnothing$; (b) $A \cap B = \varnothing$.

2. Named for John Venn, an English mathematician who used these diagrams in 1876 in a paper on Boole's algebra of logic.

DEFINITION

The set of all elements that are in either a set A or a set B (or both) is called the **union** of A and B, and is written $A \cup B$. In symbols,

$$A \cup B = \{x \mid x \in A \text{ and } x \in B\}$$

EXAMPLE 4.10

If $A = \{a, b, c\}$ and $B = \{a, c, e, g\}$, then $A \cup B = \{a, b, c, e, g\}$.

Some simple, but useful, consequences of the definitions of union and intersection are listed here for convenience.

×××

If A and B are subsets of some universal set $\mathcal{U}$, then

$$A \cap A = A \qquad\qquad A \cup A = A$$
$$A \cap \emptyset = \emptyset \qquad\qquad A \cup \emptyset = A$$
$$A \cap \mathcal{U} = A \qquad\qquad A \cup \mathcal{U} = \mathcal{U}$$

If $B \subseteq A$, then $A \cap B = B$ and $A \cup B = A$. (See *Figure 4.1*.)

×××

The complements of unions and intersections are characterized by a pair of statements known as *De Morgan's Laws*. They are named after Augustus De Morgan, a nineteenth-century British mathematician who used a version of these laws in his development of formal logic.

×××

De Morgan's Laws: If A and B are subsets of some universal set, then

$$(A \cup B)' = A' \cap B' \qquad \text{and} \qquad (A \cap B)' = A' \cup B'$$

×××

The Venn diagrams in *Figure 4.3* illustrate the first of these laws. Illustration of the second is similar.

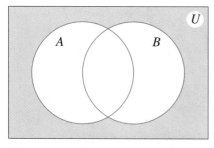

(a)

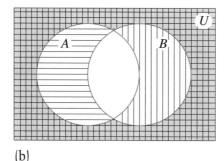

(b)

Figure 4.3 These diagrams illustrate the first of De Morgan's Laws: $(A \cup B)' = A' \cap B'$. (a) $(A \cup B)'$ (shaded in color); (b) $A' \cap B'$ (shaded in color).

The shaded portion of the left diagram of *Figure 4.3* represents $(A \cup B)'$. In the right diagram, A' is the vertically striped region and B' is the horizontally striped region. Their intersection, $A' \cap B'$, is represented by the cross-hatched area, which is precisely the same region as that shaded in the left diagram.

There is another way of combining sets, not quite as obvious as union but just as important. To this end, we must first consider the idea of an **ordered pair**, which is simply a pair of elements whose order is specified. We write the ordered pair with first element x and second element y as (x, y). Notice that, because order matters,

$$(x, y) \neq (y, x)$$

unless x and y are the same thing. In general:

DEFINITION

Two ordered pairs (a, b) and (c, d) are **equal** if $a = c$ and $b = d$.

DEFINITION

The **Cartesian product** of two sets A and B is the set of all ordered pairs whose first elements are from A and whose second elements are from B. It is denoted by $A \times B$.[3] In symbols, we write

$$A \times B = \{(a, b) \mid a \in A \text{ and } b \in B\}$$

EXAMPLE 4.11

If $A = \{a, b, c\}$ and $B = \{\star, \heartsuit, 1, 2\}$, then

$$A \times B = \{(a, \star), (a, \heartsuit), (a, 1), (a, 2), (b, \star), (b, \heartsuit),$$
$$(b, 1), (b, 2), (c, \star), (c, \heartsuit), (c, 1), (c, 2)\}$$

Example 4.11 illustrates a general fact about the number of elements in a Cartesian product. In that example, A contains 3 elements and B contains 4 elements; so $A \times B$ contains 12 $(= 3 \cdot 4)$ elements because for each of the 3 elements of A there are 4 ordered pairs containing it in the first position. In general,

×××

If A contains m elements and B contains n elements, then $A \times B$ contains $m \cdot n$ elements.

×××

EXAMPLE 4.12

If A is any set, then $A \times \varnothing = \varnothing$ because there are no elements to be used as second elements of the ordered pairs.

3. Named after René Descartes, the seventeenth-century French mathematician and philosopher who first published a description of analytic (coordinate) geometry.

4.2 EXERCISES

In Exercises 1–3, list the elements of each set. What is an obvious universal set in each case?

1. $\{x \mid x$ is an odd digit$\}$

2. $\{x \mid x$ is a month whose name contains the letter "r"$\}$

3. $\{x \mid x$ is a whole number and $x < 20\}$

For Exercises 4–13, let the universal set consist of all the letters of the alphabet, $\{a, b, c, \ldots, z\}$, and let $A = \{a, b, c, d, e\}$, $B = \{c, d, e, f, g\}$, and $C = \{b, c, d\}$. Which of the following statements are true?

4. $A \subseteq B$

5. $A = B$

6. $C \subseteq A$

7. $C \subseteq B$

8. $C \subset A$

9. $C \subset B$

10. $A' \subseteq C'$

11. $C' \subseteq A'$

12. $C' \subset A'$

13. $A' = C'$

For Exercises 14–16, let $A = \{1, 2, 3, 4, 5\}$. Describe A' in each case:

14. $\mathcal{U} = \{0, 1, 2, 3, 4, 5\}$

15. $\mathcal{U} = \{0, 1, 2, 3, 4, 5, 6, 7, 8, 9\}$

16. $\mathcal{U} = \{x \mid x$ is a positive whole number$\}$

For Exercises 17–28, let the universal set be

$$\{0, 1, 2, 3, 4, 5, 6, 7, 8, 9\}$$

and let

$$A = \{1, 2, 3\}, B = \{3, 4, 5, 6\}, \text{ and } C = \{5, 6\}$$

Find each of the following sets.

17. $A \cup B$

18. $A \cap B$

19. $A \cup C$

20. $A \cap C$

21. $B \cup C$

22. $B \cap C$

23. $A \times C$

24. $B \times C$

25. $(A \cup B)'$

26. $A' \cup B'$

27. $(A \cap B)'$

28. $A' \cap B'$

In Exercises 29–31, illustrate the given relationships among three nonempty sets A, B, and C by using a single Venn diagram.

29. $(A \cup B) \subset C$ and $A \cap B = \varnothing$

30. $A \subset B$, $A \cap C = \varnothing$, and $B \cap C \neq \varnothing$

31. A and B are disjoint, but neither A and C nor B and C are disjoint.

For Exercises 32–36, let A and B be subsets of some universal set $\mathcal{U}$. Show by some persuasive argument that each statement is true.

32. $A \cap \varnothing = \varnothing$

33. $A \cup \varnothing = A$

34. $A \cap \mathcal{U} = A$

35. $A \cup \mathcal{U} = \mathcal{U}$

36. If $A \subseteq B$, then $A \cap B = A$ and $A \cup B = B$.

For Exercises 37–42, decide whether or not the given statement is true for *all* sets. If it is not always true, give a *counterexample* (an example of sets for which the statement is false). If you say it is always true, provide a persuasive argument.

37. $A' \cap B = A \cup B'$

38. $A' \cap B = A \cap B'$

39. $(A' \cap B)' = A \cup B'$

40. $A \cup (B \cap C) = (A \cup B) \cap C$

41. $A \cap (B \cup C) = (A \cap B) \cup (A \cap C)$

42. $A \times (B \cap C) = (A \times B) \cap C$

WRITING EXERCISES

1. The definitions of *complement, intersection, union,* and *Cartesian product* appear in this section both in words and in symbols. Discuss the advantages and disadvantages of each way of stating definitions, using these as illustrative examples. Which of the two ways do you prefer, and why?

2. What aspect of the philosophy of René Descartes made it natural for him to be interested in pursuing mathematical ideas? Consult whatever sources you wish, but be sure to write most of this answer in your own words and to give appropriate source credit for any ideas that are not your own.

3. **(a)** Suppose that, for each positive integer n, the set $A_n = \left\{n - \frac{1}{3}, n + \frac{1}{3}\right\}$. (That is, we have infinitely many disjoint two-element sets of rational numbers with denominator 3.) Can you define a set consisting of *exactly one* element from each of the original sets $A_1, A_2, A_3, \ldots$? If so, do it. If not, explain why not.

(b) Suppose that, for each positive integer n, there is a set P_n containing two different objects that look exactly the same. (For instance, imagine infinitely many brand-new pairs of socks.) Can you define a set consisting of *exactly one* element from each of the original sets $P_1, P_2, P_3, \ldots$? If so, do it. If not, explain why not.[4]

(c) What is the essential difference between the questions in Parts **(a)** and **(b)**? Explain.

WHAT IS PROBABILITY?

4.3

In general terms, the field of mathematics called **probability** is the study of "randomness." Something is **random** if it is not predictable; that is, if it cannot be determined *with certainty* from prior knowledge or experience. The possible outcomes of an unfinished gambling game and the date of death of a living person are random phenomena that we discussed in Section 4.1. So are the results of flipping a coin or buying a lottery ticket or predicting next week's stock market prices. The uncertainty of these kinds of things is easier to cope with if we can assign a numerical value to the likelihood that a random situation will turn out one way rather than another. That allows us to measure the various alternatives, using numbers to determine which of several possible outcomes is most likely, least likely, and so forth.

One preliminary observation is in order here. Although problems involving games of chance make convenient examples for basic probability theory, they are by no means typical of the important current applications of probability. In fact, it is their *obvious* randomness that makes these examples atypical. Most scientific applications of probability arise in situations that are either caused or governed by physical, economic, or social principles. As such, these situations should, theoretically, be predictable. For instance, when a coin is tossed, the side it lands on is determined by its weight and center of gravity, the magnitude and direction of the force applied by the thumb that tosses it, the distance it falls, the surface it lands on, etc. However, the interaction

---

4. This question relates to an important principle in formal set theory called the Axiom of Choice. (See Section B.9 of the Appendix.)

The Bourse, Brussels. One of the world's major stock exchanges. Future stock-market prices can be treated as random phenomena.

of these principles is too complex to analyze easily,[5] so it is convenient to think of the possible results as random occurrences. In other words, the randomness of a situation is often an *assumption* made to allow a complex problem to be handled in a simple way.

Because the theory of probability is based on randomness, its applicability in particular situations depends on the validity of that assumption in each case. Such an assumption, together with other simplifying assumptions, transforms a concrete problem of physical or behavioral science into an analogous abstract problem. The setting for the problem in its abstract form is called a **model** of the original

5. In a field known as *chaos theory* it has been proposed that the laws governing phenomena such as coin tossing, wind currents, etc. are relatively simple; however *noise* or distortions obscure our understanding of these laws. This concept plays an important role in the recent Broadway play *Arcadia*; see Item 10 in the list *For Further Reading* at the end of this chapter.

situation. If the assumptions in this abstraction process eliminate only details that are truly irrelevant, then any solution of the model problem will also solve the original "real-world" problem. On the other hand, if the model problem is an oversimplification in some respects, its solution may have little to do with the original problem. For instance, when a die is thrown, the assumption that any one of its six faces is equally likely to turn up is valid only if the die has not been "loaded," or weighted off-center in some way.

The reliability of the abstraction process varies from situation to situation, sometimes depending on complex or subtle conditions. An investigation of this question is not essential to the development of basic probability theory, so we shall treat the relationship between concrete problems and their models as accurate and shall deal primarily with the models. Thus, for example, we shall generally assume that either side of a tossed coin is equally likely to turn up, that each face of a thrown die is as likely to turn up as any other, and so on.

Let us begin our formal study of probability by introducing some useful terminology. Any situation or problem involving uncertain results is called an **experiment** (regardless of whether or not we have control over the situation). The various possible results are called **outcomes**, and the set of all possible outcomes is called the **sample space** of the experiment. The sample space is usually denoted by S. In the language of sets, S is the *universal set* for its experiment, and the outcomes are all the elements of this universal set.

DEFINITION

Any subset of the sample space is called an **event**.

EXAMPLE 4.13

The tossing of a coin is an experiment with two possible outcomes, *heads* or *tails*. (Our model ignores the possibility that the coin might land on its edge.) If we denote these two outcomes by h and t, then the sample space is $\{h, t\}$. There are four possible events:

$\{h\}$:   The coin lands heads up.

$\{t\}$:   The coin lands tails up.

$\{h, t\}$: The coin lands either heads up or tails up.

$\varnothing$:   The coin lands neither heads up nor tails up.

EXAMPLE 4.14

Rolling a single die is an experiment with six possible outcomes. Its sample space can be represented by the set

$$S = \{1, 2, 3, 4, 5, 6\}$$

There are 64 events in this experiment because S has 64 (= 2^6) sub-sets. Two such events are:

{2, 4, 6}:   The die falls with an even number facing up.

{1, 2, 3, 4}: The die falls with a number less than 5 facing up.

EXAMPLE **4.15**	A raffle in which a single winning ticket is drawn from a barrel containing 1000 tickets is an experiment with 1000 possible outcomes. Thus, there are 1000 elements in its sample space, and there are 2^{1000} events.

EXAMPLE **4.16**	Tossing a coin twice is an experiment with four possible outcomes. Thus, the sample space S can be represented as {*hh, ht, th, tt*}. Here, *ht* represents heads on the first toss and tails on the second, whereas *th* represents tails, then heads. There are 2^4, or 16, events in this experiment. (What are they?)

The basic idea that underlies all of probability is this:

Somehow, we want to assign numbers to each event (subset) of a sample space to measure the likelihood that the outcome of the experiment (element of the sample space) will lie inside, rather than outside, the event.

Don't be put off by the formal language here; this is an idea that should be quite familiar from everyday life. For instance, when we say that there's a "50–50" chance of a tossed coin landing heads up, we are assigning the number 50% to the event {*heads*} and 50% to the event {*tails*}. In Example 4.15, the holder of a single raffle ticket would surely say he/she had one chance in a thousand of winning (assuming the raffle is fair), so the number attached to the event consisting of that one ticket would be $\frac{1}{1000}$.

The assumption that a raffle is fair is the same as saying that each ticket is as likely to be drawn as any other. Similarly, we say a coin is fair if either side is as likely to turn up as the other when it is tossed. These are examples of a general situation in which the common-sense approach to probability is the simplest and the most natural. We say that all the outcomes in a sample space are **equally likely** if each one has the same chance of happening as any other. (This is not a very satisfying formal definition, but it will serve our purposes well enough.) *Whenever possible, we shall try to describe the sample space of an experiment in such a way that all outcomes are equally likely* because in this case the assignment of numbers to events is easy.

Note
Sometimes, in order to obtain a better match of a model with its corresponding real-world situation, the "probabilities" (numbers) assigned to events are determined by data drawn from observations. Such an approach is called **empirical** or **experimental** (rather than **theoretical**). That approach does not alter the basic rules of probability, but, in our view, the common-sense motivation for those rules is then somewhat obscure. Thus, we make the initial assumption of equally likely outcomes for the sake of simplicity and intuitive comfort.

DEFINITION
In a sample space S of equally likely outcomes, the **probability** of an event E, denoted by $P(E)$, is the number of outcomes in E divided by the number of outcomes in S. That is,

$$P(E) = \frac{\text{number of outcomes in } E}{\text{number of outcomes in } S}$$

NOTATION
The number of outcomes in an event E is abbreviated as $n(E)$. Thus, the probability of an event E of equally likely outcomes is given by

$$P(E) = \frac{n(E)}{n(S)}$$

EXAMPLE **4.17**
If a fair coin is tossed, the probability of its landing "heads up" is the number of elements in the event $E = \{h\}$ divided by the number of elements in the sample space $S = \{h, t\}$. In symbols:

$$P(E) = \frac{n(E)}{n(S)} = \frac{n(\{h\})}{n(\{h, t\})} = \frac{1}{2}$$

because E contains 1 element and S contains 2.

EXAMPLE **4.18**
In Example 4.14, the sample space S contains 6 outcomes. Call the two events listed there E and F; that is,

$$E = \{2, 4, 6\} \quad \text{and} \quad F = \{1, 2, 3, 4\}$$

Then

$$P(E) = \frac{3}{6} = \frac{1}{2} \quad \text{and} \quad P(F) = \frac{4}{6} = \frac{2}{3}$$

In other words, if a single (fair) die is rolled, there are 3 chances in 6 of rolling an even number, and there are 4 chances in 6 of rolling a number less than 5.

EXAMPLE **4.19**
If Ms. Chance buys 25 of the 1000 tickets sold for a single-prize raffle, then the event that she wins the raffle consists of 25 outcomes. If we call this event W, then

$$P(W) = \frac{25}{1000} = \frac{1}{40}$$

That is, she has 25 chances in 1000—or 1 chance in 40—of winning the single prize.

EXAMPLE 4.20

As Example 4.16 shows, tossing a coin twice is an experiment with a four-element sample space. If we want the result to be one head and one tail (in either order), then the desired event, which we shall call E, is {*ht*, *th*}, and

$$P(E) = \frac{n(E)}{n(S)} = \frac{2}{4} = \frac{1}{2}$$

EXAMPLE 4.21

There is a tempting, but *wrong*, approach to the experiment described in Examples 4.16 and 4.20 that merits some attention. (See if you can figure out why it's wrong before we tell you.) We might describe the possible outcomes of tossing a coin twice by using the three-element sample space

$T = $ {both *heads*, one *head* and one *tail*, both *tails*}

In this case, the event $E = $ {one *head* and one *tail*} contains just one of the three outcomes in the sample space, so

$$P(E) = \frac{n(E)}{n(T)} = \frac{1}{3}$$

The probability as computed here is different from the probability of the same event in the same experiment as given in Example 4.20. Why is that approach correct and this one wrong?

The difficulty here is that *not all the outcomes in the sample space T are equally likely* to occur. In fact, the "one *head* and one *tail*" outcome is twice as likely to happen as either of the others, so a probability of $\frac{1}{3}$ will not provide a correct measure of the likelihood of the events in this experiment.

4.3 EXERCISES

The experiment for Exercises 1–7 consists of tossing a fair coin three times. Let E, F, and G be the following events:

■ E: The first two tosses are the same (that is, both heads or both tails).

■ F: The first and third tosses are the same.

■ G: The second and third tosses are tails.

1. List all the equally likely outcomes for this experiment; that is, find a suitable sample space.

2. List the outcomes in each of the events E, F, and G.

3. Find $P(E)$, $P(F)$, and $P(G)$.

4. List the elements of $E \cap F$; then find $P(E \cap F)$.

5. List the elements of $E \cup F$; then find $P(E \cup F)$.

6. List the elements of $F \cap G$; then find $P(F \cap G)$.

7. List the elements of $F \cup G$; then find $P(F \cup G)$.

The experiment for Exercises 8–12 consists of rolling a single die, then tossing a coin. Let E, F, G, and H be the following events:

■ **E: The number on the die is even.**

■ **F: The number on the die is 5.**

■ **G: The coin lands heads up.**

■ **H: The number on the die is less than 4 and the coin lands tails up.**

8. List all the equally likely outcomes for this experiment; that is, find a suitable sample space.

9. List all the outcomes in each of the events E, F, G, and H.

10. Find the probability of each of the events E, F, G, and H.

11. List the outcomes in $E \cap G$; then find $P(E \cap G)$.

12. List the elements in $F \cup H$; then find $P(F \cup H)$.

The experiment for Exercises 13–20 consists of rolling a pair of (fair) dice. Let E, F, G, and H be the following events:

■ **E: The sum of the numbers thrown is 2.**

■ **F: The sum of the numbers thrown is 7.**

■ **G: The sum of the numbers thrown is even.**

■ **H: The sum of the numbers thrown is less than 4.**

13. List all the equally likely outcomes for this experiment; that is, find a suitable sample space.

14. Does the set $\{2, 3, 4, \ldots, 12\}$ represent a suitable sample space for this experiment? Why or why not?

15. List all the outcomes in each of the events E, F, G, and H.

16. Find the probability of each of the events E, F, G, and H.

17. List the elements in $E \cap F$ and the elements in $E \cup F$.

18. Find $P(E \cap F)$ and $P(E \cup F)$.

19. List the elements in $G \cap H$ and the elements in $G \cup H$.

20. Find $P(G \cap H)$ and $P(G \cup H)$.

The experiment for Exercises 21–27 consists of rolling three fair tetrahedral dice. (Each die has four faces, numbered 1, 2, 3, 4.) Let E, F, G, and H be the following events:

■ **E: The sum of the numbers thrown is 2.**

■ **F: The sum of the numbers thrown is 7.**

■ **G: The sum of the numbers thrown is even.**

■ **H: The sum of the numbers thrown is less than 5.**

21. List all the equally likely outcomes for this experiment; that is, find a suitable sample space.

22. List all the outcomes in each of the events E, F, G, and H.

23. Find the probability of each of the events E, F, G, and H.

24. List the elements in $E \cap F$ and the elements in $E \cup F$.

25. Find $P(E \cap F)$ and $P(E \cup F)$.

26. List the elements in $G \cap H$ and the elements in $G \cup H$.

27. Find $P(G \cap H)$ and $P(G \cup H)$.

28. A single card is drawn from a regular deck of playing cards.

    (a) What is the probability of drawing an ace? Why?

    (b) What is the probability of drawing a heart? Why?

    (c) What is the probability of drawing the ace of hearts?

A jar contains seven ping-pong balls, numbered 1 through 7. Two balls are drawn from the jar, and the ball drawn first is *not* replaced before the second draw. Answer Exercises 29–32 for this experiment.

29. What is a suitable sample space for this experiment?

30. Find the probability of each of the following events:

    ■ *E*: Both balls have even numbers on them.

    ■ *F*: One ball has a 5 on it.

    ■ *G*: The sum of the numbers on the two balls is 6.

31. What is the probability that the sum of the numbers on the balls is *not* 6? Why?

32. What is the probability that neither ball drawn has a 5 on it? How is your answer related to $P(F)$?

Sometimes, the probabilities assigned to events are determined by data drawn from observations. As asserted in the text, treating outcomes as equally likely is often just a convenient simplifying assumption. Exercises 33–37 explore this idea further.

33. We *assume* that the probability of getting *heads* when tossing a fair coin is $\frac{1}{2}$.

    (a) If you toss a fair coin 50 times, how many times would you expect to get heads? Why?

    (b) Try it. Choose a coin that you think is fair, toss it 50 times, and keep track of the number of times you get *heads*. Does the result match your answer for Part (a)? If not, does this mean that your coin is not "fair"? Explain.

    (c) Try it again. Toss the coin 50 more times and keep track of the number of times you get *heads*. Does the result match your answer for Part (a)? Does it match your answer for Part (b)? Can you tell from this whether or not your coin is "fair"? Explain.

    (d) What is the total number of times you got *heads* out of the hundred tosses for Parts (a) and (b)? How does your answer compare with those of your classmates?

    (e) Based on your experiments and those of your classmates, what probabilities would you assign to the coin-toss outcomes *heads* and *tails*? Explain.

34. What is the probability that a phone number in your local phone directory will end with an odd digit? Use your phone directory to provide evidence that your answer is reasonable. (Tally the results for at least one page; then compare them with your probability estimate.)

35. (a) Do you think that observing the last digit of a phone number is an "experiment" with ten equally likely outcomes? Why or why not?

    (b) Pick a page of your local phone directory and tally the number of times each digit occurs last in a phone number. Do you think all ten outcomes are equally likely? If not, what probabilities would you assign to each?

    (c) Repeat Part (b), using a different page of your phone directory.

36. What is the probability that a phone number in your local phone directory will begin with an odd digit? Use your phone directory to provide evidence that your answer is reasonable. (Tally the results for at least one page; then compare them with your probability estimate.)

37. **(a)** Do you think that observing the first digit of a phone number is an "experiment" with ten equally likely outcomes? Why or why not?

(b) Pick a page of your local phone directory and tally the number of times each digit occurs first in a phone number. Do you think all ten outcomes are equally likely? If not, what probabilities would you assign to each?

(c) Do Part **(b)** again, using a different page of your phone directory.

WRITING EXERCISES

1. Saying that a process or experiment is random is not the same as saying that all possible outcomes are equally likely. Explain this statement, and illustrate it by giving an appropriate example.

2. **(a)** Is a random process always fair? Explain.

   (b) Is a fair process always random? Explain.

   (You probably should start by sharpening these questions!)

3. Discuss how the mathematical usage of the word "model" as described in this section is analogous to its nonmathematical usage.

4. If you have studied Chapter 3, compare and contrast the use of the word "model" in Section 3.4 with its use here.

COUNTING PROCESSES

As the examples and exercises of Section 4.3 suggest, our ability to analyze an experiment often depends on being able to count the exact number of outcomes or events in some set. In this section we discuss three problems whose solutions typify the most important basic counting techniques. These techniques are then used to illustrate some of the basic laws of probability.

EXAMPLE 4.22

Problem 1: How many three-digit numbers can be made using only the digits 0 and 1?

Solution: To form a three-digit number, we must fill three places *in order*, and in this case only the digits 0 or 1 may be used in any place. *Figure 4.4* on page 146, which is an example of a *tree diagram*, illustrates all the possible choices. There are two choices for the first digit. For each of these possibilities, there are two choices for the second digit, making a

1st choice   2nd choice   3rd choice

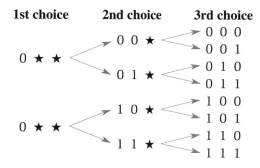

Figure 4.4 There are 2 choices for the first digit, 4 choices for the first two digits, and 8 choices for all three digits.

total of four different ways of choosing the first two digits. For each of these four ways there are two choices for the third digit. Hence, there are eight possible three-digit numbers containing only the digits 0 and 1.

Generalizing the reasoning used to solve this problem, we are led naturally to the following principle:

(4.1)

If an arrangement of objects requires r choices and if there are n possibilities for each choice, then the number of different arrangements is n^r.

An important special case of this rule gives us the formula for the number of subsets of a set. Suppose we have a set S containing r elements. Any subset of S can be formed by considering each element of S in turn and deciding whether that element should be *in* or *out* of the subset. Thus, the formation of any subset requires r choices from the two-element set

$$\{in, out\}$$

Applying Rule (4.1), we see the following:

The total number of subsets of an r-element set must be 2^r.

We can get to an even more fundamental principle by extending the reasoning of Problem 1 in several easy steps:

EXAMPLE 4.23

Problem 1a: How many three-digit numbers can be made using the digits 0 through 9?

Solution: This is a straightforward application of Rule (4.1). There are 3 choices and 10 possibilities for each choice, so there are $10 \cdot 10 \cdot 10 = 1000$ such numbers.

EXAMPLE 4.24

Problem 1b: How many three-digit numbers can be made using the digits 0 through 9 if the first digit cannot be 0?

Solution: This is almost the same as the situation in Problem 1a, except that there are only 9 choices for the first digit. Applying the reasoning behind Rule (4.1), we can see that for each of the 9 first digits there are 10 second digits—a total of 90 different ways to choose the first two digits. For each of these 90 possibilities there are 10 third digits. Thus, there are $(9 \cdot 10) \cdot 10 = 900$ such numbers.

EXAMPLE 4.25

Problem 1c: How many three-digit numbers can be made using the digits 0 through 9 if the first digit cannot be 0 and the last one must be odd?

Solution: As in Problem 1b, there are $9 \cdot 10 = 90$ possible choices for the first two digits. For each of these 90 possibilities there are 5 choices for the third digit. Thus, there are $(9 \cdot 10) \cdot 5 = 450$ such numbers.

xxxxxxxxxxxxxxxxxxxxxxxxxxxxxxxxxxxxxxxxxxxxxxxxxx

The Fundamental Counting Principle: If a first choice can be made in n ways and for each first choice a second choice can be made in m ways, then there are $n \cdot m$ possibilities for the pair of choices.

In general, suppose that there are r choices to be made. If there are n_1 possible first choices, for each first choice there are n_2 possible second choices, for each pair of first and second choices there are n_3 possible third choices, and so on, then the total number of possibilities for the r choices is

$$n_1 \cdot n_2 \cdot \ldots \cdot n_r$$

xxxxxxxxxxxxxxxxxxxxxxxxxxxxxxxxxxxxxxxxxxxxxxxxxx

EXAMPLE 4.26

Problem 2: Teams A, B, C, and D compete in a playoff tournament. If it is not possible for any two teams to tie for a place, how many different possibilities are there for the final standings of these teams?

Solution: This problem is solved by applying the Fundamental Counting Principle. We have a sequence of choices to make and a set to choose from. In this case, once an element has been chosen, it cannot be used again. (For instance, a team cannot be both first and third in the same tournament.) Thus, there are 4 possibilities for first place, and for each of them there are 3 possibilities for second place, making a total of 12 different ways the first two places can be decided. For each of these 12 situations there are 2 possibilities for third place; but once this has been decided, the remaining team must be fourth. Hence, there are

$$4 \cdot 3 \cdot 2 \cdot 1 = 24$$

possible final standings in this tournament, as shown in *Figure 4.5*.

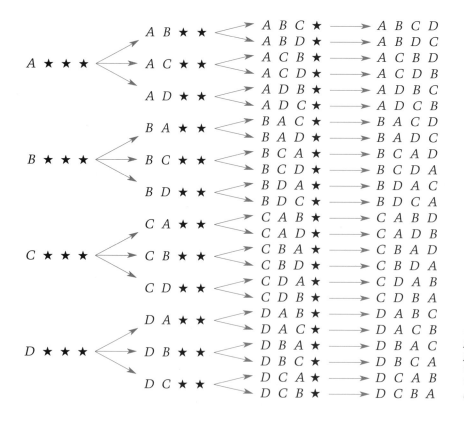

Figure 4.5 There are 4 possibilities for first place, 12 for the first two places, and 24 for the first three (and for all four) places.

In general, if an ordered arrangement of r things is to be formed from a set of n things, and if these things may *not* be used more than once in the same arrangement, there are again r choices to be made. However, although there are n possibilities for the first choice, there are only $n - 1$ possibilities for the second choice, $n - 2$ possibilities for the third, and so on. This pattern continues until the rth choice, for which there are $n - (r - 1)$ possibilities. Observing that $n - (r - 1)$ can be rewritten as $n - r + 1$, we get the following rule:

XXXXXXXXXXXXXXXXXXXXXXXXXXXXXXXXXXXXXXXXXXXXXXXXXX

(4.2)   If an arrangement of r objects is to be made from a set of n things and if none of the things can be used more than once, then the number of possible arrangements is

$$n \cdot (n - 1) \cdot (n - 2) \cdot \ldots \cdot (n - r + 1)$$

XXXXXXXXXXXXXXXXXXXXXXXXXXXXXXXXXXXXXXXXXXXXXXXXXX

DEFINITION   An ordered arrangement of r elements selected from a set of $n(> r)$ elements in such a way that no element may be used more than once is called a **permutation of n things taken r at a time**. The total number of such permutations is denoted by $_nP_r$.

NOTATION   The product of all consecutive positive integers from 1 to n inclusive is denoted by $n!$ and is read "n factorial." Thus $3! = 3 \cdot 2 \cdot 1$, $5! = 5 \cdot 4 \cdot 3 \cdot 2 \cdot 1$, etc.

In terms of this factorial notation,

$$_nP_r = n \cdot (n - 1) \cdot (n - 2) \cdot \ldots \cdot (n - r + 1)$$

$$= \frac{n \cdot (n - 1) \cdot (n - 2) \cdot \ldots \cdot (n - r + 1) \cdot [(n - r) \cdot \ldots \cdot 2 \cdot 1]}{(n - r) \cdot \ldots \cdot 2 \cdot 1}$$

$$= \frac{n!}{(n - r)!}$$

To extend this formula to the case of *all* possible permutations of n elements (that is, n things taken n at a time), we must agree that

$$0! = 1$$

Then

$$_nP_n = \frac{n!}{(n - n)!} = \frac{n!}{0!} = n!$$

EXAMPLE 4.27   **Problem 3:** If the sample space for an experiment is $\{a, b, c, d, e\}$, how many three-element events are there?

Solution: This type of problem differs from the previous ones in that the order of the elements chosen does not matter. Any two

3-element events (subsets) that contain the same elements are indistinguishable as sets, regardless of the order of the elements in them. If the order mattered, the solution would be

$$_5P_3 = \frac{5!}{2!} = 5 \cdot 4 \cdot 3 = 60$$

However, in that case the events $\{a, b, c\}$, $\{a, c, b\}$, $\{b, a, c\}$, etc., would be considered different from each other, when, in fact, they are not; that is, they are different representations of the same subset. There are exactly six possible arrangements of this subset because the number of possible permutations of a three-element set is $3! = 6$. Thus, the 60 ordered subsets can be grouped into collections according to which sets contain the same elements, and each of these collections will contain six sets. Since each collection represents a single unordered 3-element subset, the answer we seek is $\frac{60}{6} = 10$.

Here, then, is the general rule:

xxxxxxxxxxxxxxxxxxxxxxxxxxxxxxxxxxxxxxxxxxxxxxxxxxxxxxxxx

(4.3)

If we wish to count all the (unordered) r-element subsets of an n-element set, we count the number of ordered subsets of that size and then divide by the number of possible orderings of any one of them. Thus, we compute

$$\frac{_nP_r}{_rP_r} \qquad \text{or} \qquad \frac{_nP_r}{r!}$$

xxxxxxxxxxxxxxxxxxxxxxxxxxxxxxxxxxxxxxxxxxxxxxxxxxxxxxxxx

This rule is illustrated by Problem 3 (Example 4.27), in which $n = 5$ and $r = 3$.

DEFINITION

An r-element subset of an n-element set is called a **combination of n things taken r at a time**. The total number of such combinations is denoted by $_nC_r$.

By the preceding discussion,

$$_nC_r = \frac{n!}{(n - r)! \cdot r!}$$

The following examples illustrate Rule (4.1), the Fundamental Counting Principle, as well as Rules (4.2) and (4.3).

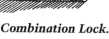

Combination Lock.

EXAMPLE 4.28

To determine how many three-digit numerals can be formed in base seven, we use Rule (4.1). There are seven single digits in base seven— 0, 1, 2, 3, 4, 5, 6—and any one of them can be used anywhere in the three places of such a numeral. Thus, $n = 7$ and $r = 3$ in this case, so there are 7^3 possibilities.

EXAMPLE 4.29

Problem: An ice cream parlor sells sundaes with any one of 12 different flavors of ice cream and any one of 5 different sauces. How many different kinds of sundaes does it sell?

Solution: This is a two-choice application of the Fundamental Counting Principle. There are 12 possible first choices and 5 possible second choices, so there are $12 \cdot 5 = 60$ possibilities in all.

Problem: If a customer may or may not choose to have whipped cream or nuts or a cherry on top, how many different kinds of sundaes are there?

Solution: Now there are three more choices to make, each one with 2 possibilities (yes or no). By the Fundamental Counting Principle, there are

$$60 \cdot 2 \cdot 2 \cdot 2 = 480$$

different kinds of sundaes.

EXAMPLE 4.30

Problem: A small club has 7 members. The club decides to elect three officers—president, secretary, and treasurer. Assuming all 7 members are candidates for each office, but nobody can hold two offices, how many possible outcomes does this election have?

Solution: To answer this, we observe that each possible outcome is an ordered triple selected from a set of 7 elements, with no element used more than once. This means that we can use Rule (4.2), with $n = 7$ and $r = 3$. Thus, the answer is $7 \cdot 6 \cdot 5$, which equals 210. In other words, there are 210 possible slates of officers for this club. This is the number of permutations of 7 things taken 3 at a time, which is symbolized by $_7P_3$. (Do you see why this situation requires permutations, rather than combinations?)

EXAMPLE 4.31

Problem: A typical poker hand is a random selection of 5 cards from a deck of 52. What is the chance of getting a hand containing a specific royal flush—say, the Ace, King, Queen, Jack, and 10 of hearts?

Solution: To answer this, we need to know how many different 5-element subsets a 52-element set has. (The order in which the cards are dealt doesn't matter.) Those subsets (the possible poker hands) form the sample space for this experiment. This means that Rule (4.3) applies, with $n = 52$ and $r = 5$. Thus, we want

$$_{52}C_5 = \frac{52!}{(52-5)! \cdot 5!} = \frac{52 \cdot 51 \cdot 50 \cdot 49 \cdot 48}{5!} = 2,598,960$$

Because the event we want consists of a single element of this large sample space, its probability is

$$\frac{1}{2,598,960}$$

Problem: *What is the probability of getting a royal flush in hearts?*

4.4   EXERCISES

For Exercises 1–6, use the Fundamental Counting Principle to calculate how many numbers fit the given description. Draw a "tree diagram" (a diagram like *Figure 4.4* or *4.5*) to illustrate your answer.

1. A 2-digit number in which the first digit is nonzero and even and the second digit is divisible by 3.

2. A 2-digit number in which the first digit is odd and the second digit is divisible by 4.

3. A 2-digit number in which the first digit is greater than 5 and the second digit is divisible by 5.

4. A 3-digit number in which the first digit is nonzero and divisible by 4, the second digit is divisible by 3, and the third digit is divisible by 5.

5. A 3-digit number in which the first digit is greater than 6, the second digit is less than 2, and the third digit is odd.

6. A 4-digit number in which the first digit is nonzero and divisible by 3, the second digit is divisible by 5, the third digit is divisible by 7, and the fourth digit is divisible by 8.

For Exercises 7 and 8, recall that a standard telephone assigns 3 letters to each number from 2 through 9; the letters Q and Z are left out.

7. How many different 7-letter strings represent the phone number 234-5678? What about the number 555-5555?

8. How many different seven-digit phone numbers have the form "D I A L _ _ _"? How many different letter triples does each of those numbers represent?

For Exercises 9–16, compute:

9. $_6P_4$ 10. $_{10}P_6$ 11. $_{10}P_3$ 12. $_8P_8$

13. $_6C_4$ 14. $_{10}C_6$ 15. $_{10}C_3$ 16. $_8C_8$

In Exercises 17 and 18, how many numbers of no more than four digits can be made by using:

17. only the odd digits?

18. only 2, 4, 6, and 8?

In Exercises 19–24, how many five-letter license plates can be made by using:

19. All the letters of the alphabet?

20. Only the letters A, B, C, D, E, F, G, H, I, J.

21. Only the vowels A, E, I, O, U?

22. All the letters of the alphabet, with no repetitions?

23. Only the letters A, B, C, D, E, F, G, H, I, J, with no repetitions?

24. Only the vowels A, E, I, O, U, with no repetitions?

In Exercises 25–30, how many ways can the letters in each word be rearranged?

25. CAT   26. WORDS

27. TICKLE   28. ARTICHOKE

29. SCATTER (*Watch out!*)

30. MISSISSIPPI (*Be careful!*)

Exercises 31–34 are about Peppy's Pizza Parlor, which sells thin-crust or thick-crust pizzas in 3 different sizes and offers a selection of 10 optional extra toppings.

31. How many varieties of plain pizza are there?

32. How many one-topping pizza varieties are there?

33. How many two-topping pizza varieties are there, not allowing a double order of the same topping?

34. If you can choose as many different extra toppings as you want, how many pizza varieties are there? Explain.

Exercises 35–44 concern a worldwide electronic-network corporation that is setting up a system of online passwords. For efficient handling, it wants the passwords to be as short as

possible, but it needs enough different access passwords to handle all possible customers.

35. If only alphabet letters are allowed and each password is 4 letters long, how many customers can be handled?

36. If only alphabet letters are allowed and each password is 5 letters long, how many customers can be handled?

37. If alphabet letters and/or digits are allowed and each password is 4 symbols long, how many customers can be handled?

38. If alphabet letters and/or digits are allowed and each password is 5 symbols long, how many customers can be handled?

39. The corporation wants to be able to handle at least 500,000,000 customer passwords. Using letters only, what is the shortest password length it can use? For that length, how many passwords will be available?

40. The corporation wants to be able to handle at least 350,000,000 customer passwords. Using letters and/or digits, what is the shortest password length it can use? For that length, how many passwords will be available?

41. The corporation considers allowing letters and/or single digits, but wants each password to begin with a letter, not a digit. How many 6-symbol passwords will be available?

42. To avoid obscene or otherwise offensive 6-symbol passwords, the company wants to require the third and fourth symbols to be digits. They also require the password to begin with a letter. The other three symbols can be either letters or digits. Will at least 350,000,000 passwords still be available? Justify your answer.

43. If the corporation uses a 7-symbol password that alternates letters and digits, beginning with a letter, will at least 350,000,000 passwords be available? What if it requires each one to begin with a digit? Justify your answers.

44. If the corporation uses a 7-symbol password that alternates letters and digits, and it allows either a letter or a digit as the first symbol, how many passwords will be available?

45. You have wandered by accident into a class in ancient Greek literature. A ten-question multiple-choice test is handed out, with each answer to be chosen from four possibilities. Having nothing better to do, you decide to take the test. Because it is written entirely in Greek, a language you know nothing about, you answer each question completely at random. What is the probability that you get all the answers right?

46. The tickets in a single-prize raffle are identified by a five-place "number" in which the first two places are letters of the alphabet and the remaining three places are single digits. All the tickets that can possibly be numbered this way are sold. You buy the tickets AB981 through AB990 (and no others). If the raffle is fair (that is, if all outcomes are equally likely), what is the probability that you will win the prize?

WRITING EXERCISES

1. Compare and contrast the mathematical meanings of the words "permutation" and "combination" with their nonmathematical meanings.

2. Write the Fundamental Counting Principle entirely in words, without using any letters or other symbols to represent arbitrary quantities. Then, reflecting on how you dealt with this question, comment briefly on the role of symbols in writing mathematical ideas.

3. Construct a clearer and/or simpler symbolic way of representing the concepts denoted in this section by $_nP_r$ and $_nC_r$. Explain why you think your way is better than the one given here. (*Note:* Some mathematical symbolism is used simply because of tradition, not necessarily because it is the clearest or simplest way to represent an idea.)

LINK: COUNTING AND THE GENETIC CODE

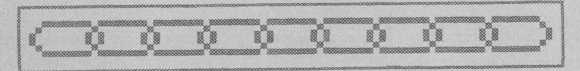

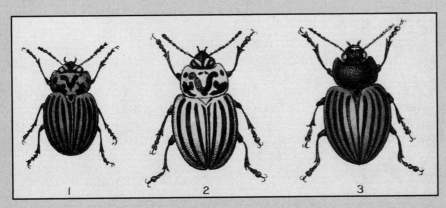

Mutations. Some of the numerous variations in *Leptinotarsa multitaeniata.* Figure 2 is the type of the species; figures 1 and 3 are extreme mutants.

4.5

Ever since mankind began to speculate about its own existence, one of the most tantalizing problems has been: How do the various life forms—people, animals, plants, even microorganisms —transmit life from generation to generation, and with it transmit the bewildering variety of characteristics that distinguish species from species and family from family? We all have physical characteristics that we say are "inherited" from our parents, but none of us is an exact copy of any ancestor. Cats breed cats and dogs breed dogs, and the kittens and pups are similar to, but not exactly like, their parents; rose bushes never blossom forth with lilies, but no two flowers are identical. There are examples almost everywhere we look.

The rise of modern science in Western culture began in the sixteenth and seventeenth centuries with the work of people such as Copernicus, Galileo, and Newton. As science sought to turn the observable into the predictable, and technology began to turn the predictable into the constructible, most natural phenomena seemed to conform to unifying principles that govern their behavior. But the propagation of life itself remained a mystery, immune to the increasingly precise mathematical techniques of science. It was almost as if nature had decreed a fundamental diversity to defy the scientists' thirst for simple logical explanations.

The first breakthrough came in the 1860s when Gregor Mendel, an Augustinian monk in Austria, observed a pattern in the inherited characteristics of peas and set down some basic principles which have since been shown to apply to all living creatures. These "Mendelian laws" described certain factors, subsequently called *genes* (from the Greek word for giving birth), whose behavior could be described precisely enough to make the occurrence of some inherited characteristics fairly predictable. The study of these inheritance patterns soon came to be called *genetics*.

Despite their predictive power, Mendel's laws did not explain exactly *how* these genetic factors are passed on. In particular, *how* do a single sperm cell and a single egg cell carry with them the complex variety of factors that govern the birth and growth of an entire human being? The answer to this question emerged in 1944, after 80 years of biochemical research, with the first direct experimental evidence that an organic acid called DNA contains all that information in its molecular structure. Each DNA molecule has the form of a *double helix*, a barber-pole spiral of two linked chemical chains. Each chain is made up of alternating sugar and phosphate molecules, and the two chains are bonded together at the sugars by linked pairs of chemical bases called *nucleotides*. There are only four nucleotides in DNA—adenine (A), cytosine (C), guanine (G), and thymine (T)—yet the sequence of these bonded pairs in a single DNA molecule carries the entire genetic "blueprint" for its organism! Moreover, the bonding occurs in such a way that the links in one of the chains determine the links in the other, so a single strand of DNA carries all the hereditary information to be passed along. Thus, the entire *genetic code* for all living things is representable as sequences of the four letters A, C, G, and T.

The marvelous simplicity of this device for passing on hereditary characteristics might seem puzzling at first. How (you might ask) is it possible for so few different chemicals to account for the seemingly endless variety of plants and animals? The answer lies in the Fundamental Counting Principle, explained in Section 4.4, and the fact that DNA chains are quite long. Let us take a simple example. A single virus particle contains only one DNA molecule. Its nucleotide chain may contain 5000 to 200,000 "links," depending on the type of virus. If we take the simplest type, a 5000-link chain of nucleotides, how many different genetic patterns are possible? There are 5000 positions in the chain and each one can be filled in 4 different ways; so the Fundamental Counting Principle says that there are 4^{5000} different possible chains. Exponential notation can be deceptively brief, so let us do a little estimating to get a better sense of the size of this number:

$$4^{5000} = (4^5)^{1000} = (1024)^{1000} > (1000)^{1000} = (10^3)^{1000} = 10^{3000}$$

This last number, if written out, would have a 1 followed by three thousand 0s. If typewritten (in 10-pitch) on a single line, this number, even without commas, would be about 25 feet long!

We humans are much more complex than viruses, so our DNA chains are much longer. A DNA chain in a typical human cell might be some five billion nucleotides long. That is, the number of various human genetic patterns is about $4^{5,000,000,000}$, somewhat larger than $1000^{1,000,000,000}$. This number, if typed on a single line, would be a 1 followed by more than 4700 *miles* of 0s; it would stretch from New York City to Moscow. With so many possible genetic variations, it is not surprising that no two people are exactly alike! (By way of comparison, the total number of people who have ever lived has recently been estimated as not much more than a mere 100,000,000,000, of whom more than 5,000,000,000 are living now.)

A modest application of the Fundamental Counting Principle played a role in "breaking" the genetic code; that is, in seeing how the information contained in the DNA chains determines the way an organism grows. Virtually all chemical reactions that take place in the growth and development of cells are controlled by the presence or absence of specific types of proteins called *enzymes*. The many, many different kinds of enzymes are formed from long chains of amino acids, but there are only 20 different amino acids. The key question is: How does a chain made up from four different things (nucleotides) determine a specific chain made up from 20 different things (amino acids)? It is clear that some combination of nucleotides is needed to determine a single amino acid, but how many? The Fundamental Counting Principle provides some important hints:

■ If we consider only ordered pairs of nucleotides, there are not enough possibilities because the total number of different ordered pairs is

$$4 \cdot 4 = 16$$

■ there are more than enough ordered triplets of nucleotides because

$$4 \cdot 4 \cdot 4 = 64$$

■ there appear to be far too many ordered quadruples because

$$4 \cdot 4 \cdot 4 \cdot 4 = 256$$

It was these *numerical* considerations that led researchers to concentrate their attention on nucleotide triplets. In 1961, experimental results began to confirm that the genetic code was indeed based on ordered triplets of nucleotides! The difference between the 64 triplets and the 20 amino acids was explained by the fact that several different triplets correspond to each acid, and within a few years a complete "dictionary" for translating the genetic code into amino acids by way of nucleotide triplets was well on its way to completion.

If you are interested in a more complete, but easily readable, explanation of this topic, you might enjoy Isaac Asimov's book, Item 2 in the list *For Further Reading* at the end of this chapter.

SOME BASIC RULES OF PROBABILITY

4.6

Now that we have some efficient ways of counting the number of outcomes in a sample space, let us return to the concept of probability as it was described in Section 4.3. Recall that we have been dealing with experiments whose sample spaces contain equally likely outcomes; in this case the probability of an event is simply the number of outcomes in the event divided by the total number of outcomes in the sample space. Because events are just subsets of the sample space, we can make some easy observations about counting sets of things that will prove to be valuable for generalizing the probability concept to other kinds of experiments. In fact, some of these observations are the *axioms* of probability; they are statements *assumed* to be true in every probability situation, from which the other general laws of probability are proved.

First of all, recall that the sample space S of an experiment plays the role of the universal set for that experiment. Because S is a subset of itself, we can ask about its probability. The common-sense answer is that the actual outcome of the experiment is certain to be somewhere in the sample space (by the definition of a sample space), and thus, the number assigned to S should indicate certainty. The definition of probability says that this number must be 1 (or 100%, if you prefer):

$$P(S) = \frac{n(S)}{n(S)} = 1$$

for any sample space S. This becomes our first general axiom.

Axiom 1 The probability of any sample space is 1.

Because any event E is a subset of S, the number of elements in E cannot exceed the number of elements in S, and the fewest elements E can have is none at all. Thus,

$$0 \le n(E) \le n(S)$$

and

$$0 = \frac{0}{n(S)} \le \frac{n(E)}{n(S)} \le \frac{n(S)}{n(S)} = 1$$

Because

$$P(E) = \frac{n(E)}{n(S)}$$

it follows that $P(E)$ must be a number between 0 and 1. This becomes our second general axiom.

Axiom 2 For any event E, $0 \le P(E) \le 1$.

The third and final axiom also follows from a common-sense observation about sets: If two sets are disjoint, then the number of elements in their union can be found by counting the elements in each set separately, then adding. That is:

xxxxxxxxxxxxxxxxxxxxxxxxxxxxxxxxxxxxxxxxxxxxxxxxxxxx

If $E \cap F = \emptyset$, then $n(E \cup F) = n(E) + n(F)$.

xxxxxxxxxxxxxxxxxxxxxxxxxxxxxxxxxxxxxxxxxxxxxxxxxxxx

If these sets are events of a sample space, then saying they are disjoint means that, if the actual outcome of the experiment falls within one of these events, then it cannot also fall within the other. In other words, if either of these events happens, the other cannot. Disjoint events are given a name to reflect this property; they are called **mutually exclusive** events.

In a sample space of equally likely outcomes, if we know the separate probabilities of two mutually exclusive events E and F, the probability of their union is easily found by applying the counting observation just made:

$$P(E \cup F) = \frac{n(E \cup F)}{n(S)} = \frac{n(E)}{n(S)} + \frac{n(F)}{n(S)} = P(E) + P(F)$$

This becomes our third general axiom.

Axiom 3 If E and F are mutually exclusive events, then

$$P(E \cup F) = P(E) + P(F)$$

These three simple axioms allow us to prove some useful rules about how probability works in *any* experiment. Before we develop them, however, it might be useful to review some interconnections among events, sets, and logic.

- An event is a subset of the sample space of an experiment. When we say that an event E *happens*, we mean that the actual outcome of the experiment lies *within* that particular subset E.

- To say an event E *does not happen*, then, means that the actual outcome lies *outside* of E (but still in the sample space, of course). Consequently, if an event E does not happen, its complement E' must happen.

■ If we have two events E and F of a sample space, then saying that one or the other of the events happens means that the actual outcome of the experiment lies within $E \cup F$, and saying that both of the events happen means that the actual outcome lies within $E \cap F$.

Thus, knowing how to compute the probabilities of complements, unions, and intersections of events whose individual probabilities are known is an important tool for analyzing experiments and drawing conclusions about them. The rest of this section is devoted to developing and applying the fundamental probability rules for complement, union, and intersection of events.

The complement of an event E, written E', is the set of all outcomes in the sample space S but not in E. Because E and E' are mutually exclusive, Axiom 3 says that

$$P(E) + P(E') = P(E \cup E')$$

But $E \cup E' = S$ by definition of complement, and Axiom 1 tells us that $P(S) = 1$; so

$$P(E) + P(E') = 1$$

Subtracting $P(E)$ from both sides of this equation, we get the rule for complements:

xxxxxxxxxxxxxxxxxxxxxxxxxxxxxxxxxxxxxxxxxxxxxxxxxxxxxxxx

(4.4)   If E is any event of a sample space S, then

$$P(E') = 1 - P(E)$$

xxxxxxxxxxxxxxxxxxxxxxxxxxxxxxxxxxxxxxxxxxxxxxxxxxxxxxxx

EXAMPLE 4.32

The probability of drawing a heart from a well-shuffled standard deck of cards is $\frac{1}{4}$. (*Hearts* is one of four 13-card suits in the 52-card deck). Thus, by Rule (4.4), the probability of not drawing a heart is

$$1 - \frac{1}{4}, \text{ or } \frac{3}{4}$$

EXAMPLE 4.33

If a pair of fair dice is thrown, there are 6^2 possible outcomes in the sample space. (Why?) Of these 36 possible outcomes, six of them add up to 7. (Which ones are they?) Thus, if we let E be the event of throwing a 7, then $P(E)$ is $\frac{6}{36}$, or $\frac{1}{6}$. This means that $P(E')$, the probability of not throwing a 7, is

$$1 - \frac{1}{6} = \frac{5}{6}$$

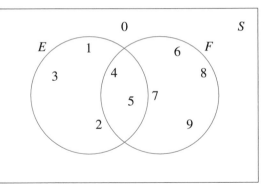

Figure 4.6 $P(E \cup F) < P(E) + P(F)$.

To compute the probability of the union of two events E and F, we need to generalize Axiom 3 to the case where E and F might have some outcomes in common. That is, we must consider the case where $E \cap F$ is not empty. Just adding the probabilities of E and F, ignoring the requirement that they be mutually exclusive, will give us incorrect information. A simple example of this is pictured in *Figure 4.6*. Suppose the sample space S contains the ten outcomes $\{0, 1, 2, \ldots, 9\}$, and the two events are

$$E = \{1, 2, 3, 4, 5\} \quad \text{and} \quad F = \{4, 5, 6, 7, 8, 9\}$$

so that

$$E \cap F = \{4, 5\}$$

Then

$$P(E) = \frac{5}{10} \quad \text{and} \quad P(F) = \frac{6}{10}$$

Therefore,

$$P(E) + P(F) = \frac{11}{10}$$

which is bigger than 1 and hence cannot be the probability of *any* event. (Why not?) Moreover, counting the outcomes in $E \cup F$ tells us that $P(E \cup F) = \frac{9}{10}$. In this case, the difference between $P(E) + P(F)$ and $P(E \cup F)$ is accounted for by the fact that the two elements in $E \cap F$ were counted in computing $P(E)$ and also in computing $P(F)$. Thus, the sum $P(E) + P(F)$ counts these same elements *twice*, although they only appear *once* in $E \cup F$.

To resolve the difficulty of counting the "overlapping" elements of E and F twice, observe that there is a way of expressing $E \cup F$ as the union of *disjoint* sets:

$$(E \cup F) = E \cup (E' \cap F)$$

where $E' \cap F$ is the set of all outcomes in F that are not in E. (See *Figure 4.7*.) We can apply Axiom 3 to this equation, and obtain

Figure 4.7
$E \cup F = E \cup (E' \cap F)$.
(a) $E \cup F$ is shaded in color; (b) E and $E' \cap F$ are each shaded in color—thus, $E \cup (E' \cap F)$ is shaded in color.

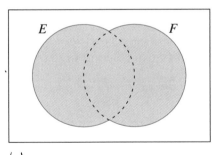

(a)

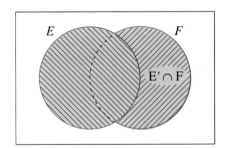

(b)

(4.5) $$P(E \cup F) = P(E) + P(E' \cap F)$$

Now, F can be described as the union of the set of all outcomes in F that are *not in E* together with the set of all outcomes in F that are *in E*, and it is easy to see that these two parts of F are mutually exclusive. Therefore,

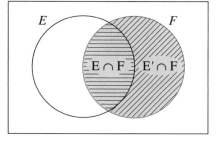

$$F = (E' \cap F) \cup (E \cap F)$$

(See *Figure 4.8*.) By Axiom 3,

$$P(F) = P(E' \cap F) + P(E \cap F)$$

This last equation can be rewritten as

Figure 4.8 $F = (E' \cap F) \cup (E \cap F)$.

$$P(F) - P(E \cap F) = P(E' \cap F)$$

Plugging the information from this last equation into Equation (4.5), we obtain the general rule for computing the probability of the union of two events:

×××

(4.6)   If E and F are any two events of a sample space, then

$$P(E \cup F) = P(E) + P(F) - P(E \cap F)$$

×××

Despite the apparent complexity of its derivation from the axioms, Rule (4.6) really states a very simple principle:

Because the sum of $P(E)$ and $P(F)$ counts the common outcomes in $E \cap F$ twice, one "copy" of $P(E \cap F)$ should be thrown away to get the proper value for $P(E \cup F)$.

EXAMPLE 4.34   Suppose a single die is thrown. Let E be the event of getting an even number and F of getting a number less than 4. Then

$$E = \{2, 4, 6\}, \quad F = \{1, 2, 3\}, \quad \text{and} \quad E \cap F = \{2\}$$

Moreover, $E \cup F$ represents getting either an even number or a number less than 4; its probability can be computed from Rule (4.6) as follows:

$$P(E \cup F) = P(E) + P(F) - P(E \cap F) = \frac{3}{6} + \frac{3}{6} - \frac{1}{6} = \frac{5}{6}$$

EXAMPLE 4.35   If a single card is drawn at random from a standard deck, the probability that it will be

- a heart is $\frac{13}{52}$,
- a face card (king, queen, jack) is $\frac{12}{52}$,
- the king, queen, or jack of hearts is $\frac{3}{52}$.

By Rule (4.6), the probability that it will be either a heart or a face card is

$$\frac{13}{52} + \frac{12}{52} - \frac{3}{52} = \frac{22}{52}$$

EXAMPLE 4.36

A survey of results from a recent mathematics quiz showed that

- 30% of the students misspelled "commutative,"
- 20% misspelled "associative,"
- 12% misspelled both words.

The probability that a student chosen at random from that class misspelled at least one of the two words is

$$30\% + 20\% - 12\% = 38\%$$

Working with the intersection of two events is a bit more complicated than handling the union or complement. Clearly, if you know the probabilities of two events E and F *and* the probability of their union, then the probability of their intersection can be found by Rule (4.6). However, if only $P(E)$ and $P(F)$ are known, there is no simple general formula for finding the probability of $E \cap F$.

One important special case, however, deserves some attention. That is the case in which E and F are "independent." A rigorous discussion of independent events requires considering *conditional probability*, which is the subject of the next section. At this point we shall content ourselves with an informal definition of independence: Two events are said to be **independent** if the occurrence or nonoccurrence of one has no effect on the likelihood of the other. In other words,

if the value of $P(E)$ is the same whether or not we know that the actual outcome of the experiment lies in F, then E and F are independent events.

The rule for computing the probability of the intersection in this case is as follows:

xxxxxxxxxxxxxxxxxxxxxxxxxxxxxxxxxxxxxxxxxxxxxxxxxxx

(4.7)   If E and F are independent events, then

$$P(E \cap F) = P(E) \cdot P(F)$$

xxxxxxxxxxxxxxxxxxxxxxxxxxxxxxxxxxxxxxxxxxxxxxxxxxx

| EXAMPLE 4.37 | An experiment consists of flipping a coin three times. If E represents obtaining heads on the first toss and F represents obtaining heads on the third toss, then these are independent events because the occurrence of E has no effect on the occurrence of F. Clearly, $P(E)$ and $P(F)$ both equal $\frac{1}{2}$. Thus, to find the probability of getting heads on both the first and third tosses, we apply Rule (4.7): |

$$P(E \cap F) = P(E) \cdot P(F) = \frac{1}{2} \cdot \frac{1}{2} = \frac{1}{4}$$

| EXAMPLE 4.38 | If a single die is thrown, then the event E of obtaining an even number has probability $\frac{1}{2}$. The event F of obtaining an odd number also has probability $\frac{1}{2}$; so |

$$P(E) \cdot P(F) = \frac{1}{2} \cdot \frac{1}{2} = \frac{1}{4}$$

However, this is not $P(E \cap F)$ because $E \cap F$ is empty and hence has probability 0. Obtaining an even number and obtaining an odd number are not independent events; whichever one happens, the probability of the other drops from $\frac{1}{2}$ to 0.

4.6   EXERCISES

In Exercises 1–8, assume that E and F are mutually exclusive events of a sample space S. If $P(E) = .45$ and $P(F) = .30$, compute:

1. $P(E \cap F)$ 2. $P(E \cup F)$

3. $P(E')$ 4. $P(F')$

5. $P(E \cap E')$ 6. $P(E \cup E')$

7. $P(E' \cap F)$ 8. $P(E' \cup F)$

In Exercises 9–14, assume that E and F are independent events of a sample space S. If $P(E) = 25\%$ and $P(F) = 60\%$, compute:

9. $P(E')$ 10. $P(F')$

11. $P(E \cap F)$ 12. $P(E \cup F)$

13. $P(E' \cup F')$ 14. $P(E' \cap F')$

In Exercises 15–19, let E and F represent events of a sample space S. Explain why each of the following situations is impossible.

15. $P(E) = .40$ and $P(E') = .75$

16. E and F are mutually exclusive, $P(E) = .5$, and $P(F) = .7$

17. $P(E) = .6$ and $P(E \cup F) = .4$

18. $P(E) = .35$, $P(F') = .7$, and E is a subset of F.

19. E and F are independent, $P(E) = .5$, $P(F) = .6$, and E is a subset of F.

In Exercises 20–25, assume that a kindergarten classroom has a box with 40 crayons in it. Five of them are red, six are blue, and eight are green; the rest are other assorted colors. A girl reaches in without looking and picks a single crayon. What is the probability that she picks:

20. A red one?   21. A blue one?

22. One that is not red?

23. One that is not green?

4. One that is neither red nor green?

5. Either a red one or a blue one?

**n Exercises 26–34, assume that a single card is
rawn from a regular deck. Consider the events**

- **D: A diamond is drawn.**
- **F: A face card is drawn.**
- **H: A heart is drawn.**

**ind the probability of each of the following
vents:**

6. D   27. F   28. H

9. D ∩ F   30. D ∩ H

31. Drawing either a heart or a diamond.

32. Drawing either a diamond or a face card.

33. Drawing a black card.

34. Drawing a black card that is not a face card.

**In Exercises 35–40, suppose that E and F are in-
dependent events of a sample space S containing 200 equally likely outcomes, and that**

$$P(E) = \frac{1}{5} \quad \text{and} \quad P(F') = \frac{3}{4}$$

Compute:

35. n(E)   36. P(F)   37. n(F)

38. P(E ∩ F)   39. P(E ∪ F)   40. P(E' ∩ F)

VRITING EXERCISES

. Without referring to coins, dice, or playing
cards, describe examples that illustrate each
of these concepts:

(a) Mutually exclusive events.

(b) For events E and F,

$$P(E \cup F) = P(E) + P(F) - P(E \cup F)$$

(c) Independent events.

. Write a clear explanation of the difference
between mutually exclusive events and independent events. In the course of your explanation, answer these questions, with
justification for your answers:

- Are mutually exclusive events always independent?
- Are mutually exclusive events ever independent?
- Are independent events always mutually
  exclusive?
- Are independent events ever mutually exclusive?

3. Suppose we changed Axiom 1 to:

   The probability of any sample space
   is 5.

How would that alter the other material
presented in this section? Would it seriously
effect the theory of probability? Discuss.

CONDITIONAL PROBABILITY

4.7

The intersection of two sets contains only those elements of one set
that are also contained in the other. If the sets are the events of an experiment, then the intersection contains only those outcomes of one
event that can occur if the other actually happens. This situation is
illustrated by the following tale drawn from life in the suburbs:

Unable to locate a volunteer to walk the family dog at six o'clock the next morning, a husband-wife couple of mathematicians and their two precocious children faced the unpleasant task of choosing an unwilling conscript from among their own ranks. The parents, sensitive to the precepts of modern psychology, were reluctant simply to appoint either child, but at the same time were unwilling to become servants to the whims of their offspring. They decided, therefore, to draw lots and let the matter be decided by chance. As they were finishing dinner, the mother cut four identical slips of paper, marked one with an x, and placed them in a hat. Each member of the family, she explained, would in turn select a slip (and not put it back); whoever drew the x would have to walk the dog.

At that point the younger child (more talented in democracy than in mathematics) protested that this was unfair, claiming that whoever drew first took the least risk. He argued that, while the probability of getting the x on the first draw was $\frac{1}{4}$, the probability of getting it on the second draw rose to $\frac{1}{3}$ because there would only be three slips left to choose from. Similarly, he said, the probability of getting it on the third draw would be $\frac{1}{2}$, and on the fourth draw it would be 1. Because the order of choice apparently affected the amount of risk, he insisted that they find a fair way of determining that order, or else scrap the whole procedure. A heated debate began immediately. Just as the mother was about to restore tranquillity by a true analysis of the probabilities involved, it was discovered that the dog had run away and would be gone for at least several days. Since the problem had disappeared with the dog, all discussion ceased and peace returned to the dinner table.

If the mother in this story had been allowed to explain the situation, she undoubtedly would have acknowledged the basis of her child's objections. It is true that, if the first slip drawn is not marked x, then there are three slips left as possible outcomes for the second draw, and hence the probability of x appearing is $\frac{1}{3}$. However, it is not certain that x will not appear on the first draw. If it does appear first, then the probability of its appearance on the second draw is 0. (In fact, there is no need for a second draw.)

If "0" is used to represent any of the unmarked slips, then the four equally likely possible outcomes of this experiment may be denoted by

$$(x, 0, 0, 0), \quad (0, x, 0, 0), \quad (0, 0, x, 0), \quad (0, 0, 0, x)$$

Call this set of outcomes S. If we denote the event of getting x on the first draw by X_1, then $X_1 = \{(x, 0, 0, 0)\}$ and

$$X_1' = \{(0, x, 0, 0), (0, 0, x, 0), (0, 0, 0, x)\}$$

The sets X_2, X_3, X_4, and their complements may be defined similarly.

Thus,

$$n(X_1) = n(X_2) = n(X_3) = n(X_4) = 1$$

and

$$n(X_1') = n(X_2') = n(X_3') = n(X_4') = 3$$

It follows that

$$P(X_1) = P(X_2) = P(X_3) = P(X_4) = \frac{1}{4}$$

and

$$P(X_1') = P(X_2') = P(X_3') = P(X_4') = \frac{3}{4}$$

The discrepancy between $P(X_2)$ as given here and the child's evaluation of it can be explained informally by saying that, if this experiment were performed many times, we could expect x to occur second one-third of the times *in which it did not appear first*. But it would not appear first only three-fourths of the total number of times, so it would appear second *one-third of three-fourths of the time*—that is, one-fourth of the time.

To generalize this line of reasoning, suppose that E and F are two events of an experiment and that we wish to find the probability that F will occur, assuming E has already occurred. Our assumption restricts the possible outcomes of the experiment to the ones in E; so the only outcomes of F that we may consider are those that are also in E.

DEFINITION

Let E and F be two events of a sample space S of equally likely outcomes. The **conditional probability of F assuming E** is

$$\frac{n(E \cap F)}{n(E)}$$

NOTATION

The conditional probability of F assuming E is denoted by $P(F \mid E)$ and is often called "the probability of F, given E."

This definition allows us to deal with the probability of the intersection of two events.

xxxxxxxxxxxxxxxxxxxxxxxxxxxxxxxxxxxxxxxxxxxxxxxxxxxxxx

(4.8)   If E and F are any two events of a sample space S, then

$$P(E \cap F) = P(E) \cdot P(F \mid E) = P(F) \cdot P(E \mid F)$$

xxxxxxxxxxxxxxxxxxxxxxxxxxxxxxxxxxxxxxxxxxxxxxxxxxxxxx

Rule (4.8) is true for any sample space. If the space consists of equally likely outcomes, it is easy to verify this statement directly from the basic definitions:

$$P(E) \cdot P(F \mid E) = \frac{n(E)}{n(S)} \cdot \frac{n(E \cap F)}{n(E)} = \frac{n(E \cap F)}{n(S)} = P(E \cap F)$$

Moreover, interchanging E and F in Statement (4.8) and using the fact that

$$F \cap E = E \cap F$$

we also have

$$P(F) \cdot P(E \mid F) = P(E \cap F)$$

EXAMPLE 4.39

In the dog-walking story at the beginning of this section, for the x to be drawn second, both X_1' and X_2 must occur. Hence, the probability is

$$P(X_1' \cap X_2) = P(X_1') \cdot P(X_2 \mid X_1') = \frac{3}{4} \cdot \frac{1}{3} = \frac{1}{4}$$

The child correctly computed $P(X_2 \mid X_1')$, but did not account for the probability of X_1'.

EXAMPLE 4.40

If two playing cards are drawn from a regular deck and the first is *not* replaced before the second is drawn, the probability that both cards are aces can be computed by using Rule (4.8). The probability that the first card is an ace is $\frac{4}{52} = \frac{1}{13}$ and, assuming this has happened, the probability that the second card is an ace is $\frac{3}{51} = \frac{1}{17}$. Hence, the probability that both cards are aces is

$$\frac{1}{13} \cdot \frac{1}{17} = \frac{1}{221}$$

Now independent events can be defined more precisely than they were at the end of Section 4.6.

DEFINITION

If E and F are two events such that $P(F) = P(F \mid E)$, then F is **independent of** E. Two events are **independent** if each is independent of the other. In general, events $E_1, E_2, \ldots, E_n$ are **independent** if every possible pairing of them consists of two independent events.

This definition can be combined with Rule (4.8) to yield a convenient proof and generalization of Rule (4.7), which appeared at the end of Section 4.6. Rule (4.8) states that

$$P(E \cap F) = P(E) \cdot P(F \mid E)$$

for *any* two events E and F. If E and F are *independent*, we can substitute $P(F)$ for $P(F \mid E)$ to obtain

$$P(E \cap F) = P(E) \cdot P(F)$$

In general:

XXX

(4.9)

If $E_1, E_2, \ldots, E_n$ are independent events, then

$$P(E_1 \cap E_2 \cap \ldots \cap E_n) = P(E_1) \cdot P(E_2) \cdot \ldots \cdot P(E_n)$$

XXXXXXXXXXXXXXXXXXXXXXXXXXXXXXXXXXXXXXXXXXXXXXX

EXAMPLE 4.41

The probability of rolling a 5 in any throw of a single (fair) die is $\frac{1}{6}$, and the fact that a 5 may have turned up before has no effect on the likelihood of its turning up again. Thus, if F_n represents the event that 5 turns up on the nth roll of a single die, then the probability that 5 turns up both times on the first two rolls is

$$P(F_1 \cap F_2) = P(F_1) \cdot P(F_2) = \frac{1}{5} \cdot \frac{1}{5} = \frac{1}{25}$$

If the die is rolled four times, the probability that 5 turns up every time is

$$P(F_1 \cap F_2 \cap F_3 \cap F_4) = P(F_1) \cdot P(F_2) \cdot P(F_3) \cdot P(F_4) = \left(\frac{1}{5}\right)^4 = \frac{1}{625}$$

EXAMPLE 4.42

The main assembly line of an automobile factory is fed by two independent auxiliary lines, one for engines and the other for prefabricated body units. Statistical studies have shown that the rate of defective engines is 1 in 400 and the rate of defective body units is 1 in 500. The studies have also shown that the appearance of these defects follows no apparent pattern; that is, they appear at random. If we denote the event that a car from this factory has a defective engine by E and the event of a defective body by B, then

$$P(E) = \frac{1}{400} \qquad \text{and} \qquad P(B) = \frac{1}{500}$$

The two auxiliary lines are independent of each other, so

$$P(E \mid B) = P(E) \qquad \text{and} \qquad P(B \mid E) = P(B)$$

Thus, the probability of a car coming off the main assembly line with both engine and body defective is given by

$$P(E \cap B) = P(E) \cdot P(B) = \frac{1}{400} \cdot \frac{1}{500} = \frac{1}{200,000}$$

To summarize and unify our discussion of the computational rules of probability, let us consider a fairly complex problem derived from one of the classic games of chance.[6] In one of the common forms of

---

6. The rest of this section is optional. The reference to binomial distributions later in this chapter is reinforced by this material, but does not depend on it.

the card game "poker," a *hand* consists of five cards dealt to a player (from a standard deck of 52 cards), with no option for exchanging any of them. The most valuable hand in the game is called a *royal flush*; it consists of the ace, king, queen, jack, and ten, all of one suit. The question we pose is:

What is the probability that a player will be dealt a royal flush exactly three times in the first ten hands of a game?

Notice that we could view the dealing of each hand as a separate experiment, or we could consider the whole situation as a single experiment with ten similar parts. To make the necessary distinction when an experimental procedure is repeated two or more times, we call each repetition a **trial**, and reserve the term "experiment" for the entire collection of trials performed. In this case, then, our experiment consists of ten trials. For each trial there are as many outcomes as there are five-card subsets (hands) of a set (deck) of 52 cards. A single outcome *of the entire experiment* is an ordered set of ten hands.

The next step in solving the problem is to find the probability of getting a royal flush in any one hand. Treating a single trial as an experiment by itself (just for the moment), let us denote the set of all possible hands by H and the event of getting a royal flush by R. If we assume that any card is as likely to be dealt as any other, the definition of probability tells us that

$$P(R) = \frac{n(R)}{n(H)}$$

It is easy to see that $n(R) = 4$ because there is just one royal flush for each suit. Now, $n(H)$ is the number of 5-element subsets of a 52-element set, so

$$n(H) = {}_{52}C_5 = \frac{52!}{(52 - 5)! \cdot 5!} = 2,598,960$$

(See Example 4.31.) Thus, the probability of getting a royal flush (of any suit) in a fair deal is

$$\frac{4}{2,598,960} = \frac{1}{649,740}$$

This says that, on the average, a poker player may expect one royal flush out of every 649,740 hands. (You might like to verify this experimentally if you are fond of long poker sessions.)

The question of getting exactly three royal flushes in ten hands involves ten independent events—getting a royal flush in the first hand, getting a royal flush in the second hand, etc. (The hands are independent because they are dealt to the same person each time, so they must occur on different deals. It is assumed that the cards are shuffled thoroughly before each hand is dealt.) If we denote these events by $R_1, R_2, \ldots, R_{10}$, then

$$P(R_1) = P(R_2) = \ldots = P(R_{10}) = \frac{1}{649,740}$$

Call this number p. Then the probability that any one of these events will *not* happen is $1 - p$, so

$$P(R_1') = P(R_2') = \ldots = P(R_{10}') = 1 - p = \frac{649,739}{649,740}$$

If we wanted the three royal flushes in three particular hands—say the first, fourth, and seventh—then Rule (4.9) tells us that the probability of this happening is

$$P(R_1 \cap R_2' \cap R_3' \cap R_4 \cap R_5' \cap R_6' \cap R_7 \cap R_8' \cap R_9' \cap R_{10}')$$
$$= p \cdot (1 - p) \cdot (1 - p) \cdot p \cdot (1 - p) \cdot (1 - p) \cdot p \cdot (1 - p) \cdot (1 - p) \cdot (1 - p)$$
$$= p^3 \cdot (1 - p)^7$$

However, we do not care which three hands are royal flushes, so *any* 3-element subset of the ten hands will satisfy our needs just as well as any other. There are $_{10}C_3$ different 3-element subsets, each of them representing a desirable event for the experiment as a whole. The probability we seek, then, is the probability of the union of these events.

Now, if any particular pattern of three royal flushes turns up, then none of the others can, so these desirable events are mutually exclusive. Therefore, by Axiom 3 on page 159, the probability of their union is simply the sum of the probabilities of the events. But each of these events has probability $p^3 \cdot (1 - p)^7$; so the probability of getting exactly three royal flushes in ten hands of poker is

$$_{10}C_3 \cdot p^3 \cdot (1 - p)^7 = 120 \cdot \left(\frac{1}{649,740}\right)^3 \cdot \left(\frac{649,739}{649,740}\right)^7$$

which is, approximately

$$\frac{1}{2,000,000,000,000,000}$$

The method of solution just used typifies the process by which in the seventeenth century Jakob Bernoulli was able to prove a general theorem about problems of this type:

✕✕

The Binomial Distribution Theorem:[7] Suppose an experiment consists of n independent trials. If the event E being considered is the same for each trial and if its probability for any single trial is p, then the probability that E will occur exactly r times is given by

$$_nC_r \cdot p^r \cdot (1 - p)^{n-r}$$

✕✕

EXAMPLE 4.43

The probability of obtaining exactly four heads in six tosses of a coin is

$$_6C_4 \cdot \left(\frac{1}{2}\right)^4 \cdot \left(\frac{1}{2}\right)^2 = 15 \cdot \frac{1}{16} \cdot \frac{1}{4} = \frac{15}{64}$$

EXAMPLE 4.44

The probability of getting exactly three 6s in five throws of a single fair die is

$$_5C_3 \cdot \left(\frac{1}{6}\right)^3 \cdot \left(\frac{5}{6}\right)^2 = 10 \cdot \frac{1}{216} \cdot \frac{25}{36} = \frac{125}{3888}$$

EXAMPLE 4.45

If a machine that makes Christmas tree ornaments produces random defects at the rate of 1 in 100, then the probability of finding exactly two defects in a package of six ornaments is obtained by letting $n = 6$, $r = 2$, and $p = \frac{1}{100}$ in the Binomial Distribution Theorem. This yields

$$_6C_2 \cdot \left(\frac{1}{100}\right)^2 \cdot \left(\frac{99}{100}\right)^4 = 15 \cdot \frac{1}{10,000} \cdot \frac{95,059,601}{100,000,000}$$

or approximately .00144. The package contains *at least* two defects if there are exactly two, exactly three, exactly four, exactly five, or exactly six defects in it; these are five mutually exclusive events. Thus, to find the probability that the package contains at least two defects, we just add the probabilities of these five separate cases, in which r takes on the values 2, 3, 4, 5, 6:

---

7. The term **binomial distribution** refers to the patterns of possible results of a re-peated-trials experiment for which the outcome of each trial is either a "success" or a "failure"—an experiment whose results can be thought of as *yes–no* lists. The royal-flush question just discussed is such an experiment because each trial (hand) is either a royal flush or it isn't. For more about binomial distributions, see Section 4.11.

$$_6C_2 \cdot \left(\frac{1}{100}\right)^2 \cdot \left(\frac{99}{100}\right)^4 + {}_6C_3 \cdot \left(\frac{1}{100}\right)^3 \cdot \left(\frac{99}{100}\right)^3 + {}_6C_4 \cdot \left(\frac{1}{100}\right)^4 \cdot \left(\frac{99}{100}\right)^2$$

$$+ {}_6C_5 \cdot \left(\frac{1}{100}\right)^5 \cdot \left(\frac{99}{100}\right)^1 + {}_6C_6 \cdot \left(\frac{1}{100}\right)^6 \cdot \left(\frac{99}{100}\right)^0$$

that is, approximately .00146.

4.7   EXERCISES

In Exercises 1–6, assume that three playing cards are dealt from a thoroughly shuffled standard deck.

1. What is the probability that the first card is a heart?

2. What is the probability that the second card is a heart?

3. If the first card is accidentally turned over and seen to be a heart, what is the probability that the second card is a heart?

4. If the first card is accidentally turned over and seen to be a club, what is the probability that the second card is a heart?

5. If neither of the first two cards is seen, what is the probability that both are hearts?

6. If none of the cards are seen, what is the probability that all three are hearts?

In Exercises 7–18, assume that one hundred slips of paper, numbered from 1 through 100, are placed in a barrel and mixed thoroughly. The even-numbered slips are green and the odd-numbered slips are red. A blindfolded person draws the slips from the barrel.

7. If a single slip is drawn, what is the probability that its number is even?

8. If a single slip is drawn, what is the probability that its number is 25?

9. If a single slip is drawn, what is the probability that its number is a multiple of 3?

10. If a single green slip is drawn, what is the probability that its number is a multiple of 3?

11. If a single slip is drawn, what is the probability that it is green and its number is a multiple of 3?

12. If a single slip is drawn, what is the probability that it is green or its number is a multiple of 3?

13. If a single slip is drawn, what is the probability that its number is a multiple of 4?

14. If a single red slip is drawn, what is the probability that its number is a multiple of 4?

15. If a single slip is drawn, what is the probability that it is red or its number is a multiple of 4?

16. If two slips are drawn in succession without being replaced, what is the probability that both numbers are multiples of 3?

17. If three slips are drawn in succession without being replaced, what is the probability that all three are red?

18. A single slip is drawn and then replaced in the barrel. If this experiment is done seven times, what is the probability that exactly four of the drawings result in red slips?

In Exercises 19–22, assume that the birth rate for boys is approximately 49%. Assume further that each birth is independent of the others and that the sex of a child is a random occurrence.

19. If a couple's first two children are boys, what is the probability that their third child will also be a boy?

20. What is the probability that a couple's first four children will all be girls?

21. If it is known that a couple's first two children are boys, what is the probability that all four of the couple's children are boys?

22. If a couple has six children, what is the probability that at least three are girls?

23. What is the probability of rolling three 7s in three consecutive tries with a pair of fair dice?

24. What is the probability of rolling exactly two 7s in three tries with a pair of fair dice?

For Exercises 25–28, presume that a person whose name is chosen at random from the Los Angeles telephone directory is as likely to have been born on any day of the year as on any other.

25. Suppose that such a name has initials T. D. H. and that this person T. D. H. was born in one of the four months September, April, June, November.

    (a) What do the months September, April, June, and November have in common? (*Hint*: Think of a grade-school rhyme.)

    (b) What is the probability that T. D. H. was born in June?

    (c) What is the probability that T. D. H. was born on the 14th of some month?

    (d) What is the probability that T. D. H. was born on June 14th?

    (e) Do Parts (b) and (c) represent independent events? Why or why not?

    (f) How are Parts (b), (c), and (d) related? Do your answers reflect this relationship?

26. Suppose that such a name has initials J. J., and assume that J. J. was not born on the leap-year day, February 29.

    (a) What is the probability that J. J. was born in January?

    (b) What is the probability that J. J. was born in March?

    (c) What is the probability that J. J. was born in June?

    (d) If J. J. was born in either January or June, what is the probability that J. J. was born on the 1st day of a month?

    (e) If J. J. was born in either January or June, what is the probability that J. J. was born on the 31st day of a month?

    (f) Which of the following terms apply to the events represented by Parts (a) and (b)? (Choose all that apply.)
        i. independent
        ii. mutually exclusive
        iii. equally likely

    (g) Which of the following terms apply to the events represented by Parts (a) and (c)? (Choose all that apply.)
        i. independent
        ii. mutually exclusive
        iii. equally likely

27. Suppose that such a name has initials U. S., and assume that U. S. was not born on the leap-year day, February 29.

    (a) What is the probability that U. S.'s birthday is in July?

    (b) What is the probability that U. S.'s birthday is in February?

    (c) What is the probability that U. S.'s birthday is the Fourth of July?

(d) If you know that U. S.'s birthday is in July, what is the probability that it is the Fourth of July?

(e) How are Parts (a), (c), and (d) related? Do your answers reflect this relationship?

(f) Do Parts (a) and (c) represent independent events? Why or why not?

(g) Do Parts (a) and (b) represent independent events? Why or why not?

8. Suppose that such a name has initials R. R.

(a) What is a "leap year"? How often do leap years occur?

(b) What is the probability that R. R.'s birthday is in February?

(c) What is the probability that R. R.'s birthday is on the 4th of some month?

(d) What is the probability that R. R.'s birthday is on the 4th of February?

(e) What is the probability that R. R.'s birthday is on the 29th of some month?

(f) What is the probability that R. R.'s birthday is on the 29th of February?

(g) Do Parts (b) and (c) represent independent events? Why or why not?

(h) Do Parts (c) and (d) represent independent events? Why or why not?

(i) Do Parts (b) and (f) represent independent events? Why or why not?

(j) Which pairs of the Parts (b), (c), (d), (e), and (f) represent mutually exclusive events?

29. Let $E_1, E_2, \ldots, E_n$ be events in a sample space S, where $n \geq 3$.

(a) Prove:
$$P(E_1 \cap E_2 \cap E_3) = P(E_1) \cdot P(E_2 \mid E_1) \cdot P(E_3 \mid E_1 \cap E_2)$$

(b) Conjecture about a formula for
$$P(E_1 \cap E_2 \cap \ldots \cap E_n)$$

30. (*Refer to Exercise 45 of Section 4.4.*) Each question on a ten-question multiple-choice test in ancient Greek literature has four possibilities. If you answer each question completely at random, what is the probability that you answer at least seven of the ten questions correctly?

WRITING EXERCISES

. (a) State in one (complete) sentence the general probabilistic idea that is illustrated by the dog-walking story at the beginning of this section.

(b) Write a different story that illustrates the same idea.

.. The definition of *independent* in this section appears to define the same word three times. Explain the differences among these three uses of the word.

3. Describe how at least three of the dozen problem-solving tactics of Section 1.2 can be used to help in the solution of Exercise 30. (Even if you can't complete Exercise 30, you should be able to do this writing exercise.)

LINK: Probability and Marketing

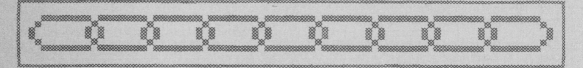

4.8

Abasic problem in marketing a product or service is how to get maximum public response for one's advertising dollar. Large companies have entire departments whose main task is to find innovative ways of solving this problem, and every day we are confronted with the results of their efforts—newspaper ads, television commercials, discount coupons, giveaway games, direct-mail flyers, and the like. Among these modern marketing and advertising techniques there are various forms of a historically popular money-making scheme called a *lottery*.

In its simplest form, a lottery is a game in which people buy tickets, and one of these tickets is chosen at random ("by lot") to determine the winner of a prize. Lotteries have been used to raise money for churches and schools, for local and state governments, and for private profit. In the 1800s, the use of lotteries to make money from the unwary was raised to a virtual art form by showmen, like P. T. Barnum (who once remarked, "There's a sucker born every minute.") Because of the widespread abuses of lottery schemes, private and public lotteries became illegal in this country by 1890. In recent years many states, starting with New Hampshire, have instituted public lotteries to raise revenue.

One of the most common lottery-type marketing techniques is the giveaway game, the awarding of prizes to a few people chosen by lot from among a much larger audience of potential consumers. Insofar as people do not have to pay (directly) for the chance to win, this game is not a lottery in the strict sense, but it operates on the same principle. In the rest of this section we shall examine a fictional, but typical, lottery-type giveaway situation, focusing on the balance between monetary cost and psychological attractiveness.

The Situation

An automobile manufacturer was bringing out a new type of light truck. The marketing department had targeted a preferred mailing list of 100,000 people as good prospective buyers, provided they could be induced to test-drive the trucks. The company allocated $150,000 of their advertising budget for a direct-mail pitch to these people. Of that money, $100,000 was needed to cover the direct costs of preparation, printing, and mailing.

Loteria Nacionale by
John Phillip, 1866.

The Problem

How could the other $50,000 best be used to induce these people to test-drive
the trucks?

After considerable discussion, it was decided that a giveaway scheme should
be set up. Each person would be mailed a certificate to be turned in at a local
dealership where he or she would test-drive a truck and then receive a reward.
However, with only $50,000 available to reward 100,000 people, equal distribu-
tion would not provide much of an incentive because $50,000 ÷ 100,000 is only
50 cents per person. Not many people would spend an hour or so of time and
travel for 50 cents.

On the other hand, the entire $50,000 could be offered as a single prize, to
be awarded by random drawing from all the certificates turned in. The chance of
winning $50,000 should be a much more attractive reward! An objection was

raised: Consumer-protection regulations require the mailing to state clearly the chances of winning, which in this case would be only 1 in 100,000 (assuming every certificate is returned). This low probability might well discourage a lot of people, who would feel that the odds of winning were too poor to bother with.

The scheme finally chosen struck a balance between high-payoff possibilities and attractive odds. Different prize values were set:

- 1st prize: $10,000 —   1 to be awarded = $10,000
- 2nd prize: $ 1,000 —   5 to be awarded =   5,000
- 3rd prize: $ 50 —   200 to be awarded =   10,000
- 4th prize: $ 5 —   5000 to be awarded =   25,000

Total: 5206 prizes $50,000

The advertising copy could then state truthfully:

> Come in for a test-drive. Return this certificate and be eligible to **WIN** one of **MORE THAN 5200 CASH PRIZES**, up to a **GRAND PRIZE OF $10,000!**
>
> Chances of winning better than 1 in 20
> (No prize less than $5)

From a mathematical viewpoint, two aspects of this $50,000 giveaway story are worth a closer look. To gain some precision in our analysis, let us add one more useful, common-sense definition to our store of probability ideas. In situations like this one, where the various outcomes are uncertain and each one results in a monetary payoff if it occurs (even though the payoff sometimes may be $0), the **mathematical expectation of an outcome** is defined to be the payoff for that outcome multiplied by its probability. [Notice that, because the probability is always a number less than or equal to 1, this product *reduces* the amount of money "expected" in accordance with the likelihood that the particular outcome will actually occur.] The **mathematical expectation** of the whole situation (game, lottery, or whatever) is just the sum of the mathematical expectations of all the different outcomes.

Let us apply this idea to the auto manufacturer's ad campaign just described. First of all, although the three proposed ways of distributing the money appear to be quite different, *their mathematical expectations are the same:*

- 50 cents to everyone — The probability of winning is 1 because everyone who turns in a certificate gets the 50 cents. So the mathematical expectation is

$$(50 \text{ cents}) \cdot 1 = 50 \text{ cents}$$

- $50,000 to one winner drawn at random — The probability of winning is $\frac{1}{100,000}$, so the mathematical expectation is

$$\$50,000 \cdot \frac{1}{100,000} = 50 \text{ cents}$$

■ Multiple-prize scheme — The mathematical expectation is

$$\$10,000 \cdot \frac{1}{100,000} + \$1000 \cdot \frac{5}{100,000} + \$50 \cdot \frac{200}{100,000} + \$5 \cdot \frac{5000}{100,000}$$

$$= \$.10 + \$.05 + \$.10 + \$.25$$

$$= 50 \text{ cents}$$

Secondly, the probability of winning a prize in the multiple-prize scheme is, of course, the total number of prizes being awarded divided by the total number of certificates, in this case

$$\frac{5006}{100,000}, \text{ about } 5.2\%$$

Now, $\frac{1}{20} = 5\%$; so the ad copy's assessment of chances is accurate. However, the probability of winning the Grand Prize is still only $\frac{1}{100,000}$, or .0001%, and the probability of winning more than \$50 is only $\frac{6}{100,000}$, or .0006%. Thus, the mathematical expectation is made to appear better than it actually is.

This persuasive technique is now used in almost every lottery or giveaway promotion. For instance, the Connecticut Weekly Lottery listed a Grand Prize of \$100,000 and other prizes of \$2000, \$500, \$50 and \$5; every ticket states: "Overall chance of winning: 1 in 25." A recent fast-food company giveaway listed prizes from thousands of dollars to a bag of french fries, and claimed that the overall chance of winning was 1 in 91. The lottery approach to marketing (and taxation) is not dishonest, but a correct understanding of it requires a grasp of the probabilistic principles upon which it is based. The lack of this understanding increases the profit margins of lotteries and enhances the effectiveness of marketing giveaway schemes. There's nothing wrong with trying to "beat the odds," but you really ought to know what odds you're up against before you try.

4.8   EXERCISES

WRITING EXERCISES

1. Find a lottery or marketing giveaway scheme currently active in your locality and determine its mathematical expectation. Discuss how its advertising statements help prospective participants to understand or hinder them from understanding their chances of winning.

2. Write a giveaway story that is analogous to the truck-advertising scheme described in this section. Use a different setting and different numbers, but illustrate all the same mathematical ideas.

WHAT IS STATISTICS?

4.9

Statistics is a word that is used in a wide variety of senses, and is often invoked in attempts to lend credibility to otherwise doubtful opinions. The word itself has two meanings, one plural and one singular. In the plural sense it denotes collections of facts that can be stated numerically, such as mortality tables or census figures. In the singular, it is the science that deals with such facts, collecting and classifying them in such a way that general predictions may be based on them. This second meaning is the theme of the rest of this chapter.

Probability and statistics are closely related fields because they investigate the same basic situation from two different viewpoints. Both approaches deal with a large collection of things and subsets of that collection. In terms of the vocabulary we have just developed, probability begins with a *sample space*, a set whose elements are known, and evaluates the likelihood that some outcome of an *event*, a subset of the sample space, will occur in an experiment. Statistics begins with a **sample**—a known subset of a larger set that is mostly unknown—and by an analysis of that sample attempts to infer the composition of the larger set, which is called the **population**.

The facts gathered in a sample are called **data**.[8] Your name, height, weight, birthday, and eye color are data. So are your grades from last year, sports scores in the newspaper, listings in the TV program guide, food prices in the local supermarket, expenditure amounts in the federal budget, populations of cities, and so on. Data are often, but not always, numbers. If this were a full course in statistics, we would spend some time discussing the different ways to handle numerical vs. nonnumerical data. However, since the purpose of these few sections is just to give you a brief overview of the subject of statistics, we shall simplify our work by considering only numerical data. As a further simplification, the data sets in some examples will be artificially small, so that the concepts being illustrated may stand out as clearly as possible.

As you can see, data are everywhere—in newspapers and magazines, in almanacs and encyclopedias, as well as on the Internet. This abundance of data leads to two problems:

1. Before collecting data about a question, it is important to understand the question well enough to know which data to collect.

2. Once you have the data, they must be organized in ways that makes them easy to understand and use.

---

8. The noun *data* is plural; its singular form (not used very much) is *datum*, a single item of information. When *data* refers to a whole collection of pieces of information, it sometimes is used with a singular verb.

The first of these two problems can be harder than you might imagine. Really understanding a question usually requires some careful thinking about the situation in which it arises. For instance, do the scores on a multiple-choice "intelligence test" really measure intelligence, or do they primarily reflect people's test-taking skills? Does a pre-election phone poll in a poor, rural county give you a reliable prediction of how the election will turn out, or does it only reflect the likely choices of people affluent enough to own telephones? Would a phone poll be reliable in a wealthy, suburban county? Would it depend on when the calls were made? Some of the problem-solving tactics of Chapter 1 may come into play here; these include checking the definitions, restating the question, and breaking it into simpler questions. There is not much more to say about this part of the data-gathering process, except that *careful thinking takes time.* We (necessarily) leave to you the task of understanding each question as it arises.

Once you have understood a question and have gathered data about it, you need to organize the data you have collected. That is, you must mold your separate bits of information into a single, useful tool for answering your question. (This step is what justifies using *data* with a singular verb as indicated in Footnote 8, page 180.) But how? One approach is to represent the data in a visually informative picture, diagram, or chart. There are several common ways to do this, some of which undoubtedly are familiar to you. Newspapers, magazines, and television often display data using pictures of some sort.

Two of the most common kinds of data-display pictures are *bar graphs* and *pie charts.* These two devices are most useful when the data refer to amounts of a fairly small number of different things. A **bar graph** represents the amount of each thing as the length of a bar, relative to some scale. A **pie chart** (or **circle graph**) represents the amount of each thing as a wedge-shaped portion (a "slice") of a whole disk (the "pie"). Because of their simplicity and visual impact, bar graphs and pie charts are preprogrammed graphics options in many packages of business, accounting, and presentation software. *Figures 4.9 and 4.10* present a simple example of each device—a bar graph and a pie chart showing the approximate popular vote totals for the major candidates in the 1992 presidential election.

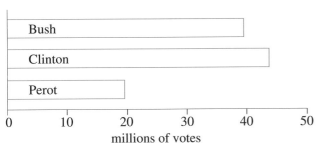

Figure 4.9 A bar graph.

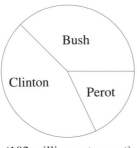

Figure 4.10 A pie chart.   (102 million votes cast)

EXAMPLE 4.46

Bar graphs and pie charts are not always equally clear or effective ways of representing data. The data shown by the bar graph of *Figure 4.11* would not be as clear if it were represented by a pie chart. The segments of all the pieces other than "Military" would be too small to show their size relationships to each other accurately. Might there be data sets for which a pie chart would be better than a bar graph? (See Exercise 6.)

Clear display is only a first step toward making data work for us. If a set of data is to be an effective tool, we have to do more than just see what it *says*. We have to analyze it carefully to understand what it *means*. The two key questions in this regard are:

▪ How are the data bunched together?
▪ How are they spread out?

The first question asks for a way of indicating some sort of "center" for a sample; any number used in this way is called a measure of the **central tendency** (or simply a **center**) of the sample. A number that describes the way a sample is spread about some central value is said to measure the **dispersion**, or the **spread**, of the sample. These two questions focus our ideas better than the wide-open invitation to "analyze the data," but they aren't very precise yet. Nevertheless, they are good enough to help us focus our examination of some typical examples.

For instance, here is a simple example of data—the scores of twelve students on a 10-point vocabulary quiz:

3, 5, 7, 7, 7, 8, 8, 8, 8, 9, 10, 10

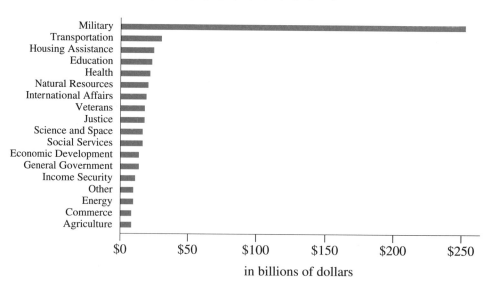

Figure 4.11 *FY1996 federal discretionary spending, as proposed by Congress.*
(Source: Center for Defense Information newsletter, Spring 1995)

One way to picture these data is shown in *Figure 4.12*. In that picture, the stacks of "boxes" represent the repeated scores in the data.

A common way to express the "center" of a set of test scores is to "average" them; that is, to divide the sum of the scores by the number of students taking the test. In this case, the sum of the twelve scores is 90, so the average is 90 ÷ 12, which is 7.5. Statisticians have a simple word for this concept:

DEFINITION

The **mean** of a collection of numerical data is the sum of the data divided by the number of data items.[9]

NOTATION

It is customary to denote the mean of a sample, $x_1, x_2, \ldots, x_n$, by $\bar{x}$. In symbols,

$$\bar{x} = \frac{x_1 + x_2 + \ldots + x_n}{n}$$

One way to picture the mean of a data set is to think of it as a "balance point" for the data. In the case of the twelve quiz scores, for instance, if a 1-ounce weight were placed at each score on a weightless number line, then the line would balance at 7.5. (See *Figure 4.13*.)

Notice that some quiz scores occur more than once. The number of times a value appears in a data set is called the **frequency** of that value. As *Figure 4.12* shows, 8 has a frequency of 4, 7 has a frequency of 3, 10 has a frequency of 2, and each of the other scores has a frequency of 1.

EXAMPLE 4.47

A group of students scored as follows on a short test:

50, 60, 60, 65, 70, 70, 70, 75, 80, 80

The sum of these ten scores is 680, so the mean score is 68. The frequency of 70 is 3; 60 and 80 have frequency 2; 50, 65, and 75 have frequency 1.

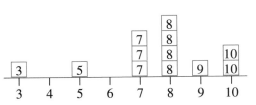

Figure 4.12 Twelve scores on a quiz.

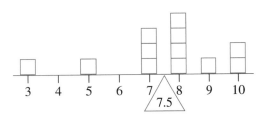

Figure 4.13 The mean as balance point.

9. More formally, this is called the **arithmetic mean** to distinguish it from the *geometric mean*, another kind of "middle" between two numbers that we shall not discuss.

A second group of ten students took the same test and scored

30, 40, 50, 65, 70, 75, 75, 80, 95, 100

The sum of these ten scores is also 680, so the mean score is again 68. The frequency of 75 is 2; all the other scores have frequency 1.

To illustrate two other common ways of displaying data, we again use test scores as a typical data set, just for simplicity and familiarity. Of course, these display methods apply equally well to sports scores, stock market quotations, marketing surveys, or just about any other sets of data. Suppose that a class of twenty students got these scores on a history exam:

83, 94, 76, 85, 97, 88, 75, 86, 91, 51,
79, 86, 90, 81, 73, 79, 85, 86, 99, 93

Computing the mean of these scores (a calculator is recommended), we get

$$1677 \div 20 = 83.85$$

This number gives us some idea of the center of the data, but it doesn't really provide any sense of how the scores are spread out or where they bunch together. To get a better sense of that, we might try making a balance picture, as was done in *Figure 4.13* for the quiz scores. (See *Figure 4.14*.)

As *Figure 4.14* shows, there aren't many repeated scores among these data; most of the values have frequency 1. In cases like this, it may be useful to summarize the data by grouping them in some way. When the data values are numbers—particularly two- or three-digit numbers—a common way to group them is by tens. In such cases, the grouping of the data is easy to see when it is organized in a form called a **stem-and-leaf plot**. *Table 4.1* shows a stem-and-leaf plot for these data. It is a table of numbers separated by a vertical line. Each number on the left side (a "stem") represents a tens group; the numbers to the right of it (the "leaves") are the last digits of the data values in that tens group. There is a leaf for each data item, even if its value repeats an earlier one. For instance, look at the third line of the table:

7 | 6 5 9 3 9   stands for   76, 75, 79, 73, 79

In this way, the frequency of each tens group can be seen by just counting the leaves for its stem.

STEMS	LEAVES
5	1
6	
7	6 5 9 3 9
8	3 5 8 6 6 1 5 6
9	4 7 1 0 9 3

Table 4.1
A stem-and-leaf plot.

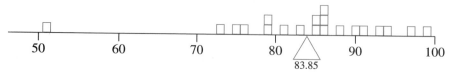

Figure 4.14 A picture of the history exam scores.

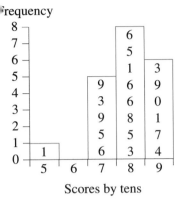

Scores by tens

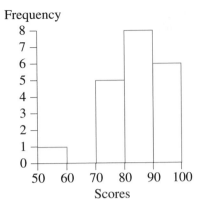

Scores

Figure 4.15 Turning a stem-and-leaf plot into a histogram.

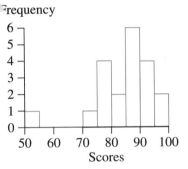

Scores

Figure 4.16 The exam scores grouped by fives.

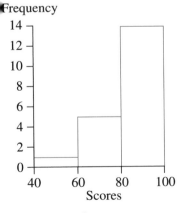

Scores

Figure 4.17 The exam scores grouped by twenties.

A handy feature of the stem-and-leaf plot is that it is easy to turn it into a **histogram**, a vertical bar graph. We really do *turn* it—90° counterclockwise—so that the stems are on the bottom and the leaves form columns going up. The bottom line of the histogram shows the range of possible values of the data; for the stem-and-leaf plot of *Table 4.1*, the values are from 50 to 100. The bottom line is divided into sections of equal length. The vertical bars represent the frequencies of the data in each section. When turned up like this, the stems of the stem-and-leaf plot of *Table 4.1* show the tens intervals on the bottom and the leaves show the heights of the vertical bars. The resulting histogram appears in *Figure 4.15*, first with the stem-and-leaf information displayed, and then without it.

A histogram provides a graphic illustration of how a set of data is spread out and how it is bunched together. Making the size of the groupings (along the bottom line of the histogram) smaller gives the picture more detail; making the grouping size bigger provides less detail. *Figure 4.16* shows a histogram for the history-exam scores grouped by fives, with the 0, 1, 2, 3, 4 leaves of each stem in one (narrower) bar and the 5, 6, 7, 8, 9 leaves in another. *Figure 4.17* shows a histogram for the same set of scores grouped by twenties.

A histogram is such a common and useful way of presenting data that some graphics calculators (the TI-82, for example) have a built-in histogram display feature. After storing your data in a memory list, you can select the maximum and minimum values and the interval size to be displayed. In that way, each histogram display can be "customized" to enhance the effectiveness of the picture on the screen.

4.9   EXERCISES

Exercises 1–5 provide sets of data. For each data set, draw a bar graph and a pie chart (if possible). Try to make your pictures clear and accurate, and make a note of anything that keeps you from being as accurate as you would like. After you finish drawing the pictures, answer these questions for each data set:

(a) Which was easier to make, the bar graph or the pie chart? Why?

(b) Which do you think would be easier for someone else to understand, the bar graph or the pie chart? Why?

(c) If you could only use one of these two depictions of this set of data, which would you choose? Why?

1. In an effort to show how little money is devoted to its work, NASA compiled the data shown in *Table 4.2*, concerning spending amounts from the 1992 federal budget.

CATEGORY	BILLIONS OF DOLLARS
Fixed spending	727
Defense/International	333
Domestic	202
Interest on debt	199
NASA	14

Table 4.2 The 1992 federal budget.

COUNTRY	DOLLAR AMOUNT
Canada	$3.26
United Kingdom	$3.24
Ireland	$3.09
New Zealand	$1.92
Australia	$1.50
United States	$0.46

Table 4.4 Cigarette taxes.

2. The Big Umbrella Insurance Co. reported its latest annual corporate travel expenses, a total of $82 million, as shown in *Table 4.3*.

3. A survey of cigarette taxes per pack in developed countries where English is the predominant language yielded the data in *Table 4.4*. The amounts are expressed in 1992 U. S. dollars. (Data from Public Citizen Health Research Group *Health Letter*, 1992.)

4. *Table 4.5* shows approximate 1993 U. S. purchased beverage consumption, in gallons per person. (Data adapted from a chart in *Nutrition Action*, June 1995.)

5. *Table 4.6* shows the approximate heating value of various types of air-dry firewood, in millions of BTUs per cord. (From D. Havens, *The Woodburners Handbook*. Brunswick, ME: Harpswell Press, 1973.)

ITEM	PERCENTAGE
Automobile	27%
Airfare	20%
Sundry	17%
Lodging	13%
Relocation	11%
Meals	7%
Entertainment	5%

Table 4.3 An annual travel-expense report.

BEVERAGE	GALLONS
Soda (pop)	47
Beer	32.5
Coffee	26
Milk	25
Bottled water	9
Fruit juices	8.5
Tea	7

Table 4.5 1993 U. S. beverage consumption.

6. Find or construct an example of data that you think is better represented by a pie chart than by a bar graph. Explain.

For Exercises 7–12, find the mean of the specified data set. Then answer these questions:

a) What does this mean represent?

b) Is the mean a useful statistic in this case? Justify your opinion.

c) Is the mean a good measure of center for this data set? Why or why not?

7. The spending amounts in *Table 4.2*.

8. The expense figures in *Table 4.3*.

9. The cigarette taxes in *Table 4.4*.

10. The beverage figures in *Table 4.5*.

11. The heating values in *Table 4.6*.

12. The spending amounts illustrated by *Figure 4.11* on page 182. The corresponding numbers, from bottom to top (in billions of dollars) are:

4, 4, 6, 6, 12, 13, 13, 16, 17, 19, 19, 21, 22, 23, 26, 27, 38, 261

13. *Figure 4.13* on page 183 displays the relationship between twelve quiz scores and their mean, 7.5.

(a) Suppose that the quiz grader made an error and everybody lost another point. What would the new mean be? Why?

(b) Suppose, instead, that the grader's error resulted in only six of the twelve students losing a point. What would the new mean be? Why?

14. Make up two data sets that satisfy all these conditions:
- Each data set contains 9 items.
- Both data sets have the same mean.
- Every value in the first set has frequency 1.
- Every value in the second set has frequency 2 or 3.

15. List the individual data items contained in the stem-and-leaf plot of *Table 4.7*. Draw a histogram to represent these data. Then find the mean.

16. An observer of a state lottery compiled this list of payoff numbers between 1 and 40, which was selected during six televised drawings:

4, 7, 8, 21, 34, 39
1, 5, 10, 11, 23, 29
2, 10, 16, 19, 22, 36
4, 7, 10, 23, 26, 35
3, 11, 15, 17, 25, 36
2, 3, 6, 18, 23, 34

(a) Make a stem-and-leaf plot for the data.

(b) Make a histogram for the data.

Species	Millions of BTU
Ash	20.0
Aspen	12.5
Beech, American	21.8
Birch, yellow	21.3
Elm, American	17.2
Hickory, shagbark	24.6
Maple, red	18.6
Maple, sugar	21.3
Oak, red	21.3
Oak, white	22.7
Pine, Eastern white	13.3

Table 4.6 *Firewood heating values.*

Stems	Leaves
2	3 1 5 0
3	7 7 2 1 6
4	
5	3 9
6	0 4 4 2 4 8
7	2 1 2

Table 4.7 *A stem-and-leaf plot for Exercise 15.*

(c) Do you think that the mean would be a useful statistic to have in this case? Why or why not?

17. The histogram in *Figure 4.18* shows the results of a 100-point biology exam.

(a) How many students took the exam?

(b) How many students scored between 70 and 79 (inclusive)?

(c) How many students scored between 70 and 74 (inclusive)?

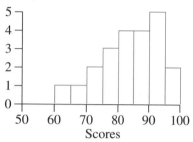

Figure 4.18 The histogram for Exercise 17.

(d) How many students scored between 80 and 89 (inclusive)?

(e) How many students scored between 80 and 84 (inclusive)?

(f) Can you say exactly what the scores were? Why or why not?

(g) Find an approximate mean ("class average"); then explain how you obtained it.

18. To help plan its orders for frozen chicken, a restaurant did a two-week survey of the number of customers ordering chicken dinners. The daily totals were:

47, 36, 55, 43, 24, 46, 38,

47, 52, 50, 38, 41, 45, 32

(a) Estimate the sum and the mean of the data.

(b) Find the mean, rounding it to two decimal places. (A calculator might help here.) Was your estimate pretty close?

(c) Make a stem-and-leaf plot for the data.

WRITING EXERCISES

1. Write a brief comparison of the four data-display methods described in this section: bar graphs, pie charts, stem-and-leaf plots, and histograms. What are the advantages of each? What are the shortcomings of each? Is any one of them *always* better than the others? Give reasons to support your opinions.

2. Under what circumstances do you think the mean might be a misleading measure of the "center" of a data set? Justify your opinion. (Feel free to refer to examples in this section, but don't limit your thinking just to them.)

CENTRAL TENDENCY AND SPREAD

4.10

The mean of a data set is the most common measure of its center, and often the most reliable. However, its use may obscure certain essential characteristics of the central tendency of the data. In Example 4.47, for instance, both groups contain ten students, and the mean scores of both groups is 68. Yet even a cursory glance at the individual scores shows that the data "bunches" very differently in the two cases.

Here's a more striking example of how the mean glosses over differences:

Suppose I offer to sell you a $1 ticket for a special lottery. I say (truthfully), "Only 1000 tickets will be sold and there will be 100 winning tickets. That gives you 1 chance in 10 of winning, and the mean value of the cash prizes is $5.99."

Is that a pretty good deal? Do you have about 1 chance in 10 of winning $5.99? Maybe, maybe not. Here are three possible lottery setups, each with 100 winners and a mean prize value of $5.99:

- *Lottery 1*: All 100 prizes are exactly $5.99.
- *Lottery 2*: First prize $500, second prize $50, and 98 prizes of 50 cents each.
- *Lottery 3*: First prize $100, 2 prizes of $50, 22 prizes of $10, 26 prizes of $5, and 49 prizes of $1.

First of all, let us check that the mean of each of these prize patterns really is $5.99. For *Lottery 1*, that's easy: The sum of 100 prizes of $5.99 each is

$$100 \times \$5.99 = \$599.00$$

Dividing by 100, we get back $5.99. In *Lottery 2*, the sum of the prizes is

$$\$500 + \$50 + (98 \times \$.50) = \$599.00$$

Dividing by 100, we get $5.99 again. In *Lottery 3*, the sum of the prizes is

$$\$100 + (2 \times \$50) + (22 \times \$10) + (26 \times \$5) + (49 \times \$1) = \$599$$

Again, the mean of the 100 prizes is $5.99. Thus, the mean value of the prizes doesn't tell you enough to know how the prizes are distributed. Only in the first lottery does the mean represent the value of the prize a winner is most likely to get. In the second one, the most likely prize is 50 cents; 98 of the 100 prizes are worth only 50 cents. In the third one, the most likely prize is $1, the value of 49 of the 100 prizes.

These lottery examples illustrate a measure of center that is useful when some data values show up fairly often:

DEFINITION   A **mode** of a set of data is a data value with the highest frequency.

We say "*a* mode," rather than "*the* mode," because data may have more than one value with the same highest frequency. If there are two values with the same highest frequency, the data set is called **bimodal**. When we say "*the* mode," we are saying that there is only one mode for the data we are dealing with.

In *Lottery 2*, the mean is not a good measure of center because a few data numbers (two, in this case) are far away from the rest. The mode (50 cents) is better than the mean at predicting what you are most likely to win—*if* you win at all—because 98 prizes out of 100 are worth only 50 cents.

If you use the mode as the "typical" prize in *Lottery 3*, however, it doesn't work as well. The $1 prize is the mode because it has the highest frequency (49), but *51 of the prizes are worth at least $5*. This means that a winner is more likely to win at least $5 than to win only $1. In other words, $5—which is in the middle of the data when they are arranged in size order—is a more typical prize in this case. This illustrates another important measure of central tendency:

DEFINITION   The **median** of a set of numerical data is the middle data value when the data are arranged in size order. (If there is an even number of data items, the median is the mean of the two middle ones.)

In *Lottery 3*, the median is the mean (the average) of the 50th and 51st prizes, arranged in size order. Since both of them are $5, the median is $5. *Table 4.8* shows the three different measures of central tendency for each of the three lotteries.

The mean is the most commonly used measure of central tendency for a numerical data set. However, a few data items that are far away from the rest can shift the mean enough to make it misleading at times. The median (the middle number), however, is not much affected by a few data items far to one extreme or the other, making it a more informative indicator of central tendency in such cases. Example 4.48 describes a simple instance in which the median is a better measure of center than either the mean or the mode.

Lottery	Mean	Median	Mode
1	$5.99	$5.99	$5.99
2	5.99	.50	.50
3	5.99	5.00	1.00

Table 4.8 Centers for the three lotteries.

EXAMPLE 4.48   You are the mayor of a village with eight small businesses and a shoe factory. The Census Bureau asks you how many people are employed by a typical company in your village. You do a quick phone survey and get these numbers:

1, 1, 2, 3, 4, 5, 6, 7, 79

What number do you tell the Census Bureau? Would the mean, 12, tell them what they wanted to know? Would the mode, 1, tell them what they wanted to know? It makes better sense to choose the middle number, 4, to represent the number of employees of a typical company in your village.

Note   Sometimes people use the word "average" to refer to a "typical" value or characteristic, which may be the mean, the median, or the mode. If you really mean *mean*, say so, in order to avoid confusion.

The second question that we asked about analyzing data was

How are the data spread out?

The mean, median, and mode of a collection of data may not help us much in answering this question. For instance, these three data sets,

Data set 1: 1, 1, 4, 6, 9, 13, 15
Data set 2: 2, 5, 6, 6, 7, 7, 16
Data set 3: 2, 2, 2, 6, 10, 11, 16

have exactly the same mean, 7, and median, 6, but they are spread out very differently, as you can see from *Figure 4.19*.

What we want is a way of summarizing the spread of a data set by a single number, just as we have used numbers to represent the center. That is, we want to find some number that measures how the data are spread out around some central point. The simplest measure of the spread of data is the difference between the largest and smallest values; this is called the **range** of the data.

EXAMPLE 4.49   You are going on a vacation trip to a place where the average daily temperature is 75° (Fahrenheit). What clothes would you pack? Would it make a difference if you knew that this average came from the fact that it hits a low of 35° at night and a high of 115° in the daytime? Would you pack differently if you knew that this average came from a low of 65° and a high of 85°?

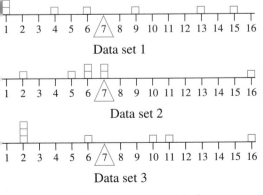

Data set 1

Data set 2

Data set 3

Figure 4.19 Three data sets with the same mean.

Example 4.49 describes a situation in which the range is a useful, simple measure of spread. However, its very simplicity makes it an inadequate descriptor for many sets of data. For instance, the three data sets of *Figure 4.19* all have the same range (what is it?), but they are spread out very differently. Thus, the range is not a good enough measure of spread for most data sets. We have to do better.

Another natural way to analyze the spread is to look at the difference between each data value and the mean. Recall that $\bar{x}$ (*read "x bar"*) represents the mean of a data set. If we let x represent a data value, each such difference has the form

$$x - \bar{x}$$

How about using the average of these differences to describe the spread of the data set? Let us try this for data sets (1) and (2) of *Figure 4.19*, both of which have mean 7. *Table 4.9* shows these differences and their sums.

As *Table 4.9* illustrates, the sum of all the differences $x - \bar{x}$ is *always* zero. (In fact, if you think about the mean as the "balance point" for the data, isn't this what you would expect?) Not a very promising result!

But looking at those differences is not a bad idea. Observing how data values are spread out means considering the distances between data points, and that's a lot like looking at the distances between individual data points and some central point. If we think of each difference as the *distance* between the data item and the mean, and if we recall that a distance is positive or zero, then we get something worthwhile. *Table 4.10* shows these computations for the three data sets of *Figure 4.19*. (Remember that the nonnegative value of an expression is called its *absolute value*.)

DATA SET 1		DATA SET 2	
x	$x - \bar{x}$	x	$x - \bar{x}$
1	−6	2	−5
1	−6	5	−2
4	−3	6	−1
6	−1	6	−1
9	2	7	0
13	6	7	0
15	8	16	9
Sum:	0	Sum:	0

Table 4.9 Differences from the mean, $\bar{x}$, which is 7 for both data sets.

DEFINITION

The mean of all the distances between the individual data items and the mean of these data is called the **average deviation** of the data set. (Its more formal name is the **mean absolute deviation** of the data set.)

You can see from *Table 4.10* that the average deviation is a good enough measure of spread to distinguish between data set 2 and the other two, but not between data sets 1 and 3. Thus, it is a fairly good measure of spread, but it doesn't always tell you what you might need

DATA SET 1		DATA SET 2		DATA SET 3	
x	$\lvert x - \bar{x} \rvert$	x	$\lvert x - \bar{x} \rvert$	x	$\lvert x - \bar{x} \rvert$
1	6	2	5	2	5
1	6	5	2	2	5
4	3	6	1	2	5
6	1	6	1	6	1
9	2	7	0	10	3
13	6	7	0	11	4
15	8	16	9	16	9
Sum: 49	32	Sum: 49	18	Sum: 49	32
Mean: 7	4.57	Mean: 7	2.57	Mean: 7	4.57

Table 4.10 Distances from the mean, $\bar{x}$, which is 7 in each case.

to know. Moreover, its use of absolute value makes the average deviation unwieldy for algebraic manipulation because of the sign changes that are required sometimes, but not always. There is another measure of spread that has all the advantages of average deviation without the drawbacks. Although its definition seems a little complicated at first, a second look at what we have just done should show you why the pieces are put together as they are.

Let us look at *Figure 4.19* more carefully. It appears that the main difference between *Data sets 1* and *3* is that some of the "weights" are a little farther away from the balance point (the mean) in *set 3* than in *set 1*. We need some kind of measurement that emphasizes such differences in distance from the mean. That's not a very precise statement, but it helps to focus the discussion.

One process that emphasizes differences in quantities is squaring. Moreover, the square of a number is always nonnegative. In this case, the mean of the *squares* of the distances between the data numbers and their mean distinguishes between *Data sets 1* and *3* (as well as between each of these sets and *Data set 2*). *Table 4.11* shows these calculations for all three data sets.

For practical purposes, the mean of the squares of the individual differences is an important measure of spread called the *variance* of the data set. We say "for practical purposes" because a small adjustment needs to be made to make this idea conform to the formal definition: Instead of dividing the sum of squares by the number of data items, the sum is divided by one less than the number of items. This adjustment makes the theoretical analysis work better (because it corrects a consistent underestimation relative to the population as a whole), but it makes very little difference in the numerical result when the data sets are even moderately large. The formal definition is as follows.

Table 4.11 Squared distances from the mean, $\bar{x}$, which is 7 for all three data sets.

DATA SET 1		DATA SET 2		DATA SET 3	
x	$(x - \bar{x})^2$	x	$(x - \bar{x})^2$	x	$(x - \bar{x})^2$
1	36	2	25	2	25
1	36	5	4	2	25
4	36	6	1	2	25
6	1	6	1	6	1
9	4	7	0	10	9
13	36	7	0	11	16
15	64	16	81	16	81
Sum: 49	213	Sum: 49	112	Sum: 49	182
Mean: 7	30.43	Mean: 7	16	Mean: 7	26

DEFINITION   The **variance** of a data set $x_1, x_2, \ldots, x_n$ is

$$\frac{(x_1 - \bar{x})^2 + (x_2 - \bar{x})^2 + \ldots + (x_n - \bar{x})^2}{n - 1}$$

The square root of the variance is one of the most important measures in statistics, the *standard deviation*, which will be discussed in the next section. Formally:

DEFINITION   The **standard deviation** of a data set $x_1, x_2, \ldots, x_n$ is

$$\sqrt{\frac{(x_1 - \bar{x})^2 + (x_2 - \bar{x})^2 + \ldots + (x_n - \bar{x})^2}{n - 1}}$$

NOTATION   The standard deviation of a data set is commonly denoted by s or by σ. In particular, many calculators use σ for the standard deviation.

EXAMPLE 4.50

Example 4.47 (in Section 4.9) calculated the mean test scores of two groups of ten students; in both cases the mean was 68. Here are the scores of the two groups:

Group 1: 50, 60, 60, 65, 70, 70, 70, 75, 80, 80

Group 2: 30, 40, 50, 65, 70, 75, 75, 80, 95, 100

The variance of the first set of scores is

$$\frac{(-18)^2 + 2 \cdot (-8)^2 + (-3)^2 + 3 \cdot 2^2 + 7^2 + 2 \cdot 12^2}{9} = \frac{810}{9} = 90$$

The standard deviation is $\sqrt{90} \approx 9.5$ for this first set.[10] For the second set of scores a similar computation yields a variance of 506.67 and a standard deviation of approximately 22.5. (Check it for yourself.) The difference between these two standard deviations provides a clear measure of the difference in spread between these two sets of scores.

Although the standard deviation is—as its name implies—a standard way of describing the spread of a set of data, its computational complexity makes it a bit awkward for some everyday applications of statistics. Often a much more "rough-and-ready" approach to the spread of data, called a *five-number summary*, will suffice. A **five-number summary** shows the high, low, and median numbers of the data and it also shows two other numbers, called "quartiles." Just as the median divides the data in half, the **quartiles** divide the data into

---

10. The symbol $\approx$ means "approximately equal to" in this context.

quarters. Here is a step-by-step way to get a five-number summary from a set of data:

1. Arrange the data in size order, from smallest to largest.
2. Find the median. This is the **second quartile**.
3. Look at the set of numbers *before* the median and find its median. This is the **first quartile**.
4. Look at the set of numbers *after* the median and find its median. This is the **third quartile**.
5. Write down the five numbers in this order:

   smallest, first quartile, median, third quartile, largest

EXAMPLE 4.51

Here are two simple cases to illustrate that the idea underlying quartiles and the five-number summary is just the common-sense notion of dividing a data set (approximately) into quarters.

The first, second, and third quartiles of the following 15-element data set appear in boldface:

1, 2, 3, **4**, 5, 6, 7, **8**, 9, 10, 11, **12**, 13, 14, 15

The quartiles of the following 8-element data set are each halfway between two data values. They appear here in parentheses:

1, 2, (2.5) 3, 4, (4.5) 5, 6, (6.5) 7, 8

EXAMPLE 4.52

Coach Cliffhanger's high school basketball team is getting ready for a state championship game. The other team, the Aces, comes from a small town far across the state and not much is known about them. However, statistics recently published in the newspaper gave this five-number summary of the Aces' regular season scores:

33, 49, 56, 66, 78

This says that the Aces' highest game score during the season was 78 points; their lowest scoring effort was a dismal 33 points. Moreover, the scores for half of their games are between 49 (the first quartile) and 66 (the third quartile), with no more than half of those above 56.

The graphical counterpart of the five-number summary is the **box-plot**, sometimes called a "box-and-whisker diagram." It is a five-number summary in picture form. A (narrow) box, cut in two by a vertical line at the median, shows where the middle 50% of the data lie. Horizontal lines on each side of the box (the "whiskers") show the range of the lowest and highest quartiles.

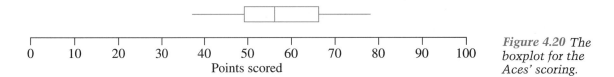

Figure 4.20 The boxplot for the Aces' scoring.

A single example, drawn from the Aces' scoring summary in Example 4.52, should suffice to explain how this process works. Their five-number summary is 33, 49, 56, 66, 78, so the box, drawn above a numerical scale, stretches from 49 to 66 with a vertical dividing line at 56. This "middle half" of the data gives you a good idea of how well the Aces usually score, so this part of the range is more important, in some sense. The two whiskers, representing the lowest and highest quarters of the data, stretch from 33 to 49 and from 66 to 78, respectively. The boxplot is shown in *Figure 4.20*.

EXAMPLE **4.53**

Look back at the three data sets of *Figure 4.19* on page 191. They are almost too small to bother with, but they provide some easy practice on making boxplots. *Data set 1* is

$$1, \ 1, \ 4, \ 6, \ 9, \ 13, \ 15$$

Its median is 6. The first quartile—the median of 1, 1, 4—is 1. The third quartile—the median of 9, 13, 15—is 13. Thus, the five-number summary for data set 1 is

$$1, \ 1, \ 6, \ 13, \ 15$$

Following this example, find the five-number summaries for *Data sets 2* and *3*. Then compare your results with the boxplots for all three sets, which are shown in *Figure 4.21*. In particular, notice how the boxplots illustrate the differences in the spreads of these three sets.

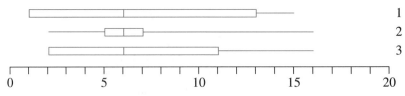

Figure 4.21 Boxplots for data sets 1, 2, and 3.

EXAMPLE **4.54**

A statewide educational research study examined the IQ scores of 500 children. The boxplot in *Figure 4.22* illustrates these scores. It corresponds to the five-number summary

$$53, \ 90, \ 102, \ 111, \ 142$$

Figure 4.22 The statewide IQ scores.

From this information, you can conclude that the scores for 250 children (50% of the sample) are between 90 and 111. (The word "between" is used loosely here to give you a quick, intuitive idea of what's going on. Scores right at the dividing points should be interpreted in accordance with the formal definitions of the quartile ranges.) Since 102 is the median, 125 scores are between 90 and 102, and 125 scores are between 102 and 111. The two "whiskers" of the boxplot indicate that 125 children have scores between 53 and 90, and the remaining 125 have scores between 111 and 142. The boxplot does *not* indicate how the scores are distributed within each of these four ranges.

Some popular graphics calculators, including the TI-82, provide five-number summaries and boxplots as part of their calculations and displays for any data set you enter. (To use these features to their best advantage, you may have to adjust the range settings for the graphics window.) If you ever use such a calculator to handle a set of data, the foregoing descriptions should allow you to take full advantage of the numerical and visual information provided by these automatic features.

4.10   EXERCISES

You are encouraged to use a calculator for these exercises. They may also be done without a calculator, but you'll need a little patience.

Exercises 1–8 provide examples for comparing the various measures of center and spread. For each set of data,

a) calculate the mean, the median, and the mode(s);

b) write a five-number summary and draw a boxplot;

c) calculate the average deviation, the variance, and the standard deviation.

Round all answers to one decimal place.

1. *Data*: 3, 1, 5, 10, 8, 1, 5, 2, 1

2. *Data*: 13, 10, 2, −5 , 7, 10, −2

3. *Data*: 3.9, 8.05, 5.2, .11, 10.6, 2.0

4. A class of twelve students scored as follows on an exam:

$$75, 60, 60, 72, 70, 86$$
$$93, 54, 72, 50, 72, 85$$

5. The daily totals of shoppers at a market during a certain week are:

Mon. 700   Tues. 450   Wed. 630
Thurs. 520   Fri. 810   Sat. 1050

6. A college hockey team achieved these scores during a fifteen-game season:

2, 0, 3, 3, 1, 5, 4, 7, 0, 2, 3, 12, 6, 0, 1

7. One family's monthly long-distance phone bills for a year were

$15.16, $32.22, $17.30, $20.21
$60.26, $ 5.99, $ 3.43, $ 1.96
$12.34, $30.51, $47.52, $82.27

8. While considering improvement to a stretch of road, the Highway Commission decided to determine the amount of traffic that used the road. It made a daily count of the number of cars using the road during a 30-day sampling period. The data from their survey, rounded to the nearest hundred cars, appear in *Table 4.12.*

Number of Days	Number of Cars per Day
4	1100
6	1600
3	1900
2	2000
6	2300
6	2500
3	2900

*Table 4.12
30-day traffic
survey.*

9. A class of 15 students took a calculus exam and got these scores:

78, 82, 95, 86, 73
99, 95, 89, 90, 85
23, 92, 80, 84, 85

(a) Find the mean and the median of these scores.

(b) Construct a five-number summary and a boxplot for these scores.

(c) The professor found out that the student who got the 23 was sick during the exam, so she dropped this score. How much did this change the class mean? How much did it change the class median? Draw a new boxplot.

10. This puzzle demonstrates how changing one or two data values affects the mean and the median.

(a) Make up a data set of twelve numbers between 0 and 10 (inclusive) with these properties:

■ No number appears more than three times.

■ The mean of the data is 7.

■ The median of the data is also 7.

(b) By changing no more than two numbers in your data set, make it come out with the same mean, 7, but with median 8. (You may have to redo Part (a) to make this work.)

(c) By changing no more than two numbers in your original data set, make it come out with the same median (7), but with mean 6. (You may have to redo Parts (a) and/or (b) to make this work.)

11. A census-taker surveys a (very peculiar) neighborhood and finds only two income levels: 20 families make $100,000 a year and 19 families make $10,000 a year.

(a) What are the mean, median, and mode of this sample?

(b) Rechecking his data, the census-taker finds that he made a tallying error. The true figures show that 19 families make $100,000 a year and 20 families make $10,000. Compare the mean, median, and mode of this sample with the figures you found in Part (a). Comment.

12. *Figure 4.23* shows two boxplots of runs scored in each game of the season for two Texas baseball teams, the Armadillos (A) and the Broncos (B).

(a) Write a five-number summary for each one.

(b) Which team scored at least 6 runs in half of their games?

(c) Which team scored at least 9 runs in one-quarter of their games?

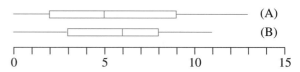

Figure 4.23 The Armadillos and the Broncos.

3. Here is a five-number summary of a set of 40 test scores, all of which are whole numbers.

$$53, 65.5, 78.5, 88.5, 97$$

(a) Make a boxplot for this set of data.

(b) In which quarter of the data set is there a score of 82? Mark this point (approximately) on your boxplot.

(c) How many scores can you guarantee to be less than 82? Do you think there may be more than that? Why or why not?

(d) How many data items lie between 65 and 79 (not inclusive)?

(e) How many data items are less than 89?

4. (a) Compute the mean and all the individual differences, $x - \bar{x}$, between the individual items of this data set and its mean:

$$1, 3, 7, 4, 5, 4, 9, 1, 2$$

(b) Verify that the sum of all the differences in Part (a) is 0.

(c) Prove: For *any* data set, the sum of the differences $x - \bar{x}$ for all x in the set must be 0.

5. Two boatloads of tourists went sport fishing. There were 11 tourists in each boat. In the first boat, every tourist caught 3 fish. In the second boat, five tourists caught 6 fish each, one caught 3 fish, and five caught none. We can think of these numbers of fish caught as two 11-number data sets, one for each boat.

(a) What is the mean and the median of each data set?

(b) Construct a boxplot for each data set. Do these boxplots represent the data sets well? Why or why not?

(c) Calculate the range and the standard deviation for each set.

(d) In your opinion, which of these measures of central tendency and spread best distinguishes between these two data sets. Explain.

16. Construct two seven-element data sets with a mean of 5 and a range of 10, but such that the standard deviation of the first set is at least 1.5 more than that of the second.

17. (This exercise illustrates that in some instances, the mean, median, and mode of a data set are not enough to give you a very good picture of the data, and even boxplots may not be very informative.) Three government data collectors took surveys of net household incomes in a housing development. Each collector recorded 25 incomes. Their results, rounded to the nearest thousand dollars, are as follows:

Collector 1: 5 households had net incomes of $10,000; 15 were at the $20,000 level; the remaining 5 made $30,000.

Collector 2: All 25 households had net incomes of $20,000.

Collector 3: 23 households had $20,000 incomes, one made only $1000, and one made $39,000.

(a) Find the mean, median, and mode of each of these three data sets.

(b) Construct five-number summaries and boxplots for each data set.

(c) Calculate the standard deviation of each data set.

(d) Do your results support the parenthetical claim at the beginning of this exercise? Why or why not?

18. (A calculator or computer spreadsheet is strongly recommended for this exercise.) You are a rich executive looking for a warm place to spend Thanksgiving Day. Your travel agent tells you of a remote island where the mean low temperature on Thanksgiving for the last ten years has been 70°, with an average deviation of less than 4°. Before buying your airplane ticket, you decide to use some mathematics to see how low the temperature is likely to get during your trip.

(a) Verify that this is one possible set of low Thanksgiving temperature readings for the last ten years:

$$73°, \ 71°, \ 68°, \ 66°, \ 71°$$
$$70°, \ 75°, \ 72°, \ 66°, \ 68°$$

(b) Show that it is possible for the island to have had a low Thanksgiving temperature of 55° sometime during the last ten years. (*Hint*: Experiment by changing some of the data from Part **(a)**.) Justify your answer by making up a set of ten temperatures that includes 55°, that has a mean of 70°, and that has an average deviation of less than 4°.

(c) Judging from this temperature data, would you take the trip? Why or why not?

WRITING EXERCISES

1. Discuss the relative advantages and disadvantages of using the mean or the median as the primary measure of the central tendency of a sample. Describe a situation in which the mean would be better than the median; then describe a situation in which the median would be better than the mean. Explain what "better" means in each case.

2. **(a)** Write the definition of *variance* of a sample completely in words. Do not use any symbols, not even letters to represent numbers.

(b) Comment in general on the role of notation in mathematics. In particular:

   i. Is mathematical notation a useful "shorthand" for expressing quantitative ideas?

   ii. Is mathematical notation *more than* just a shorthand? Does it play any other role in the subject? If so, what?

   iii. What are the drawbacks to using notation?

DISTRIBUTIONS

4.11

As one passes from the purely descriptive part of statistics to the uncertainties of making predictions from a sample, the mathematical waters suddenly deepen. The background required to discuss predictive statistics rigorously is extensive, and any pretense of completeness in a treatment as brief as this section and the next would be far less than honest. We therefore limit our exploration to a general outline of some fundamentals of **statistical inference**, methods by which the characteristics of a sample are used to obtain a description of the population it represents.

   The fundamental idea is simple: If the size of the population makes it impossible to examine all of its members, then we select a sample (a subset of those members) and, by analyzing the sample, try to infer something about the entire population. For instance:

- A newspaper that wants to know how all the registered voters in the country are going to vote in an upcoming election will poll a relatively small sample of voters and use the results to predict how the election will turn out.

- A manufacturer of shotgun shells who wants to check that the shells will fire properly can only afford to test relatively few of them.

- A meteorologist who wants to know the average snowfall in a region during a storm cannot measure the total amount of snow that falls and divide by the area it covers. Instead, samples taken at various spots around the region are used to infer the average for the region as a whole.

It would be ideal if we could say that the characteristics of any sample exactly reflect the characteristics of the population from which it was drawn. Unfortunately, this is hardly true. We know from our study of probability that, even in the case of a known population (sample space), the samples drawn from it (the results of an experiment) are not always the same. To get truly reliable information about entire populations, we must interpret the results of samples very carefully in relation to the limitations of the sampling process. Our study of this process begins with *distributions*.

Informally, the idea is this: Suppose we are interested in some numerical characteristic of a large population—the gas mileage of each automobile, the rainfall at each place in the world on a particular day, the height of every living person, the number of copies of the *New York Times* sold at each newsstand in New York City yesterday, etc. Such a numerical characteristic is called a **variable** (because it can vary from member to member in the population). The number associated with a particular member of the population is called the **value** of the variable for that member.[11] Sometimes the values of a variable can be found explicitly; for example, we can measure someone's height. Sometimes, however, the values cannot be determined. For instance, the amount of rainfall in your hometown a year from today, which will be a specific number *then*, cannot be stated with certainty *now*. A variable whose numerical values cannot be determined with certainty is called a **random variable**.[12]

Whether or not a variable is random, its values over a large population are hard to comprehend if they are just viewed as a big collection of numbers. Thus, it is natural to seek some pattern that describes in

11. In other mathematical words, a variable is a *function* from the population to some set of numbers, and its value for a particular element of the population is the *image* of that element.
12. See the beginning of Section 4.3 for a discussion of randomness.

summary form how the numerical values are spread out; such a pattern is called the **distribution** of the variable. A distribution pattern may be represented visually by a graph or algebraically by a formula. The most important kinds of distributions can be represented both ways, as we shall see. Although an algebraic formula is essential for drawing numerical conclusions about the underlying population, a graphical picture is often easier to understand.

EXAMPLE **4.55**	Ten slips of paper, numbered 1 through 10, are placed in a hat, and then a number is drawn at random. The histograms in *Figures 4.24 and 4.25* show the frequencies of each number drawn in 100 repetitions and in 1000 repetitions of this experiment, respectively.[13]

In Example 4.55, the uncertain outcome of the experiment makes the variable (the number to be drawn) random, at least before the drawings occur. Each repetition of the drawing yields a value of the random variable. Of course, after all the drawings have occurred, each of the 100 or 1000 repetitions has its assigned number, so these outcomes are no longer random. Hence, these results can be represented by graphs.

However, if we think about these 100 or 1000 repetitions as being *samples* of the larger population of all possible repetitions of this experiment, then we are still dealing with a random variable. Theoretically, this variable assigns each possible drawing a whole number from 1 to 10, but we cannot know with certainty each one of all these possible outcomes. If asked to predict (or bet on) the outcome of the next

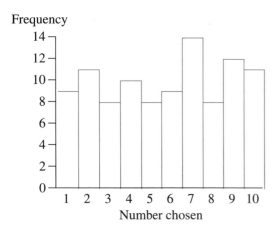

Figure 4.24 100 repetitions.

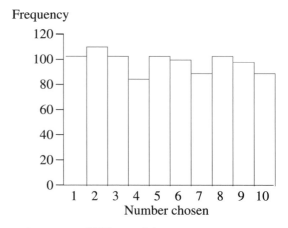

Figure 4.25 1000 repetitions.

13. These experiments were simulated by using a random-number-generating computer program.

drawing, what would your common sense say? Think about it for a moment, then see if the next few sentences come close to "reading your mind."

> Since there are ten numbers in the hat, there is only 1 chance in 10 that any particular number will be drawn. Hence, no matter what I say, I have a 90% chance of being wrong. That's a pretty risky bet!

In other words, our common-sense understanding of the "law of averages" would tell us that the expectation of the outcome of this experiment is closely tied to the probability of choosing each number, which (assuming all the slips of paper are the same size, are thoroughly mixed, etc.) is $\frac{1}{10}$. "In the long run," you might say, "each number should turn up about $\frac{1}{10}$ of the time." If you were to guess "7" over and over again for many drawings, for instance, you would expect to be right about 10% of the time.

The mathematical counterpart of this common-sense idea is Bernoulli's Law of Large Numbers. Roughly speaking, this law says that the more times an experiment is repeated, the closer the ratio of successes to total trials will approximate the probability of success on a single trial. Here, essentially, is how the law applies to statistical sampling:

×××

The Law of Large Numbers: The larger the sample taken, the closer the distribution of the sample will approximate the distribution of the population.

×××

(Of course, the statement of the Law of Large Numbers may be made much more precise than this, but the mathematics required would take us far afield.[14]) In the case of drawing the slips of paper from the hat, this law tells us that larger and larger numbers of repetitions of this experiment should be distributed more and more uniformly, with each of the ten values tending toward a frequency that is $\frac{1}{10}$ of the total number of trials. *Figures 4.24–4.26* on pages 202 and 204 illustrate this idea. (A frequency value expressed as a fraction or percentage of the total number of trials is called a **relative frequency**.)

The validity of the Law of Large Numbers depends to a great extent on the condition that the sample be chosen randomly. In other

14. A readable discussion of the Law of Large Numbers appears in Chapter XI of Item 9 in the list *For Further Reading* at the end of this chapter.

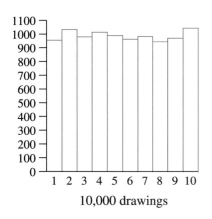

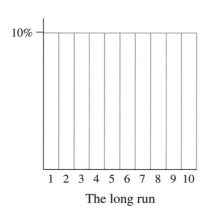

Figure 4.26 Repeated drawings of numbered slips.

10,000 drawings

The long run

words, the elements must be selected without regard to any predetermined pattern, in such a way that each element of the population has an equal chance of being chosen. This type of sample is called a **random sample**; these are the samples governed by the laws of probability and statistics. *From now on, any sample discussed is assumed to be random.*

Generalizing from a sample involves two questions:

> What statements can be made about the characteristics of the population as a whole?

and

> How reliable are these statements?

The key to answering both of these questions may be found by taking another look at distributions. In Example 4.55, we described a distribution in terms of the frequencies of the numerical values assigned to the underlying population. Such distributions are called, naturally enough, **frequency distributions**. However, when we extend the number-drawing experiment of Example 4.55 to a consideration of *all possible* outcomes, the distribution can no longer be a frequency distribution because the actual numbers of outcomes of each kind cannot be determined. Instead, using the Law of Large Numbers, we arrive at a uniform distribution based on the *probability* of each possible outcome, as illustrated by the second graph in *Figure 4.26*. Such a distribution is called (you guessed it!) a **probability distribution**.

EXAMPLE 4.56

Here is a probability distribution that is not uniform (and hence is more interesting). Suppose we toss a fair coin five times and record the total number of *heads*. There are 32 equally likely possible outcomes of this experiment (why?), and they may be classified conveniently in terms of six mutually exclusive events:

> 0 *heads*, 1 *head*, 2 *heads*, 3 *heads*, 4 *heads*, 5 *heads*

By counting the outcomes in the various events, we can compute their respective probabilities (as in Section 4.3):

$$\frac{1}{32}, \quad \frac{5}{32}, \quad \frac{10}{32}, \quad \frac{10}{32}, \quad \frac{5}{32}, \quad \frac{1}{32}$$

These six events and their probabilities constitute a probability distribution for the set of all possible five-toss outcomes (the underlying population in this case). The "$n = 5$" graph in *Figure 4.27* illustrates this distribution.

   In general, if the experiment consists of tossing a coin n times, the outcomes can be stated in terms of obtaining exactly r heads, where r can be any integer from 0 to n, inclusive. The corresponding probabilities will be

$$_nC_r \cdot \left(\frac{1}{2}\right)^n$$

for each r from 0 to n, inclusive. (The factor $\left(\frac{1}{2}\right)^n$ is constant for an experiment of n trials; it accounts for the total number of elements in the sample space.) The $n + 1$ mutually exclusive events and their corresponding probabilities form a probability distribution for the set of all possible n-toss outcomes, which is the underlying population. *Figure 4.27* also contains the graphs for the $n = 3$ and $n = 10$ distributions.

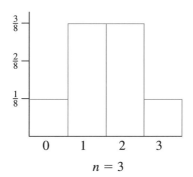

$n = 3$

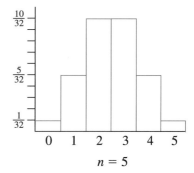

$n = 5$

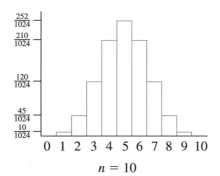

$n = 10$

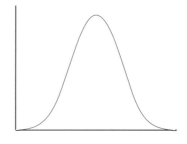

Normal curve

Figure 4.27 Some binomial distributions and a related normal curve.

Example 4.56 typifies a **binomial distribution**, a probability distribution with an underlying experiment that can be analyzed in terms of two alternatives—*success* or *failure, yes* or *no, heads* or *tails,* etc. Binomial distributions are determined by the numbers

$$_nC_r \cdot p^r \cdot (1 - p)^{n-r}$$

which depend on the numbers $_nC_r$ for a particular probability p and some fixed number n of independent trials.[15] If either p or n is altered, the binomial distribution will also change. Distributions of this kind are useful because many statistical problems can be structured in terms of two-alternative questions, and also because they have a relatively simple computational form. In Example 4.56, the fact that $p = 1 - p = \frac{1}{2}$ makes the distributions for each n especially simple to calculate.

The final graph in *Figure 4.27* is a curve that approximates the binomial distribution for $p = \frac{1}{2}$ and large values of n. This curve is known as a **normal curve**; it represents a common and very important type of distribution for infinite populations. Although normal curves may be taller or shorter, wider or narrower than the one pictured here, each one is "bell-shaped" and is symmetric about the vertical line through its highest point, curving downward toward the bottom axis more and more steeply at first, then leveling out as it approaches the axis on either side. The shape of the curve represents the distribution of the population, in much the same way as the shape of a bar graph or histogram does. For large populations, you might visualize lots of thin relative-frequency rectangles under the curve, generalizing the pictures in *Figure 4.27*. This description can be made much more precise if we relate each normal curve to the numerical properties of the distribution it represents.

The two keys to describing a **normal distribution**—a distribution described by a normal curve—are the mean and the standard deviation of the values of the random variable on the population being studied. We saw how to compute these two numbers for relatively small sets of numbers in Section 4.10. The ideas are exactly the same here, except that the populations may be very large, so it usually is not practical actually to compute these numbers.[16] Nevertheless, the numbers exist in theory, at least, and their properties are fundamental to most of statistics. It is traditional to denote this mean and this standard deviation by the lower-case Greek letters μ ("mu") and σ ("sigma"), respectively.

15. Compare this with the statement of the Binomial Distribution Theorem, near the end of Section 4.7.

16. Section 4.12 discusses how the mean and standard deviation of a large population may be determined by looking at samples.

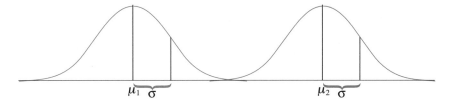

Figure 4.28 Two normal curves with different means, but with the same standard deviation.

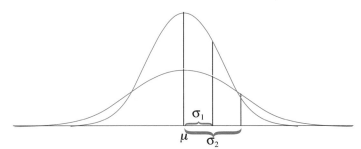

Figure 4.29 Two normal curves with the same mean, but with different standard deviations.

(These are the Greek counterparts of our letters *m* and *s*, respectively.) To distinguish them from the mean and standard deviation of a particular sample, μ and σ are usually called the **population mean** and **population standard deviation**, respectively.

An important fact about a normal curve is that it is *completely determined by the numbers μ and σ*. Pictorially, the highest point of the curve occurs at μ and the curvature changes from becoming steeper to becoming shallower (that is, from *concave downward* to *concave upward*) at the points that are exactly σ distance units away from μ on either side. *Figure 4.28* shows two normal curves with different means, μ_1 and μ_2, but with the same standard deviation, σ; *Figure 4.29* shows two normal curves with the same mean, μ, but with different standard deviations, σ_1 and σ_2.

Regardless of its mean μ and standard deviation σ, a normal curve has the following remarkable and useful properties:

- ▪ Although the curve extends infinitely far in both directions, the entire area between the curve and the horizontal axis is considered to be exactly 1, which equals 100%.[17]

- ▪ About 68% of that area lies between the vertical lines at $\mu - \sigma$ and $\mu + \sigma$.

- ▪ About 95% of that area lies between the lines at $\mu - 2\sigma$ and $\mu + 2\sigma$.

- ▪ About 99.7% of that area lies between the lines at $\mu - 3\sigma$ and $\mu + 3\sigma$.

17. Rigorous explanations of how to compute the area under a normal curve and how the area under an infinite curve can be finite require the methods of integral calculus, which are beyond the scope of this book. For our purposes, it is enough to know that such computations can be done and that, beyond some finite distance from μ in either horizontal direction, the area under the normal curve is negligible.

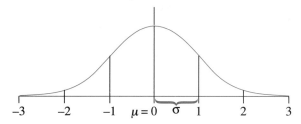

Figure 4.30 The standard normal curve.

Thus, it is often convenient to "standardize" the graphing of normal distributions so that the mean lies at 0 on the horizontal axis and the horizontal unit of measure is σ; that is, $\mu = 0$ and $\sigma = 1$. *Figure 4.30* is a picture of the standard normal curve.

These four area properties of the standard normal curve can be translated into statements about the data items in a normally distributed population:

- The area under the curve represents 100% of the population.
- About 68% $\left(\text{slightly more than } \frac{2}{3}\right)$ of the items are within 1 standard deviation of the mean.
- About 95% of the items are within 2 standard deviations of the mean.
- About 99.7% of the items are within 3 standard deviations of the mean.

Thus, for any normally distributed set of data, the standard deviation provides a precise, efficient way of describing how the data "clusters" around the mean. In fact, the term *"standard* deviation" is suggested by these facts, in that *all normally distributed populations can be standardized (and hence compared) by using their standard deviations as the common unit of measure.* This property makes standard deviation a powerful tool for predicting the reliability of statistical results.

Note   If you are curious about the precise formula that defines a normal curve, here it is:

$$f(x) = \frac{1}{\sigma\sqrt{2\pi}} \cdot e^{\frac{-(x-\mu)^2}{2\sigma^2}}$$

where e, which is approximately 2.718, is the base for natural logarithms. Standardizing to $\mu = 0$ and $\sigma = 1$ gives us the somewhat less formidable expression

$$f(x) = \frac{1}{\sqrt{2\pi}} \cdot e^{\frac{-x^2}{2}}$$

but it, too, takes us beyond the scope of this book. Some calculus is required to derive the properties described above from these formulas. We shall not use either formula in the rest of this chapter; they are included just to satisfy your curiosity.

"So what?" we hear you say. "Why all this fuss about one particular kind of distribution?"

Consider this: The concept underlying all of probability and statistics is *randomness*; that is, the behavior of random phenomena is the cental theme of this area of mathematics. The fact is that

if the outcome of *any* random phenomenon is averaged over many repetitions, its distribution tends to be normal.

This powerful statement is a consequence of the Central Limit Theorem, which we shall discuss at the beginning of the next section. For now, suffice it to say that the normal distribution truly reflects the "normal" behavior of randomness.

4.11   EXERCISES

For Exercises 1–8, make a histogram to represent the given data.

1. A single die was thrown 100 times. The frequency that each number appeared is:

    1: 14 times;   2: 20 times;   3: 16 times
    4: 18 times;   5: 13 times;   6: 19 times

2. A single die was thrown 500 times. The frequency that each number appeared is:

    1: 78 times;   2: 90 times;   3: 82 times
    4: 80 times;   5: 89 times;   6: 81 times

3. A single die was thrown 1000 times. The frequency that each number appeared is:

    1: 159 times;   2: 165 times;   3: 173 times
    4: 164 times;   5: 161 times;   6: 178 times

4. A pair of dice was thrown 100 times. The frequency of the sums of numbers that appeared is:

    2: 4 times;   3: 6 times;   4: 11 times
    5: 8 times;   6: 15 times;   7: 13 times

    8: 9 times;   9: 14 times;   10: 12 times
    11: 8 times;   12: 1 time

5. A pair of dice was thrown 500 times. The frequency of the sums of numbers that appeared is:

    2: 17 times;   3: 27 times;   4: 45 times
    5: 62 times;   6: 65 times;   7: 84 times
    8: 62 times;   9: 51 times;   10: 42 times
    11: 32 times;   12: 13 times

6. A pair of dice was thrown 1000 times. The frequency of the sums of numbers that appeared is:

    2: 31 times;   3: 49 times;   4: 84 times
    5: 114 times; 6: 141 times;   7: 166 times
    8: 138 times; 9: 114 times;   10: 87 times
    11: 53 times; 12: 23 times

7. The grades of 30 students on a 10-point quiz are:

    8, 6, 9, 10, 0, 6, 7, 9, 8, 4
    7, 10, 8, 6, 2, 9, 9, 8, 5, 7
    8, 8, 2, 9, 6, 10, 4, 6, 8, 10

8. The grades of 30 students on a 100-point exam are:

   79, 81, 94, 76, 75, 87, 57, 55, 93, 62
   87, 86, 86, 91, 83, 78, 53, 91, 84, 98
   65, 82, 39, 67, 87, 81, 64, 80, 74, 66

   (*Hint:* You might want to group the grades in some way so that the visual representation is informative.)

9. Relate the data sets in Exercises 1–3 to the Law of Large Numbers.

10. Relate the data sets in Exercises 4–6 to the Law of Large Numbers. (*Hint:* What is the probability of the occurrence of each number? Why? See Example 4.33 in Section 4.6.)

11. Consider the experiment of tossing a fair coin four times, and suppose that we are interested in the number of *heads* that turn up.

    (a) Compute the probabilities of getting each possible number of heads.

    (b) Draw a histogram to represent the probabilities you computed in Part (a).

    (c) Take a coin and actually do this experiment 10 times; then draw a histogram to represent your results.

    (d) (For people with patience.) Repeat this experiment 20 more times, and draw a histogram to represent the results of these twenty trials.

    (e) Combining your data for Parts (c) and (d), draw a histogram representing the results of the thirty trials.

12. A computerized advertising display contains a row of 8 lights which are programmed to switch on and off as follows: At 2-second intervals, the computer assigns 0 or 1 to each light at random with a probability of $\frac{1}{2}$ for either choice. The lights assigned "1" go on (or stay on) and the lights assigned "0" go off (or stay off).

    (a) What is the probability that exactly 4 lights will be on at some particular time?

    (b) What is the probability that exactly 6 lights will be on at some particular time?

    (c) What is the probability that exactly 0 lights will be on at some particular time?

    (d) Describe the binomial distribution that represents the probabilities that for each possible integer n from 0 to 8, exactly n lights will be on at any given time.

    (e) Represent the distribution of Part (d) by a histogram.

    Suppose that the computer is reprogrammed so as to be twice as likely to assign a 1 as to assign a 0 to to each of the lights.

    (f) What is the probability that exactly 4 lights will be on at some particular time?

    (g) What is the probability that exactly 6 lights will be on at some particular time?

    (h) What is the probability that exactly 0 lights will be on at some particular time?

    (i) Describe the binomial distribution that represents the probabilities that for each possible integer n from 0 to 8, exactly n lights will be on at any given time.

    (j) Represent the distribution of Part (i) by a histogram.

13. Open your local phone book to a page in the residential listings (usually, the white pages), and choose two full columns of phone numbers on that page.

    (a) Make a histogram showing the distribution of the *last* digits of the phone numbers in the two columns. (This histogram should have ten bars.)

    (b) Make a histogram showing the distribution of the *first* digits of the phone numbers in the two columns.

    (c) Do the last digits appear to be randomly distributed? Do the first digits appear to be randomly distributed? Relate your answers to your histograms.

    (d) If you were to make histograms for the first-digit and last-digit distributions of *all* the numbers in your phone book, what do you think their shapes would look like? Why?

14. (For students who have access to a computer with BASIC and can use it.) The BASIC program given at the end of this exercise will choose a random sample of five single-digit numbers, compute the mean of the sample, and print it. (*Note*: Each time the program is run, your computer will ask:

Random number seed (–32768 to 32767)?

Just choose a number in the given range and type it in.[18] Each time you run the program, choose a different number.)

(a) Run the program 30 times, and represent your results by a histogram.

(*Note*: If you are comfortable with programming in BASIC, feel free to modify the program so that it automatically repeats the process 30 times and records all 30 means.)

(b) Run the program 30 more times, and represent your results by a histogram.

(c) Combine your results from Parts (a) and (b), and represent them by a single histogram.

(d) If you think of these samples as being drawn from a very large population of single-digit numbers, with equally many of each digit, what do you think the population mean would be? Why? How well does your conjecture fit with your results in Parts (a)–(c)?

(e) Modify the program to handle samples of ten single-digit numbers; then repeat Parts (a) through (d).

```
10 RANDOMIZE
20 LET A = 0
30 FOR I = 1 TO 5
40 LET B = INT(RND*10)
50 A = A + B
60 NEXT I
70 PRINT A/5
80 END
```

WRITING EXERCISES

1. (*Note*: Situations used as examples in this section are "off limits" as responses.)

(a) Describe a situation in everyday life that is an example of a normal distribution of a random variable. Explain.

(b) Describe a situation in everyday life that is an example of a random-variable distribution which is *not* normal. Explain.

2. Is it appropriate to grade a class by using a normal distribution curve? What characteristics of a class might make this grading procedure more or less meaningful? Discuss.

3. (For students who are comfortable with programming.)

(a) Write a computer program that will simulate tossing a coin 5 times and record the total number of heads.

(b) Modify your program from Part (a) so that it will simulate tossing a coin *n* times,

where *n* is a number chosen and entered by the user, and will record the total number of heads.

(c) Write instructions for your program of Part (b) so that it can be used to generate data by someone who is a beginner in using computers.

4. (For students who are familar with graphics calculators or spreadsheets.) Many graphics calculators (such as the TI-82) and electronic spreadsheets (such as Microsoft Excel and Lotus 1-2-3) have built-in random-number functions. Rewrite the instructions to Exercise 14 so that it can be done using one of these calculators or spreadsheets. Check the clarity of your instructions by seeing if a classmate or friend can use them to produce a random sample of five single-digit numbers and calculate its mean.

18. Some advanced forms of BASIC (such as True BASIC) will randomize without your help.

GENERALIZATION AND PREDICTION

4.12

Because the idea of distribution applies both to sample spaces and to populations (known and unknown collections), it forms a natural link between probability and statistics. "Wait a minute," you say. "If we don't know the elements of a population, how do we know anything about its distribution?" That is a perceptive natural question at this point (we're glad you thought of it!) and its answer actually determines how statistics is applied in the real world. The key to the answer is **sampling**—selecting small subsets of the population at random and using the characteristics of these samples to infer the makeup of the population as a whole. Two major results, mentioned in Section 4.11, provide the theoretical foundation for sampling: The Law of Large Numbers and the Central Limit Theorem. It is time to take a closer look at how they work.

Roughly speaking, the Law of Large Numbers (as applied to sampling) says that, the larger the sample taken, the more likely it is that the characteristics of the sample will approximate the characteristics of the population from which it was drawn. For instance, suppose a big-city newspaper wants to predict the outcome of an upcoming election between two mayoral candidates, Abbott and Baker (whom we shall sometimes call A and B, for short) by polling registered voters in its area. Common sense and the Law of Large Numbers tell us that asking 10 voters how they will vote is better than asking only one, and asking 100 voters is better than asking 10. In fact, the best sample would be the *entire* set of voters; but clearly, the constraints of time and money make this option impossible. There are, say, 500,000 eligible voters! For the editors of the newspaper, then, knowing that "bigger is better" is not sufficient; they also need to know "How large is large enough?" That is, when will the results of the poll they take approximate the actual voting preferences of the entire population within some reasonable margin of error, and how large is that margin of error? Answering these questions depends on the Central Limit Theorem.

To understand the main idea of the Central Limit Theorem, we must distinguish between two populations—the population P we are sampling and the population S of all possible random samples of a particular size.[19] Here is some notation to help keep the necessary distinctions straight:

- We shall use x as the random variable representing the property of P that interests us. In our election example, for instance, we

---

19. Since there is a population S for each possible sample size n, it would be more precise to denote these populations by S_n. However, this notational refinement is not essential to understanding the main idea.

can let x take on the value 1 when the voter chooses Abbott, say, and 0 when the voter chooses Baker. (To keep the example simple—at the expense of some realism—we shall assume that every eligible voter votes, and that every vote cast is either for A or for B.)

■ Recall from Section 4.11 that the mean and standard deviation of x in the population P are denoted by μ and σ, respectively. In our election example, μ is the proportion of the votes to be cast for Candidate A. This number is what the newspaper wants to determine; it specifies both the winner of the election and the relative margin of victory.

■ In Section 4.9, we denoted the mean of a sample by $\bar{x}$. In the language of Section 4.11, $\bar{x}$ is a random variable that takes on values from the population S. That is, each possible random sample (member) of the population S has a mean $\bar{x}$, a numerical value that depends on the actual random choice of the elements of the sample. (The means of different random samples would, presumably, be different.)

The Central Limit Theorem relates the distribution of the random variable $\bar{x}$ over S to the distribution of the random variable x over P. Loosely stated, it says that, for "large enough" sample sizes, $\bar{x}$ is approximately normally distributed around the mean μ of the original population P—regardless of how P itself is distributed—and the standard deviation of $\bar{x}$ can be approximated in terms of the standard deviation σ of P. More precisely:

xxxxxxxxxxxxxxxxxxxxxxxxxxxxxxxxxxxxxxxxxxxxxxxxxxxx

The Central Limit Theorem:[20] Let P be a population with mean μ and standard deviation σ, and let S be the population of all n-element samples of P. Then the distribution of the random variable $\bar{x}$ of the means of these n-element samples is approximated by a normal distribution that has mean μ and standard deviation $\frac{\sigma}{\sqrt{n}}$.

xxxxxxxxxxxxxxxxxxxxxxxxxxxxxxxxxxxxxxxxxxxxxxxxxxxx

Thus, the random variable $\bar{x}$ for the population S and the random variable x for the population P have the same mean—μ—and the standard deviation of $\bar{x}$, which we shall call $\bar{\sigma}$, is found by dividing the standard deviation of x by $\sqrt{n}$:

20. The first general version of this theorem was proved in 1810 by Pierre Simon Laplace, an outstanding mathematician and astronomer who was also a prominent political figure in France during the Napoleonic Era.

$$\bar{\sigma} = \frac{\sigma}{\sqrt{n}}$$

Notice that the effect of the denominator $\sqrt{n}$ is to make $\bar{\sigma}$ smaller as the sample size n gets bigger. This provides a numerical measure of our intuition that, the larger the sample, the less its mean, $\bar{x}$, should deviate from the the true population mean, μ.

The Central Limit Theorem holds for "reasonably large" samples in a "large enough" population. The detail involved in making these quantifying phrases exact is unnecessary for our purposes; the statisticians' working "rule of thumb" will serve us well enough. They consider the theorem to be reliable in situations where the sample size is at least 30 and the population is at least 20 times as large as the sample size. In such a situation, the Central Limit Theorem tells us:

1. The mean μ of the population is in some sense the "most likely" value for the mean $\bar{x}$ of any sample.

2. The means of the samples are symmetrically distributed about μ; any particular value of $\bar{x}$ is as likely to be smaller than μ as it is to be greater than μ.

3. About 68% of the sample means will fall between $\mu - \frac{\sigma}{\sqrt{n}}$ and $\mu + \frac{\sigma}{\sqrt{n}}$; about 95% will fall between $\mu - 2\frac{\sigma}{\sqrt{n}}$ and $\mu + 2\frac{\sigma}{\sqrt{n}}$; about 99.7% will fall between $\mu - 3\frac{\sigma}{\sqrt{n}}$ and $\mu + 3\frac{\sigma}{\sqrt{n}}$.

Let us apply these ideas to our election example. The newspaper plans to ask a randomly chosen sample of voters whether they will vote for Abbott or for Baker. Of course, it would be foolish for the editors to predict that the proportions of all the votes cast in the election will be *exactly* the same as the proportion they find in their sample. For instance, if they were to question 30 voters and find that 18 (= 60%) choose Abbott and 12 (= 40%) choose Baker, it would be very unwise to predict that Abbott will get *exactly* 300,000 votes (60% of the total). However, the principles of the preceding paragraph can be used to provide some idea of a likely range of results, along with a numerical estimate of the probability that the election results will fall within the predicted range.

A numerical mean $\bar{x}$ for any voter-choice sample can be found by assigning 1 to each of the a choices for Abbott and 0 to each of the b choices for Baker; that is,

$$\bar{x} = \frac{a \cdot 1 + b \cdot 0}{a + b}$$

This is the ratio of a, the number choosing Abbott, to $a + b$, the total number of voters surveyed.[21] Now, Statement 3 implies that the mean of the voting population—the number the editors want—has

21. Compare this with the definition of *mean* in Section 4.9.

(4.10)

▪ about a 68% chance of being within $\dfrac{\sigma}{\sqrt{n}}$ of $\bar{x}$;

▪ about a 95% chance of being within $2\dfrac{\sigma}{\sqrt{n}}$ of $\bar{x}$;

▪ about a 99.7% chance of being within $3\dfrac{\sigma}{\sqrt{n}}$ of $\bar{x}$.

Thus, if they knew σ, the standard deviation of the voting population, then they could predict a likely range for the outcome of the election and have some measure of the confidence to place in their prediction.

Now, recall from Section 4.10 that the standard deviation is the square root of the variance v, which, in turn, is the "average" of the squares of the differences between the individual values of x and the mean, μ. In this case, there is an easy way to approximate the variance.[22] Since each value of x is either 1 or 0, the only two differences between them and the mean are $1 - \mu$ and μ, respectively. Now, μ also represents the proportion of the 500,000 voters who will choose Abbott, and consequently, $1 - \mu$ represents the proportion of voters who will choose Baker; that is, $500,000\mu$ votes will be cast for Abbott and $500,000(1 - \mu)$ votes will be cast for Baker. Thus, the variance is

$$v = \frac{(1 - \mu)^2 \cdot 500,000\mu + \mu^2 \cdot 500,000(1 - \mu)}{500,000}$$

Cancelling the factor 500,000 from both the numerator and the denominator, we get

$$\begin{aligned}
v &= (1 - \mu)^2 \cdot \mu + \mu^2 \cdot (1 - \mu) \\
&= \mu(1 - \mu)(1 - \mu + \mu) \\
&= \mu(1 - \mu)
\end{aligned}$$

The standard deviation, σ, is the square root of this value:

(4.11)

$$\sigma = \sqrt{\mu(1 - \mu)}$$

But this means that the editors need to know μ in order to find σ, so they seem to be back where they started!

Fortunately, they don't need to know σ *exactly* in order to estimate how good their prediction is; an approximation will do. Moreover, the form of Equation (4.11) is particularly nice in this situation because it provides a pretty close approximation for σ even when the estimate of μ is relatively crude. For instance, suppose their survey of 30 voters came out as described before, with a sample mean of .6. Then, using .6 as an estimate for μ, they would get

$$\sigma \approx \sqrt{.6(1 - .6)} = \sqrt{(.6)(.6)} = \sqrt{.24} \approx .49$$

---

22. This is where the simplifying assumption that there are only two possible choices pays off. We have a binomial distribution, so the computations are much simpler.

Now, suppose that this sample mean is pretty far off from the actual value of μ; let's say the actual value is .5. In that case, using $\mu = .5$, we would obtain

$$\sigma \approx \sqrt{.5(1 - .5)} = \sqrt{(.5)(.5)} = \sqrt{.25} \approx .50$$

a difference of only .01. Thus, if they use the sample mean as an approximation for μ in Equation (4.11), the resulting value for σ will be close enough to be used for determining just how good an approximation it really is.

To see what the value $\sigma = .49$ would tell them about the reliability of this poll, refer to the three parts of Statement (4.10). They say that

- about a 68% chance of being within $\frac{\sigma}{\sqrt{n}}$ of μ;

- about a 95% chance of being within $2\frac{\sigma}{\sqrt{n}}$ of μ;

- about a 99.7% chance of being within $3\frac{\sigma}{\sqrt{n}}$ of μ.

Now, because the distance between two points is the same, regardless of the point from which you measure, an interval of a particular size centered at μ will contain $\bar{x}$ if and only if an interval of the same size centered at $\bar{x}$ contains μ. Thus, we can use the sample mean, .6, as the midpoint of each interval and, recalling that the sample size n in this case is 30, we can compute these value ranges:

$$\frac{\sigma}{\sqrt{n}} \approx \frac{.49}{\sqrt{30}} \approx .09$$

Therefore,

- $\approx$ 68% confidence: μ is between .51 and .69;
- $\approx$ 95% confidence: μ is between .42 and .78;
- $\approx$ 99.7% confidence: μ is between .33 and .87.

This means that newspaper editors have a problem. Remember that the winner of this election will get more than 50% of the votes cast, so μ must be larger than .5 in order for their prediction that Abbott will win to be correct. But only in the first of these three cases is the entire interval of possible values for μ greater than .5, and the confidence for such a prediction is only about 68%. A prediction that has about 1 chance in 3 of being wrong isn't very reliable. A prediction that has a 95% chance of being correct would be good enough for the editors, but the interval of possible values for μ is too big (because it includes values below .5). The only way to shrink the size of the interval is to gather a bigger sample.

How big is big enough? Well, we want to be sure that $\bar{x} - 2\frac{\sigma}{\sqrt{n}}$, the lower end of the 95% confidence interval, is greater than .5; so a precise answer to that question depends on knowing in advance what the mean of this new sample will turn out to be. In this case, if the new sample mean is not very far from .6, then a sample of 100 or so will do, because

$$.6 - 2 \times \frac{.49}{\sqrt{100}} \approx .6 - 2 \times \frac{.49}{10} = .502$$

Here is a summary of the ideas typified by this polling example.

- The general problem it represents is that of trying to estimate the mean of a large population by examining a relatively small random sample.

- As we saw, it is impossible to determine the exact population mean with certainty. The best we can hope for is a range of values along with a numerical measure of the "confidence" we can have in our prediction that the actual mean lies within that range. The range of values is called a **confidence interval**; the measure of confidence is called a **confidence coefficient**.

- Common confidence coefficients are (approximately) 95% and 99.7%. These choices are natural consequences of the characteristics of normal distributions and the fact that the distribution of sample means tends to be approximately normal for "large enough" samples. (This is the main point of the Central Limit Theorem).

- The size of a confidence interval depends on the desired confidence coefficient. For the two most common coefficients, the corresponding confidence intervals are $(\bar{x} - 2\bar{\sigma}, \bar{x} + 2\bar{\sigma})$ and $(\bar{x} - 3\bar{\sigma}, \bar{x} + 3\bar{\sigma})$.

- Since $\bar{\sigma} = \frac{\sigma}{\sqrt{n}}$, the standard deviation $\bar{\sigma}$ gets smaller as the sample size n gets larger. That is, the distribution of the sample means tends to be more tightly grouped around the population mean. Therefore, for a fixed confidence coefficient, the size of the confidence interval can be made smaller by making the size of the sample larger.

- Using this description of the standard deviation $\bar{\sigma}$ of the sample means in terms of the standard deviation σ of the population P depends on being able to approximate σ in some way.

  - If P is *binomially distributed*—that is, if there are only two possible outcomes, say c and d, for the random variable x as it ranges over P—then a working approximation of σ is easy to describe. Assigning 1 and 0 to the outcomes c and d, respectively, we observe that μ and $1 - \mu$ are the respective proportions of c and d outcomes for the entire population and hence that $\sigma = \sqrt{\mu(1 - \mu)}$. (See the discussion leading up to Equation

(4.11) in the election example.) As we have seen, this formulation shows that changes in the value of μ result in much smaller changes in σ; so we can approximate σ fairly well by using the mean, $\bar{x}$, of a reasonably large sample in place of μ. That is,

$$\sigma \approx \sqrt{\bar{x}(1-\bar{x})}$$

❑ If P is not binomially distributed, then σ must be determined in some other way. Further discussion of finding σ for other distributions would quickly take us well beyond the intended scope of this book.[23] In some of the examples and exercises that follow, we shall simply presume that a given value of σ has been found or approximated appropriately.

EXAMPLE 4.57

Problem: A large supply of old bean seeds is found in the back storeroom of an agricultural supply store. Before selling them (at a reduced price), the store manager wants to be able to tell his customers what percent of the seeds are likely to germinate. He chooses 200 seeds at random and plants them in his greenhouse. After two weeks (a reasonable maximum germination time), 167 of the seeds have sprouted. What is the likely germination rate of the supply of bean seeds?

Solution: The information sought is a confidence interval for the likely proportion of viable seeds (that is, seeds that will germinate). If we assign 1 and 0 to viable and nonviable seeds, respectively, then this proportion is the mean μ of a binomial distribution with $n = 1$ from which a 200-element sample was drawn and tested. (Note that, as stated in the Central Limit Theorem, the mean proportion of viable seeds for all 200-element samples is also μ.) The first step in the process is to decide on an acceptable confidence coefficient; let us assume that the manager decides that a 95% confidence coefficient is good enough. Then, using $\bar{x}$—the proportion of sample seeds that germinated—as an estimate for μ, the endpoints of the interval $(\bar{x} - 2\bar{\sigma}, \bar{x} + 2\bar{\sigma})$ are computed as follows:

$$\bar{x} = \frac{167}{200} = .835$$

$$\sigma \approx \sqrt{\bar{x}(1-\bar{x})} = \sqrt{(.835)(.165)} \approx .371$$

$$\bar{\sigma} = \frac{\sigma}{\sqrt{n}} \approx \frac{.371}{14} = .0265$$

23. Detailed examinations of various types of distributions may be found in most standard introductory texts on statistics, such as Items 1 and 8 in the list *For Further Reading* at the end of this chapter.

$$(\bar{x} - 2\bar{\sigma}, \bar{x} + 2\bar{\sigma}) \approx (.835 - 2(.0265), .835 + 2(.0265))$$
$$= (.782, .888)$$

Of course, this interval either contains μ or it doesn't. However, the fact that this is a 95% confidence interval means, roughly speaking, that the likelihood of getting such a sample when the actual germination rate of the seeds falls *outside* the interval (.782, .888) is only about 5%. But a normal distribution is symmetric about its mean; so any outcome outside this interval would be equally likely to be above it as below it. Thus, there is only about a 2.5% chance that such a sample would occur if the actual germination rate is *below* .782. This means that the manager can predict with at least 97.5% confidence that the germination rate for these bean seeds is at least 78%.

EXAMPLE 4.58

A breakfast-food company wants to know if the mean net weight of the boxes of corn flakes filled by its machinery is within .1 oz. of the stated net weight of 12 oz.—that is, between 11.9 oz. and 12.1 oz. It carefully weighs a random sample of the contents of 10 boxes and finds a mean net weight $\bar{x} = 11.97$ oz. and a standard deviation $s = .11$ oz. Wanting to be quite sure, the company decides to use 99.7% as the confidence coefficient. Using s as an estimate of the population standard deviation σ, it computes the appropriate confidence interval:

$$(\bar{x} - 3\bar{\sigma}, \bar{x} + 3\bar{\sigma}) \approx \left(11.97 - 3\frac{.11}{\sqrt{10}}, 11.97 + 3\frac{.11}{\sqrt{10}}\right)$$
$$\approx (11.87, 12.07)$$

Since the lower end of this confidence interval is more than .1 oz. from the target weight of 12 oz., the company cannot declare with 99.7% confidence that its machinery is working properly. Rather than stopping production to make adjustments, however, it decides to check a larger sample. Weighing a random sample of the contents of 50 boxes, it gets a mean net weight of 11.96 oz. and a standard deviation of .13 oz. This is initially worrisome, since the new sample mean is farther off target than the old sample mean and the new standard deviation suggests a slightly broader dispersion than the old one. However, computing the 99.7% confidence interval, the company observes that

$$(\bar{x} - 3\bar{\sigma}, \bar{x} + 3\bar{\sigma}) \approx \left(11.96 - 3\frac{.13}{\sqrt{50}}, 11.96 + 3\frac{.13}{\sqrt{50}}\right)$$
$$\approx (11.905, 12.015)$$

Since this entire interval lies between 11.9 and 12.1, the company can conclude with 99.7% confidence that the machinery is functioning properly.

Example 4.58 shows how increasing the size of the sample shrinks the size of the confidence interval. Since the confidence interval determines the "margin of error" (loosely speaking), increasing the size of the sample increases the reliability of predictions based on that sample.

This example further illustrates the idea of **quality control**, an application of statistical theory to industrial production that uses regular sampling to check acceptable product characteristics. Other examples of applied statistics abound. The life insurance industry depends heavily on statistical predictions, as do advertisers and broadcasting companies. Since most sciences are experimental by nature and proceed from a limited number of observations to the enunciation of general laws, statistics plays a major role in deriving these laws. For instance, the normal curve is an essential tool in genetics, and the application of the normal distribution to the movement of particles explains the physical phenomenon of Brownian motion. The social sciences and education depend on statistics for the proper formulation and interpretation of surveys and tests. The list goes on and on.

The peculiar strength of statistics is its ability to deal quantitatively with uncertainty, thereby bridging the gap between theoretical results and practical applications. But this strength is also its weakness. Many applications of statistical theory depend upon the *assumption* that a normal distribution is right for a situation as well as upon the choice of an appropriate confidence coefficient and several other choices. These assumptions and choices, if not made properly, render the statistical conclusions based upon them invalid and thus useless or misleading. The LINK section that ends this chapter describes a particular application of statistics. Whether this application represents a strength or a weakness is left for you to decide.

4.12   EXERCISES

For Exercises 1–12, you are given a sample mean $\bar{x}$, a standard deviation $\bar{\sigma}$, and a confidence coefficient. Specify the confidence interval determined by these values.

1. $\bar{x} = 0$, $\bar{\sigma} = 1$; 68% confidence

2. $\bar{x} = 0$, $\bar{\sigma} = 1$; 95% confidence

3. $\bar{x} = 0$, $\bar{\sigma} = 1$; 99.7% confidence

4. $\bar{x} = 100$, $\bar{\sigma} = 10$; 68% confidence

5. $\bar{x} = 100$, $\bar{\sigma} = 10$; 95% confidence

6. $\bar{x} = 100$, $\bar{\sigma} = 10$; 99.7% confidence

7. $\bar{x} = 8.1$, $\bar{\sigma} = 1.25$; 68% confidence

8. $\bar{x} = 8.1$, $\bar{\sigma} = 1.25$; 95% confidence

9. $\bar{x} = 8.1$, $\bar{\sigma} = 1.25$; 99.7% confidence

10. $\bar{x} = .32$, $\bar{\sigma} = .2$; 68% confidence

11. $\bar{x} = .32$, $\bar{\sigma} = .2$; 95% confidence

12. $\bar{x} = .32$, $\bar{\sigma} = .2$; 99.7% confidence

For Exercises 13–18, you are given a sample mean $\bar{x}$, a standard deviation $\bar{\sigma}$, and a confidence interval. Specify the confidence coefficient for the given interval.

13. $\bar{x} = 5$, $\bar{\sigma} = 1$; interval $(3, 7)$

14. $\bar{x} = 34$, $\bar{\sigma} = 3$; interval $(31, 37)$

15. $\bar{x} = 7.33$, $\bar{\sigma} = .5$; interval $(5.83, 8.83)$

16. $\bar{x} = 1.71$, $\bar{\sigma} = .01$; interval $(1.68, 1.74)$

17. $\bar{x} = .4$, $\bar{\sigma} = .25$; interval $(-.1, .9)$

18. $\bar{x} = 74$, $\bar{\sigma} = 8$; interval $(66, 82)$

For Exercises 19–22, you are given the standard deviation σ for a large population from which an n-element random sample is to be drawn. Determine a minimum size for n so that there is 95% confidence that the mean of the sample is less than .5 unit away from the actual population mean μ.

19. $\sigma = 3.5$

20. $\sigma = 4.98$

21. $\sigma = 2.2$

22. $\sigma = 7.1$

For Exercises 23–28, $\sigma = 5.2$ is the standard deviation of a large population from which an n-element random sample is to be drawn. In each case, approximate with 95% confidence the maximum "margin of error" between the sample mean and the actual population mean. (Round your answer to 2 decimal places.)

23. $n = 1$

24. $n = 2$

25. $n = 5$

26. $n = 10$

27. $n = 30$

28. $n = 50$

For Exercises 29–32, assume that σ is the standard deviation of a large binomially distributed population (with $n = 1$) from which a 50-element sample has been drawn. In each case, compute an approximate value for σ using the given sample mean $\bar{x}$. (Round your answer to 2 decimal places.)

29. $\bar{x} = .7$

30. $\bar{x} = .62$

31. $\bar{x} = .45$

32. $\bar{x} = .18$

33. The label on a package of 200 corn seeds states that the germination rate is 94% ± 3%. Assuming that the stated variation represents a 95% confidence interval,

(a) What is the standard deviation for the germination rate for the seed population?

(b) Predict with 99.7% confidence the maximum number of seeds from this package that will germinate.

34. In an attempt to get an early indication of the outcome of a citywide bonding referendum vote on election day, a Kansas City television station conducted an exit poll of 300 voters. They found that 168 voted *yes* and 132 voted *no*. With what level of confidence can the station predict that the referendum will pass? Justify your answer.

35. A candidate in a primary needs to get more than 50% of the votes cast to win without a runoff election. On the evening before the election, a random telephone survey of 50 voters indicated that 29 of them would vote for her.

(a) The candidate's campaign manager, seeking at least 95% assurance of victory, was not content with this survey. Why not?

(b) The campaign staff phoned 100 more voters. Again, the same proportion of voters contacted—58 of the 100—said they would vote for the candidate. This time, the campaign manager was satisfied. Why?

36. A standardized nationwide examination is known to have a mean score of 72 and a standard deviation of 8.

(a) A group of 30 students took the exam and got a mean score of 75. How likely is it that these students' scores represent a random sample of all scores on the exam? Justify your answer.

(b) A group of 100 students took the exam and got a mean score of 74.5. Compared with the group in Part (a), is it more or less likely that these students' scores represent a random sample of all scores on the exam? Justify your answer.

37. The machinery of a toy manufacturer cuts small brass axles to a mean length of 15 cm with an allowable variation of ±.1 cm. Every two hours, the quality-control department takes a random sample of 9 axles, measures them, and records the mean and standard deviation.

   (a) One such sample has a mean of 15.04 cm and a standard deviation of .06 cm. How confident can the company be that the machinery is working properly?

   (b) Another such sample has a mean of 14.95 cm and a standard deviation of .12 cm. How confident can the company be that the machinery is working properly?

WRITING EXERCISES

1. In your own words, explain the meaning of the word "confidence" as it is used in this section.

2. (a) Which idea of this section interests you the most? Why?

   (b) Which idea of this section interests you the least? Why?

   (c) Which idea of this section confuses you the most? In what way is it confusing to you?

3. Do some library research on the subject of quality control; then write a nontechnical description (about 1 page long) of how sampling is used to monitor quality in industrial production. Be sure to provide a bibliography that lists any sources you used, and be careful that you do not simply copy or paraphrase the material from your sources.

LINK: STATISTICS IN THE PSYCHOLOGY OF LEARNING

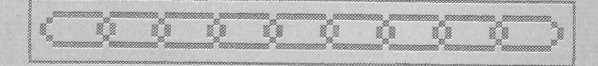

4.13

The IQ test is one of the most widely known measurement instruments in educational psychology. Almost all of us have had our IQs measured at least once, and some of us may even recall the score(s). Yet relatively few people know what an IQ score really means (do you?), except for some vague sense that it measures intelligence level in some way. In fact, the IQ test is a psychological testing tool whose meaning depends almost entirely on the concepts you studied in this chapter. Here we shall explain briefly the statistical basis for IQs, then tie this to current public policy in a particular area of education.

Testing.

"IQ" is an abbreviation for *Intelligence Quotient*. A quotient is, of course, a number formed from the division of one number by another. In this case the quotient is formed by dividing a person's mental age by his or her chronological age. It is standard practice to multiply the resulting fraction by 100 and then round after the decimal point, so that IQ scores appear as whole numbers. Thus,

$$IQ = \frac{\text{mental age}}{\text{chronological age}} \times 100$$

rounded to the nearest whole number. For example, a nine-year-old child with a mental age of 10 would have an IQ of 111 because

$$\frac{10}{9} \times 100 = 111.11 \ldots$$

The obvious question is: What is *mental age*? This is where statistics enters the scene. In order to measure a child's mental age, educational researchers devise a test that can be administered to children over a broad span of ages.[24]

24. The most common IQ tests in use today are the Wechsler Intelligence Scale for Children (WISC), spanning ages 6 through 15, and the Stanford-Binet Intelligence Scale, spanning all ages from preschool through high school.

To be standardized, these tests are administered to a large random sample of children of every age group, and the mean score in each age group is used to represent the mental age. Thus, a child whose individual test score matches the mean score for twelve-year-old children has a mental age of 12, regardless of that child's actual (chronological) age.

For instance, suppose the mean score for ten-year-old children on a particular intelligence test is 92. Then

- an eight-year-old child who scores 92 on that test has a mental age of 10 and, therefore, has an IQ of 125 $\left[\text{that is, } \frac{10}{8} \times 100\right]$;

- a twelve-year-old child who scores 92 also has a mental age of 10 and thus has an IQ of 83 [obtained from $\frac{10}{12} \times$ by rounding];

- a ten-year-old child who scores 92 has a mental age of 10 and an IQ of 100 $\left[= \frac{10}{10} \times 100\right]$.

In general, a person with an IQ of 100 is someone whose intelligence, or intellectual ability, tests out at the mean for his/her age group.

Notice that the computation process just described has actually *defined* mental age in a completely statistical way. That's fine, *if* mental age is understood solely in that way. Unfortunately, even the term itself suggests something more, as if mental age were some sort of fixed characteristic of individual minds that is discoverable by testing. It is quite clear, however, that an eight-year-old child with a mental age of 10 and a fifteen-year-old child with a mental age of 10 have very different kinds of minds.

Moreover, the IQ formula implies that an eight-year-old girl with an IQ of 125 is two years ahead in mental age $\left[\text{because } \frac{125}{100} \times 8 = 10\right]$, but when that same child reaches age twelve (presumably with the same IQ), she must be three years ahead in mental age $\left[\text{because } \frac{125}{100} \times 12 = 15\right]$. According to the formula, on her 24th birthday this adult with an IQ of 125 would be fully 6 years ahead in mental age! The obvious extension of this idea leads to absurd conclusions, so the interpretation of variance from the norm is adjusted to mean different things at different ages. Such difficulties with the conceptual validity of "mental age" have led in recent years to IQ testing based on a more formal statistical foundation. Some of the major intelligence tests have now abandoned "mental age" in favor of scoring tables designed so that the (standardized) mean for each age group is 100 and the standard deviation is 15 or 16.

This little exercise in applied statistics can affect people's lives in many ways. Let us look at one important example. Since about 1970, many school systems around the country have begun to provide supportive educational services to a special class of handicapped children known as "learning disabled."

These special services are now required (and partially funded) by the federal government. But children qualifying for these services are vaguely defined by federal regulations, which say only that such a child must have "a severe discrepancy between achievement and intellectual ability" in oral and/or written language or mathematics.[25] The question is: What is a "severe discrepancy"? The answer to this question has profound implications for parents and children, school boards and teachers, towns and taxpayers. Who qualifies for this help? How much service must be provided? How much will it cost? All these answers flow from the definition of "severe discrepancy," a definition the federal government does not provide.

Among the states attempting to provide legal definitions for themselves is Connecticut, whose answer is a striking illustration of the effect of statistics on public policy:

> [A] severe discrepancy is said to exist whenever the difference between academic achievement (AA) and intellectual functioning (IQ) (each expressed as a standard score) is equal to or greater than one-and-one-half standard deviations—that is,[26]

$$IQ - AA = 1\tfrac{1}{2} SD$$

In other words, to qualify for special services in this area, children must satisfy a statistical criterion that rests, in turn, on statistically based test measurements. Their intellectual potential is measured by an IQ test and then their achievement is measured by other tests whose scores are standardized for purposes of comparison. (The Connecticut guidelines even provide a formula for converting other scoring scales to a mean of 100 and a standard deviation of 15.) Only if the achievement scores fall $1\tfrac{1}{2}$ standard deviations (22 points) below the IQ scores do the Connecticut guidelines declare these children to be learning disabled. Thus, the concept of "learning disabled" is defined in Connecticut in a fundamentally statistical way.

The foregoing is but one illustration of how statistical concepts may be so thoroughly intermixed with ideas from another discipline that there is no way to separate them. As you encounter the behavioral and social sciences in your education, you might find it helpful to look for concepts that are defined fundamentally by statistics. Recognizing them is a critical first step in truly understanding the theories that use them and the implications of those theories in our lives and in our world.

25. *Federal Register*, 42 (240): 65083, Dec. 29, 1977.

26. *Guidelines for Identification and Programming of Learning Disabilities*. Hartford, CT: Connecticut Department of Education, 11/27/81 draft, p. 30.

TOPICS FOR PAPERS

1. This is an independent research project to illustrate how the statistical methods of this chapter are used. It is suitable for either an individual or a small group.

   (a) Formulate a *yes–no* question that can be asked of all the students at your institution. (Before going any further, check with your instructor to be sure that your question is suitable as a basis for this paper.)

   (b) Choose a random sample of students and ask them the question.

   (c) Determine with 95% confidence the proportion of the entire student body that would answer *yes* to your question.

   (d) Write a detailed description of the methods and conclusions of your survey, beginning with a clear statement of the survey question. Be sure to cover *at least* the following points in your paper:
   - How did you decide on the size of your random sample?
   - How did you choose the people in your sample? From what population are they drawn? Why is the choice process you used random?
   - Did any unexpected complications arise when you were conducting your survey? If so, what were they and how did you handle them?
   - What is the proportion of the population that you predict would say *yes* to your question? What is the confidence interval for your result? (Include the computations that justify your answers.)
   - Now that you have completed the project, what stands out as the most interesting thing you learned from it?

2. Write a research paper elaborating on or extending the ideas in one of the three LINK sections of this chapter. Consider your reader to be an anonymous student in the class, a person who knows the material of the chapter. Be sure to provide a bibliography that lists any sources you used in writing the paper, and be careful that you do not simply copy or paraphrase the material from your sources. Any material not explicitly quoted should have been mentally digested by you, so that your own words describe what really are your own ideas when you write them down.

FOR FURTHER READING

1. Alder, Henry L., and Edward B. Roessler. *Introduction to Probability and Statistics*, 3rd ed. San Francisco: W. H. Freeman and Co., 1964.

2. Asimov, Isaac. *The Genetic Code*, New York: The Orion Press, 1962.

3. COMAP (Solomon Garfunkel, Project Director). *For All Practical Purposes*, Part II. New York: W. H. Freeman and Co., 1988.

4. Diamond, Solomon. *The World of Probability*. New York: Basic Books, Inc., 1964.

5. Freund, John E. *Modern Elementary Statistics*, 4th Ed. Englewood Cliffs, NJ: Prentice-Hall, Inc., 1973.

6. Johnson, Robert. *Elementary Statistics*, 4th Ed. Boston: Duxbury Press, 1984.

7. Kac, Mark. "Probability," *Scientific American* 211 (3): 92–108, 1964.

8. Moore, David S., and George P. McCabe. *Introduction to the Practice of Statistics*. New York: W. H. Freeman and Co., 1989.

9. Robbins, Herbert. "Theory of Probability," *NCTM 23rd Yearbook, Insights into Modern Mathematics*. Washington, DC: National Council of Teachers of Mathematics, 1957.

10. Stoppard, Tom. *Arcadia*. London: Faber and Faber, 1993.

11. Weaver, Warren. *Lady Luck*. Garden City, NY: Doubleday & Co., 1963.

12. Willerding, Margaret. *A Probability Primer*. Boston: Prindle, Weber & Schmidt, Inc., 1968.

MATHEMATICS OF INFINITY: CANTOR'S THEORY OF SETS

WHAT IS SET THEORY?

Mathematics is many things to many people. To some it is just a routine tool; to others, a convenient language; to still others, a strict science. And to a surprisingly large number of people, mathematics is mainly an art, pursued for its own sake out of curiosity and with an appreciation of abstract beauty, much as when a chess grandmaster seeks an elegant checkmate of a respected opponent. This chapter treats some mathematics developed from that last point of view.

As soon as it was introduced by the German mathematician Georg Cantor in 1872, set theory caused intense controversy in mathematical, philosophical, and theological circles. That controversy contributed to Cantor's eventual mental breakdown, caused severe divisions of opinion among European theologians and mathematicians, and resulted in a continuing three-way philosophical split with regard to the foundations of mathematics. But Cantor's work also affected mathematics in a decidedly positive way. His basic set theory provided a simple *unifying* approach to many different areas of mathematics, including probability and statistics, modern geometry, and abstract algebra. Moreover, the strange paradoxes encountered in some early extensions of his work encouraged mathematicians to put their logical house in order, so to speak. Their careful examination of the logical foundations of mathematics led to many new results in that area and paved the way for even more abstract unifying ideas.

The key to the universality of Cantor's work is the simplicity of its starting point. This fact is especially convenient for us because it allows us to get a good look at some of his most important results without requiring much preliminary material. In fact, we need assume only a few elementary ideas from your prior mathematical experience. Specifically, we assume you know (or are willing to believe) that two points determine exactly one line, and that each point of a line can be matched with exactly one real number, and vice versa. (A more detailed description of the real numbers appears in Section 5.4.) We also assume you have a little experience with various kinds of numbers—integers, fractions, decimals.

We begin with a brief summary of the basic terms of the language of set theory. You may or may not have seen these terms before (in your pre-college mathematics). If you have, this summary will serve as a compact review. If not, you might find it helpful to study the more detailed description of the language of sets that appears in Section 4.2. Here, then, are the basic terms of set theory:

■ Any collection of objects whatsoever is called a **set**; the objects themselves are called **elements** of that set. Sets are usually denoted by capital English letters, such as A, B, or S; elements often are denoted by small (often italicized) letters, like a, b, c, etc. If a set is described by listing its elements, that list is enclosed in braces. Thus, the set of the first four letters of the alphabet is written {a, b, c, d}. The set of all letters of the alphabet may be written {a, b, c, . . . , z}, where the ellipsis ". . ." is an abbreviated way of showing that all the letters between c and z are included. In general, the ellipsis indicates that a pattern is continued. If it is followed by a specific number, as in

$$\{-1, -2, -3, \ldots, -25\}$$

the ellipsis indicates that the pattern ends with that final number. If it is not followed by a number, as in

$$\{-1, -2, -3, \ldots\}$$

it indicates that the pattern continues without end. Often letters or other symbols are used with ellipses to help describe the pattern, as in

$$\{-1, -2, -3, \ldots, -n, \ldots\}$$

where n is understood to be any natural number.

■ Sometimes it is inconvenient or impossible to list all the elements in a set, or to establish enough of a pattern so that the ellipsis notation can be used without danger of confusion. The natural thing to do in such a case is to specify some property or properties that describe all the elements in the set and nothing else. In that way it becomes possible to "build" the set from its description. In these cases we use a standard form of notation:

$$\{x \mid x \text{ has a certain property}\}$$

This is read "the set of all x such that x has a certain property." Such a set is **well-defined** if the statement "x has a certain property" always is unambiguously true or false, no matter what x is.

■ If the defining condition for a set is *always* false regardless of what is substituted for the variable, then the set has no elements in it; but it is considered to be a set, just the same. The set containing no elements is called the **empty set** (or **null set**) and is denoted by $\varnothing$.

■ A set A is a **subset** of a set B if every element of A is also an element of B. We write this as "$A \subseteq B$." Two sets A and B are **equal** if $A \subseteq B$ and $B \subseteq A$; in this case we write "$A = B$." A set A is a **proper subset** of a set B if A is a subset of B but A does not equal B; here we write "$A \subset B$."

■ Sometimes it is necessary to specify the set of all elements that are considered appropriate for a particular discussion. This set is called the **universal set** for the discussion, and is usually denoted by $\mathcal{U}$. The set of all elements in the universal set that are not in a particular set A is called the **complement** of A, and is denoted by A'.

■ The set of all elements in both a set A and a set B is called the **intersection** of A and B, and is symbolized by $A \cap B$. If A and B are sets such that $A \cap B = \varnothing$, we say that A and B are **disjoint**. The set of all elements that are either in a set A or in a set B, possibly in both, is called the **union** of A and B, and is symbolized by $A \cup B$.

■ Another way of combining sets is based on the concept of an **ordered pair**, which is simply a pair of elements in a specified order. We write the ordered pair with first element a and second element b as (a, b). The **Cartesian product** of two sets A and B is the set of all ordered pairs whose first elements are from A and whose second elements are from B; it is denoted by $A \times B$.

Listed here for reference are some infinite sets of numbers that will be used frequently throughout the chapter. The first few are familiar to you, no doubt. A fuller description of the others, as well as a discussion of what is meant by saying they are "infinite," will appear shortly.

$\mathbf{N} = \{1, 2, 3, 4, 5, \ldots\}$, the **natural numbers**

$\mathbf{E} = \{2, 4, 6, 8, 10, \ldots\}$, the **even natural numbers**

$\mathbf{D} = \{1, 3, 5, 7, 9, \ldots\}$, the **odd natural numbers**

$\mathbf{I} = \{\ldots, -3, -2, -1, 0, 1, 2, 3, \ldots\}$, the **integers**

Q denotes the **rational numbers**, the set of all numbers that can be written as fractions with integer numerators and nonzero integer denominators. In symbols,

$$\mathbf{Q} = \left\{\frac{p}{q} \,\middle|\, p, q \in \mathbf{I}, q \neq 0\right\}$$

R denotes the **real numbers**. It is the set of all numbers that can be written as finite or infinite decimals, and it contains **Q** as a subset. (An explanation of finite and infinite decimals appears in Section 5.4.) This set can be used to represent all the points on a single line.

[0, 1] denotes the set of all real numbers between 0 and 1, inclusive. It is called the **unit interval**.

[a, b] denotes the set of all real numbers between a and b, inclusive. It is called the **(closed) interval** a, b. (The term "closed" refers to the fact that the endpoints a and b are included in the set, which is also denoted by the use of the square brackets. If it is clear from the context that the endpoints are included, the word "closed" is often omitted.)

$\mathbf{R} \times \mathbf{R}$ denotes the set of all ordered pairs of real numbers, which can be used to represent the set of all points of a plane.

5.1   EXERCISES

or these exercises, refer to the sets described t the end of this section. In Exercises 1–38, nswer true or false:

1. **E** is a subset of **N**.

2. **D** is a proper subset of **N**.

3. **E** is a subset of **D**.

4. **E** equals **D**.

5. **E** and **D** are disjoint.

6. **E** ∪ **D** = **N**

7. **E** ∪ **D** is a subset of **N**.

8. **E** ∪ **D** is a proper subset of **N**.

9. **E** and **N** are disjoint.

10. **E** ∪ **N** = **N**

11. **E** ∪ **N** is a subset of **N**.

12. **E** ∪ **N** is a proper subset of **N**.

13. **E** ∩ **D** = **N**

14. **E** ∩ **D** is a subset of **N**.

15. **E** ∩ **N** = **N**

16. **E** ∩ **D** is a proper subset of **N**.

17. **E** ∩ **N** is a subset of **N**.

18. **E** ∩ **N** is a proper subset of **N**.

19. $N \times N$ is a subset of N.

20. $N \times N$ and N are disjoint.

21. I is a subset of N.

22. N is a subset of I.

23. Q is a proper subset of I.

24. I is a proper subset of Q.

25. Q is a subset of Q.

26. $[0, 1]$ is a subset of R.

27. $[0, 1]$ is a subset of $[-1, 5]$.

28. $[0, 1]$ equals $[1, 2]$.

29. $[0, 1]$ is a subset of $[2, 4]$.

30. $[0, 1] \cup [1, 2] = [0, 2]$

31. $[0, 1] \cap [1, 2] = [0, 2]$

32. .75 is an element of $[0, 1]$.

33. 1.02 is an element of $[1, 2]$.

34. $\frac{1}{3}$ is an element of $[0, 1]$.

35. $\frac{3}{5}$ is an element of $[3, 5]$.

36. The ordered pair $(3, 5)$ is an element of $R \times R$.

37. $[3, 5]$ is a subset of $R \times R$.

38. $[0, 1] \times [0, 1]$ is a subset of $R \times R$.

39. **(a)** List all the integers in $[5, 8]$.

    (b) List ten rational numbers in $[5, 8]$ other than the integers of Part **(a)**.

40. **(a)** List all the integers in $[-2, 4]$.

    (b) List ten rational numbers in $[-2, 4]$ other than the integers of Part **(a)**.

41. List five elements of the set $I \times I$.

42. List five elements of the set $Q \times I$ that are not in $I \times I$.

43. List five elements of the set $Q \times Q$ that are not in $Q \times I$.

44. List five elements of the set $[0, 1] \times [1, 2]$.

In Exercises 45–54, use the notation of this section to represent the given set in symbols.

45. The set of all real numbers between 4 and 9, inclusive.

46. The set of all integers between 4 and 9, inclusive.

47. The set of all integers between -2 and 2, inclusive.

48. The set of all real numbers between -2 and 2, inclusive.

49. The set of all integers between -6 and -2, inclusive.

50. The set of all real numbers between -6 and -2, inclusive.

51. The set of all ordered pairs of even natural numbers.

52. The set of all ordered pairs of natural numbers in which the first one is odd and the second is even.

53. The set of all ordered pairs of real numbers in which the second number is between 1 and 8, inclusive.

54. The set of all ordered pairs of real numbers in which both numbers are between -1 and 3.75, inclusive.

WRITING EXERCISES

1. Compare the definitions of *union*, *intersection*, and *Cartesian product* as they are given in this section with the symbolic versions of these definitions as they appear in Section 4.2. Which form do you prefer, and why?

2. What do you think we should mean by the "size" of a set? Can you think of more than one way to define this concept? Use some specific sets, including some of the sets of numbers listed in this section, to illustrate your definition(s).

3. How is the ellipsis (". . .") used in ordinary English? How is that similar to its mathematical usage. How is it different?

INFINITE SETS

In its most literal sense, infinite means "unbounded, endless, without limits of any kind"; but these definitions are not much more informative than the word itself. The confusion is compounded by the fact that nothing in the world around us seems to possess this property, with the possible exceptions of space and time, which appear to be limitless, but certainly are finite as far as the experience of any individual is concerned. Even a quantity as vast as the total number of electrons and protons in the universe can be calculated, at least approximately, and is thereby bounded by some large but finite number. (In 1938, the British astronomer Sir Arthur Eddington actually proposed a number for this: $2 \times 136 \times 2^{256}$.)[1]

The first outright encounter with infinity usually comes from a consideration of numbers themselves. Very early in their mathematical experience, children discover that there is no largest number because 1 can be added to any number to get a larger one. The natural numbers are thus envisioned as an endless chain stretching out "to infinity," a place just beyond the farthest cloud and off limits to sane, stable people. This pathway to infinity has a definite starting point and is composed of discrete numerical steps. But it operates like a treadmill: you can walk as far as you like, but your goal will be as far away when you stop as it was when you began.

Contrasted with this discrete, step-by-step idea of infinity is the predominantly geometric idea of continuous infinity. Although a straight line can be shown to be infinite because one can "step off" intervals in either direction as far as one likes, there is no need to "step" at all. Because there are no gaps in the line, it is possible to proceed as far as is desired without singling out specific points along the way.

"But," you say, "this is really the same kind of process. One simply glides from one point to the next, thus taking very small but definite steps."

The Greeks considered a line (and time) to be composed of a succession of adjacent points (or moments) until Zeno showed how that assumption led to absurd results. (See Section B.3 of the Appendix.) There is no "next" point to any given point and there are no gaps anywhere; so the set of points that make up a line appears to constitute a kind of infinity that is somehow different from that of the natural numbers. One might even speculate that these two infinite sets are of different "sizes," so to speak.

---

1. See p. 1069 of Item 8 (Volume 2) in the list *For Further Reading* at the end of this chapter.

If, just for the sake of argument, we suppose that there are, indeed, at least two different sizes of infinity, are there more? What about the set **I** of integers or the set **Q** of rational numbers? Clearly, **I** contains no smallest number, so this infinity appears two-sided, in some sense. Between any two rational numbers there are infinitely many others. Does this mean that **I** and/or **Q** somehow represent larger sizes of infinity than **N**? If infinity is truly limitless, can there be "larger" or "smaller" infinite sets, or different sizes in any sense? To answer these questions (and others that probably are starting to occur to you), we need to define precisely what we mean by an "infinite set," and we must also define "same size."

For instance, consider the sets

$$A = \{a, b, c, d\} \quad \text{and} \quad B = \left\{ \square, \triangle, \bigcirc, \square \right\}$$

Are these two sets the same size? Before this question can be answered reasonably, we must specify what we mean by "same size." Clearly, if we are talking about the area of the page each occupies, the answer is *no*. However, the size question, as it concerns us in this chapter, is more accurately translated by asking if *A* and *B* have the same number of elements, and in this case the answer is *yes*. But what does "same number" mean, especially in the case of infinite sets? Must we have particular numbers of elements before we can decide whether or not two sets are the same size in this sense? We certainly can use numbers to count the elements of *A* and *B*; but saying that *A* and *B* are the same size just means we ended up with the same number in each case. Now, counting is nothing more than matching things up with part of a known set (the natural numbers 1, 2, 3, . . .). If we are going to match up sets, there is no need to use numbers at all; we can directly match the sets we want to compare. Thus, we can pair off the elements of *A* and *B* as in *Figure 5.1* and conclude that *A* and *B* are the same size.

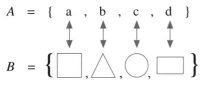

Figure 5.1 Corresponding elements in sets A and B.

As we prepare to investigate infinite sets, where counting all the elements is a hopeless task, it is useful to have a formal definition of this simpler comparison process.

DEFINITION   A **one-to-one** (or **1-1**) **correspondence** between two sets *A* and *B* is any rule or process by which each element of *A* is associated with exactly one element of *B* and each element of *B* is associated with exactly one element of *A*.

DEFINITION   Two sets *A* and *B* are **equivalent** if there exists a one-to-one correspondence between them. We shall write this as $A \leftrightarrow B$.

When we say that two sets "have the same size," we mean that they are equivalent.

EXAMPLE 5.1

There are many different 1-1 correspondences between the sets

$$A = \{a, b, c, d\} \quad \text{and} \quad B = \{w, x, y, z\}$$

Two of them are indicated here by means of arrows:

$$A = \{a, \quad b, \quad c, \quad d\} \qquad\qquad A = \{a, \quad b, \quad c, \quad d\}$$
$$\updownarrow \quad \updownarrow \quad \updownarrow \quad \updownarrow \qquad \text{and} \qquad \diagdown\!\!\diagup \quad \diagdown\!\!\diagup$$
$$B = \{w, \quad x, \quad y, \quad z\} \qquad\qquad B = \{w, \quad x, \quad y, \quad z\}$$

EXAMPLE 5.2

The matching

$$A = \{a, \quad b, \quad c, \quad d\}$$
$$\updownarrow \quad \nwarrow\!\updownarrow \quad \updownarrow$$
$$B = \{w, \quad x, \quad y, \quad z\}$$

is not a 1-1 correspondence (even though the two sets are equivalent). Although each element of A corresponds to exactly one element of B, the element x in B is not matched with anything in A, whereas y is matched with two elements of A.

EXAMPLE 5.3

There is no way to put the sets $\{a, b, c, d\}$ and $\{x, y, z\}$ in 1-1 correspondence. Therefore, these two sets are not equivalent.

EXAMPLE 5.4

A 1-1 correspondence between the set of all natural numbers and the set of all negative integers is given by

$$\{ 1, \quad 2, \quad 3, \quad 4, \quad \ldots, \quad n, \quad \ldots\}$$
$$\updownarrow \quad \updownarrow \quad \updownarrow \quad \updownarrow \qquad\quad \updownarrow$$
$$\{-1, \quad -2, \quad -3, \quad -4, \quad \ldots, \quad -n, \quad \ldots\}$$

We cannot list every pairing in the correspondence; but the pattern given allows us to determine specifically how each element in either set is matched with an element of the other. For instance:

The natural number 37 is matched with the negative integer -37;
the negative integer -45 is matched with 45; etc.

DEFINITION

A set A is **finite** if it is empty or if there is a natural number n such that A is equivalent to $\{1, 2, 3, \ldots, n\}$.

Thus, a set is **infinite**—that is, not finite—if its elements *cannot* be counted completely, no matter how fast we count or how much time we take. This fits well with the intuitive sense of infinity described at the beginning of this section and at the same time it makes good mathematical sense. So, let us move on to the first major question we face:

Are all infinite sets equivalent?

In other words, can two infinite sets *always* be put in 1-1 correspondence? If so, then all infinite sets are the same size (in this sense). If not, then there are actually different sizes of infinity!

Intuitive reactions ("gut feelings") about this question vary widely. Since we human beings have never encountered actually infinite collections of things in our material experience, all of our attempts to deal with them must involve projecting our finite experience into an area in which its applicability is unknown. Therefore, we must rely on logical reasoning to guarantee the validity of any statements we make about infinity. In particular, we must apply the formal definitions carefully, and then be prepared to accept the consequences of our reasoning, regardless of whether or not they conform to intuition.

If we try to apply the previous description of *infinite set* to a specific set, a difficulty arises almost immediately. To verify that a set is infinite by that description, we must prove that *no* natural number will determine (the number of elements of) an equivalent set. The logical argument required to prove that something does not exist is often delicate and somewhat obscure. Fortunately, there is another property of infinite sets that serves to distinguish them from finite sets. Consider this impossible, but informative, example from the folklore of mathematics.

> An owner of a 100-room motel in a busy convention center is often completely booked up and has to turn away customers. In planning to enlarge his establishment, this innkeeper faces a dilemma—if he doesn't make it large enough, he will still have to turn away business; but if he makes it too big he will lose money on the vacant rooms. Based on past experience he decides that even on the slowest days he can fill half the rooms. After pondering the problem awhile, he decides that the perfect solution is to build a motel with infinitely many rooms, which he will number 1, 2, 3, ... He reasons as follows:
>
> > On the nights when I have filled every room and an extra customer arrives, I can still accommodate him or her. I'll give the new guest Room 1 and move the guest in Room 1 to Room 2, the guest in Room 2 to Room 3, the guest in Room 3 to Room 4, and so on. Because there is no last room, everyone will have a place to stay.
> >
> > On the slow nights, when only half the rooms are filled, I'll just make sure I've assigned all the even-numbered rooms first. Then, before everyone settles in for the night, I'll move all the guests to the rooms whose numbers are half the numbers of the rooms they were originally assigned—2 will move to 1, 4 will move to 2, 6 will move to 3, and so on. Then *all* the rooms will be filled for the night!

We might suspect that this innkeeper is a better mathematician than businessman. Although he might never finish building his infinite motel, his mathematical conclusions about it are correct, and they illustrate a characteristic property of infinite sets. Since all infinite sets have this property, and no finite sets do, we have a useful definition:

DEFINITION   A set is **infinite** if it is equivalent to a proper subset of itself.

This definition of infinite set and the description given earlier yield the same results—a set that fits either description also fits the other. (The proof of this fact is somewhat difficult and involves a subtle axiomatic consideration; it is therefore omitted.) Hence, the two definitions may be used interchangeably, according to which is more convenient in a given situation.

EXAMPLE 5.5   We can easily show that the set $\mathbf{N}$ of natural numbers is infinite by using the definition of infinite set that appears just before this example. Indeed, $\mathbf{E}$ is a proper subset of $\mathbf{N}$, and we can establish the equivalence of the two sets by the 1-1 correspondence

$$\mathbf{N} = \{1,\ 2,\ 3,\ 4,\ \ldots,\ n,\ \ldots\}$$
$$\updownarrow\ \updownarrow\ \updownarrow\ \updownarrow\ \qquad \updownarrow$$
$$\mathbf{E} = \{2,\ 4,\ 6,\ 8,\ \ldots,\ 2n,\ \ldots\}$$

EXAMPLE 5.6   The easiest way to show that $\{a, b, c, d, e\}$ is *finite* is to count its elements; that is, to apply the definition of finite set. Accordingly,

$$\{a,\ b,\ c,\ d,\ e\}$$
$$\updownarrow\ \updownarrow\ \updownarrow\ \updownarrow\ \updownarrow$$
$$\{1,\ 2,\ 3,\ 4,\ 5\}$$

5.2   EXERCISES

1. Find three more 1-1 correspondences between $\{a, b, c, d\}$ and $\{w, x, y, z\}$ besides the ones in Example 5.1. How many such correspondences are there? Why?

2. Find three more 1-1 correspondences between the set of all natural numbers and the set of all negative integers besides the one in Example 5.4.

3. Find three sets that can be put in 1-1 correspondence with $\{0, 1, 2, 3, 4\}$, and find two sets that cannot be.

4. Are the sets
   $$\{1, 2, 3, \ldots, 50\} \quad \text{and} \quad \{1, 2, 3, \ldots\}$$
   equivalent? Why or why not? Are they equal? Why or why not?

5. Are the sets

   $\{1, 2, 3, \ldots\}$ and $\{1, 2, 3, 4, \ldots\}$

   equivalent? Why or wht not? Are they equal? Why or why not?

6. If two sets are equivalent, must they be equal? Why or why not?

7. If two sets are equal, must they be equivalent? Why or why not?

8. Show that **E** is equivalent to **D**.

9. Show that **E** is equivalent to

   $\{5, 10, 15, \ldots, 5n, \ldots\}$

10. If two sets are both equivalent to a third set, must they be equivalent to each other? Why or why not? Find an example to illustrate your answer.

In Exercises 11–14, find specific examples of finite sets of the indicated sizes. How would you show that each of these sets is finite?

11. 7 elements

12. 10 elements

13. more than 100 elements

14. more than 1000 elements

For Exercises 15–19, let $A = \{a, b, c, d, e\}$, and recall that N denotes the set of all natural numbers. Find an example of each of the following:

15. A set that is not equal to A, but that is equivalent to A.

16. A set that is not equal to **N**, but that is equivalent to **N**.

17. A proper subset of A that cannot be put in 1-1 correspondence with A.

18. A proper subset of **N** that cannot be put in 1-1 correspondence with **N**.

19. A proper subset of **N** that can be put in 1-1 correspondence with **N**.

In Exercises 20–23, find an example of each of the following:

20. An infinite set of positive numbers all less than 5.

21. An infinite set of numbers between 2 and 3.

22. An infinite set containing the alphabet as a proper subset.

23. A 1-1 correspondence between **E** and one of its proper subsets. (By doing this you are proving that **E** is an infinite set.)

For each of the sets given in Exercises 24–29, find a proper subset that is equivalent to it; then find one that is not.

24. **D** (the set of odd natural numbers)

25. $\{3, 6, 9, 12, \ldots, 3n, \ldots\}$

26. $\{1, \frac{1}{2}, \frac{1}{3}, \frac{1}{4}, \ldots, \frac{1}{n}, \ldots\}$

27. $\{5, 6, 7, 8, \ldots, n + 4, \ldots\}$

28. $\{-2, -4, -6, -8, \ldots, -2n, \ldots\}$

29. $\{1, 4, 9, 16, \ldots, n^2, \ldots\}$

30. Use the definition of infinite set (given on page 237) to prove that the set **I** of integers is infinite.

Exercises 31–34 are related. If you intend to do any of these questions, you should read the statements of all of them.

31. Describe how to subdivide the set **N** of natural numbers into three nonoverlapping infinite sets. That is, find three sets, A, B, and C, such that

    (1) the intersection of any two of these sets is empty, and

    (2) $A \cup B \cup C = $ **N**

32. Describe how to subdivide the set **N** of natural numbers into six nonoverlapping infinite sets. That is, find six sets such that

    (1) the intersection of any two of these sets is empty, and

    (2) the union of all six sets equals **N**.

33. (This exercise generalizes Exercises 31 and 32.) Describe how to subdivide the set **N** of natural numbers into k nonoverlapping infinite sets; here, k can be any natural number.

34. (This exercise is an extension of Exercise 33.) Describe how to subdivide the set **N** of natural numbers into infinitely many nonoverlapping infinite sets. (Be careful! Your solution for Exercise 33 may not generalize easily.)

35. Can you put the set $\{3, 6, 9, \ldots, 3n, \ldots\}$ of all multiples of 3 into 1-1 correspondence with the set of all natural numbers that are *not* multiples of 3? If you can, do it. Write a general rule that describes your correspondence and use your rule to find the number that corresponds to 300. If you cannot, explain why you think it can't be done.

36. Can you put the set $\{2, 4, 8, \ldots, 2^n, \ldots\}$ of all powers of 2 into 1-1 correspondence with the set of all natural numbers that are *not* powers of 2? If you can, do it. Write a general rule that describes your correspondence and use your rule to find the number that corresponds to 2^{50}. If you cannot, explain why you think it can't be done. If you think it can be done, but can't find a general rule, describe in words some pattern that you think might work.

WRITING EXERCISES

1. Do infinite sets actually exist in the physical world? In the real world? Are your answers to these two questions different? Why or why not?

2. Describe at least two other ways, besides 1-1 correspondences, of interpreting the statement that two things are "the same size." Can either of these other ways be applied generally to "things" that are sets? If so, how; if not, why not?

3. Do you think that *any* two infinite sets of natural numbers can be put in 1-1 correspondence? Give reasons to justify your opinion. Think a little about Exercises 31–36 (even if you didn't actually do them) before answering this question.

4. Using complete sentences, write a half-page summary that identifies the main theme of this section and outlines how that theme is carried out.

THE SIZE OF N

5.3

One way to rephrase the definition of infinite set is to say that a set is infinite if we can throw away some of it and still be left with a set of the same size. Example 5.5 of Section 5.2 showed that **N** and **E** are the same size, even though **E** was obtained by throwing away "half" of **N** (that is, by throwing away all the odd natural numbers). It is also easy to see that **D**, the set of odd natural numbers, is equivalent to **N**. These two correspondences can be used to extend our size comparison of the sets listed in Section 5.1 one step further.

Saying that some infinite sets can be put in 1-1 correspondence with the set of natural numbers is the same as saying that these sets can be ordered sequentially, starting with a first element, then a second element, then a third, and so on. Now, when we deal with sets of numbers—like **I** or **Q** or **R**—we tend to think of them as being arranged in their usual size order, as if they were on a number line. However, they are still the same sets even if we jumble the order or rearrange them to fit some other pattern. (If a box of registration cards for everyone in this class, arranged in alphabetical order, were spilled on the floor and then picked up in haste, the cards would probably not remain in the same order; but they would still be the same collection of cards. If they were rearranged according to the numerical order of the social security numbers on them, they would also still be the same set of cards.)

If we ignore the usual ordering of the integers, we can show that **I** is equivalent to **N** by choosing a first element of **I**, then a second element, then a third, and so on. The only tricky part is that we must be sure that every integer gets matched with some natural number. The matching between **N** and **E** in Example 5.5 provides a clue for an easy way to match **I** with **N**; it tells us we can match up all the positive integers with just the even natural numbers, like this:

$$\mathbf{N} = \{1, \; 2, \; 3, \; 4, \; 5, \; 6, \; 7, \; 8, \; 9, \; \ldots\}$$
$$\mathbf{I^+} = \{ \quad 1, \qquad 2, \qquad 3, \qquad 4, \ldots\}$$

The odd natural numbers are then still available to be matched up with zero and the negative integers, like this:

$$\mathbf{N} = \{1, \; 2, \; 3, \; 4, \; 5, \; 6, \; 7, \; 8, \; 9, \; \ldots\}$$
$$\mathbf{I^-} \cup \{0\} = \{0, \qquad -1, \qquad -2, \qquad -3, \qquad -4, \ldots\}$$

Combining both match-ups, we obtain:

$$\mathbf{N} = \{1, \quad 2, \quad 3, \quad 4, \quad 5, \quad 6, \quad 7, \quad 8, \quad 9, \; \ldots\}$$
$$\mathbf{I} = \{0, \quad 1, \; -1, \quad 2, \; -2, \quad 3, \; -3, \quad 4, \; -4, \; \ldots\}$$

It should be easy to see that the pattern established by the first few terms of this correspondence can be continued as far as we please. For instance, the positive integer 10 corresponds to the natural number 20, the natural number 51 corresponds to the negative integer -25, etc. A general description of this matching is:

Any even natural number n corresponds to the integer $\frac{n}{2}$;

any odd natural number n corresponds to the integer $\frac{1-n}{2}$.

Thus, we have shown that:

xx

N and **I** are the same size.

xxxxxxxxxxxxxxxxxxxxxxxxxxxxxxxxxxxxxxxxxxxxxxxxxx

At this point it is convenient to make a useful observation:

xxxxxxxxxxxxxxxxxxxxxxxxxxxxxxxxxxxxxxxxxxxxxxxxxx

(5.1)   For any sets *A*, *B*, and *C*, if *A* is equivalent to *B* and *B* is equivalent to *C*, then *A* and *C* are also equivalent.

xxxxxxxxxxxxxxxxxxxxxxxxxxxxxxxxxxxxxxxxxxxxxxxxxx

The required 1-1 correspondence between *A* and *C* may be obtained by "patching together" the *A*-to-*B* correspondence with the *B*-to-*C* correspondence. For example, if the sets

$$A = \{x, y, z\}, \quad B = \{p, q, r\}, \quad \text{and} \quad C = \{5, 6, 7\}$$

are matched as in *Figure 5.2*, then we automatically have a 1-1 correspondence between *A* and *C* that matches *x* with 6 (by way of *q*), *y* with 7 (by way of *p*), and *z* with 5 (by way of *r*).

Using Statement (5.1), our work, so far, has shown that the first four infinite sets in the list of Section 5.1 are all the same size. Now let us deal with **Q**, the set of all rational numbers. By again ignoring the usual ordering of the numbers, we can devise a method for putting the positive integers and the positive rationals in 1-1 correspondence. Arrange all the positive rationals in an infinite rectangular array, listing all the fractions with numerator 1 in the first row, all the fractions with numerator 2 in the second row, all the fractions with numerator 3 in the third row, and so on, as shown in *Figure 5.3* on page 242. Then, starting at the upper-left corner and omitting all fractions that are not in lowest terms, take the positive rationals in a zigzag diagonal order and match them with successive positive integers.

As the arrows in *Figure 5.3* indicate, this 1-1 correspondence

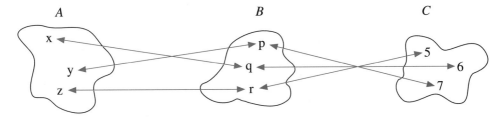

Figure 5.2 *A is in 1-1 correspondence with B and B is in 1-1 correspondence with C. Thus, A is in 1-1 correspondence with C.*

between **Q⁺**, the set of positive integers, and **I⁺**, the set of positive rationals, begins:

$$\mathbf{Q^+} = \{1, \tfrac{1}{2}, 2, 3, \tfrac{1}{3}, \tfrac{1}{4}, \tfrac{2}{3}, \tfrac{3}{2}, 4, 5, \tfrac{1}{5}, \ldots\}$$

$$\updownarrow \; \downarrow \; \downarrow \; \downarrow \; \downarrow \; \downarrow \; \downarrow \; \downarrow \; \downarrow \; \updownarrow \; \downarrow$$

$$\mathbf{I^+} = \{1, \; 2, \; 3, \; 4, \; 5, \; 6, \; 7, \; 8, \; 9, \; 10, \; 11, \ldots\}$$

The same process also gives us a 1-1 correspondence between all the negative integers and all the negative rationals, just by labeling all the numbers in both sets as negative and using the same matching. The 1-1 correspondence between **I** and **Q** is completed by assigning the integer 0 to the rational number 0. Consequently:

xxxxxxxxxxxxxxxxxxxxxxxxxxxxxxxxxxxxxxxxxxxxxxxxxxxxxxxxx

The sets **I** and **Q** are the same size.

xxxxxxxxxxxxxxxxxxxxxxxxxxxxxxxxxxxxxxxxxxxxxxxxxxxxxxxxx

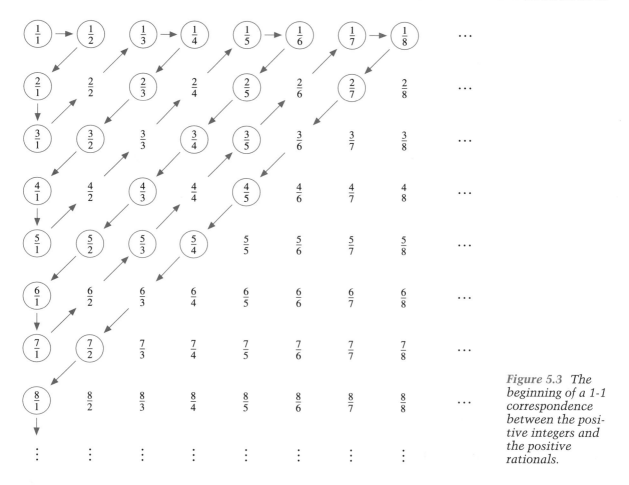

Figure 5.3 The beginning of a 1-1 correspondence between the positive integers and the positive rationals.

5.3   EXERCISES

In Exercises 1–5, describe three different orders for the elements of each set.

1. {1, 2, 3, 4, 5}
2. {a, b, c, d, . . . , y, z}
3. **N**, the set of natural numbers
4. **I**, the set of integers
5. **Q**, the set of rational numbers

6. Set up a 1-1 correspondence that proves that the set **D** of all odd natural numbers is equivalent to **N**. Try to write a general formula for that correspondence; then use this formula to find which natural numbers correspond to the odd numbers 99 and 243.

Use the 1-1 correspondence between N and I given in this section to answer Exercises 7–12.

7. Which integer is matched with the natural number 100?

8. Which natural number is matched with the integer −100?

9. Which integer is matched with the natural number 75?

10. Which natural number is matched with the integer −30?

11. If p denotes a positive integer, what is its corresponding natural number?

12. If $-p$ denotes a negative integer, what is its corresponding natural number?

13. Set up a 1-1 correspondence between **N** and the set
$$\{30, 60, 90, 120, . . .\}$$
Try to write a general formula for this correspondence; then use this formula to find which numbers correspond to the natural numbers 17 and 50.

14. Use your answers to Exercises 6 and 13 to set up a 1-1 correspondence between **D** and
$$\{30, 60, 90, 120, 150, . . .\}$$
Try to write a general formula for this correspondence; then use this formula to find which numbers correspond to the odd numbers 17, 99, and 243.

Use the correspondence between I and Q described in this section to answer Exercises 15–23.

15. Which rational number corresponds to the integer 15?

16. Which rational number corresponds to the integer 30?

17. Which rational number corresponds to the integer −15?

18. Which rational number corresponds to the integer −30?

19. Which integer corresponds to the rational number $\frac{5}{2}$?

20. Which integer corresponds to the rational number $\frac{4}{7}$?

21. Which integer corresponds to the rational number $\frac{-5}{3}$?

22. Which integer corresponds to the rational number $\frac{-3}{8}$?

23. Why were the fractions that are not in lowest terms omitted in the correspondence process?

For Exercises 24–32, prove that the given subset of Q is equivalent to the set N of natural numbers. In each case, write a specific description of the 1-1 correspondence. (A formula is preferable, but not necessary.) Check the clarity of your description by finding the rational numbers that correspond to the natural numbers 20 and 21 in each case.

24. $\left\{\frac{2}{1}, \frac{2}{2}, \frac{2}{3}, \frac{2}{4}, \cdots\right\}$

25. $\left\{\frac{2}{1}, \frac{2}{3}, \frac{2}{5}, \frac{2}{7}, \cdots\right\}$

26. $\left\{\frac{2}{1}, \frac{4}{3}, \frac{8}{5}, \frac{16}{7}, \cdots\right\}$

27. $\left\{\cdots, \frac{-5}{3}, \frac{-5}{2}, \frac{-5}{1}, 0, \frac{5}{1}, \frac{5}{2}, \frac{5}{3}, \cdots\right\}$

28. $\left\{\cdots, \frac{-5}{16}, \frac{-5}{8}, \frac{-5}{4}, \frac{-5}{2}, 0, \frac{5}{2}, \frac{5}{4}, \frac{5}{8}, \frac{5}{16}, \cdots\right\}$

29. The set of all positive rational numbers that can be written with a numerator of 1 or 2.

30. The set of all rational numbers that can be written with a numerator of 1 or 2.

31. The set of all positive rational numbers that can be written with single-digit numerators.

32. The set of all rational numbers with numerator some positive integral power of 2 and denominator some positive integral power of 3.

33. Describe a 1-1 correspondence between **N** and **N** × **N** (the set of all ordered pairs of natural numbers).

WRITING EXERCISES

1. Does every infinite set have a subset that can be put in 1-1 correspondence with the set **N** of natural numbers? Explain.

2. Describe how to set up a 1-1 correspondence between **N** and the set of all rational numbers whose denominators are powers of 10. If you find this question difficult, describe how you might use at least two of the dozen problem-solving tactics of Section 1.2 to help you answer it.

3. If the rational numbers are arranged in size order, then between any two of them there are always infinitely many others. Mathematicians refer to this property by saying that **Q** is a **dense** set. Write a brief explanation of why the word "dense" is a good label for this property. Think of at least one other English word which would be just as good (or better) in evoking an intuitive sense of this property? Explain your choice(s).

RATIONAL AND IRRATIONAL NUMBERS

5.4

(For those already familiar with the characterization of rational and irrational numbers by means of their decimal expansions, this section may be omitted without loss of continuity.)

How large is the set **R** of all real numbers, both rational and irrational? Before we can answer this, we must learn a little more about rational and irrational numbers. To begin with, the set **R** can be described as the set of all decimals, provided that we allow decimals to have infinitely many digits. For brevity, we shall call decimals with infinitely many digits **infinite decimals** and decimals with only finitely many digits **finite decimals**.

Every rational number can easily be converted to a decimal by dividing its numerator by its denominator. Some rational numbers, such as

$$\frac{1}{3} = .33333 \ldots$$

require infinitely many digits in their decimal expansions. If we agree to attach an infinite "tail" of zeros to an ordinary decimal, then every rational number can be written as an infinite decimal. For example,

$$\frac{3}{4} = .75000\ldots \qquad\qquad \frac{7}{3} = 2.33333\ldots$$

$$\frac{13}{8} = 1.62500\ldots \qquad\qquad \frac{2}{11} = .18181\ldots$$

and so forth. In other words, **Q** is a subset of **R**.

However, not every infinite decimal represents a rational number. To distinguish those decimals that represent rationals from those that do not, consider first an example of the division process just mentioned.

EXAMPLE 5.7

To convert $\frac{47}{22}$ to a decimal, we divide 47 by 22 as follows:

```
          2.136
   22 ) 47.0000   . . .
        44
         3 0
         2 2
           80      ←
           66
          140
          132
            8      ←
```

The remainder, 8, at the fourth subtraction step is the same as the one at the second step, so the third and fourth digits of the quotient will repeat again and again from there on, without interruption. Thus,

$$\frac{47}{22} = 2.1363636\ldots$$

Must such a repetition of remainders happen with any division example, or is this just a carefully chosen special case? The answer is *yes* to both parts of that question. A repetition must always occur *eventually*; but the fact that it happens so conveniently soon in this example is the result of careful choice.

Any case in which the numerator is larger than the denominator can be simplified by first taking care of the integer part of the quotient. For instance, we could have begun Example 5.7 by writing

$$\frac{47}{22} = 2 + \frac{3}{22}$$

then converting $\frac{3}{22}$ to the decimal .1363636 . . . afterwards. In this way we can safely confine our attention to fractions in which the numerator is smaller than the denominator.

In general, when one integer is divided by another (larger) one using long division, the remainder at any subtraction step must be less than the divisor. Thus, if we are converting the rational number $\frac{p}{q}$ to a decimal, the largest possible remainder is $q - 1$; that is,

(5.2)   The number of different nonzero remainders in the division problem $p \div q$ can be no larger than $q - 1$.

Statement (5.2) indicates that, once the nonzero digits of p have all been used, after at most q more steps in this division problem the remainder *must* repeat an earlier one. Therefore, the digits in the quotient that occur in the steps between the two equal remainders will repeat over and over from there on without interruption. Sometimes this repetition occurs early, as in Example 5.7. Sometimes all possible nonzero remainders are used; for instance,

$$\frac{4}{7} = .571428 \ 571428 \ 571428 \ \ldots$$

Thus, we have established an important fact:

XXXXXXXXXXXXXXXXXXXXXXXXXXXXXXXXXXXXXXXXXXXXXXXXXXXXX

(5.3)   Every rational number can be expressed as an infinite decimal in which, from some specific digit on, a finite sequence of digits repeats again and again in the same order, without interruption.

XXXXXXXXXXXXXXXXXXXXXXXXXXXXXXXXXXXXXXXXXXXXXXXXXXXXX

Such decimals are called **repeating decimals**. To denote the repeating sequence of digits without having to write it several times over, we just put a bar over it. For instance, the results of the preceding two examples are written

$$\frac{47}{22} = 2.1\overline{36} \qquad \text{and} \qquad \frac{4}{7} = .\overline{571428}$$

The finite decimal $\frac{2}{5} = .4000\ldots$ can be written $.4\overline{0}$, but usually the repeated zeros are just omitted; in this case, we would write $\frac{2}{5} = .4$.

Two natural questions arise at this point:

1. Does *every* repeating decimal represent a rational number?

2. Are there really any (or many) such things as infinite *non*repeating decimals?

It is surprisingly easy to show that the answer to the first question is *yes*. To do this, we use a simple method for converting any repeating decimal to fractional form. The next two examples illustrate how this method is used to convert repeating decimals to fractions.

EXAMPLE 5.8

Suppose $d = .\overline{35}$. Since there are two digits in the repeating sequence, we multiply d by 100 and then subtract d from this product:

$$100d = 35.353535\ldots$$
$$-\ d =\ \ \ .353535\ldots$$
$$99d = 35.000000\ldots$$

$$d = \frac{35}{99}$$

EXAMPLE 5.9

Suppose $d = 2.8\overline{473}$. Since there are three digits in the repeating sequence, we multiply d by 1000 and then subtract d from this product:

$$1000d = 2847.3473473473\ldots$$
$$-\ \ \ d =\ \ \ \ \ \ 2.8473473473\ldots$$
$$999d = 2844.5000000000\ldots$$

$$d = \frac{2844.5}{999}$$

Strictly speaking, a rational number is the quotient of two *integers*; to put the answer in appropriate form, then, we should eliminate the decimal point from the numerator. This is easily done by multiplying both numerator and denominator by 10. This answer can be reduced to lowest terms, if desired, by eliminating all common factors of the numerator and denominator. Thus:

$$d = \frac{28{,}445}{9990} = \frac{5689}{1998}$$

EXAMPLE 5.10

A finite decimal, such as .375, can be handled in the same way by treating it as $.375\overline{0}$, a repeating decimal with a single-digit sequence. (Multiply by 10, subtract, and divide.) It is *much* simpler, however, to observe that a finite decimal is just another notation for a fraction whose denominator is a power of 10 and to write the fractional form immediately:

$$.375 = \frac{375}{1000} = \frac{3}{8}$$

The general method for converting infinite decimals to fractions may be described like this:

Suppose d is a repeating decimal with n digits in its repeating sequence.

■ If we multiply d by 10^n and then subtract d from $10^n \cdot d$, we will get a *finite* decimal that equals $(10^n - 1) \cdot d$.

■ Then d is the fraction obtained by dividing that finite decimal by $10^n - 1$.

■ To insure that the numerator is an integer, multiply both numerator and denominator by the same power of ten, if necessary.

Perhaps this description reads more like a magic incantation than a logical procedure. It was stated in general terms to show that there is a procedure that works all the time. Look back at the preceding examples, then reread the general method; you should find it much clearer.

Since every rational number can be expressed as a repeating decimal, and vice versa, we have just characterized all those real numbers (in decimal form) that are rational—they are just the repeating decimals. The other real numbers are called **irrational numbers**; they are the infinite nonrepeating decimals. But are there any such numbers? Certainly—here is one:

$$.101001000100001\ldots$$

where the ellipsis implies that the continuing pattern requires one more 0 after each successive 1. Because there is no finite sequence of digits that repeats again and again in the same order without interruption, this is not a repeating decimal. (Any other pair of digits in place of 0 and 1 would work equally well, of course.)

To construct other examples of infinite nonrepeating decimals, we can choose the sequence of successive multiples of some natural number, such as

$$3, 6, 9, 12, 15, 18, 21, 24, \ldots$$

and form decimals using these digits in order:

$$.3691215182124\ldots, \quad 3.691215182124\ldots, \quad 36.91215182124\ldots$$

and so forth. (Can you convince yourself that the digit sequence in such cases is nonrepeating?) These examples should suggest how to construct many other irrational numbers.

Sometimes irrational numbers arise in other contexts, and when they do, it often is difficult to prove that they are irrational. Some familiar irrational numbers are $\sqrt{2}$ (the length of the diagonal of a unit square), π (the ratio of the circumference of a circle to its diameter), and e (the base of the natural logarithms). The proof that $\sqrt{2}$ cannot be expressed as a rational number was known to the early Greeks; a version of it appears in Section B.3 of the Appendix. The proofs that π and e are irrational are much more difficult; they were only established within the last 200 years.

The decimal characterization of rational and irrational numbers is important and useful; we summarize it here for emphasis:

xxx

The real numbers that are rational are the repeating decimals; the real numbers that are irrational are the (infinite) nonrepeating decimals.

xxxxxxxxxxxxxxxxxxxxxxxxxxxxxxxxxxxxxxxxxxxxxxxxxxx

Note   There is a small problem of ambiguity of representation in Statement (5.3). Decimals ending in repeated 9s represent the same numbers as the corresponding decimals ending in repeated 0s. For example, .4999. . . = .5000. . . . (To see that this is true, convert these two decimals to fractional form.) If we discard one or the other of these representations in each case, then there is a 1-1 correspondence between the set of all repeating decimals and the set of all rational numbers. For our purposes, it suffices to assume that this has been done in a way that allows us to use whichever form is more useful in each particular situation.

5.4   EXERCISES

Before performing the divisions in Exercises 1–9, indicate the maximum number of different nonzero remainders that can occur. Then do the computations to find out how many different remainders actually do occur. Is the number of different remainders the same as the number of digits in the repeating sequence?

1. $3 \div 5$ 2. $2 \div 3$ 3. $11 \div 37$

4. $7 \div 10$ 5. $3 \div 7$ 6. $14 \div 6$

7. $70 \div 14$ 8. $451 \div 999$ 9. $7832 \div 9990$

Convert each fraction in Exercises 10–21 to a repeating decimal. (Find the entire repeating sequence of digits.)

10. $\frac{1}{9}$ 11. $\frac{2}{5}$ 12. $\frac{13}{6}$ 13. $\frac{15}{22}$

14. $\frac{5}{11}$ 15. $\frac{2}{13}$ 16. $\frac{21}{9}$ 17. $\frac{9}{21}$

18. $\frac{10}{7}$ 19. $\frac{22}{7}$ 20. $\frac{112}{37}$ 21. $\frac{5}{111}$

Convert each decimal in Exercises 22–33 to fractional form.

22. $1.\overline{4}$ 23. $.0\overline{2}$ 24. $.\overline{9}$ 25. $.\overline{02}$

26. $.\overline{15}$ 27. $.1\overline{5}$ 28. $.0\overline{15}$ 29. $62.1\overline{75}$

30. $62.\overline{175}$ 31. $.\overline{40}$ 32. $6.85\overline{14}$ 33. $6.8\overline{514}$

34. Write five irrational numbers in infinite decimal form.

For Exercises 35–38, arrange the given numbers in size order, from smallest to largest.

35. $.3, .33, .03, .\overline{3}, .0\overline{3}, .\overline{03}$

36. $.1, .01, .\overline{10}, .\overline{01}, .\overline{010}, .1010010001. . .$

37. $.567, .\overline{6}, .\overline{567}, .5\overline{67}, .56\overline{7}, .567566756667. . .$

38. $\pi, 3.14, 3.1416, \frac{22}{7}, 3.1\overline{4}, 3.1416181101. . .$

WRITING EXERCISES

1. Describe at least two ways in which the term *repeating decimal* can be misleading, if it is not understood carefully.

2. The word *numerator* literally means "a person or thing that numbers or counts something." The word *denominator* literally means "namer"; it is closely related to *denomination*—the name of a specific kind of thing—as in a religious denomination or a denomination of currency. With this in mind, explain why it is appropriate to use the words *numerator* and *denominator* to refer to the top and bottom numbers of a fraction, respectively.

3. Discuss the relative merits of using the upper-bar notation versus the ellipsis ("...") notation to represent repeating decimals.

4. In Section 5.1, we said that **R** could be regarded as the set of all points on a single line. Explain how decimals are related to locations on a line. As an example in your explanation, describe the location of the decimal $d = 7.135$ in steps, as follows:

(a) What does the digit 7 tell you about the location of d?

(b) What does the digit 1 tell you about the location of d?

(c) What does the digit 3 tell you about the location of d?

(d) What does the digit 5 tell you about the location of d?

(e) What does the fact that there are no more digits say about the location of d?

Similarly, describe the locations of the *repeating decimals* $.\overline{3}$ and $.\overline{9}$.

5. Do you *really* believe that $.\overline{9}$ is actually equal to 1? Do you really believe that $.\overline{3}$ is actually equal to $\frac{1}{3}$? Discuss.

A DIFFERENT SIZE

5.5

The equivalence of such different sets as **I** (the integers) and **Q** (the rationals) suggests that any two infinite sets may be equivalent. Let us assume for the moment that this assumption is true and see where it leads us. In particular, let us suppose that **N**, the set of natural numbers, is equivalent to [0, 1], the set of all real numbers between 0 and 1, inclusive. Because the real numbers are all the infinite decimals, the real numbers between 0 and 1 can be described as all infinite decimals that only have digits to the right of the decimal point. (The number 1 is represented by the infinite decimal .999... in this case.) Then any 1-1 correspondence between **N** and [0, 1] will be a sequential listing of these infinite decimals, one for each natural number. For example, such a correspondence might begin as illustrated in *Figure 5.4*.

In any such 1-1 correspondence there is an infinite decimal corresponding to each natural number. Let us use this correspondence to define a new infinite decimal, according to the following rule:

N		[0, 1]
1	⟷	0 . ③ 0 1 2 5 9 4 ...
2	⟷	0 . 1 ⑥ 6 5 2 1 8 ...
3	⟷	0 . 4 1 ① 2 1 0 7 ...
4	⟷	0 . 2 0 5 ⓪ 9 6 3 ...
5	⟷	0 . 0 0 0 1 ① 1 1 ...
6	⟷	0 . 8 5 7 3 0 ⑨ 9 ...
⋮		⋮ ⋱

Figure 5.4 The beginning of a proposed 1-1 correspondence between N and [0, 1].

For each natural number n, look at the nth (decimal) digit of its corresponding decimal. If that digit is 1, let 2 be the nth (decimal) digit of the new decimal; otherwise, let this digit be 1.

Looking at the encircled digits in *Figure 5.4*, we see that the new decimal would begin .112121. . . . Notice that this new number differs from the first number in the [0, 1] listing in at least the first decimal place; it differs from the second number in at least the second place; it differs from the third number in at least the third place; and so on. Because this new number differs from each number of [0, 1] in the list, it does not correspond to any natural number; yet it is clearly a real number between 0 and 1. In other words,

we have found a number in [0, 1] that was left out of the alleged 1-1 correspondence between all of N and all of [0, 1].

Therefore, the proposed correspondence cannot exist!

But *any* proposed 1-1 correspondence between N and [0, 1] presents the same problem. Despite the specific choice of the first few decimals to illustrate the correspondence, the method given for constructing the "new" decimal number is general enough to apply *to any proposed listing*. Thus, the process just described—called **Cantor's Diagonalization Process**—shows that any proposed 1-1 correspondence between N and [0, 1] must necessarily leave out at least one element of [0, 1]. Therefore,

xxxxxxxxxxxxxxxxxxxxxxxxxxxxxxxxxxxxxxxxxxxxxxxxxxxxx

N and [0, 1] cannot be equivalent.

xxxxxxxxxxxxxxxxxxxxxxxxxxxxxxxxxxxxxxxxxxxxxxxxxxxxx

(In fact, any attempted 1-1 correspondence between N and [0, 1] actually will leave out many, many of the numbers in [0, 1]; but to prove the correspondence invalid, all we need to show is that at least one number in [0, 1] is skipped.) This means that the infinite sets N and [0, 1] are *not* the same size. Therefore, *there are at least two different sizes of infinity*!

Having used the unit interval [0, 1] to discover that there are different sizes of infinite sets, it seems natural to ask whether the length of an interval is related to its size as a set of points. For example, can we get an infinite set that is not equivalent to [0, 1] by choosing a longer interval, such as [0, 5] or [−36, 247], or a shorter interval, such as $\left[\frac{1}{2}, \frac{3}{4}\right]$ or [.00001, .00002]? It would seem reasonable to assume that an interval of length 5 or 50 or 1000 should contain too many points to be equivalent to an interval of length 1. However, if $a < b$, a surprisingly simple geometric argument proves that:

xxxxxxxxxxxxxxxxxxxxxxxxxxxxxxxxxxxxxxxxxxxxxxxxxxx

Any interval $[a, b]$, of any length whatsoever, can be put in 1-1 correspondence with $[0, 1]$.

xxxxxxxxxxxxxxxxxxxxxxxxxxxxxxxxxxxxxxxxxxxxxxxxxxx

To show how this can be done, place the interval $[0, 1]$ perpendicular to $[a, b]$, with point 0 lying on point a. Draw the line determined by the two points 1 and b, and choose a point P somewhere on the extension of that line from b past the point 1, as shown in *Figure 5.5*. Now, if we choose any point x in $[0, 1]$, the line through P and x must intersect $[a, b]$ at some point y. Also, if we choose a point y' in $[a, b]$, the line through P and y' necessarily intersects $[0, 1]$ at some point x'. This process gives us a correspondence between the points of $[0, 1]$ and the points of $[a, b]$.

Notice that different points of $[0, 1]$ determine different lines through P, so these lines must intersect $[a, b]$ at different points, and vice versa (because two points determine exactly one line). Therefore, the correspondence is one-to-one, and consequently $[0, 1]$ and $[a, b]$ are equivalent sets. In fact, the presence or absence of endpoints does not affect this result, as Exercise 38 shows. This means that *all* intervals, regardless of length, are equivalent to $[0, 1]$ and hence to each other. Thus, *the length of an interval does not affect its size as a set of points!*

A somewhat similar argument shows that the interval $[0, 1]$ is equivalent to the entire number line **R**. (We omit the details of that argument.) Coupling this information with the first result of this section, we arrive at a startling conclusion about the number line and its intervals, *considered as sets of points*:

xxxxxxxxxxxxxxxxxxxxxxxxxxxxxxxxxxxxxxxxxxxxxxxxxxx

R is the same size as $[0, 1]$ or any other interval.

xxxxxxxxxxxxxxxxxxxxxxxxxxxxxxxxxxxxxxxxxxxxxxxxxxx

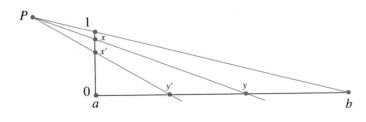

Figure 5.5 A 1-1 correspondence between $[0, 1]$ and $[a, b]$.

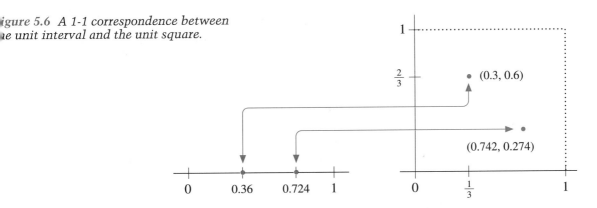

That is, the *length* of a line or line segment and its *size* as a set of points are totally different mathematical concepts.

What if we change dimension? Do we get a different-sized set? It seems plausible that the set of all points in a plane, for example, is far too large to be matched up with the set of points on a single line. Once again, however, we are in for a surprise; a 1-1 correspondence between these two sets is possible. We shall look at a special case—a correspondence between the unit interval [0, 1] and the set of all points in a 1-by-1 square.

First, recall from high school algebra that a **coordinate system** can be placed on the plane by using two copies of the real line **R**, one horizontal and one vertical, intersecting at their respective zero points. Each of these two lines is called a **coordinate axis**. Using this system, any point P on the plane can be labeled by an ordered pair of real numbers (x, y), where x is the number on the horizontal axis directly above or below P and y is the number on the vertical axis directly to the left or right of P. The numbers x and y are called the **coordinates** of the point P. (If P is on an axis, one of its coordinates is 0.) Thus, the set of all points in the plane can be treated as the set $\mathbf{R} \times \mathbf{R}$ of all ordered pairs of real numbers.

If we take all the points whose coordinates lie within the [0, 1] interval on each axis, we have a square 1 unit on a side. Set up a 1-1 correspondence between this square and a single unit interval, roughly, as shown in *Figure 5.6*.[2] Each point in the square has coordinates of the form

$$(.x_1 x_2 x_3 \dots, \ .y_1 y_2 y_3 \dots)$$

where $.x_1 x_2 x_3 \dots$ and $.y_1 y_2 y_3 \dots$ are decimals between 0 and 1. Match

---

2. We say "roughly" because the repeated-0/repeated-9 ambiguity described in the *Note* at the end of Section 5.4 causes some trouble. Certain decimal forms have to be omitted initially, then "patched" back in by an argument similar to the one needed for Exercise 38. Nevertheless, the main argument works, as outlined here.

each such point with the decimal formed by "interweaving" the x-digits and y-digits in an alternating pattern, starting with x_1:

$$.x_1 y_1 x_2 y_2 x_3 y_3 \ldots$$

This number is also a decimal between 0 and 1. For example, the point $\left(\frac{1}{3}, \frac{2}{3}\right)$ in the square, which can be written

$$(.333 \ldots, \ .666 \ldots)$$

corresponds to the point .363636. . . in the unit interval. Conversely, the point $.\overline{724}$ (= .724724. . .) in the unit interval corresponds to the point $(.\overline{742}, .\overline{274})$ in the square.

The construction of a 1-1 correspondence between **R** and **R** × **R** is an extension of the technique shown here. We omit its details as well as the proof that this matching leads to a 1-1 correspondence. The fact that such a correspondence can be constructed is the main point here because it establishes the equivalence of **R** and **R** × **R**. If we disregard the spatial properties of lines and planes and consider them just as sets of points, this equivalence means:

×××

The set of all points in a plane (a two-dimensional geometric object) is the same size as the set of all points on a single line (a one-dimensional geometric object).

×××

The trick of interweaving the digits of decimal representations of numbers can also be used for three-dimensional space. For instance, each point in the "unit cube" $[0, 1] \times [0, 1] \times [0, 1]$ is determined by three coordinates of the form

$$(.x_1 x_2 x_3 \ldots, \ .y_1 y_2 y_3 \ldots, \ .z_1 z_2 z_3 \ldots)$$

Each such point can be matched with the single infinite decimal

$$.x_1 y_1 z_1 x_2 y_2 z_2 x_3 y_3 z_3 \ldots$$

in $[0, 1]$. The extension of this technique to all triples of real numbers can be used to establish the equivalence of the sets **R** and **R** × **R** × **R**. Thus:

×××

The set of all points on a single line is the same size as the set of all points in three-dimensional space.

×××

5.5   EXERCISES

1. Find five rational numbers between 0 and 1.

2. Find five irrational numbers between 0 and 1.

3. Find five rational numbers between .5 and .53.

4. Find five irrational numbers between .5 and .53.

5. Suppose that an alleged 1-1 correspondence between **N** and [0, 1] begins as follows:

$$
\begin{array}{ccl}
\mathbf{N} & & [0, 1] \\
1 & \leftrightarrow & .040404\ldots \\
2 & \leftrightarrow & .111111\ldots \\
3 & \leftrightarrow & .257346\ldots \\
4 & \leftrightarrow & .101001\ldots \\
5 & \leftrightarrow & .763988\ldots \\
6 & \leftrightarrow & .121121\ldots \\
\vdots & & \vdots
\end{array}
$$

(a) Construct the first six decimal places of a number in [0, 1] that cannot be matched with any natural number in this correspondence.

(b) Construct three more such numbers, all different from each other and from the number you found in Part (a). (*Hint*: There is nothing special about the digits 1 and 2 in Cantor's diagonalization process.)

6. Answer Parts (a) and (b) of Exercise 5 for the following proposed correspondence:

$$
\begin{array}{ccl}
\mathbf{N} & & [0, 1] \\
1 & \leftrightarrow & .111111\ldots \\
2 & \leftrightarrow & .101010\ldots \\
3 & \leftrightarrow & .321321\ldots \\
4 & \leftrightarrow & .765432\ldots \\
5 & \leftrightarrow & .151617\ldots \\
6 & \leftrightarrow & .222222\ldots \\
\vdots & & \vdots
\end{array}
$$

Using a ruler, draw a 2-inch (2″) line segment T perpendicular to a 5-inch line segment *F*, with an endpoint of one segment coinciding with an endpoint of the other. Then, mimicking the argument given in conjunction with *Figure 5.5*,

answer Exercises 7–16 as precisely as you can by measuring.

7. Which point of *F* corresponds to the point at 1″ on *T*?

8. Which point of *F* corresponds to the point at $\frac{1}{2}$″ on *T*?

9. Which point of *F* corresponds to the point at $\frac{3}{2}$″ on *T*?

10. Which point of *T* corresponds to the point at 1″ on *F*?

11. Which point of *T* corresponds to the point at 2″ on *F*?

12. Which point of *T* corresponds to the point at 3″ on *F*?

13. Which point of *T* corresponds to the point at 4″ on *F*?

14. Which point of *T* corresponds to the point at 5″ on *F*?

15. Which point of *T* corresponds to the point at 0″ on *F*?

16. Which point of *F* corresponds to the point at 2″ on *T*?

In Exercises 17–24, decide whether or not the two sets given are equivalent. Give a reason to justify each answer.

17. **I** and the unit interval, [0, 1]

18. [0, 1] and **Q**

19. [0, 1] and the set of all points in a plane

20. **R** × **R** and **R** × **R** × **R**

21. [0, 1] and the set of all points in three-dimensional space

22. **Q** and the set of all points in a plane

23. **I** and the set of all points in three-dimensional space

24. The interval [3, 8] and the set of all points in three-dimensional space

The 1-1 correspondence between $[0, 1] \times [0, 1]$ and $[0, 1]$ described in this section matches ordered pairs of real numbers with single real numbers by "interweaving" decimals. In Exercises 25–32, a point in one of these two sets is given; find the corresponding point in the other set.

25. $(.555 \ldots, .111 \ldots)$ 26. $(.85, .222 \ldots)$

27. $\left(\dfrac{3}{4}, \dfrac{1}{3}\right)$ 28. $(.\overline{43}, .\overline{918})$

29. $.777 \ldots$ 30. $\dfrac{2}{3}$

31. $\dfrac{1}{2}$ 32. $.\overline{73125}$

33. A 1-1 correspondence between two intervals of different lengths can be defined without pictures by using a little elementary algebra. For instance, the interval $[0, 1]$ can be matched with $[0, 3]$ by "stretching" it according to the formula $y = 3x$. That is, each point x in $[0, 1]$ is matched with the point $3x$ in $[0, 3]$. By this formula:

   (a) Which points of $[0, 3]$ correspond to the following points of $[0, 1]$?

   $$\dfrac{1}{4}, \dfrac{1}{2}, \dfrac{3}{4}, \dfrac{1}{3}, .4, 0, 1$$

   (b) Which points of $[0, 1]$ correspond to the following points of $[0, 3]$?

   $$0, 1, 2, 3, \dfrac{1}{2}, \dfrac{1}{3}, .243, \dfrac{5}{2}, 2.7$$

34. Using the method described in Exercise 33, write an algebraic formula for a 1-1 correspondence between the intervals $[0, 1]$ and $[0, 100]$. Then answer these questions, which refer to your formula:

   (a) Which points of $[0, 100]$ correspond to the following points of $[0, 1]$?

   $$\dfrac{1}{4}, \dfrac{1}{2}, \dfrac{3}{4}, \dfrac{1}{3}, .4, 0, 1$$

   (b) Which points of $[0, 1]$ correspond to the following points of $[0, 100]$?

   $$0, 1, 10, 50, 100, \dfrac{1}{2}, \dfrac{57}{3}, 75.243$$

35. Using the method described in Exercise 33, write an algebraic formula for a 1-1 correspondence between the intervals $[0, 1]$ and $[5, 23]$. Then answer these questions, which refer to your formula:

   (a) Which points of $[5, 23]$ correspond to the following points of $[0, 1]$?

   $$\dfrac{1}{4}, \dfrac{1}{2}, \dfrac{3}{4}, \dfrac{1}{3}, .4, 0, 1$$

   (b) Which points of $[0, 1]$ correspond to the following points of $[5, 23]$?

   $$5, 10, 15, 20, \dfrac{13}{2}, \dfrac{57}{3}, 17.357$$

36. Write an algebraic formula for a 1-1 correspondence between the intervals $[0, 1]$ and $[0, b]$, where b can be any positive real number. (*Hint*: Look back at Exercises 33 and 34.)

37. Write an algebraic formula for a 1-1 correspondence between the intervals $[0, 1]$ and $[a, b]$, where a and b can be any real numbers such that $a < b$. (*Hint*: Look back at Exercise 35.)

38. These questions lead to a proof that the unit interval $[0, 1]$ is equivalent to the "open" unit interval

   $$(0, 1) = \{x \mid x \in \mathbf{R} \text{ and } 0 < x < 1\}$$

   (a) Suppose that sets A and B have a subset S in common and that for some sets C and D,

   ■ $A = C \cup S$ and $B = D \cup S$;
   ■ $C \cap S = \varnothing$ and $D \cap S = \varnothing$; and
   ■ there is a 1-1 correspondence between C and D.

   Use this information to describe a 1-1 correspondence between A and B.

   (b) Describe a 1-1 correspondence between the sets

   $$\left\{0, 1, \dfrac{1}{2}, \dfrac{1}{3}, \dfrac{1}{4}, \ldots\right\} \text{ and } \left\{\dfrac{1}{2}, \dfrac{1}{3}, \dfrac{1}{4}, \ldots\right\}$$

   (c) Use Parts (a) and (b) to prove that the intervals $[0, 1]$ and $(0, 1)$ are equivalent sets.

WRITING EXERCISES

1. Describe how at least three of the dozen problem-solving techniques of Section 1.2 can be used to help in the solution of Exercise 38. (Even if you can't complete Exercise 38, you should be able to do this writing exercise.)

2. Consider the subset H of [0, 1] consisting of all decimals that are one hundred digits long.

   (a) Is H a finite set or an infinite set? Justify your answer.

   (b) If you tried to apply Cantor's diagonalization process to H, what would go wrong? Be specific.

3. Consider the subset V of [0, 1] consisting of all the infinite decimals that have only even digits. Does Cantor's diagonalization process still work for this set? Explain.

CARDINAL NUMBERS

5.6

We have now examined all the sets listed at the end of Section 5.1 and have found that they can be categorized by two distinct sizes:

- Those that are equivalent to N.
- Those that are equivalent to R.

It is convenient to have names for these sizes; so we shall adopt for them the names Cantor used.

DEFINITION   The collection of all sets that are equivalent to **N** is denoted by $\aleph_0$, and is read **aleph null**. ($\aleph$ is the first letter of the Hebrew alphabet; the significance of the subscript 0 will appear later.)

DEFINITION   The collection of all sets that are equivalent to **R** is denoted by c, and is called the **cardinality of the continuum**. ("Continuum" is the name chosen by mathematicians to convey the idea of an unbroken line.)

The sizes $\aleph_0$ and c can be regarded as numbers, much like the usual counting numbers 1, 2, 3, 4, 5, ... The number 3, for example, can be considered as the common property possessed by all sets that can be put in 1-1 correspondence with {a, b, c}. But the idea of "common property" (or "threeness," if you prefer) is not precise enough to work with. Instead, we consider 3 to be the collection of all sets that are equivalent to {a, b, c}. In general:

DEFINITION

A **cardinal number** is the collection of all sets that are equivalent to a particular set. Any particular set used to describe a cardinal number is called a **reference set** for that number. The cardinal number to which a set belongs is called the **cardinality** of that set.

Thus, 3 is a cardinal number and {a, b, c} is a reference set for it. Of course, we could use {★, ◇, ♡} or any other three-element set as a reference set for 3 just as well. Similarly, the definitions of $\aleph_0$ and c indicate that they are also cardinal numbers. Those definitions also specify the most commonly used reference sets for these two cardinal numbers—

$$\mathbf{N} \text{ for } \aleph_0 \quad \text{and} \quad \mathbf{R} \text{ for } c$$

DEFINITION

The **finite** cardinal numbers are those with finite reference sets; the **transfinite** cardinal numbers are those with infinite reference sets.

The whole numbers 0, 1, 2, 3, . . . , n, . . . are the finite cardinals; $\aleph_0$ and c are transfinite cardinals.

Some natural questions arise at this point. One that fairly cries out for an answer is:

Are there any other transfinite cardinal numbers?

Another is:

Does it make sense to ask which of the numbers $\aleph_0$ and c is bigger?

Let us take on the second question first, using our experience with finite numbers as a guide.

What does it mean to say that 5 is larger than 3? Because we are dealing with cardinal numbers, the behavior of their reference sets should somehow determine the behavior of the numbers. In this sense, we would say that 5 is larger than 3 because any attempted 1-1 correspondence between a reference set for 5 and a reference set for 3 results in having some elements of 5's reference set "left over." For example, the letters d and e are not used in the following correspondence:

$$\{a, \ b, \ c, \ d, \ e\}$$
$$\updownarrow \ \ \updownarrow \ \ \updownarrow$$
$$\{x, \ y, \ z\}$$

This observation can be generalized to provide a definition of *larger than* that applies to any pair of cardinal numbers:

DEFINITION

Let $\mathcal{A}$ and $\mathcal{B}$ be cardinal numbers, with reference sets A and B, respectively. We say the number $\mathcal{A}$ is **larger than** the number $\mathcal{B}$ (or $\mathcal{B}$ is **smaller than** $\mathcal{A}$) if

1. there exists a 1-1 correspondence between all of set B and a proper subset of set A, and

2. there does not exist a 1-1 correspondence between B and all of A.

NOTATION   We symbolize "A is larger than B" by writing $A > B$ (or $B < A$).

Condition 2 might seem redundant at first, and in fact for finite numbers it is; but it is necessary in order for the definition to make sense when applied to transfinite numbers. For example, consider two different reference sets for $\aleph_0$, namely **N** and **E**. There obviously is a 1-1 correspondence between **E** and a proper subset of **N** (**E** itself), so Condition 1 of the definition is satisfied. Thus, if we were to use Condition 1 alone as the definition of *larger than*, we would find ourselves in the ridiculous position of having to say that $\aleph_0$ is larger than itself! Condition 2 allows us to avoid this absurdity because there is also a 1-1 correspondence between **E** and all of **N**. Therefore, the given definition of *larger than* does not apply to $\aleph_0$ and itself.

Now that we have a workable definition of *larger than*, it is easy to prove that c is larger than $\aleph_0$ by using the results of our previous work. Choose **N** as the reference set for $\aleph_0$ and $[0, 1]$ as the reference set for c. We have already proved that there is no 1-1 correspondence between **N** and all of $[0, 1]$, so Condition 2 of the definition is satisfied. To verify Condition 1, observe that

$$\{1, \frac{1}{2}, \frac{1}{3}, \frac{1}{4}, \ldots, \frac{1}{n}, \ldots\}$$

is a proper subset of $[0, 1]$ and it can be put in 1-1 correspondence with **N** by matching each fraction of the form $\frac{1}{n}$ with the natural number n that is its denominator. Thus, we have shown that

xxxxxxxxxxxxxxxxxxxxxxxxxxxxxxxxxxxxxxxxxxxxxxxxxxxxxx

$$\aleph_0 < c$$

xxxxxxxxxxxxxxxxxxxxxxxxxxxxxxxxxxxxxxxxxxxxxxxxxxxxxx

5.6   EXERCISES

1. Find three different sets that can be used as reference sets for the cardinal number 5.

2. Find three different sets that can be used as reference sets for the cardinal number 8.

3. The set {a, b, c, . . . , y, z} is a reference set for which cardinal number?

4. Determine a reference set for the cardinal number 0.

5. Set up a 1-1 correspondence between

$$\{\spadesuit, \heartsuit, \diamondsuit, \clubsuit\}$$

and a proper subset of

$$\{a, b, c, d, e, f, g\}$$

Is there more than one way to do this?

6. Let $\mathcal{A}$ be a cardinal number with reference set $\{x, y, z\}$. Which specific number is $\mathcal{A}$? Why?

Each statement in Exercises 7–14 is a typical error made by students having difficulty distinguishing between cardinal numbers and reference sets. For each one, write a phrase or sentence explaining what is wrong with the statement. Then write a correct version of what you think the student is trying to say.

7. $\{a, b, c, d\} = 4$ 8. $\{a, b\} < \{c, d, e\}$

9. $\mathbf{N} < \mathbf{R}$ 10. $\aleph_0 = \mathbf{N}$

11. $[0, 1] = c$ 12. $\mathbf{N} = \mathbf{Q}$

13. $\aleph_0 \subset c$ 14. $\mathbf{Q} < [0, 1]$

15. Suppose that $\mathcal{B}$ is a cardinal number with reference set $\{p, q\}$ and that C is a cardinal number with reference set $\{t, y\}$. What can you say about $\mathcal{B}$ and C? Why?

16. Suppose that $\mathcal{A}$ is a cardinal number with reference set A and that $\mathcal{B}$ is a cardinal number with reference set B. If A can be put in 1-1 correspondence with B, how are the numbers $\mathcal{A}$ and $\mathcal{B}$ related? Why?

17. If two cardinal numbers are equal, must a randomly chosen reference set for one be equal to a randomly chosen reference set for the other? Must these reference sets be equivalent? Why?

18. Use the concept of reference sets to explain why there is no cardinal number between 4 and 5.

19. Suppose there were a cardinal number X strictly between $\aleph_0$ and c; that is, suppose that

$$\aleph_0 < X < c$$

If X were a reference set for such a number, how would it compare with $\mathbf{N}$? with $\mathbf{R}$? Explain briefly.

20. Use the definition given in this section to prove that 7 is larger than 4.

21. Prove that $\aleph_0$ is larger than 3.

22. Prove that c is larger than 5.

23. Prove that $\aleph_0$ is larger than any whole number n.

24. Find an example to illustrate the following general statement:

For any three cardinal numbers $\mathcal{A}$, $\mathcal{B}$, and C, if $\mathcal{A} < \mathcal{B}$ and $\mathcal{B} < C$, then $\mathcal{A} < C$.

25. Use the general statement given in Exercise 24 and the statement proved in Exercise 23 to show that c is larger than any whole number n.

WRITING EXERCISES

1. (a) What is the distinction between *cardinal numbers* and *ordinal numbers*? (You might find it helpful to consult a dictionary.)

   (b) Write a paragraph defending the use of the adjective "cardinal" in this numerical sense as compatible with its nonmathematical usage in English.

2. Explain in your own words the distinction between a *cardinal number* and a *reference set*. Provide both finite and infinite examples to illustrate your explanation.

CANTOR'S THEOREM

Up to now, our investigation of particular infinite sets has led us to classify all the infinite sets we have seen into two sizes: the "discretely" infinite size $\aleph_0$ of the set of natural numbers and the "continuously" infinite size c of the real number line. It is tempting, especially after finding that even a change of dimension does not produce a set of larger size than c, to guess that there are no other infinite sizes. But this is a deceptive temptation. In proving his most important general statement, the theorem that bears his name, Georg Cantor showed that no matter what size we consider, there is a way to find a set of a larger size. Thus,

there are infinitely many different sizes of infinity!

Cantor's Theorem is based on a simple, almost trivial, observation about finite sets: A set has more subsets than it has elements. The surprise is that, unlike many other observations about finite sets, this one also holds for infinite sets. We look first at a finite example. The three-element set $\{1, 2, 3\}$ has eight subsets:

$$\varnothing,\ \{1\},\ \{2\},\ \{3\},\ \{1, 2\},\ \{1, 3\},\ \{2, 3\},\ \{1, 2, 3\}$$

This collection of subsets cannot be put in 1-1 correspondence with the original set $\{1, 2, 3\}$; but there are many ways to put $\{1, 2, 3\}$ in 1-1 correspondence with some subcollection of its subsets. One obvious way is to match each number with the single-element subset containing it:

$$\{\ 1,\quad 2,\quad 3\ \}$$
$$\updownarrow\quad \updownarrow\quad \updownarrow$$
$$\{\,\{1\},\ \{2\},\ \{3\}\,\}$$

Clearly, such a correspondence can be established between *any* set and all its single-element subsets. For instance, the infinite set **N** of all natural numbers can be matched with its one-element subsets in this way:

$$\{\ 1,\quad 2,\quad 3,\quad 4,\quad \ldots,\quad n,\quad \ldots\}$$
$$\updownarrow\quad \updownarrow\quad \updownarrow\quad \updownarrow\qquad \updownarrow$$
$$\{\,\{1\},\ \{2\},\ \{3\},\ \{4\},\ \ldots,\ \{n\},\ \ldots\}$$

Thus, **N** can be matched with a proper subset of the set of all of its subsets.

It is not obvious, however, that an infinite set *cannot* be put in 1-1 correspondence with the set of all of its subsets. That was the ingenious part of Cantor's proof. He used a method analogous to his diagonalization process to show that no such matching is ever possible. In particular, he showed that, given any alleged 1-1 correspondence between an infinite set and all of its subsets, one could construct a specific subset, step by step, so that it could not possibly be matched with any element of the original set. The details of his proof are given below. The steps of that proof are not difficult; but you may have to ponder them awhile to understand how they establish a contradiction, and why this contradiction is the key to the entire logical argument.

xxxxxxxxxxxxxxxxxxxxxxxxxxxxxxxxxxxxxxxxxxxxxxxxxxxxxxx

Cantor's Theorem: Let $\mathcal{A}$ be any cardinal number, with reference set A, and let S be the set of all subsets of A. If S is the cardinal number that represents the size of S, then $\mathcal{A}$ is smaller than S.

xxxxxxxxxxxxxxxxxxxxxxxxxxxxxxxxxxxxxxxxxxxxxxxxxxxxxxx

(Informally, this says that the size of any set is smaller than the size of the set of all of its subsets.)

Proof [OPTIONAL]

As stated in the theorem, A is a reference set for the cardinal number $\mathcal{A}$, and S (the set of all subsets of A) is a reference set for the cardinal number S. By the definition of *larger than*, to prove that S is larger than $\mathcal{A}$ we must show that

1. A can be put in 1-1 correspondence with a proper subset of S, and
2. A cannot be put in 1-1 correspondence with all of S.

The first part is easy; we shall dispose of it quickly. The second part is a bit more complicated; but a little concentrated study should give you not only an understanding of the logical steps, but an appreciation of the elegance of Cantor's work, as well.

Condition 1. (For simplicity, we assume $A \neq \varnothing$; the special case where $A = \varnothing$ is handled by a logical "trick" which need not concern us here.) Among all the subsets of A are the single-element subsets; that is, for each element x in A there is the subset $\{x\}$ in S. The matching of each x with $\{x\}$ is obviously a 1-1 correspondence between all the elements of A and some, but not all, of the elements of S. (In particular, $\varnothing$ is an element of S not used in this correspondence.) Thus, Condition 1 is satisfied.

Condition 2. To prove that there cannot exist a 1-1 correspondence between A and all of S, we use an indirect argument. We assume that such a correspondence does exist and derive a contradiction from this assumption.[3]

3. This indirect method of proof is quite common in mathematics and in classical philosophy. It is called *proof by contradiction*, or *reductio ad absurdum*. See the end of Appendix A for a brief description of its logical structure.

Thus, assume that each subset of A corresponds to exactly one element of A, and vice versa. Because each element of A is matched with a subset of A by this presumed correspondence, it makes sense to ask whether or not an element is actually contained in the set with which it is matched. We select each element of A that is *not* in the set with which it is matched, and denote the set of all such elements by W. Clearly, W is a subset of A, and hence it is an element of S.

Because A and S are in 1-1 correspondence, there must be some element z in A that is matched with the set W. Now, either z is contained in W or it is not; let us examine both of these alternatives:

■ If z is contained in W, then z is contained in the set with which it is matched. But then, the definition of W implies that z *cannot* be contained in W, which is an outright contradiction.

■ If z is not contained in W, then z is not contained in the set with which it is matched. Then the definition of W implies that z *must* be in W, again a contradiction.

There are no other alternatives, so we are forced to conclude that our argument contains a flaw somewhere. But each step follows logically from the one before it, except for our initial supposition that a 1-1 correspondence between A and S exists. Hence, this supposition must be false, so Condition 2 of the definition of *larger than* is satisfied.

Thus, we have proved that S, the size of the set of all subsets of A, is larger than $\mathcal{A}$, the size of A.

5.7   EXERCISES

1. Write out the set of all subsets of {a, b, c}.

2. Write out the set of all subsets of {2, 4, 6, 8}.

In Exercises 3–6, find a proper subset of the set N **of natural numbers that has size:**

3. 3   4. 10   5. 500   6. $\aleph_0$

In Exercises 7–10, find a proper subset of the interval [0, 1] that has size:

7. 3   8. 100   9. $\aleph_0$ 10. c

11. Let A = {a, b}. Write out the set S of all subsets of A; then write out the set T of all subsets of S. What size is S? What size is T?

12. Let A = {p, q, r, s, t} and let S be the set of all subsets of A.

(a) Find a three-element subset of A.

(b) Find two elements of S.

(c) Find a three-element subset of S.

(d) What cardinal number represents the size of S? Why?

13. (a) Let A = {u, v, w, x, y, z}. Set up a 1-1 correspondence between A and a set consisting of six of its subsets. (You may choose any six subsets.)

(b) In the proof of Cantor's Theorem we constructed a set W of all elements of A that are not in their corresponding subsets. Using the set A and your correspondence from Part **(a)** of this exercise, list the elements of such a set W.

(c) Do Parts **(a)** and **(b)** again, choosing a different collection of six subsets.

14. An alleged 1-1 correspondence between **N** and its set of subsets begins:

$$1 \leftrightarrow \emptyset$$
$$2 \leftrightarrow \mathbf{D}$$
$$3 \leftrightarrow \{p \mid p \text{ is prime}\}$$
$$4 \leftrightarrow \mathbf{E}$$
$$5 \leftrightarrow \{1, 4, 7, 10, 13, 16, \ldots\}$$
$$6 \leftrightarrow \{2, 4\}$$
$$\vdots \qquad \vdots$$

(a) Describe briefly how the "contradiction" set W of Cantor's Theorem is constructed.

(b) Following the process you described in Part **(a)**, indicate which of the numbers 1, 2, 3, 4, 5, 6 are in W.

Exercises 15–19 describe five different 1-1 correspondences between N and (some of) its subsets. In each case:

(a) Write out the sets that correspond to the natural numbers 1 through 5.

(b) Specify the "contradiction" set W of Cantor's Theorem that is not included in the 1-1 correspondence.

(c) Say whether or not 32 is an element of W.

(d) Verify that W cannot be the set that corresponds to 32 by identifying at least one number that is in one set, but not in the other.

15. Each natural number n is matched with the set $\{1, 2, \ldots, n\}$.

16. Each natural number n is matched with the set $\{1, 3, \ldots, 2n - 1\}$.

17. Each natural number n is matched with the set $\{1, 4, \ldots, n^2\}$.

18. Each natural number n is matched with the set $\{2, 4, \ldots, 2^n\}$.

19. Each natural number n is matched with the set $\{n + 1, n + 2, \ldots, 2n\}$.

20. **(a)** Describe a set of a larger size than size c.

(b) Describe a set of a larger size than the set you chose for Part **(a)**.

(c) Describe a set of a larger size than the set you chose for Part **(b)**.

WRITING EXERCISES

1. Referring to the statement of Cantor's Theorem, discuss the notational distinctions between the script capital letters ($\mathcal{A}$ and $\mathcal{S}$) and the italic capital letters (A and S). Why is this distinction made? Is the distinction necessary? Do you think this notation is helpful or confusing in conveying its message? How would you improve upon it? (In what way is your approach better than the one in the text?)

2. Reread the proof of Cantor's Theorem. What is the *first* thing in the proof that you don't understand? Write out two questions (in complete sentences) about that difficulty, as specifically as you can.

3. Explore the analogy between Cantor's diagonalization process and the proof of Cantor's Theorem. How are they similar? How are they different?

THE CONTINUUM HYPOTHESIS

5.8

Successive applications of Cantor's Theorem, first to a set A, then to the set S of all subsets of A, then to the set of all subsets of S, and so on, provides a way of getting larger and larger sizes of infinite sets, thus guaranteeing an infinite string of different transfinite numbers. However, the theorem does *not* guarantee that the numbers obtained in this way are *successive*; that is, if a set A represents a particular cardinal number, Cantor's Theorem does not tell us whether the set of all subsets of A represents the next cardinal number, or even whether the idea of "next" makes sense in this context. We do not get successive numbers in the finite case (except for 0 and 1)—a 2-element set has 4 subsets, a 4-element set has 16 subsets, and so on. However, our previous experience with expecting infinite sets to behave like finite sets should warn us against assuming the situations are alike.

Basing his work on some commonly accepted axioms,[4] Cantor established several useful facts about transfinite numbers. We state them here without proof:

- $\aleph_0$ is the smallest transfinite cardinal number.

- Each of the transfinite cardinal numbers has an immediate successor, a "next larger" cardinal number. Thus, we may begin to list the transfinite cardinals as

$$\aleph_0 < \aleph_1 < \aleph_2 < \ldots < \aleph_n < \ldots$$

with no other cardinal numbers in between.

- c is the size of the set of all subsets of the natural numbers.

All three of these statements are interesting and important to mathematicians, but one of Cantor's most intriguing statements about transfinite numbers is an assertion he could *not* prove.

Having observed that a set of the smallest infinite size, $\aleph_0$, has a set of all subsets whose size is c, Cantor conjectured that c is the next smallest infinite size; that is, he guessed that $c = \aleph_1$. However, he was unable to prove his conjecture, and by 1900, this question had become one of the most famous unsolved mathematical problems. Known as the "Continuum Hypothesis," it may be stated more formally as follows:

---

4. One of Cantor's implicit assumptions was the Axiom of Choice, which was not recognized explicitly as an axiom until about 25 years later. See page 572 for a brief description of the Axiom of Choice.

XXX

The Continuum Hypothesis: If a set has size $\aleph_0$, then the set of all its subsets has size $\aleph_1$. More generally, if a set has size $\aleph_n$ for some n, then the set of all its subsets has size $\aleph_{n+1}$.

XXXXXXXXXXXXXXXXXXXXXXXXXXXXXXXXXXXXXXXXXXXXXXXXXXXXX

(Properly speaking, the Continuum Hypothesis is just the first of these two sentences.)

During the first half of this century there were many fruitless efforts to prove the Continuum Hypothesis. The elusiveness of this problem was made even more frustrating by the fact that set theory had rapidly become respectable in most areas of mathematics and by 1930, had been put on a solid axiomatic foundation. Many people felt that such a rigorously logical approach to set theory would quickly provide a solution for the Continuum Hypothesis. But the problem was more difficult than they anticipated, and the eventual solution came in a surprising form.

In 1940, Kurt Gödel, an Austrian logician, proved that the Continuum Hypothesis is consistent with the axioms of set theory. In other words, he proved that the *assumption* that the Continuum Hypothesis is *true* will not lead to any contradictions within set theory. Of course, this did not prove the Continuum Hypothesis itself; rather, it showed only that the Continuum Hypothesis *could not be proven false* by a logical argument based on any of the commonly accepted axiom systems for set theory.

The problem remained unresolved until 1963, when Paul Cohen, of Stanford University, proved that the *assumption* that the Continuum Hypothesis is *false* also will not lead to any contradictions within set theory! This means that the Continuum Hypothesis can neither be proved nor disproved within set theory. In other words, we may treat it as a separate axiom that provides information not found in the other axioms of set theory. If we assume it to be true, we get one kind of set theory; if we assume it to be false, we get another kind of set theory, different from the first, but equally valid. (Compare this with the discussion of Euclid's Parallel Postulate in Chapter 3, especially in Section 3.6.)

This discovery provides a striking illustration to support the modern view of mathematics as a study that is independent of the physical world. If there were a single "true" theory of sets waiting for discovery, we could not have conflicting, but equally consistent, theories of sets. It appears, then, that mathematics is invented by human beings and is then applied to the world around them. This imposes on the universe a convenient order useful for explaining observed phenomena; but the universe is not determined by those phenomena. In the words of Cantor himself,

> mathematics is entirely free in its development and its concepts are restricted only by the necessity of being noncontradictory and coordinated to concepts previously introduced by precise definitions. . . . The essence of mathematics lies in its freedom.[5]

5.8   EXERCISES

For the purpose of these exercises, assume that the general form of the Continuum Hypothesis is true.

1. **(a)** Give an example of a set A of size $\aleph_1$.

   (b) List three specific elements of A.

   (c) Find a subset of A that has size 5.

   (d) Find a subset of A that has size $\aleph_0$.

2. **(a)** Give an example of a set B of size $\aleph_2$.

   (b) List three specific elements of B.

   (c) Find a subset of B that has size 5.

   (d) Find a subset of B that has size $\aleph_0$.

   (e) Find a subset of B that has size $\aleph_1$.

In Exercises 3–8, let K be a set of size $\aleph_5$ and let S be the set of all of its subsets.

3. What is the size of S?

4. Is K a subset of S or is it an element of S?

5. Is $\{K\}$ a subset of S or is it an element of S? What is its size?

6. What is the size of the set of all subsets of S?

7. Find a subset of S that can be put in 1-1 correspondence with K.

8. Find a subset of S that is infinite but that cannot be put in 1-1 correspondence with K.

9. How many transfinite cardinal numbers are smaller than $\aleph_3$? Give an example of a set of each of those sizes.

10. Give a persuasive argument to explain why there should be no size of infinity smaller than $\aleph_0$.

WRITING EXERCISES

1. What idea in this section do you find most confusing or difficult to understand? If, after rereading the section, you still do not understand this idea, write three specific questions that you think might help you. If you now understand it, write a brief explanation that you think might help a classmate who is having trouble with the idea.

2. Assume that the (first part of the) Continuum Hypothesis is false. In this case, what can be said about sets of size $\aleph_1$? Write as many characteristics of these sets as you can find.

3. If, as this section suggests, mathematical truth is not determined by reality but, rather,

5. Georg Cantor, 1883, as quoted on page 1031 of Item 7 in the list *For Further Reading* at the end of this chapter.

simply depends on assumptions you choose to begin with, how is it that mathematics works so well in describing the physical world? (After all, mathematics is fundamental to the construction and operation of almost all the modern machinery we have—automobiles, televisions, airplanes, satellites, CD players, etc.) Think about this for a little while; then summarize your thoughts in one or two paragraphs.

4. In his essay in Appendix C, physiologist James O. Bullock says,

Mathematics differs from other languages . . . because of its complete detachment from the complications of what we experience by direct observation.

Write a paragraph or two relating this comment to the quotation from Georg Cantor that appears at the end of this section.

THE FOUNDATIONS OF MATHEMATICS

Although Cantor's theory of sets was well received in many parts of the mathematical community, acceptance was by no means universal. Cantor's set-theoretic treatment of infinity generated heated opposition from some of his foremost contemporaries, notably Leopold Kronecker, a prominent mathematics professor at the University of Berlin. Kronecker based his approach to mathematics on the premise that a mathematical entity does not exist unless it is actually constructible in a finite number of steps. From this point of view, infinite sets do not exist because it is clearly impossible to construct infinitely many elements in a finite number of steps. The natural numbers are "infinite" only in the sense that the finite collection of natural numbers constructed to date may be extended as far as we please, but "the set of all natural numbers" is not a legitimate mathematical concept. To Kronecker and those who shared his views, Cantor's work was a dangerous mixture of heresy and alchemy that introduced potentially lethal dosages of fantasy into the bloodstream of mathematics.

Kronecker's fears for the safety of mathematical consistency were at least partially justified by the appearance of several paradoxes in set theory. Among the most renowned of these is the self-contradictory notion of the "set of all sets." Cantor's concept of set was extremely general:

By a set we are to understand any collection into a whole of definite and separate objects of our intuition or our thought.

Now, a set is itself a "definite and separate object of our thought," so it would seem sensible to consider the set S of all sets. By its very nature, S would have to contain at least as many elements as any other set; that is, the cardinal number of S would have to be greater than or equal to the cardinal number of any other set. But Cantor's Theorem states that there is no greatest cardinal number; so there must be a set whose cardinal number is greater than that of S (namely, the set of all subsets of S)!

Perhaps the most famous set-theoretic paradox of all was formulated by Bertrand Russell in 1902, and is now known as **Russell's Paradox**. It depends solely on the notion of set, thus striking at the very heart of set theory. The paradox begins by observing that all sets may be classified according to whether or not they are elements of themselves. For example, the set of abstract ideas is an abstract idea and hence is an element of itself; but the set of all elephants is hardly an elephant. Let us call a set that is not an element of itself *normal*, and consider the set $\mathcal{N}$ of all normal sets. In symbols,

$$\mathcal{N} = \{S \mid S \notin S\}$$

Question: Is $\mathcal{N}$ normal?

If we answer *yes*, then $\mathcal{N}$ is in the set of all normal sets; that is, $\mathcal{N}$ is an element of itself, implying that it is not normal. If we answer *no*, then $\mathcal{N}$ is an element of itself, and hence must be normal because it is thereby contained in the set of all normal sets. Thus, either choice leads to a contradiction.[6]

Dilemmas such as this, resulting from an unrestricted use of the seemingly harmless concept of set, forced mathematicians of the late nineteenth and early twentieth centuries to undertake a thorough reappraisal of the foundations of mathematics in an attempt to free it from the dangers of self-contradiction. This, in turn, led to the formulation of several different philosophies of mathematics.

A philosophy of mathematics may be described as a viewpoint from which the various bits and pieces of mathematics can be organized and unified by some basic principles. There have been many philosophies of mathematics throughout history. As the body of mathematical knowledge grew and was changed by the results of new investigations, its philosophies underwent similar mutations. The advent of set theory in all its

Picture Collection, The Branch Libraries, The New York Public Library

The Beardless Barber. He shaves all those who do not shave themselves. Who shaves the barber?

6. There are many popularized versions of Russell's paradox. Russell himself gave one in 1919: A (beardless) barber in a certain village claims that he shaves all those villagers who do not shave themselves, and that he shaves no one else. If his claim is true, who shaves the barber?

unifying simplicity and then the discovery of serious flaws in its fundamental structure, all within less than half a century, brought about a violent upheaval in mathematical philosophy. Its development may be separated into three branches, each attempting in its own way to safeguard mathematics from internal contradictions. In the brief space remaining we summarize a few of the basic tenets that characterize each of these schools of thought.

Logicism

Logicism regards mathematics as a branch of logic, claiming that all mathematical principles are completely reducible to logical principles. Several attempts have been made to reduce all of mathematics to a symbolic logical system, culminating in Bertrand Russell and Alfred North Whitehead's *Principia Mathematica* (1910). The *Principia* bases all of mathematics on a logical system derivable from five primitive logical statements whose truth is founded on basic intuition. Refinements of Russell and Whitehead's work, made by a number of people during the first half of this century, have succeeded in ironing out many of the minor difficulties in the *Principia*.

Nevertheless, there are some fundamental objections to the logistic viewpoint as a whole. It is claimed that some primitive mathematical ideas must be used to develop the system in an orderly fashion, and thus the system is not completely self-contained. Some have also objected that logicism implies that all mathematical ideas are contained in the five initial statements and that the rest of mathematics is merely an exercise in redundancy, a formalized restatement of these five principles, involving no additional information at all.

Intuitionism

Intuitionism proceeds from the premise that mathematics must be based solely on the intuitively given notion of a succession of things, exemplified by the sequence of natural numbers. The intuitionists claim that mathematics is dependent neither on language nor on classical logic. The symbols in mathematics are used for communication only and they are incidental, because mathematics is essentially an individual matter and need not be communicated in order to exist. Moreover, the rules of mathematical reasoning are arrived at intuitively from a logic system that differs from classical logic and is applicable only to mathematics. Like Kronecker, the intuitionists reject the idea that infinitely many elements can be treated as a single thing (a set). They define a set as a *law* that generates a succession of elements. Thus, the "law"

$$\frac{1}{n} \text{ for any } n \in \mathbf{N}$$

is a set whose elements are 1, $\frac{1}{2}$, $\frac{1}{3}$, . . . ; but the collection of its elements cannot be completed by proceeding in this way.

Intuitionist logic does not accept the Law of the Excluded Middle (the principle that a statement that is not false must be true). Thus, for them, proof by contradiction is an invalid procedure. The existence of a mathematical object can be proven only by establishing a method for constructing it in a finite number of steps. These restrictions successfully eliminate the contradictions stemming from set theory, but they also eliminate sizable portions of generally accepted classical mathematics. It remains to be seen whether the intuitionists can refine their approach sufficiently to obtain all of classical mathematics while retaining their advantage of working in a contradiction-free system.

Formalism

Formalism claims that mathematics is concerned solely with the development of systems of symbols. A formal mathematical system is a collection of abstract statements expressing relationships among undefined terms that are subject to a variety of interpretations. Because these systems have no necessary relation to reality, the formalists must guard against potential contradictions by proving that their systems are internally consistent.

Attempts were made to find one provably consistent axiom system for all of mathematics, but in 1931 Kurt Gödel proved this goal to be unattainable. Formalism is thus restricted to a piecemeal verification of the consistency of separate parts of mathematics,[7] and is forced to be content with the tentative assurance that any one mathematical system is as likely to be consistent as any other.

It would be inaccurate and misleading to suggest that all or even most mathematicians adhere rigidly to one of these three mathematical philosophies. The fact is that most mathematicians work in their respective fields "doing mathematics" and concern themselves very little with questions of philosophy. They have formulated opinions about what constitutes mathematics that are sufficient to guide them in their research, and these opinions are often mixtures of the

7. In many instances, a mathematical system is proved to be consistent relative to the assumed consistency of some other, more familiar system, such as the arithmetic of the integers.

viewpoints expressed here. Even those who do concern themselves directly with mathematical philosophy seldom agree in every detail.

The classifications that we have described represent general trends of thought regarding the foundations of mathematics, especially the mathematical treatment of infinity. However, they are at best incomplete views of a subject far more complex than any of these three philosophies seems willing to admit. A truly adequate philosophy of mathematics, if such is possible, invites the efforts of a future generation.

5.9   EXERCISES

WRITING EXERCISES

1. Summarize Russell's Paradox in your own words, without using *any* symbols. Try to capture the main idea as briefly as you can.

2. **(a)** Rewrite this section's description of logicism in a different style; that is, rewrite it in such a way that someone reading the book's description and yours side by side would see all the same main ideas, but would not think that the two descriptions had been written by the same person.

   (b) Do the same as in Part (a) for the description of intuitionism.

   (c) Do the same as in Part (a) for the description of formalism.

3. Reflect on what you know about mathematics so far; then decide which of the three philosophies of mathematics described in this section best fits your sense of the nature of mathematics. State and explain your choice. What do you find wrong or uncomfortable about the other two philosophies? How (if at all) does the one you chose *fail* to match what you think?

4. In his essay reprinted in Appendix C, physiologist James O. Bullock says,

   . . . mathematics is comprehensible precisely *because* it is abstract. We are not dependent on observation to know that lines are completely straight and parallel lines never meet. It is possible to make these assertions with confidence because mathematics was not discovered, it was invented.

   Into which of the three philosophies of mathematics does Bullock's viewpoint fit best? Does it fit all of them equally well? Write one or two paragraphs explaining your answers to these questions.

5. Keeping in mind what you have seen in this chapter, answer the opening question of the book: What is mathematics?

LINK: SET THEORY AND METAPHYSICS

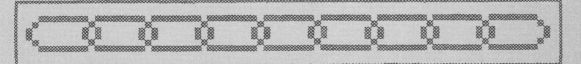

Picture Collection, The Branch Libraries, The New York Public Library

Neo-Thomism at Catholic Universities. Some Neo-Thomists adopted Cantor's theory to defend the existence of infinities. Pictured is the University of Pisa in the latter part of the nineteenth century.

5.10

In very general terms, terms that might well make a professional philosopher bubble over with disclaimers and qualifying phrases, philosophy is the search for a true understanding of reality by means of logical reasoning. Traditionally, the field has been broken down into a number of major subfields, including

- logic—the study of reasoning processes;
- aesthetics—the study of beauty;
- ethics—the study of the principles of human conduct;
- politics— the study of the principles of social organization;
- epistemology—the study of the limits and validity of human knowledge;
- metaphysics—the study of being and reality.

Philosophy as a study distinguishable from religious activity began in Greece in the sixth century B.C., along with deductive mathematics. In fact, sometimes it was impossible to distinguish mathematical questions from philosophical ones, especially when it came to ideas such as motion, time, and infinity. (See, for example, the material on the Pythagoreans and on Zeno's Paradoxes in Section B.3 of the Appendix.)

Bertrand Russell has described philosophy as lying somewhere between the dogmatic pronouncements of religion and the experimental verifications of science. This distinction, while largely accurate today, has not always been valid. In the past, philosophy has often been regarded as encompassing both religion and science, borrowing from or lending to those fields the authority to defend the correctness of some particular theory or thesis. One example of this attempted merger of philosophy, theology, and science occurred in the late nineteenth century, just as Cantor's set theory was emerging into the intellectual daylight.

In 1879, Pope Leo XIII issued the encyclical *Aeterni Patris*, in which he instructed the Catholic Church to revitalize its study of scholastic philosophy. This type of philosophy, also called Thomism, is based on the *Summa Theologica* of Saint Thomas Aquinas (c. 1225–1274), a truly monumental work that attempted to integrate traditional Roman Catholic doctrine with Aristotle's metaphysics. *Aeterni Patris* gave rise to neo-Thomism, a school of philosophical thought that proceeded in part from the view that atheism and materialism result from an incorrect philosophical interpretation of modern scientific achievements and that scholastic philosophy is the key to a "proper" (i.e., orthodox Roman Catholic) understanding of science. With the urging and guidance of the Vatican and the educational influence of various teaching orders of priests (notably the Jesuits), neo-Thomism soon became a major form of philosophy taught and studied in Catholic colleges and universities throughout the world.

When Cantor's work on the mathematics of infinity became known in the 1880s, it generated considerable interest among neo-Thomist philosophers. There was some initial concern that the assertion of the actual existence of infinite sets would lead logically to the heresy of pantheism. Cantor, a devoutly religious man, nevertheless maintained the view that his mathematics of infinite sets indeed dealt with actual reality. For instance, he stated in an 1896 letter:

> The establishing of the principles of mathematics and the natural sciences is the responsibility of metaphysics. . . . The general theory of sets . . . belongs entirely to metaphysics.[8]

Thus, to Cantor, actually infinite collections of numbers had a real (though not necessarily material) existence. In patient, persistent correspondence with some of the leading Catholic theologians, he successfully distinguished his position from heretical views and gained semiofficial acceptance. Some neo-Thomistic philosophers in Germany even adopted his theories to defend the existence of actual infinities. For example, they argued that, because the Mind of God is all-knowing, It must know *all* numbers; hence, not only do all the natural numbers actually exist in the Mind of God, but so do all rationals, all infinite decimals, and so forth.

The most important effect of set theory in philosophy goes far beyond the arguments of the neo-Thomists, however. The attempts of Cantor and his successors to rid set theory of contradictions and thereby make it metaphysically sound led to profound investigations into the foundations of mathematics. Those introspective investigations early in this century that led to the three major schools of mathematical thought described in Section 5.9 have also led to clarifications of logical forms, methods of proof, and errors of syntax, which have, in turn, been used to refine the arguments of philosophy. Modern mathematics has provided philosophy with some explicit, formal guidelines for admissible kinds of reasoning and possible logical constructions. The boundaries between religion, philosophy, and science have been brought into sharper focus as a result.

In particular, a major effect of these efforts has been the removal of mathematics from the realm of metaphysics! Thus, mathematics is now seen as a subject that is independent of reality. In the words of David Hilbert:

> It is not truth, but only "certainty," which is at stake.[9]

Thus, the saga of set theory has a final ironic twist: Cantor, working to establish the metaphysical reality of set theory, began the process that has freed all of mathematics from metaphysics. Perhaps Cantor would have seen this as a tragedy; most modern mathematicians and philosophers see it as a giant forward stride in the progress of human thought.

---

8. Quoted in Herbert Meschkowski, *Ways of Thought of the Great Mathematicians* (trans. by John Dyer-Bennet). San Francisco: Holden-Day, Inc., 1964.

9. Ibid., p. 129.

TOPICS FOR PAPERS

*General
Instructions*

In doing these papers, you may use any outside sources you like. However, please remember that you must properly credit ideas and statements drawn from those other sources. Be careful that you do not simply copy or paraphrase the material from your sources. Any material not explicitly quoted should have been mentally digested by you, so that your own words describe what really are your own ideas when you write them down.

1. This paper assignment asks you to compare and contrast the mathematical idea of infinity with the idea of infinity as it occurs in some other area or as it is seen from some other point of view. Your title should be something like

    "Infinity in Mathematics and Infinity in _____ ,"

    or

    "Mathematical Infinity vs. _____ Infinity."

    Consider the reader of your paper to be an anonymous student in your class—someone who knows the material of Chapter 5 fairly well, but who knows little or nothing about the other area you choose to discuss.

    (a) Choose some area of thought or activity, other than mathematics, in which some idea of infinity arises. Define the term *infinite* as it is used in that area. (Consider: Is the term formally defined in that area? If not, then how do we arrive at an understanding of it? From where does it derive its meaning?)

    > *Note*: Before going further, it might be wise to check with your instructor to be sure that the area and idea you have chosen are suitable for this project.

    (b) Give an example to illustrate the definition you stated in Part **(a)**. (*Consider*: What is the connection between this example and your abstract definition or description?)

    (c) Compare and contrast the idea of infinity you have chosen with the mathematical idea of infinity described in this chapter. How are they similar? How are they different? Apply these questions specifically to the definitions and examples in both areas; then try to formulate some general assessment of the similarities and differences.

    (d) How do the similarities and differences you described in Part **(c)** reflect more general differences between mathematics and the other field you have chosen to discuss?

2. The twentieth-century distinction in mathematics between truth and consistency, noted at the end of this chapter's LINK section, raises many paradoxical questions. Consider these statements:

- As we saw in Section 5.10, Georg Cantor said: "The establishing of the principles of mathematics and the natural sciences is the responsibility of metaphysics. . . ." However, he also said: "The essence of mathematics lies in its freedom."

- Albert Einstein said: "As far as the laws of mathematics refer to reality, they are not certain; and as far as they are certain, they do not refer to reality." Nevertheless, scholars in many areas of thought —from physics and chemistry to psychology, sociology, medicine, and even literary criticism—work hard to make their areas more mathematical in order that their conclusions will be accepted as reliably true.

- David Hilbert said about mathematics: "It is not truth, but only 'certainty,' which is at stake." But many people indicate their conviction that something is *true* by saying that it is "as sure as $1 + 1 = 2$," or words to that effect.

Pretend that you are writing an article about truth and consistency for the *New York Times* Magazine Section. (That is, think of your reader as a fairly intelligent person who is widely read, but who has no special background in mathematics beyond high school.) Write a carefully organized paper in which you

(a) clearly describe the distinction between truth and consistency;

(b) analyze each of the three given statements with respect to the truth/consistency distinction;

(c) use your analysis as the basis for a general discussion of the relationship between mathematics and reality;

(d) apply your general discussion to the concept of mathematical infinity as it has been described in this chapter. (If you covered Chapter 3, you might also find it worthwhile to make connections with some of the ideas presented there.)

You may include your own opinions, but you must give arguments to justify any position you take.

3. In his essay, "Literacy in the Language of Mathematics," which appears in Appendix C, physiologist James O. Bullock makes a number of related assertions about the nature of mathematics:

  - "Mathematics can more properly be regarded as a form of language, developed by humankind in order to converse about the abstract concepts of numbers and space."

  - "The scientist uses the language of mathematics to construct *metaphors* which represent insights into the workings of nature."

  - "Metaphors do not have answers, they have implications."

■ "What distinguishes the mathematical metaphor is the extraordinary power of this language to uncover implications of the underlying idea."

■ "In interpreting mathematical language in the scientific literature, we tend to become confused by the Platonistic view that mathematical objects themselves actually exist. . . . [T]he use of this language does not imbue one's ideas about nature with an independent existence. The use of mathematical objects is still a metaphorical device. Mathematical metaphors are not different from other forms of human understanding. They may provide significant insights, but they are not a holy grail of ultimate truth."

Write an essay relating Bullock's view of mathematics as a language of metaphors to

(a) Cantor's theory of sets and

(b) the philosophies of mathematics described in Section 5.9.

In particular, comment on Bullock's implicit assertion that mathematical objects do not exist. Do you agree or disagree? For what reasons? Does the existence or nonexistence of mathematical objects affect the usefulness of mathematics as a tool for understanding reality? Why or why not? There are no universally correct answers to these questions. The quality of the paper you write will depend on the clarity and persuasiveness of the arguments you present in support of your opinions.

FOR FURTHER READING

1. Bell, E. T. *Men of Mathematics*. New York: Simon and Schuster, Inc., 1937, Chapter 29.

2. Dauben, Joseph Warren. *Georg Cantor*. Cambridge, MA: Harvard University Press, 1979.

3. Davis, Philip J., and Reuben Hersh. *The Mathematical Experience*. Boston: Birkhäuser Boston, Inc., 1981, Part 7.

4. Devlin, Keith. *Mathematics: The New Golden Age*. London: Penguin Books, 1988, Chapter 2.

5. Dunham, William. *Journey Through Genius: The Great Theorems of Mathematics*. New York: John Wiley and Sons, Inc., 1990, Chapters 11 and 12.

6. Eves, Howard, and Carroll V. Newsom. *An Introduction to the Foundations and Fundamental Concepts of Mathematics*. New York: Holt, Rinehart and Winston, 1965, Chapters 8 and 9.

7. Kline, Morris. *Mathematical Thought from Ancient to Modern Times.* New York: Oxford University Press, 1972, Chapters 41, 43, and 51.

8. Newman, James R., ed. *The World of Mathematics.* New York: Simon and Schuster, Inc., 1956, Part X.

9. Reid, Constance. *From Zero to Infinity*, 3rd Ed. New York: Thomas Y. Crowell Company, 1964, Chapter ... .

MATHEMATICS OF SYMMETRY: FINITE GROUPS

WHAT IS GROUP THEORY?

The historical roots of group theory lie in early nineteenth-century Europe. At that time mathematicians in several different countries made largely independent efforts to free algebra from its dependence on numbers. As they worked to strip their algebraic tools down to the essential underlying ideas, they were led to remarkably similar patterns and concepts. This process of stripping down an idea to its essential features is called "abstraction"; it will be discussed more a little later in this section. For now, let us just say that the abstraction process is reminiscent of the Cheshire Cat of Wonderland, who faded slowly away before Alice's very eyes, until only its grin remained.

The grin that remained in algebra after the numbers disappeared is a mathematical theory of symmetry known as *group theory*. Symmetry is one of the most basic concepts of human thought. We observe balance and proportion in nature and we try to capture it in art; we observe symmetrical structure in science and repeat its harmony in music. Group theory's importance to us today stems from the fact that it is the mathematical tool used to describe symmetry in all these areas. Groups occur in the geometry of artistic perspective, in the design of wallpaper patterns, in the theory of modern computer codes, and in the patterns of traditional bell ringers. Chemists use group theory to classify and analyze molecules somewhat as they use the periodic table to relate the chemical properties of the elements. Physicists use group theory as a basis for theories of nuclear structure. Mathematical group theory is even beginning to turn up in sociology and psychology. And, despite all these applications to the "real world," some mathematicians study group theory for its own sake, to extend their understanding and appreciation of its internal symmetries and harmonies.

The theory of groups is part of a mathematical field called *abstract algebra*. Each of these two words has been known to cause discomfort; together they can be positively terrifying. Let us put these fears to rest by examining exactly what this phrase means. Many people consider the word abstract to be a synonym for "vague" or "hard to understand"; but, in fact, its meaning is quite different. Its literal meaning is derived from Latin and is better reflected in our use of the verb *to abstract*, which means "to pull out of" or "to separate from." (It has the same root as "tractor.") Thus, the adjective *abstract* is used to indicate some property of a thing that is considered apart from that thing's other characteristics. The abstraction process is a way of simplifying a situation by focusing directly on a specific aspect of it, separating its essential features from other facts that might confuse the issue.[1]

Let us turn to arithmetic for an example. Among the many common arithmetic facts at our disposal, we know that

$$6 \cdot 4 = 24, \quad 7 \cdot 3 = 21, \quad 8 \cdot 2 = 16, \quad 9 \cdot 1 = 9, \quad 10 \cdot 0 = 0$$

This list of data has many features, some interesting from one point of view, some from another. For instance, we might observe that every digit except 5 was used, or that the products are alternately even and odd, or that all the numerals are printed with the same-size type, or that the two factors add up to 10 in each case. All these observations are true, but they may distract us from some more important ones. In this example, let us rearrange the information given, by starting with the observation that in each case 5 is "in the middle of" the two numbers being multiplied:

$$(5 + 1)(5 - 1) = \ 6 \cdot 4 = 24 = 25 - \ 1$$
$$(5 + 2)(5 - 2) = \ 7 \cdot 3 = 21 = 25 - \ 4$$
$$(5 + 3)(5 - 3) = \ 8 \cdot 2 = 16 = 25 - \ 9$$
$$(5 + 4)(5 - 4) = \ 9 \cdot 1 = \ 9 = 25 - 16$$
$$(5 + 5)(5 - 5) = 10 \cdot 0 = \ 0 = 25 - 25$$

A close look at these facts in this form suggests a general observation about the behavior of numbers, which can be written in the following *abstract* form:

---

1. For a related view of mathematical abstraction, read the first paragraph of "Mathematics is a Language" in James O. Bullock's essay, "Literacy in the Language of Mathematics," in Appendix C.

If two numbers are equidistant from and on opposite sides of 5, then their product equals 25 minus the square of their distance from 5; that is,

$$(5 + d) \cdot (5 - d) = 5^2 - d^2$$

The use of d here in place of the various specific distances is an essential abstraction step; it unifies the five specific cases of the one pattern and brings the pattern itself into focus. Moreover, this last equation suggests that the number 5 might not be crucial to the behavior of such products. This leads us to an even more general form, which you may remember from high school:

For any numbers x and y, $(x + y)(x - y) = x^2 - y^2$

This example typifies much of what is called *algebra* in high school. High school algebra, and indeed, the only algebra studied and practiced by mathematicians until fairly modern times, is little more than symbolized arithmetic. Various methods for manipulating numbers and the basic arithmetic operations $+$, $-$, $\times$, and $\div$ are abstracted (pulled out) by using letters to stand for numbers. This makes it easier for us to focus on general rules that describe the behavior of our number system. That area of mathematics is sometimes called "classical algebra"; it might appropriately be called "abstract arithmetic."

Abstract algebra takes a further step in generality (and thus also in simplicity). Rather than confining itself to the numbers and operations of our usual number system, this form of algebra looks at various collections (*sets*) of objects and discusses operations on them. Because the objects considered need not be numbers (for instance, they could be letters, or motions, or just about anything), the usual operations of arithmetic may not apply to them. Hence, the fundamental properties of the "operation" concept must be abstracted from their numerical settings and put in a general form to permit application to other collections of things. This shift of attention from the concept of number to the concept of operation is what distinguishes modern algebra from classical algebra.

The conceptual unity that results from the abstraction process is one of the most significant features of modern mathematics as a whole. Let us look at a simple example of this unifying effect. Consider these three situations:

1. A 3-way light has four switch positions: *low, medium, high,* and *off.* Each time the light switch is turned, the light goes from one of these settings to the next, in order. Each "click" of the switch signals a movement from one level of brightness to the next; with the fourth click the light is *off,* with the fifth it is at *low* again, and so on.

Watches. Since the days of sailing ships, a watch has been four hours long.

2. The duty periods for crew members aboard ship are called *watches*. By a tradition dating back to the early days of sailing ships, each watch is four hours long and the passing of every hour is marked by pairs of bells—2 bells for the first hour, 4 for the second, 6 for the third, and 8 bells to signal the end of a watch (and the beginning of another).

3. A square tile works loose from its position in a bathroom floor. Its shape allows it to be put back in any of four ways: exactly as it was originally, or rotated (clockwise) 90°, or 180°, or 270°. (A 360° rotation puts it back to its original position.) Even if it has been kicked around for a while, whatever way it is put back *must* correspond to one of these four "quarter turns" relative to its original (but perhaps forgotten) position.

Let us put the abstraction process to work. Some common features of these three very different situations are obvious: Each involves four "things"—brightness levels, watch hours, tile positions—and we can go from each to the next in succession until we return to a "starting point"—*off*, beginning of watch, original position. Now, if we focus not on the things themselves but on the way of getting from one to another, a common structure emerges. In each case, we go from one "thing" to another by steps in a cyclic pattern; moreover, if we take more than three steps we will have "gone around" the cycle completely and will have started over. Thus, we might as well ignore all "strings" of four successive steps. For example:

A Square Tile in a Bathroom Floor. The square shape of a loosened tile allows it to be put back in any of four ways.

- 5 switch clicks from *off* gets us to *low* again—so it's the same as 1 click;

- 6 hours from the beginning of a watch is 2 hours into a watch (the next one, of course);

- 7 quarter turns of the tile puts it in the same position as 3 quarter turns;

and so on.

The numbers of the steps that take us to the different positions in each situation, then, form the set $\{0, 1, 2, 3\}$. If we follow any of these step numbers by any other, we get the same result as some one of these four numbers (2 steps following 3 steps is equivalent to 1 step, etc.). If we denote by f the process "followed by," we can tabulate all possible combinations of these four step numbers, as in *Table 6.1*.

If we further simplify this tabulation by listing just the four columns of answers, we obtain a table whose patterns and symmetry almost beg for further exploration! (See *Table 6.2*.)

Further examination of this pattern is best left until we get some basic definitions and structural ideas under our belts. For now, suffice it to say that this pattern is an example of something called a *group*, and this particular group describes the operational behavior of the three different situations that we have described. From this viewpoint, then, the three situations have been unified into one, and *the study of the single table representing this one pattern (Table 6.2) tells us about all three of the situations from which it was derived.*

This example illustrates in a simple way the value of abstraction. The abstraction process unifies seemingly different situations by

$$
\begin{array}{llll}
0\ f\ 0 = 0 & 0\ f\ 1 = 1 & 0\ f\ 2 = 2 & 0\ f\ 3 = 3 \\
1\ f\ 0 = 1 & 1\ f\ 1 = 2 & 1\ f\ 2 = 3 & 1\ f\ 3 = 0 \\
2\ f\ 0 = 2 & 2\ f\ 1 = 3 & 2\ f\ 2 = 0 & 2\ f\ 3 = 1 \\
3\ f\ 0 = 3 & 3\ f\ 1 = 0 & 3\ f\ 2 = 1 & 3\ f\ 3 = 2
\end{array}
$$

Table 6.1 *Combinations of the four step numbers.*

)	1	2	3
(	2	3	0
2	3	0	1
3	0	1	2

Table 6.2 The common pattern of situations 1, 2, and 3.

focusing on (abstracting) some particular properties those situations have in common, so that a single study of the abstract properties can give us information about all the "different" situations at the same time. Thus, abstraction is a powerful tool in applied science; it reduces the amount of effort required to analyze things with similar characteristics and sometimes suggests unsuspected connections among seemingly different phenomena.

6.1 — EXERCISES

For Exercises 1–4, write three specific instances of the given abstract numerical formula. (Assume x, y, and z represent numbers of some kind.) Work out the arithmetic to verify that your examples are true statements.

1. $3x + 3y = 3(x + y)$

2. $(x + y) + (x - y) = 2x$

3. $(x + y) + z = x + (y + z)$

4. $(-x)(-y)(-z) = (-x)yz$

For Exercises 5–12, write a general expression using letters to abstract a general form from the given set of numerical data.

5. $2 \cdot 3 = 3 + 3$
   $2 \cdot 5 = 5 + 5$
   $2 \cdot 14 = 14 + 14$
   $2 \cdot 27 = 27 + 27$

6. $2 + 3 = 3 + 2$
   $5 + 8 = 8 + 5$
   $7 + 19 = 19 + 7$
   $43 + 88 = 88 + 43$

7. $2 \cdot 3 + 2 \cdot 5 = 2 \cdot 8$
   $7 \cdot 12 + 7 \cdot 8 = 7 \cdot 20$
   $5 \cdot 40 + 5 \cdot 2 = 5 \cdot 42$
   $3 \cdot 1 + 3 \cdot 1 = 3 \cdot 2$

8. $25 - 9 = 8 \cdot 2$
   $36 - 25 = 11 \cdot 1$
   $100 - 4 = 12 \cdot 8$
   $81 - 49 = 16 \cdot 2$

9. $2 \cdot 2^2 = 2^3$
   $2 \cdot 2^5 = 2^6$
   $2 \cdot 2^9 = 2^{10}$
   $2 \cdot 2^{37} = 2^{38}$

10. $2 \cdot 2^2 = 2^3$
    $2^2 \cdot 2^5 = 2^7$
    $2^{-1} \cdot 2^9 = 2^8$
    $2^{10} \cdot 2^{10} = 2^{20}$

11. $3 \cdot 3^2 = 3^3$
    $5 \cdot 5^5 = 5^6$
    $7 \cdot 7^9 = 7^{10}$
    $3.45 \cdot 3.45^{37} = 3.45^{38}$

12. $3 \cdot 3^2 = 3^3$
    $6^2 \cdot 6^5 = 6^7$
    $(-12)^{-1} \cdot (-12)^9 = (-12)^8$
    $17^{10} \cdot 17^{10} = 17^{20}$

For Exercises 13 and 14, recall that Situation 3 (described on page 283), which led to the operation described by *Tables 6.1 and 6.2*, was that of a loose square tile.

13. What if the tile were in the form of an equilateral triangle? Describe the resulting set and operation, and write them in the two forms exemplified by *Tables 6.1 and 6.2*.

14. What if the tile were in the form of a regular hexagon (as many old bathroom tiles are)? Describe the resulting set and operation, and write the pattern in the form exemplified by *Table 6.2*.

15. Construct a table with 5 rows and 5 columns that has the same pattern as *Table 6.2* on page 285. Describe the pattern in words. Then translate the information in your table into the form of *Table 6.1* on page 284.

16. The hour counter of a digital clock displays only the numbers 1, 2, . . . , 12. Represent by ⊕ the process of adding hours to some displayed number of hours and displaying the result. Thus, for example,

$$10 \oplus 2 = 12, \quad 10 \oplus 3 = 1, \quad 10 \oplus 4 = 2$$
and so forth.

(a) What is displayed for 3 ⊕ 5? for 7 ⊕ 8? for 9 ⊕ 10?

(b) Display all the possible ⊕ combinations as a table.

(c) Compare the pattern of your table with the pattern of *Table 6.2*. How are they similar? How are they different?

WRITING EXERCISES

1. Examine *Table 6.2* on page 285 and describe patterns (of any sort) that you see in it. Explain how each pattern you find might be translated to say something about each of the three different situations the table describes.

2. Find and describe another real-world situation (besides the three given in the text) that fits the pattern summarized by *Table 6.2*.

3. *Table 6.2* expresses an underlying structure that is common to three real-world situations, each of which was decribed by a short paragraph. In like manner, find and describe three real-world situations with a common underlying structure that is described by *Table 6.3*.

0	1
1	0

Table 6.3 Another pattern common to some real-world situations.

OPERATIONS

6.2

The first step in our investigation of groups will be to isolate the "operation" idea in some general form. If we look at the familiar arithmetic operations of addition, subtraction, multiplication, and division, we see that they share a common characteristic: They all are ways of assigning an "answer" number to two numbers. For instance, addition assigns 8 to 3 and 5 (because $3 + 5 = 8$), whereas multiplication assigns 15 to 3 and 5 (because $3 \cdot 5 = 15$). To abstract this idea from its numerical setting, we begin with the idea of an **ordered pair**, a pair of elements (of any sort) in which the first element is distinguishable from the second. We write (x, y) to denote the ordered pair in which x is the first element and y is the second. If x and y are different, the ordered pairs (x, y) and (y, x) are different. (By way of contrast, the set $\{x, y\}$ is the same as the set $\{y, x\}$ because both sets contain exactly the same elements.) Now it becomes easy to state a useful general definition of operation.

DEFINITION	An **operation** on a set S is any rule or process that assigns to each ordered pair of elements of S exactly one element of that set S.[2]

NOTATION	We shall generally use $*$ as the symbol for an operation on a set, although other symbols will be used as well. The element of the set that the operation $*$ assigns to the ordered pair (x, y) will be written

$$x * y$$

EXAMPLE 6.1	As noted above, the four arithmetic operations are examples of operations. The first three are defined on the set of all real (or complex) numbers; division is an operation on the set of all nonzero real (or complex) numbers.

- Addition assigns to each ordered pair (x, y) the sum $x + y$;
- subtraction assigns to (x, y) the difference $x - y$;
- multiplication assigns to (x, y) the product $x \cdot y$;
- division assigns to (x, y) the quotient $x \div y$.

EXAMPLE 6.2	The tabulation at the end of Section 6.1 defines an operation, "f," on the set $\{0, 1, 2, 3\}$ because it assigns to each of the 16 ordered pairs of those four numbers some number from that set—$(2, 1)$ is assigned 3, $(1, 3)$ is assigned 0, $(0, 1)$ is assigned 1, and so forth.

EXAMPLE 6.3	As we saw in Section 6.1, a particular case of the preceding example is the set of the four rotations of a square—rotations of $0°$, $90°$, $180°$, and $270°$. Because any of these rotations followed by any other gives the same result as if one of the original four rotations had occurred, the *followed by* process is an operation on this set of rotations.

A similar operation can be defined for rotations of any regular polygon, such as an equilateral triangle or a regular pentagon or hexagon. To find the numbers of degrees for the appropriate rotations, just divide 360 by the number of sides of the figure; then take successive multiples of that number until you get back to 360. For instance, the appropriate rotations for a triangle are

$$0°, 120°, \text{ and } 240°$$

For a hexagon they are

$$0°, 60°, 120°, 180°, 240°, \text{ and } 300°$$

2. Strictly speaking, such operations are usually called *binary* operations because they combine two elements at a time. This is the only kind of operation we shall consider, so the word *binary* has been omitted for simplicity.

In each case the result of following one of these rotations by another gives the same result as one of the original rotations in the set. We will have more to say about movements of geometric figures in Section 6.9.

EXAMPLE 6.4

Let S be the set of all students in a class. If we assign to each ordered pair of students the taller one, we have an operation on S. If we pick out a particular student and assign that student to each ordered pair, we have another operation on S (even though in this case all the "answers" are the same). However, if we assign to each ordered pair of students in S the average of their grades on the last exam, we do *not* have an operation on S, because the "answers" are not in the set S of students.

EXAMPLE 6.5

The process that assigns to each pair (x, y) of numbers the *mean*, $(x + y) \div 2$, is an operation on the set of all rational numbers, but not on the set of whole numbers. (Why not?)

EXAMPLE 6.6

Let S be the set of letters {n, d, q}, and define an operation * on S by listing all possible ordered pairs and assigning elements to them at random. An example of this is shown in *Table 6.4*. Because every ordered pair is assigned exactly one element of S, this listing specifies an operation on S (even though there is no apparent pattern to the assignments).

Listing the assignments for each ordered pair is sometimes the only way a particular operation can be written down because it is not always possible to give a succinct general rule for it. We can make the job shorter and the result clearer, however, by organizing the information in an **operation table**. An operation table works like the grid used for reading a road map and is perhaps most easily explained by example. To put the operation of Example 6.6 into tabular form, we list the elements of S in a row and in a column, as shown in *Table 6.5*. This arrangement creates a square array of nine boxes, as indicated by the dotted lines, which are to be filled. If we "name" each box by the letter indicating its row and then the letter for its column, we have a box for

$n * n = q$	$n * d = d$	$n * q = n$
$d * n = d$	$d * d = n$	$d * q = n$
$q * n = n$	$q * d = n$	$q * q = d$

*Table 6.4 A table for an operation, *, on S = {n, d, q}.*

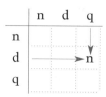

*Table 6.5 The form of an operation table, showing the entry d * q = n.*

*	n	d	q
n	q	d	n
d	d	n	n
q	n	n	d

Table 6.6 The information of Table 6.4 in a more compact form.

each ordered pair of elements of S, and we can fill each one with the element assigned to that pair by the operation. The box filled in *Table 6.5* shows the entry that corresponds to d * q. *Table 6.6* shows the complete table for this operation (with the dotted lines omitted). The information supplied is the same as that given in *Table 6.4*, but it is displayed much more compactly. The symbol for the operation is noted in the upper-left corner for convenience.

EXAMPLE 6.7

Let $S = \{a, b, c\}$, and define the operation * on S by means of *Table 6.7*, which says that a * a = b, a * b = a, a * c = c, etc.

EXAMPLE 6.8

Table 6.8(a) and (b) defines the two operations * and ∘ on the set $\{0, 1\}$. (How many other operations can you define on this set?)

Let us denote the set $\{\ldots, -3, -2, -1, 0, 1, 2, 3, \ldots\}$ of all integers by $\mathbf{I}$ and the set $\{1, 2, 3, \ldots\}$ of all positive integers by $\mathbf{I}^+$. Ordinary addition and multiplication are operations on both $\mathbf{I}$ and $\mathbf{I}^+$. However, subtraction is an operation on $\mathbf{I}$ but not on $\mathbf{I}^+$ because, for example, subtraction assigns to the ordered pair of positive integers (3, 4) the number -1, which is in $\mathbf{I}$ but not in $\mathbf{I}^+$. (Recall that the definition of operation requires the assigned elements to come from the original set.)

In general, an operation defined on a set S may not be an operation on every subset[3] of S, because the operation on S may assign to an ordered pair of subset elements some element of S that is not in the subset. If the operation does assign to each pair of subset elements an element from that subset, we say that the subset is **closed** under the operation (or the operation is closed on the subset). This property is called **closure**.

*	a	b	c
a	b	a	c
b	c	b	a
c	a	c	b

*Table 6.7 A table for the operation * on S = {a, b, c}.*

*	0	1
0	0	1
1	1	0

(a)

∘	0	1
0	0	0
1	0	1

(b)

*Table 6.8 Tables for the operations * and ∘ on the set {0, 1}.*

---

3. A subset of a set S is any set whose elements all are also elements of S.

EXAMPLE 6.9

The even integers are closed under addition; that is, the sum of two even integers is always even. The odd integers are not closed under addition; for example, $3 + 5 = 8$, so the sum of two odd integers is not always (in fact, is never) odd. Both the even integers and the odd integers are closed under multiplication. (Can you justify these statements?)

EXAMPLE 6.10

Table 6.9 defines an operation, *, on the set {a, b, c, d}. The subset {a, b} is closed under *, but the subsets {c, d} and {a, b, c} are not. (Can you justify these statements?)

*	a	b	c	d
a	a	b	c	d
b	b	a	d	c
c	c	d	a	b
d	d	c	b	a

*Table 6.9 The operation * on the set {a, b, c, d}.*

6.2   EXERCISES

1. Referring to the operation * defined by *Table 6.9*, find a * b, b * c, d * d, and c * a.

Table 6.10 defines an operation ◇ on the set
{2, 3, 4, 5, 6}
Use this table for Exercises 2–11.

◇	2	3	4	5	6
2	4	2	6	3	5
3	5	5	2	5	2
4	3	6	2	2	3
5	4	4	4	4	4
6	2	6	3	4	5

Table 6.10 The operation ◇ on the set {2, 3, 4, 5, 6}.

2. $2 \diamond 4 =$

3. $3 \diamond 5 =$

4. $4 \diamond 2 =$

5. $6 \diamond 6 =$

6. $5 \diamond 5 =$

7. $6 \diamond 3 =$

8. $(2 \diamond 3) \diamond 4 =$

9. $2 \diamond (3 \diamond 4) =$

10. $(5 \diamond 5) \diamond 5 =$

11. $5 \diamond (5 \diamond 5) =$

12. Let $S = \{q, r, s, t\}$ and define an operation, *, on S as follows:
    - If q is in a pair, that pair is assigned q;
    - If r is in a pair, that pair is assigned the other element in the pair, and r * r = r;
    - $s * t = t * s = t$;
    - $s * s = t * t = s$.

    Describe * by means of an operation table.

13. Let $S = \{1, 2\}$. Construct tables describing *all* possible operations on S. (This exercise will be used in Exercise 2 of Section 6.3.)

For Exercises 14 and 15, let $S = \{1, 2, 3, 4, 5\}$ and let x and y represent arbitrary elements of S.

14. Define an operation, ⋆, on S by $x \star y = y$ for all elements x and y in S. Construct an operation table for ⋆ on S.

15. Define an operation, ○, on S by:
    - If $x \neq y$, then $x \circ y$ is the larger number;
    - if $x = y$, then $x \circ y = 1$.

    Construct an operation table for ○ on S.

For Exercises 16–19, let E be the set of all English words and let S be the set of all "letter strings" (finite lists of letters, possibly repeated) of our alphabet. Notice that E is a subset of S. In each exercise a process, $*$, that makes sense on these two sets is described. The Greek letters α and β represent arbitrary elements of these sets.) Answer these questions for each exercise:

a) Give two more examples to show how $*$ works.

b) Is $*$ an operation on S? on E? Justify each answer with a reason or a counterexample.

16. $\alpha * \beta$ is formed by putting α and β next to each other, forming a single string. For example,

moon $*$ glow = moonglow

(This is like the password process used by CompuServe and other electronic network services.)

17. $\alpha * \beta$ is the word or string that in alphabetical order comes first. For example,

trip $*$ trap = trap

(This is a basic part of any word-sorting algorithm.)

18. $\alpha * \beta$ is the string of letters, including repeated letters, that are common to α and β, arranged in alphabetical order. For example,

beaker $*$ knee = eek

19. $\alpha * \beta$ is the number of letters in the longer word or string, or if they have the same number of letters, it's that number. For example,

movie $*$ theater = 7

20. Let $T = \{r, s, t\}$.

(a) Construct five different operation tables for T.

(b) How many operations are possible on T? Why?

(c) Which of the operations that you defined in Part (a) are closed on the subset $\{s, t\}$ of T?

21. Find two examples to show that the odd integers are not closed under addition.

22. Find three examples to illustrate that the even integers are closed under addition.

23. Abstract from the examples given in Exercise 22 to make a general argument showing that the sum of two even integers must be even. (*Hint:* Any even integer is of the form $2n$, where n is an integer.)

24. Keeping in mind the discussion in Example 6.3, list the appropriate rotations for a regular pentagon, and show that any one of them followed by any other must give the same result as one of the original rotations. (*Hint:* There are five of them.)

WRITING EXERCISES

1. We have seen that an operation on a finite set can be described either by stating some sort of succinct procedural rule or by filling in a table; many operations can be described both ways. Discuss the relative advantages and disadvantages of each approach.

2. The footnote to the definition of *operation* observes that this definition actually only refers to *binary* operations because it is stated in terms of pairs.

(a) By analogy, define "ternary operation."

(b) Construct and describe at least one interesting example of such an operation. (An interesting example should not be simply the repeated application of a single binary operation.)

(c) What would an operation table for a ternary operation look like? If possible, construct a table for a ternary operation on the set 0, 1 and describe how it works.

SOME PROPERTIES OF OPERATIONS

The abstract definition of an operation is so general that the list of all possible operations even on a small set would be far too long to permit a meaningful examination of each one. For instance, because any operation on a three-element set can be written as a 3-by-3 table and because each of the nine entries in that table can be any of the three elements in the set, there is a total of 3^9 such operation tables. Now, $3^9 = 19{,}683$; so on any three-element set we can define 19,683 different operations! To organize this bewildering variety of things in some manageable way, we look at several properties that make some operations "nicer" than others. The properties discussed here should already be familiar to you; they are just generalizations of arithmetic concepts.

The most obvious "nice" properties are suggested by a close look at the definition of *operation* itself. The definition says that the operation must deal with ordered pairs, implying that the element assigned to a pair in one order might differ from the one assigned to that pair taken in the reverse order. Indeed, this is what happens with the usual arithmetic operations of subtraction and division:

$$5 - 3 = 2 \quad \text{but} \quad 3 - 5 = -2$$

$$10 \div 2 = 5 \quad \text{but} \quad 2 \div 10 = \frac{1}{5}$$

However, addition and multiplication assign the same number to a pair of numbers in both orders:

$$5 + 3 = 8 \quad \text{and} \quad 3 + 5 = 8$$

$$10 \cdot 2 = 20 \quad \text{and} \quad 2 \cdot 10 = 20$$

Thus, addition and multiplication are easier to work with. This suggests the following general property:

DEFINITION

An operation $*$ on a set S is **commutative** if

$$a * b = b * a$$

for all elements a and b in S.

EXAMPLE 6.11

The operation on the set of letters {a, b, c} defined in *Table 6.7* of Section 6.2 is not commutative because, for example, a $*$ b = a, but b $*$ a = c. However, the operations given in *Tables 6.6, 6.8, and 6.9* are all commutative.

Table 6.11
Commutativity in an operation table .

An easy way to see whether or not a table defines a commutative operation is to imagine folding it along the diagonal from the upper-left corner of the table to its lower-right corner, thereby matching up each $a * b$ with $b * a$. If the table entries that match by this folding are the same, then the operation is commutative. *Table 6.11* illustrates this.

If an operation is not expressed by a table, then a proof that it is commutative might be difficult because a general argument is needed to show that *every* pair is assigned the same element in both orders. On the other hand, since the definition of commutativity requires the reversibility to hold in *every* case, to prove that an operation is *not* commutative, it suffices to find *one* pair that is assigned different elements in its two orders.

Returning to the definition of an operation, let us now focus on the requirement that operations assign elements to *pairs* of elements. It is natural to ask, then, how three or more elements can be combined. Any attempt to apply a (binary) operation to three elements must proceed two at a time, and (because not all operations are commutative) the order in which the elements are considered is important. However, even if the order of the elements is fixed, there are still two ways to apply the operation. If a, b, and c are three elements (taken in that order), we could

combine a and b, and then combine $a * b$ with c

or

combine b and c, and then combine a with $b * c$

In symbols, we could consider either

$$(a * b) * c \quad \text{or} \quad a * (b * c)$$

For instance, if asked to *subtract* the numbers 9, 5, and 2, we might compute

$$(9 - 5) - 2 = 4 - 2 = 2$$

or

$$9 - (5 - 2) = 9 - 3 = 6$$

On the other hand, *adding* 9, 5, and 2 is independent of this grouping process:

$$(9 + 5) + 2 = 14 + 2 = 16$$

and

$$9 + (5 + 2) = 9 + 7 = 16$$

This suggests another property that some operations possess.

DEFINITION

An operation $*$ on a set S is **associative** if

$$(a * b) * c = a * (b * c)$$

for all elements a, b, and c in S.

EXAMPLE 6.12	The operation defined by *Table 6.7* on page 289 is not associative because, for example,

$$(b * b) * c = b * c = a \qquad \text{but} \qquad b * (b * c) = b * a = c$$

EXAMPLE 6.13	Both operations in *Tables 6.8* on page 289 are associative, as is the operation defined by *Table 6.9* on page 290. Unfortunately, there is no easy way to observe this from the tables (as there is for commutativity); a proof of these facts would require checking all possible cases. We give here just one example for each operation from Tables 6.8; you may verify the other cases at your leisure.

$$(1 * 1) * 0 = 0 * 0 = 0 \qquad \text{and} \qquad 1 * (1 * 0) = 1 * 1 = 0$$
$$(1 \circ 1) \circ 0 = 1 \circ 0 = 0 \qquad \text{and} \qquad 1 \circ (1 \circ 0) = 1 \circ 0 = 0$$

6.3   EXERCISES

1. Give an example to show that ordinary division (on sets of nonzero numbers) is an operation that is not associative.

2. (This exercise refers back to Exercise 13 of Section 6.2.) In all the examples we have examined so far, operations either have been both associative and commutative or have had neither of these properties. It is natural to ask whether these properties always exist together. Show that such is not the case by finding among the 16 operations on the set {1, 2} an example of

   (a) an operation that is commutative but not associative;

   (b) an operation that is associative but not commutative.

For Exercises 3–8: Is the operation defined by the given table commutative? Is it associative? Give reasons for your answers.

3. The operation * on the set {1, 2, 3} defined by *Table 6.12(a)*.

4. The operation ∘ on the set {1, 2, 3} defined by *Table 6.12(b)*.

5. The operation * on the set {5, 9} defined by *Table 6.13(a)*.

6. The operation ∘ on the set {5, 9} defined by *Table 6.13(b)*.

7. The operation * on the set {2, 4, 6} defined by *Table 6.14(a)*.

8. The operation ∘ on the set {2, 4, 6} defined by *Table 6.14(b)*.

*	1	2	3
1	3	3	3
2	3	3	3
3	3	3	3

(a)

∘	1	2	3
1	1	3	2
2	3	3	1
3	2	1	2

(b)

*Table 6.12 The operations * and ∘ on the set {1, 2, 3}.*

*	5	9
5	5	9
9	5	9

(a)

∘	5	9
5	9	9
9	5	9

(b)

*Table 6.13 The operations * and ∘ on the set {5, 9}.*

9. Construct an operation on the set {p, q, r} that is both commutative and associative.

10. Construct an operation on the set {p, q, r} that is commutative, but not associative.

11. Construct an operation on the set {p, q, r} that is associative, but not commutative.

12. Construct an operation on the set {p, q, r} that is neither commutative nor associative.

Exercises 13–16 refer back to Exercises 16–18 of Section 6.2. Those exercises described the following three operations on the set S of all "strings" (finite lists) of letters of the alphabet. In each case, say whether or not the given operation is associative and/or commutative, supporting your answers with reasons or counterexamples. As before, the Greek letters α and β represent arbitrary elements of S.

13. $\alpha * \beta$ is formed by putting α and β next to each other, forming a single string.

14. $\alpha * \beta$ is the word or string that in alphabetical order comes first.

15. $\alpha * \beta$ is the string of letters, including repeated letters, that are common to α and β, arranged in alphabetical order.

16. Define another operation on S, one that is neither associative nor commutative. Give counterexamples to show how each property fails.

17. Example 6.3 on page 287 explained that, for each regular polygon, the *followed by* process is an operation on an appropriate set of rotations of the figure. Is this operation associative? Is it commutative? Support your answers with reasons or counterexamples.

18. The process of "raising to a power" is an operation on the set $\mathbf{I}^+$ of positive integers. That is, the process "$\wedge$" defined by
$$a \wedge b = a^b$$
is an operation called **exponentiation**.
   (a) Give two specific examples of how exponentiation works.
   (b) Is exponentiation commutative? Justify your answer.
   (c) Is exponentiation associative? Justify your answer.

19. On a number line, the distance between two points, a and b, is the absolute value of their difference. We can regard this as an operation, *, on the set of all real numbers:
$$a * b = |a - b|$$
Is * commutative? Is * associative? Justify your answers.

Exercises 20–25 present six different operations on the set I of integers. For each operation, answer these questions:
(a) Is it commutative? If so, give an example; if not, give a counterexample.
(b) Is it associative? If so, give an example; if not, give a counterexample.

20. $a * b = a + b - 1$ 21. $a * b = a - b + a$
22. $a * b = a + b + a$ 23. $a * b = a + b - a$
24. $a * b = (a - b)^2$ 25. $a * b = a^2 - b^2$

26. Give a general argument to prove that *Table 6.15* defines an associative operation on the set {1, 2, 3, 4, 5}. (A case-by-case verification is a permissible but unattractive strategy; there are 125 cases to be checked.)

*	1	2	3	4	5
1	1	1	1	1	1
2	2	2	2	2	2
3	3	3	3	3	3
4	4	4	4	4	4
5	5	5	5	5	5

Table 6.15
*The associative operation * on the set {1, 2, 3, 4, 5}.*

*	2	4	6
2	2	6	4
4	4	2	6
6	6	4	2

(a)

∘	2	4	6
2	6	4	4
4	4	6	4
6	4	4	6

(b)

Table 6.14
*The operations * and ∘ on the set {2, 4, 6}.*

WRITING EXERCISES

1. Choose one of the following parts:

   (a) Explain why the words "commutative" and "associative" are appropriate choices to represent the properties described by their definitions. (Look at the nontechnical meanings of the roots of these two words.)

   (b) If you do not think that "commutative" and "associative" are appropriate choices to represent the properties described by their definitions, suggest two words that you think would be better choices, and explain your preference.

2. Subtraction and division fail to be commutative in much the same way. That is, for nonzero rational numbers a and b,

$$a - b = -(b - a) \text{ and } a \div b = (b \div a)^{-1}$$

   (a) Write a definition for the general property suggested by these two examples; call it "anticommutativity."

   (b) Find or construct another operation that is anticommutative. (You might try making up an operation table on some set.)

   (c) Explain how you might tell from looking at an operation table whether or not the operation it describes is anticommutative.

THE DEFINITION OF A GROUP

Of the four elementary arithmetic operations, the two "better behaved" ones seem to be addition and multiplication because both of them are commutative and associative. Addition and multiplication are also well-behaved in other ways. If a and b are integers, then any equation of the form $a + x = b$ has an integer solution. For example,

$$\text{if } 7 + x = 2, \text{ then } x = -5$$

Similarly, if c and d are positive rational numbers, then any equation of the form $c \cdot x = d$ has a positive rational solution. For example,

$$\text{if } \frac{2}{3} \cdot x = \frac{1}{2}, \text{ then } x = \frac{3}{2} \cdot \frac{1}{2} = \frac{3}{4}$$

Now, if we examine the ways in which equations like these are solved, we find a common pattern that can be abstracted and made into a definition with surprisingly far-reaching consequences. Let us look at two examples, solving each equation in painfully inefficient, but enlightening, detail.

To solve $3 + x = 5$, we add -3 to both sides of the equation, regroup by associativity, and let -3 "cancel" the 3 on the left, leaving us with the desired solution:

$$-3 + (3 + x) = -3 + 5$$
$$(-3 + 3) + x = 2$$
$$0 + x = 2$$
$$x = 2$$

Similarly, to solve $3 \cdot x = 5$, we multiply both sides by $\frac{1}{3}$, regroup by associativity, and let $\frac{1}{3}$ "cancel" the 3 on the left, leaving us with the desired solution:

$$\frac{1}{3} \cdot (3 \cdot x) = \frac{1}{3} \cdot 5$$

$$\left(\frac{1}{3} \cdot 3\right) \cdot x = \frac{5}{3}$$

$$1 \cdot x = \frac{5}{3}$$

$$x = \frac{5}{3}$$

In both cases we needed a number to "cancel" the number paired with x. Here, "cancel" means that, after regrouping by associativity, the resulting number, when combined with x, left x unchanged. (Notice that commutativity was *not* needed.) These observations can be translated into specific properties of an operation on a set that guarantee the solvability of simple equations like the ones shown here.

DEFINITION

Let G be a nonempty set and let $*$ be an operation on G with the following properties:

1. $*$ is associative on G;

2. there is an element z in G such that

$$z * a = a \qquad \text{and} \qquad a * z = a$$

for any element a in G;

3. for each element a in G, there is an element a' in G such that

$$a * a' = z \qquad \text{and} \qquad a' * a = z$$

Then G together with the operation $*$, which we shall sometimes abbreviate as $(G, *)$, is called a **group**. The element z is called an **identity** element, and a' is called an **inverse** of a.

EXAMPLE 6.14

As we have already implied, the system $(\mathbf{I}, +)$ of integers under addition is a group. Its identity element is 0, and the inverse of any integer a is $-a$. Similarly, the set of positive rational numbers under multiplication is a group. Its identity element is 1 and the inverse of any number a is its reciprocal, $\frac{1}{a}$.

Note	*Tables 6.7, 6.8, and 6.9,* cited in the following examples, are on pages 289–90.

EXAMPLE **6.15**	*Table 6.8(a)* represents a group in which 0 is the identity, $0' = 0$, and $1' = 1$. *Table 6.8(b)* does not represent a group; there is an identity, 1, but there is no inverse for 0.

EXAMPLE **6.16**	*Table 6.7* does *not* represent a group table because there is no identity element. (Can you see an easy way to recognize when there is an identity element in a table?)

EXAMPLE **6.17**	*Table 6.9* represents a group. The identity element is a, and each element is its own inverse.

Note	As you can see from Example 6.17, an identity element is easy to recognize in a table because its row must be identical with the row at the top of the table and its column must be identical with the column on the left.

Many other examples of groups will turn up in the course of our work, but for now, the foregoing examples should be enough to illustrate the definition. Notice that a group operation does not have to be commutative. A group with a commutative operation will be specifically called a **commutative group**.[4] So far, all the examples we have seen are commutative groups, but noncommutative ones will begin to appear shortly.

Groups are of fundamental importance in many areas of modern mathematics. Although the defining properties of a group are simple, mathematicians have unearthed a wealth of information in their exploration of group theory since its beginnings in the nineteenth century. This chapter is too brief to do more than scratch the surface of such a vast field of knowledge, but even here we shall see some of the power of the group axioms as we investigate some specific problems.

The first of these problems can be stated immediately; it is the focus of our attention in the next section. We saw at the beginning of Section 6.3 that there are 19,683 possible operations on a three-element set. Which of these define groups?

---

4. These groups are also called *abelian groups*. They are named after Neils Henrik Abel, a Norwegian mathematician of the early nineteenth century.

6.4   EXERCISES

Exercises 1–5 refer to *Table 6.16*.

1. Determine $4 * 8$.

2. Is the set $\{2, 4, 6, 8\}$ closed under $*$? Why or why not?

3. Which element, if any, is the identity?

4. Check one case of the associative law for $*$.

5. Which element, if any, is the inverse of 8?

$*$	2	4	6	8
2	4	8	2	6
4	8	6	4	2
6	2	4	6	8
8	6	2	8	4

Table 6.16
The operation $*$ on the set $\{2, 4, 6, 8\}$.

Exercises 6–18 refer to *Tables 6.17 and 6.18*.

6. $t \circ r = $ _____

7. $3 \diamond 4 = $ _____

8. $(q \circ r) \circ p = $ _____

9. $1 \diamond (3 \diamond 2) = $ _____

10. Is there an identity element in *Table 6.17*? If so, what is it?

11. Is there an identity element in *Table 6.18*? If so, what is it?

12. Does q have an inverse in *Table 6.17*? If so, what is it?

13. Does 1 have an inverse in *Table 6.18*? If so, what is it?

14. Solve for x: $r \circ x = t$.

15. Solve for x: $2 \diamond x = 3$.

16. Find an example to show that one of these operations is not commutative.

17. Find an example to show that one of these operations is not associative.

18. One of these tables does not describe a group. Which is it, and which group properties fail?

For Exercises 19–23, let $S = \{a, b, c\}$. In each of these exercises, construct an operation table for S that satisfies the given condition or explain why this is not possible.

19. S is commutative, but has no identity element.

20. S has an identity element, but not every element has an inverse.

21. S has an identity element, but no element has an inverse.

22. S is a commutative group.

23. S is a group that is not commutative.

24. Fill in *Table 6.19* so that you have an operation $\diamond$ on the set $\{a, b, c, d\}$ with the following three properties:
    ■ c is the identity element,
    ■ b is the inverse of d, and
    ■ $\diamond$ is commutative.

$\circ$	p	q	r	s	t
p	s	r	t	p	q
q	t	s	p	q	r
r	q	t	s	r	p
s	p	q	r	s	t
t	r	p	q	t	s

Table 6.17 The operation $\circ$ on the set $\{p, q, r, s, t\}$.

$\diamond$	0	1	2	3	4
0	0	1	2	3	4
1	1	2	3	4	0
2	2	3	4	0	1
3	3	4	0	1	2
4	4	0	1	2	3

Table 6.18 The operation $\diamond$ on the set $\{0, 1, 2, 3, 4\}$.

$\diamond$	a	b	c	d
a				
b				
c				
d				

Table 6.19 An operation, $\diamond$, on the set $\{a, b, c, d\}$.

25. Fill in *Table 6.20* so that you have an operation, *, on the set {1, 2, 3, 4, 5} with the following five properties:

   ■ $4 * 3 = 1$,
   ■ 2 is the identity element,
   ■ 1 is the inverse of 5,
   ■ 3 is its own inverse, and
   ■ * is not commutative.

*	1	2	3	4	5
1					
2					
3					
4					
5					

*Table 6.20 An operation, *, on the set {1, 2, 3, 4, 5}.*

26. Find an example of a commutative group containing six elements. (Call this group *A* for future reference.)

WRITING EXERCISES

1. Évariste Galois, generally acknowledged to be the founder of group theory, was one of the more colorful, tragic figures of mathematical history. Read about him in at least two different mathematical history books; then write a short human-interest piece about some aspect of his life. (Be sure to list the sources you use, and be careful that you do not simply copy or paraphrase the material from your sources.)

2. Commutative groups are commonly called "Abelian" groups, in recognition of a mathematician whose last name was Abel. Write a paragraph that states who Abel was (what his full name was, when and where he lived, etc.) and explains why it is appropriate to name this kind of group after him. (List the sources you use, and be careful that you do not simply copy or paraphrase material.)

3. This section begins by observing that the defining properties of a group can be seen as an abstraction of the pattern for solving some simple equations.

   (a) Discuss the extent to which you think the ideas of this chapter *depend on* the standard algebraic notation for solving equations.

   (b) Section B.5 of the Appendix observes that the use of even such simple algebraic symbols as =, +, and − did not occur until the sixteenth century. With this in mind, expand on your answer to Part (a) to comment on whether or not group theory could have developed sooner than it did in our culture. (When and where did it start?)

SOME BASIC PROPERTIES OF GROUPS

6.5

The problem we confront in this section is:

Which of the 19,683 possible operations on a three-element set define groups?

This problem can be approached in many ways. The most obvious one is also the most tedious and the least enlightening: We could check

	1	2	3	4
1			1	
2			2	
3	1	2	3	4
4			4	

Table 6.21
Recognizing an
identity element.

each of the 19,683 different three-element tables to see if they had the properties required by the definition of group.

"There must be an easier way!" you say, and you are right. Instead of taking the direct, dull approach, let us first try to establish some easily recognizable properties that any group table must have, thereby eliminating many ineligible candidates at once.

The most easily recognizable group property in a table is the existence of an identity element. The row and column of an identity element must be exactly the same as the reference row and column given at the top and side of the table, respectively. For instance, an operation table on the set {1, 2, 3, 4} with 3 as its identity element must fit the pattern of *Table 6.21*, regardless of how the operation behaves elsewhere on the set.

Moreover:

xxxxxxxxxxxxxxxxxxxxxxxxxxxxxxxxxxxxxxxxxxxxxxxxxxxxx

(6.1)   Any group has only one identity element.

xxxxxxxxxxxxxxxxxxxxxxxxxxxxxxxxxxxxxxxxxxxxxxxxxxxxx

We prove this fact by showing that if y and z both represent identity elements in a group $(G, *)$, then they cannot be different. Consider the element $y * z$. If y is an identity, then

$$y * z = z$$

On the other hand, if z is an identity, then

$$y * z = y$$

Hence, we have

$$y = y * z = z$$

so y must equal z, and thus Property (6.1) has been proved.

Our experience with addition and multiplication of numbers suggests another useful general property. We know, for example, that for numbers x and y,

$$3 + x = 3 + y \qquad \text{implies} \qquad x = y$$

and also

$$3x = 3y \qquad \text{implies} \qquad x = y$$

In both cases the 3 can be "canceled." We abstract from such numerical examples a general property shared by all groups.

xxxxxxxxxxxxxxxxxxxxxxxxxxxxxxxxxxxxxxxxxxxxxxxxxxxxx

(6.2)   **The Cancellation Laws:** Let a, b, and c be any elements of a group $(G, *)$.

  1. If $a * b = a * c$, then $b = c$
  2. If $b * a = c * a$, then $b = c$

xxxxxxxxxxxxxxxxxxxxxxxxxxxxxxxxxxxxxxxxxxxxxxxxxxxxx

We prove the first of these laws here; the proof of the second is almost exactly the same and is left as an exercise.

Because $(G, *)$ is a group, the element a has an inverse a' in G. Combining a' with both sides of the given equation, we obtain

$$a' * (a * b) = a' * (a * c)$$

and thus

$$(a' * a) * b = (a' * a) * c$$

by associativity. Denoting the identity of $(G, *)$ by z, we then have

$$z * b = z * c$$

by the definition of inverse, implying

$$b = c$$

by the definition of identity.

Cancellation is an important property of groups which, in turn, allows us to prove other basic properties. For example:

xxxxxxxxxxxxxxxxxxxxxxxxxxxxxxxxxxxxxxxxxxxxxxxxxxxx

(6.3)   Any element of a group has only one inverse.

xxxxxxxxxxxxxxxxxxxxxxxxxxxxxxxxxxxxxxxxxxxxxxxxxxxx

As in the proof of Property (6.1), the approach here is to show that any two inverses of the same element must be equal. Let $(G, *)$ be a group with identity element z, and let g be an arbitrary element of G. Suppose that x and y both represent inverses of g. Then, by the definition of an inverse,

$$g * x = z \qquad \text{and} \qquad g * y = z$$

implying

$$g * x = g * y$$

Applying the first of the Cancellation Laws (6.2), we get $x = y$; so Property (6.3) is proved.

Besides providing a powerful tool for proving other general facts about groups, the Cancellation Laws (6.2) are easily recognizable properties of any operation expressed in tabular form. Suppose, for example, that *Table 6.22* represents an operation, $*$, on the set $\{1, 2, 3, 4, 5\}$, and that two entries in the third row are the same, as shown. In this case we have

$$3 * 2 = 4 \qquad \text{and} \qquad 3 * 5 = 4$$

implying

$$3 * 2 = 3 * 5$$

$*$	1	2	3	4	5
1					
2					
3		4			4
4					
5					

Table 6.22 A violation of left cancellation.

But $2 \neq 5$, so Cancellation Law 1 is violated. Thus, Cancellation Law 1 implies that there can be no repetitions in any row.

Similarly, Cancellation Law 2 prohibits repetitions in any column. In other words:

xxxxxxxxxxxxxxxxxxxxxxxxxxxxxxxxxxxxxxxxxxxxxxxxxxxxxx

(6.4)   In each row and column of any table that represents a group, every element of the set must appear exactly once.

xxxxxxxxxxxxxxxxxxxxxxxxxxxxxxxxxxxxxxxxxxxxxxxxxxxxxx

	1	2	3
1	1	2	3
2	2		
3	3		

Table 6.23
The identity 1.

Now we are ready to solve the problem of deciding which of the 19,683 three-element tables are groups. We shall use the general tools just developed to construct group tables for the set {1, 2, 3}. Notice first that every group must have an identity element, and that the choice of an element to play this role immediately determines five of the nine entries of a potential group table. For instance, the choice of 1 as identity requires that the row and column for 1 be identical with the reference row and column, as shown in Table 6.23.

There are only four remaining entries, each of which can be chosen in three ways, so there are only 81 ($= 3^4$) ways to fill in the rest of the table. Now, there are only three possible identity elements and 81 ways to complete the table for each choice of identity, implying that there are only $3 \cdot 81$ ($= 243$) ways to construct a three-element table containing an identity element. Hence, the identity requirement alone reduces the number of possible three-element group tables from 19,683 to a mere 243.

The Cancellation Laws and the consequent Statement (6.4) allow us to go even further. Referring to Table 6.23, it is not hard to see that the nonrepetition restriction on the rows and columns *requires* that the middle entry of the table be 3, since a 2 there would give us an obvious repetition in the second row and column, and a 1 there would lead to repeated 3s in the third row and column. This middle entry of 3, in turn, requires 1s at the end of the second row and column, and finally a 2 in the lower-right corner. That is, the *only way* to complete Table 6.23 without row or column repetitions is as shown in Table 6.24. Hence, there is *at most one* three-element group that has 1 as its identity element.

	1	2	3
1	1	2	3
2	2	3	1
3	3	1	2

Table 6.24 The only possible group for this set with identity 1.

Reasoning in precisely the same way, we can verify that there is at most one three-element group table with identity 2 and at most one with identity 3. We have now reduced the 243 possibilities of three-element group tables to *at most three*.

Notice that we have not yet proved that these tables actually represent groups. There are operations that satisfy the Cancellation Laws but are not associative (see Exercise 26); so we still must check these tables for associativity. We do know, however, that no other three-element tables need to be checked because none of the others have identities and satisfy cancellation.

It is not hard to check each of the three tables directly for the group properties. For instance, referring to *Table 6.24*, it is easy to see that

- the operation is closed,
- 1 is the identity, and
- $1' = 1$, $2' = 3$, and $3' = 2$.

Checking associativity is somewhat tedious (27 cases), but it can be done; in fact, associativity holds in every case. Hence, *Table 6.24* is actually a group table. The group properties for the other two tables can be verified in exactly the same way. So the problem has been solved:

Of the 19,683 different three-element tables, exactly three represent groups!

6.5   EXERCISES

For Exercises 1–4, construct an operation table for the set {1, 2, 3, 4} that satisfies the given condition.

1. Both Cancellation Laws (6.2).

2. Cancellation Law 1, but not 2.

3. Neither Cancellation Law.

4. Cancellation Law 2, but not 1.

5. Check that the operation shown in *Table 6.24* on page 303 is associative. (Work out all cases.)

6. *Table 6.25* starts to describe an operation on the set {1, 2, 3, 4} with identity 1. There

are at least two different ways to complete the table so that the Cancellation Laws hold; complete it *in two ways*. Do both define group operations? Justify your answer.

For Exercises 7–14, fill in each given table to define a group, if possible. If this is not possible, explain why not.

7. See *Table 6.26*.   8. See *Table 6.27(a)*.

9. See *Table 6.27(b)*.   10. See *Table 6.28*.

11. See *Table 6.29*.   12. See *Table 6.30*.

13. See *Table 6.31*.   14. See *Table 6.32*.

*	1	2	3	4
1	1	2	3	4
2	2			
3	3			
4	4			

*Table 6.25 An operation, *, on the set {1, 2, 3, 4} with identity 1.*

*	r	s
r	s	
s		

*Table 6.26 An operation, *, on the set {r, s}.*

*	1	2	3
1	1	2	3
2	2	4	
3	3		

(a)

∘	1	2	3
1	2		
2		2	
3			

(b)

*Table 6.27 Two operations, * and ∘, on the set {1, 2, 3}.*

For Exercises 15 and 16, set up a table for an operation ∗ on the set {a, b, c, d, e}, listing the elements across and down in that order.

15. Fill in as much of your table as is determined by this information:

    ■ The table describes a group.

    ■ The identity element is e.

    ■ a and c are inverses of each other.

    ■ d is not its own inverse.

16. Suppose also that c ∗ c = d. Fill in the rest of your table. Is this group commutative? Why or why not?

For Exercises 17–20, recall that, for each regular polygon, the *followed by* process is an operation on an appropriate set of rotations of the figure. (See Example 6.3, in Section 6.2.) This operation, which we shall denote by "∙," is associative. (See Exercise 17 of Section 6.3.)

17. Construct an operation table for ∙ and the 0°, 90°, 180°, and 270° rotations of a square. Then verify that this table represents a group by specifying the identity rotation and the inverse of each rotation and by

checking to see that cancellation holds. Is this group commutative? Why or why not?

18. Construct an operation table for ∙ and the appropriate rotations of an equilateral triangle. Then verify that this table represents a group, following the instructions of Exercise 17. Where have you seen the pattern of this table before?

19. Construct an operation table for ∙ and the appropriate rotations of a regular pentagon. Then verify that this table represents a group, following the instructions of Exercise 17. Is this group commutative? Why or why not?

20. Construct an operation table for ∙ and the appropriate rotations of a regular octagon. Then verify that this table represents a group, following the instructions of Exercise 17. Is this group commutative? Why or why not?

21. Describe in words the pattern that is common to all the tables for Exercises 17–20. Then give a persuasive argument that a table of *any* size with this pattern must represent a group.

∗	2	5	8
2			5
5		5	
8	5		

Table 6.28 An operation, ∗, on the set {2, 5, 8}.

∗	2	4	6	8
2	4	2		
4	2	4	6	8
6		6		
8		8		6

Table 6.29 An operation, ∗, on the set {2, 4, 6, 8}.

∗	a	b	c	d
a	b			
b		c		
c			d	
d				a

Table 6.30 An operation, ∗, on the set {a, b, c, d}.

∗	w	x	y	z
w	w			
x		x		
y			y	
z				z

Table 6.31 An operation, ∗, on the set {w, x, y, z}.

∗	0	1	2	3
0	0	1	2	3
1				0
2				
3		0		

Table 6.32 An operation, ∗, on the set {0, 1, 2, 3}.

22. **(a)** Notice that the pattern of *Table 6.33* is very similar to the pattern of the group tables for Exercises 17–20. (For instance, compare it with your table for Exercise 19.) Describe in words at least two ways in which these patterns are alike and at least one way in which they differ.

   (b) Does *Table 6.33* satisfy the Cancellation Laws? Why or why not?

   (c) *Table 6.33* does *not* represent a group. Why not?

•	1	2	3	4	5
1	1	2	3	4	5
2	5	1	2	3	4
3	4	5	1	2	3
4	3	4	5	1	2
5	2	3	4	5	1

Table 6.33 The operation • on the set {1, 2, 3, 4, 5}.

23. Supply a reason for each step in the proof of the statement:

   If a and b are elements of a commutative group $(G, *)$ with identity z, then
   $$((a * b) * a')' = b'$$

   Proof:

   a' and b' exist in G.

   _____

   Hence, $((a * b) * a') * b'$ is an element of G.

   _____

   $((a * b) * a') * b' = ((b * a) * a') * b'$

   _____

$$= (b * (a * a')) * b'$$

   _____

$$= (b * z) * b'$$

   _____

$$= b * b'$$

   _____

$$= z$$

   _____

   Therefore, $((a * b) * a')' = b'$.

   _____

24. Prove Cancellation Law 2.

25. Let $(G, *)$ be a finite group. Prove that if the number of elements contained in G is even, then at least one element besides the identity is its own inverse. (*This result is used later.*)

26. *Table 6.34* satisfies the Cancellation Laws. (Why?) Moreover, there is an identity element (what is it?) and each element has an inverse (what are they?). Prove that this table does *not* define a group.

*	a	b	c	d	e
a	c	d	a	e	b
b	e	c	b	a	d
c	a	b	c	d	e
d	b	e	d	c	a
e	d	a	e	b	c

*Table 6.34 The operation * on the set {a, b, c, d, e}.*

WRITING EXERCISES

1. Using complete sentences, write a half-page summary that identifies the main theme of this section and outlines how that theme is carried out.

2. Describe how at least three of the dozen problem-solving techniques of Section 1.2 can be used to help in the solution of Exercise 25. (Even if you can't complete the proof for Exercise 25, you should be able to do this writing exercise.)

3. Write two questions that extend the ideas of this section but are analogous in some way to questions answered in the section. Explain the analogy in each case.

SUBGROUPS

6.6

One way to proceed with our investigation of finite groups would be to continue the approach of Section 6.5, trying to examine all four-element operation tables, then all five-element tables, and so on. But this route is tiresome and inefficient; the rapid growth of the number of possibilities supplies too many cases for even the powerful Cancellation Laws to dispose of. Checking associativity in a five-element table can be tedious, and by the time tables with ten elements come into view, we would be approaching the limits of human patience. (There are 1000 associativity cases to be checked for a ten-element table—not much for a computer, but a truly annoying chore if done by hand! Tables of several thousand elements involve billions of cases; this is very time-consuming even for computer-assisted checking.)

We choose, instead, to seek another general insight into group theory which, like cancellation, will allow us to dispose of many particular cases all at once. This path will lead to a major result in group theory, Lagrange's Theorem. We begin by defining an important basic idea.

DEFINITION

Let G be a group with operation $*$. If S is a subset of G that is itself a group with respect to the same operation $*$ (restricted to the elements of S), then we call $(S, *)$ a **subgroup** of $(G, *)$.

EXAMPLE 6.18

The integers form a group under addition. You know from elementary arithmetic that the sum of two integers is an integer and that addition is associative; the identity element is 0; the inverse of each element is its negative. The set of even integers (under addition) forms a subgroup of this group. The sum of two even integers is even; addition is (still) associative; 0 is an even number; the negative of any even number is even.

It is important to note that the definition of a subgroup requires that *the same operation* be used for the subset as was used for the original group. Otherwise, *any* subset could be made into a group under a totally unrelated operation; this would render the subgroup concept worthless. This necessary restriction simplifies the language and notation a little: Because the operation in question is necessarily the same, we can simply say that a *subset S* of $(G, *)$ is a subgroup, instead of having to specify that $(S, *)$ is a subgroup of $(G, *)$.

Moreover, because a group operation $*$ is necessarily associative and because the definition of associativity requires that

$$(a * b) * c = a * (b * c)$$

for *any* three elements of the group, this equality surely holds whenever a, b, and c all happen to be in a particular subset. Thus, a group operation is automatically associative on every subset. Hence, in checking to see if a subset is a subgroup, we may ignore the tiresome associative property. In other words:

XXXXXXXXXXXXXXXXXXXXXXXXXXXXXXXXXXXXXXXXXXXXXXXXXXXXXXX

(6.5)

To verify that a subset S of a group $(G, *)$ is a subgroup, we need only check three properties:

1. S is closed under $*$.
2. S contains the identity element of $(G, *)$.
3. Each element of S has its inverse contained in S.

XXXXXXXXXXXXXXXXXXXXXXXXXXXXXXXXXXXXXXXXXXXXXXXXXXXXXXX

EXAMPLE 6.19

Table 6.35 defines a group with identity element 1. Let us use Statement (6.5) to find all of its subgroups. Since 1 must be in every subgroup, we need to check only the subsets containing 1 to see if they are closed and if every element has an inverse. These subsets are:

$$\{1\}, \{1, 2\}, \{1, 3\}, \{1, 4\}, \{1, 2, 3\}, \{1, 2, 4\}, \{1, 3, 4\}$$

and, of course, the entire set $\{1, 2, 3, 4\}$, which is automatically a subgroup of itself.

- The set $\{1\}$ forms a subgroup because $1 * 1 = 1$; that is, $\{1\}$ is closed under $*$, and 1 is its own inverse.
- $\{1, 2\}$ is not closed because $2 * 2 = 3$, which is outside the subset.
- $\{1, 3\}$ is a subgroup; it is closed under $*$, $1' = 1$, and $3' = 3$.
- $\{1, 4\}$ is not closed because $4 * 4 = 3$, which is outside the subset.

(See *Table 6.36*.)

We can dispose of the other cases similarly, or we can use the following slightly more efficient argument: By closure, any subgroup containing 2 or 4 must contain 3 ($= 2 * 2 = 4 * 4$). But $2 * 3 = 4$ and $4 * 3 = 2$; so any group containing either 2 or 4 must contain the other three elements, as well. Thus, the only subgroups of this group are

$$\{1\}, \{1, 3\}, \{1, 2, 3, 4\}$$

*	1	2	3	4
1	1	2	3	4
2	2	3	4	1
3	3	4	1	2
4	4	1	2	3

Table 6.35 The operation $*$ on the set $\{1, 2, 3, 4\}$.

*	1
1	1

*	1	2
1	1	2
2	2	3

*	1	3
1	1	3
3	3	1

*	1	4
1	1	4
4	4	3

Table 6.36 Some subset tables.

6.6   EXERCISES

1. *Table 6.24* on page 303 represents a three-element group with identity 1. Give a convincing argument to verify that this group has no two-element subgroups.

2. *Table 6.37* describes a five-element group with identity 1. Find all of its subgroups.

*	1	2	3	4	5
1	1	2	3	4	5
2	2	3	4	5	1
3	3	4	5	1	2
4	4	5	1	2	3
5	5	1	2	3	4

*Table 6.37 The operation * on the set {1, 2, 3, 4, 5}.*

3. Find all subgroups of the six-element group *A* that you constructed for Exercise 26 of Section 6.4.

4. *Table 6.38* defines a group; call it (*B*, ∘).

   (a) Find all of the subgroups of *B*.

   (b) Count the number of elements in each of the subgroups. How do these numbers relate to the total number of elements in *B*?

   (c) How does this group differ from your group *A* in Exercise 3?

∘	j	f	g	h	i	k
j	j	f	g	h	i	k
f	f	j	i	k	g	h
g	g	h	j	f	k	i
h	h	g	k	i	j	f
i	i	k	f	j	h	g
k	k	i	h	g	f	j

Table 6.38 The operation ∘ on the set {j, f, g, h, i, k}.

In previous sections you saw that the rotations of a regular polygon form a group under the *followed by* operation. (See, for example, Exercises 17–20 of Section 6.5.)

5. (a) Write out the operation table for the group of rotations of a regular hexagon. What is the smallest nonzero rotation in this group?

   (b) Find all the subgroups of the group you constructed in Part (a). For each one, find its smallest nonzero rotation. How are these smallest rotations related to the complete rotation, 360°?

6. (a) List all the rotations in the group for a regular decagon (10-sided polygon). Do *not* write out the operation table.

   (b) Thinking about closure and the rotation process, choose the subsets that you think form subgroups in the group of Part (a). Explain your choices.

   (c) Find the smallest nonzero rotation for each subgroup you listed in Part (b). How are these smallest rotations related to the complete rotation, 360°?

7. The solutions of the equation $x^2 + 1 = 0$ are the (complex) numbers whose squares equal -1. These numbers are usually denoted by i and $-i$; that is,

$$i = \sqrt{-1} \quad \text{and} \quad -i = \sqrt{-1}$$

   (a) Make an operation table for the set

$$S = \{1, i, -1, -i\}$$

   using ordinary multiplication and the fact that $i \cdot i = -1$.

   (b) Verify that (*S*, ·) is a group by checking all the defining properties for a group. (You may remember or assume that this multiplication is associative.)

   (c) Find all the subgroups of (*S*, ·).

   (d) What are the similarities between (*S*, ·) and the group defined in *Table 6.9* on page 290? What are the differences?

   (e) What are the similarities between (*S*, ·) and the group given in Example 6.19? What are the differences?

For Exercises 8–13, let ⊕ be the operation defined on the set C = {0, 1, 2, 3, 4, 5, 6, 7, 8, 9, 10, 11} by

$$x \oplus y = \text{the remainder when } x + y \text{ is divided by 12}$$

For example, 6 ⊕ 10 = 4 because 4 is the remainder when 16 is divided by 12.

8. $9 \oplus 7 = $ ____   9. $(6 \oplus 8) \oplus 11 = $ ____

10. $6 \oplus (8 \oplus 11) = $ ____

11. It can be shown that ⊕ is an associative operation. Assuming this is true, verify the rest of the properties needed to make $(C, \oplus)$ a group.

12. Find all of the subgroups of C.

13. Count the number of elements in each of the subgroups. How do these numbers relate to the total number of elements in C?

14. Prove that the identity element of any subgroup must be the same as the identity element of the group containing it. (*This result will be used later.*)

In Example 6.18, you saw that the set of integers under addition is a group, $(I, +)$. Exercises 15–18 refer to this group.

15. Prove that the set of multiples of 3 forms a subgroup of $(I, +)$.

16. Prove that the set of nonnegative integers under addition is *not* a subgroup of $(I, +)$.

17. Prove that the set of all multiples of 6 forms a subgroup of $(I, +)$. Then prove that it also forms a subgroup of the group of Exercise 15.

18. (a) Let k be any fixed integer. Prove that the set of all multiples of k forms a subgroup of $(I, +)$.

    (b) Suppose that the set of all multiples of some integer n is a subgroup of the group of Part (a). How are n and k related? Justify your answer.

19. (a) Let G be a commutative group with the operation ∗. Prove that for any elements a and b in G,
$$(a * b)' = a' * b'$$

    (b) In the noncommutative group $(B, \circ)$ defined by *Table 6.38*, find an instance of two elements for which the property stated in Part (a) does *not* hold.

    (c) If $(G, *)$ is a group but is not commutative, how can you write $(a * b)'$ in terms of a' and b'? Justify your answer.

20. Exercise 18 asserts that
    i. the set of all multiples of a fixed integer k forms a subgroup of $(I, +)$;
    it also implies that
    ii. if $k > 1$, then the subgroup formed does not contain all of I.

    (a) Justify Statement ii.

    (b) What are the analogous two statements for the group $(C, \oplus)$ described by Exercises 8–13? Are either or both of these analogous statements true? Justify your answers.

WRITING EXERCISES

1. In mathematical terminology, the prefix "sub-" is used to signify that a set or structure of some sort is included within another one of the same kind (as in *subset* and *subgroup*). How well does this conform to the use of "sub-" outside of mathematics? Explain, using appropriate examples from ordinary English.

2. Write three simpler questions that you think might provide helpful steppingstones for solving Exercise 14. Be as specific as possible.

3. Write three simpler questions that you think might provide helpful steppingstones for solving Exercise 19. Be as specific as possible.

4. Describe as specifically as you can an approach to one part of Exercise 19 that uses one (or more) of the problem-solving techniques listed in Section 1.2.

LAGRANGE'S THEOREM

To illustrate the variety of subgroups within a group, we consider the set of digits

$$D = \{0, 1, 2, 3, 4, 5, 6, 7, 8, 9\}$$

and define an operation $\oplus$ on D by

$$x \oplus y = \text{the remainder when } x + y \text{ is divided by } 10$$

Now, any number less than 10 when divided by 10 has quotient 0 and itself as remainder; so for pairs of numbers whose sum is less than 10 this operation is the same as addition. For pairs of numbers whose sum is greater than or equal to 10, however, the remainder feature guarantees that the results of the $\oplus$ process are always back in the set D. Thus, for example,

$$3 \oplus 5 = 8, \qquad 6 \oplus 8 = 4, \qquad 9 \oplus 2 = 1, \qquad 7 \oplus 3 = 0$$

It is not hard to verify that $(D, \oplus)$ is a group with identity element 0 and inverses as follows:

$0' = 0$	$1' = 9$	$2' = 8$	$3' = 7$	$4' = 6$
$5' = 5$	$6' = 4$	$7' = 3$	$8' = 2$	$9' = 1$

(Notice that x' is the inverse of x if $x + x' = 10$.)

Let us try to find all the subgroups of $(D, \oplus)$. We know from Subgroup Property 2 of Statement (6.5) and Exercise 14, both in Section 6.6, that any subgroup of D must contain 0, and that, in fact, $\{0\}$ itself is a subgroup. We can determine the other subgroups by trial and error; but a little ingenuity can save a lot of effort. For example, any subgroup containing 1 must also contain $1 \oplus 1$, which is 2. But then it must also contain $2 \oplus 1 = 3$, and hence $3 \oplus 1 = 4$, and so on. Thus, any subgroup of D containing 1 also contains every other element of D, and hence must be D itself.

So far we have located a one-element subgroup and a ten-element subgroup (D itself). Clearly, $\{0\}$ is the only possible one-element subgroup. (Why?) Are there any two-element subgroups? If so, each must contain 0, which is its own inverse; so the other element must also be its own inverse. The list of inverses shows that 5 is the only element besides 0 that is its own inverse; thus, the only two-element subgroup of D is $\{0, 5\}$.

How about three-element subgroups? In this case, Subgroup Property 3 of Statement (6.5) requires that the two nonzero elements be inverses of each other. Let us try the various combinations:

- {0, 1, 9} is not a subgroup because it does not contain $1 \oplus 1 = 2$;
- {0, 2, 8} does not contain $2 \oplus 2 = 4$;
- {0, 3, 7} does not contain $3 \oplus 3 = 6$;
- {0, 4, 6} does not contain $4 \oplus 4 = 8$.

Hence, there are no three-element subgroups of D.

These computations also ensure that there are no four-element subgroups because any such subgroup must contain 0 and 5 (by Exercise 25 of Section 6.5) and some other element with its inverse. Putting 5 into each of the three-element sets does not make them closed under $\oplus$.

A five-element subgroup must contain 0 and two pairs of elements that are inverses of each other (why?), and it cannot contain 1, as we have seen. Moreover, it cannot contain the pair consisting of 3 and 7 because $(7 \oplus 7) \oplus 7$ equals 1. The only possibility, then, is

$$\{0, 2, 4, 6, 8\}$$

The sum of even numbers is even, and division by 10 preserves evenness in these cases; so this is indeed a subgroup.

Similar arguments may be used to rule out the possibility of six-, seven-, eight-, and nine-element subgroups; therefore, the only subgroups of D are:

$$\{0\}, \{0, 5\}, \{0, 2, 4, 6, 8\}, \text{ and } D \text{ itself}$$

Notice that *the number of elements in each subgroup of this ten-element group is a divisor of the number 10.* Looking back at the results of Exercises 1 through 13 of Section 6.6, we can see the same phenomenon, as shown in *Table 6.39*. This pattern is not just a curious coincidence; in fact, the assertion that subgroups are always

Joseph Louis Lagrange. His work on the theory of equations led to basic theorems in group theory.

Exercise(s)	Group Size	Subgroup Sizes
1	3 elements	1, 3
2	5 elements	1, 5
3–5	6 elements	1, 2, 3, 6
6	10 elements	1, 2, 5, 10
7	4 elements	1, 2, 4
8–13	12 elements	1, 2, 3, 4, 6, 12

Table 6.39 The pattern of subgroup sizes in Exercises 1–13 of Section 6.6.

related to their "parent" group in this way is the first major theorem of group theory.

xxxxxxxxxxxxxxxxxxxxxxxxxxxxxxxxxxxxxxxxxxxxxxxxxxx

Lagrange's Theorem:[5] The number of elements in any subgroup of a finite group must be a divisor of the total number of elements in the group.

xxxxxxxxxxxxxxxxxxxxxxxxxxxxxxxxxxxxxxxxxxxxxxxxxxx

(Recall that a whole number a is a **divisor** of a whole number b if there is a whole number c such that $a \cdot c = b$.)

The proof of Lagrange's Theorem is not trivial, but it is within the scope of the mathematical tools now at our disposal. A detailed explanation of that proof appears in the next (optional) section. The proof, however, is not necessary for an understanding of the role of Lagrange's Theorem in group theory. In the rest of this section we look at a few consequences of the theorem itself to illustrate its power both in handling specific groups and in proving other general results.

For a first instance of Lagrange's Theorem at work, we turn to an example just like the one that began this section. Consider the set

$$E = \{0, 1, 2, 3, 4, 5, 6, 7\}$$

and define an operation $\oplus$ on E by

$$x \oplus y = \text{the remainder when } x + y \text{ is divided by 8}$$

$(E, \oplus)$ is a group with identity element 0 and inverses as follows:

$$0' = 0 \quad 1' = 7 \quad 2' = 6 \quad 3' = 5$$
$$4' = 4 \quad 5' = 3 \quad 6' = 2 \quad 7' = 1$$

Here, x' is the inverse of x if $x + x' = 8$. (Can you justify these statements about the identity and the inverses?)

Problem

Find all possible subgroups of $(E, \oplus)$.

Solution

Since E contains eight elements, Lagrange's Theorem tells us that it can only have subgroups containing one, two, four, or eight elements. The only one-element subgroup is $\{0\}$ because every subgroup must

5. Named after Joseph Louis Lagrange (1736–1813), an outstanding French mathematician and a friend of Napoleon.

contain 0. Clearly, the only eight-element subgroup is E itself. Now, any two-element subgroup must contain a nonzero element that is its own inverse (by Exercise 25 of Section 6.5); we can see from the list of inverses that the only two-element group is {0, 4}. Finally, any four-element subgroup must contain 0, 4, and two elements that are inverses of each other. (Why?) Checking the available inverse pairs, we must discard 1 and 7 because $1 \oplus 1 = 2$, and closure would require a fifth element in the subgroup. Similarly, the pair consisting of 3 and 5 will not work because $3 \oplus 3 = 6$. However, 2 and 6 fit with 0 and 4 to form the only possible four-element subgroup. Thus, we have found *all* the subgroups of $(E, \oplus)$.

Compare this solution with the long arguments used in finding the subgroups of $(D, \oplus)$ earlier in this section. Lagrange's Theorem allowed us to avoid a lot of tedious work!

Here is an even more striking example of the same sort. Define the operation $\oplus$ on the set $F = \{0, 1, 2, \ldots, 96\}$ by

$$x \oplus y = \text{the remainder when } x + y \text{ is divided by 97}$$

Again, $(F, \oplus)$ is a group.

Problem

Find all subgroups of this 97-element group $(F, \oplus)$.

Solution

The only divisors of 97 are 1 and 97 (that is, 97 is prime).[6] Thus, Lagrange's Theorem tells us that the only possible subgroups are of these two sizes. But {0} is the only possible one-element subgroup and F itself is the only 97-element subgroup. Therefore, we have found *all* subgroups of the group $(F, \oplus)$.

This last problem is a particular case of an obvious, but important, general consequence of Lagrange's Theorem.

xxxxxxxxxxxxxxxxxxxxxxxxxxxxxxxxxxxxxxxxxxxxxxxxxxxxxxxxxxxx

If the number of elements in a group is prime, then there are only two subgroups—the one-element subgroup consisting of the identity alone, and the entire group itself.

xxxxxxxxxxxxxxxxxxxxxxxxxxxxxxxxxxxxxxxxxxxxxxxxxxxxxxxxxxxx

---

6. **A prime number** is a whole number greater than 1 whose only divisors are 1 and the number itself.

We close this section with one more interesting corollary of Lagrange's Theorem. This result is a twin of the often used Exercise 25 of Section 6.5.

xxxxxxxxxxxxxxxxxxxxxxxxxxxxxxxxxxxxxxxxxxxxxxxxxxx

Let G be a finite group. If the number of elements contained in G is odd, then no element besides the identity can be its own inverse.

xxxxxxxxxxxxxxxxxxxxxxxxxxxxxxxxxxxxxxxxxxxxxxxxxxx

To prove this, let z denote the identity element of G, and suppose there is some other element g in G that is its own inverse. Then $\{z, g\}$ is a two-element subgroup of G (why?); so, by Lagrange's Theorem, 2 must divide the number of elements in G. But this is impossible because the number of elements in G is odd. Therefore, there cannot be an element besides z that is its own inverse.

6.7   EXERCISES

Table 6.40 represents a group that we shall call G. Use this table to answer Exercises 1–10.

*	a	b	c	d	e	f
a	a	b	c	d	e	f
b	b	c	a	e	f	d
c	c	a	b	f	d	e
d	d	f	e	a	c	b
e	e	d	f	b	a	c
f	f	e	d	c	b	a

Table 6.40 The operation * on the group $G = \{a, b, c, d, e, f\}$.

1. What is the identity element of G?

2. List the inverse of each element of G.

3. Is $\{a, b\}$ a subgroup of G? Why or why not?

4. Find a two-element subgroup of G.

5. How many two-element subgroups does G have? How do you know?

6. Is $\{a, b, c, d\}$ a subgroup of G? Why or why not?

7. Is $\{d, e, f\}$ a subgroup of G? Why or why not?

8. Is $\{a, b, c\}$ a subgroup of G? Why or why not?

9. Is $\{a, c, e\}$ a subgroup of G? Why or why not?

10. Is $\{a, d, f\}$ a subgroup of G? Why or why not?

11. Let $S = \{0, 1, 2, 3, 4, 5, 6, 7, 8\}$ and define $\oplus$ on S by

$x \oplus y =$ the remainder when $x + y$ is divided by 9

This operation makes $(S, \oplus)$ a group. Find all its subgroups.

12. Tables 6.41 and 6.42 represent groups. Find all their subgroups.

13. Consider the following information concerning groups W, X, Y, and Z:

> Group W contains 22 elements.
> Group X contains 23 elements.
> Group Y contains 24 elements.
> Group Z contains 25 elements.

(a) Which group has the most different sizes of possible subgroups, and which has the fewest?

(b) What sizes of subgroups are possible in group W? In group Z?

(c) Which of the four groups must contain an element besides the identity that is its own inverse? Which cannot contain such an element?

For Exercises 14–19, let n be some fixed positive integer and let $I_n = \{0, 1, 2, \ldots, n - 1\}$. Define $\oplus$ on I_n by

$$x \oplus y = \text{the remainder when } x + y \text{ is divided by } n$$

(For instance, the group $(D, \oplus)$ described at the beginning of this section is $(I_{10}, \oplus)$.)[7]

14. Assuming that $\oplus$ is associative (which it is), verify that $(I_n, \oplus)$ is a group, for any positive integer n.

15. (a) Without writing out an operation table, find all the subgroups of $(I_{15}, \oplus)$. (List the elements of each one.)

(b) Is there a "chain" of three different subgroups, S_1, S_2, S_3, such that
$$S_1 \subset S_2 \subset S_3 \ ?$$
If so, find one. If not, explain why not.

(c) Is there a "chain" of four different subgroups, S_1, S_2, S_3, S_4, such that
$$S_1 \subset S_2 \subset S_3 \subset S_4 \ ?$$
If so, find it. If not, explain why not.

16. (a) Without writing out an operation table, find all the subgroups of $(I_{20}, \oplus)$. (List the elements of each one.)

(b) Find a "chain" of four different subgroups, S_1, S_2, S_3, S_4, such that
$$S_1 \subset S_2 \subset S_3 \subset S_4$$

(c) Find two more subgroup chains like the one you found for Part (b).

17. (a) Find a number n such that $(I_n, \oplus)$ contains *exactly one* "chain" of five different subgroups, S_1, S_2, \ldots, S_5, such that
$$S_1 \subset S_2 \subset \ldots \subset S_5$$
Write out the elements of each subgroup in the chain.

*	p	q	r	s	t	u	v
p	p	q	r	s	t	u	v
q	q	r	s	t	u	v	p
r	r	s	t	u	v	p	q
s	s	t	u	v	p	q	r
t	t	u	v	p	q	r	s
u	u	v	p	q	r	s	t
v	v	p	q	r	s	t	u

*Table 6.41 The operation * on the group $G = \{p, q, r, s, t, u, v\}$.*

*	1	2	3	4	5	6	7	8
1	1	2	3	4	5	6	7	8
2	2	3	4	1	8	7	5	6
3	3	4	1	2	6	5	8	7
4	4	1	2	3	7	8	6	5
5	5	7	6	8	1	3	2	4
6	6	8	5	7	3	1	4	2
7	7	6	8	5	4	2	1	3
8	8	5	7	6	2	4	3	1

*Table 6.42 The operation * on the group $G = \{1, 2, 3, 4, 5, 6, 7\}$.*

7. This operation $\oplus$ is commonly known as "addition modulo n" or "+ (mod n)."

(b) Find another n for which Part **(a)** can be done.

(c) What is the smallest n for which Part **(a)** can be done? Justify your answer.

(d) Is there a largest n for which Part **(a)** can be done? Why or why not?

18. **(a)** Find a number n such that $(\mathbf{I}_n, \oplus)$ contains at least one "chain'" of six different subgroups, $S_1, S_2, \ldots, S_6$, such that

$$S_1 \subset S_2 \subset \ldots \subset S_6$$

Specify the elements of each subgroup in the chain.

(b) Is there more than one number n for which Part **(a)** can be done? If so, specify at least one more such number. If not, explain why not.

(c) What is the smallest n for which Part **(a)** can be done? Justify your answer.

19. If you are given an arbitrary natural number k (that is, a number whose choice you cannot control), can you *always* find a group $(\mathbf{I}_n, \oplus)$ containing a "chain" of k different subgroups, $S_1, S_2, \ldots, S_k$, such that

$$S_1 \subset S_2 \subset \ldots \subset S_k?$$

If so, how? If not, explain why not.

20. Let G be a finite group. Prove that the only subgroup of G that contains more than half of the total number of elements in G is G itself.

21. (This easy exercise foreshadows the proof of Lagrange's Theorem.) Let $T = \{a, b, c, \ldots, t\}$, the set of the first 20 letters of the alphabet, and let $U = \{b, d, g, j\}$.

(a) Split T into a collection of subsets such that

   i. Every element of T is in exactly one subset.

   ii. U is one of the subsets in the collection.

   iii. All subsets in the collection contain the same number of elements.

(b) Is there more than one way to choose a collection of subsets as described in Part **(a)**? Why or why not?

(c) Find an example of a subset V of T such that the splitting-up process described in Part **(a)** *cannot* be carried out if V is substituted for U.

(d) Is it possible to find a set W containing a different number of elements than U such that the same kind of splitting-up process as described in Part **(a)** can be carried out if W is substituted for U? If such a set can be found, what are the possible numbers of elements it could contain? If such a set W cannot be found, why not?

WRITING EXERCISES

1. Explain how three of the dozen problem-solving techniques of Section 1.2 were used in this section to lead up to the statement of Lagrange's Theorem. (You may be able to see more than three techniques used here; just explain three of them.)

2. (A research problem in mathematical history.) Group theory is generally regarded as having its origins in a long letter from Évariste Galois to a friend, written in 1832 on the night before Galois was killed. However, Lagrange's Theorem, a fundamental result of group theory, is named after Joseph Louis Lagrange, who died in 1813 (when Galois was about two years old). Explain this historical paradox. Be sure to cite the sources you use.

LAGRANGE'S THEOREM PROVED [OPTIONAL]

The proof of Lagrange's Theorem gives us a chance to see how all the basic machinery developed so far interacts to produce a major mathematical result.[8] It is hardly surprising that the argument is neither short nor trivial; rather, it is remarkable that such an elegant and far-reaching result can be proved at all from the few facts we have established! For an idea of how the proof will proceed, we begin by examining the general version of Exercise 21 of Section 6.7.

Suppose that a set G contains n elements and a subset H of G contains r elements. Suppose, further, that G can be split up into a collection of subsets such that the following conditions hold:

(6.6)

1. Every element of G is in exactly one subset.
2. H is one of the subsets in the collection.
3. All subsets in the collection contain the same number of elements.

In such a case, **Conditions 2** and **3** imply that each subset contains r elements. Then, because each of the n elements of G is in exactly one subset, the total number of elements in G can be found by counting the number of subsets in the collection and multiplying that number by r. That is, r must be a divisor of n.

To prove Lagrange's Theorem, then, we need only show that, for any group G and any sub*group* H of G, there is a way of splitting G up into a collection of subsets satisfying **Conditions 1**, **2**, and **3** listed in Statement (6.6). To do this, we need one new concept.

DEFINITION

Let H be a subgroup of a group G with operation $*$. If g is some fixed element of G, then the set of all elements $g * h$ for each element h of H is called a **coset** of H in G and is denoted by $g * H$.

Notice that the fixed element g is always on the left when it is combined with the elements of H. It is important to keep this in mind when forming cosets because $*$ may not be a commutative operation. Such cosets are usually called *left cosets*, but since they are the only kind we shall use, we omit the word "left."

EXAMPLE 6.20 (a)

(This example is continued throughout the section to illustrate the proof.)

---

8. This section may be omitted without disturbing the continuity of the chapter. To get just an informal idea of the proof, read up to the subheading "Proof of Lagrange's Theorem" on page 320, and skip the rest of the section.

Recall the ten-element group $(D, \oplus)$ defined in Section 6.7, where

$$D = \{0, 1, 2, 3, 4, 5, 6, 7, 8, 9\}$$

and

$$x \oplus y = \text{the remainder when } x + y \text{ is divided by 10}$$

As we have seen before, one of the subgroups of $(D, \oplus)$ is $\{0, 2, 4, 6, 8\}$; call this subgroup S. Then the coset of S determined by 3 is

$$3 \oplus S = \{3 \oplus 0, 3 \oplus 2, 3 \oplus 4, 3 \oplus 6, 3 \oplus 8\}$$
$$= \{3, 5, 7, 9, 1\}$$

Similarly, the coset of S determined by 4 is

$$4 \oplus S = \{4 \oplus 0, 4 \oplus 2, 4 \oplus 4, 4 \oplus 6, 4 \oplus 8\}$$
$$= \{4, 6, 8, 0, 2\}$$

A complete list of cosets of S in D is as follows:

$0 \oplus S = \{0, 2, 4, 6, 8\}$	$1 \oplus S = \{1, 3, 5, 7, 9\}$
$2 \oplus S = \{2, 4, 6, 8, 0\}$	$3 \oplus S = \{3, 5, 7, 9, 1\}$
$4 \oplus S = \{4, 6, 8, 0, 2\}$	$5 \oplus S = \{5, 7, 9, 1, 3\}$
$6 \oplus S = \{6, 8, 0, 2, 4\}$	$7 \oplus S = \{7, 9, 1, 3, 5\}$
$8 \oplus S = \{8, 0, 2, 4, 6\}$	$9 \oplus S = \{9, 1, 3, 5, 7\}$

Notice that all the sets in the left column are equal, as are all the sets in the right column. (Remember that two sets are equal if they contain the same elements, regardless of order.) Thus, there are only two different cosets of S in D.

EXAMPLE 6.21

One of the subgroups of the group $(B, \circ)$ defined by *Table 6.38* on page 309 for Exercise 4 of Section 6.6 is $\{j, f\}$; call this subgroup T. Then the coset of T in B determined by the element h is

$$h \circ T = \{h \circ j, h \circ f\} = \{h, g\}$$

A complete list of cosets of T in B is:

$j \circ T = \{j, f\}$	$g \circ T = \{g, h\}$	$i \circ T = \{i, k\}$
$f \circ T = \{f, j\}$	$h \circ T = \{h, g\}$	$k \circ T = \{k, i\}$

Notice that there are only three different cosets of T in B.

A closer look at these two examples can provide more insight into the arguments at the beginning of this section. In Example 6.21, notice that the three different cosets $\{j, f\}$, $\{g, h\}$, and $\{i, k\}$ form a set consisting of subsets of B that satisfy **Conditions 1**, **2**, and **3** of Statement

(6.6) with respect to subgroup T. Notice further that 2, the number of elements of subgroup T, multiplied by 3, the number of different cosets, gives us 6, the total number of elements of the original group.

The same observations can be made about Example 6.20. The two different cosets $\{0, 2, 4, 6, 8\}$ and $\{1, 3, 5, 7, 9\}$ form a set of subsets of D satisfying Statement (6.6) with respect to the subgroup S, and 5, the number of elements of S, multiplied by 2, the number of cosets, is 10, the total number of elements of D. It seems, then, that a reasonable way to proceed with proving Lagrange's Theorem is to show that these three conditions hold for the set of different cosets of *any* subgroup in *any* finite group. The rest of the section is devoted to a formal argument that does exactly this.

Proof of Lagrange's Theorem

Let $(G, *)$ be a finite group containing n elements and let H be a subgroup of G containing r elements. We want to show that the set of all the different cosets of H in G satisfies **Conditions 1, 2,** and **3** of Statement (6.6).

Condition 1. The identity element, z, must be in the subgroup H; so each element g of G is in at least one coset, namely, the one it determines. (That is, g is in $g * H$ because z is in H and $g * z = g$.) To show that no element of G can be in two different cosets, we prove that any two cosets containing a common element are necessarily the same; the argument proceeds as follows:

Suppose two cosets $x * H$ and $y * H$ both contain some particular element g. Then g in $x * H$ implies $g = x * h_1$ for some h_1 in H, and g in $y * H$ implies $g = y * h_2$ for some h_2 in H. Thus,

$$(6.7) \qquad x * h_1 = y * h_2$$

(because they both equal g). Since H is a group, h_1', the inverse of h_1 is in H; so we may combine h_1' with both sides of Equation (6.7) to get

$$(6.8) \qquad (x * h_1) * h_1' = (y * h_2) * h_1'$$

Now, by associativity, this equation becomes

$$x * (h_1 * h_1') = y * (h_2 * h_1')$$

and because $h_1 * h_1'$ is the identity element (why?), we get

$$(6.9) \qquad x = y * (h_2 * h_1')$$

We use Equation (6.9) to show that *every* element of $x * H$ must also be in $y * H$, as follows. Suppose $x * h$ is an arbitrary element of $x * H$. By Equation (6.9), we have $x * h$ in the form

$$x * h = (y * (h_2 * h_1')) * h$$

and, by associativity, this can be rewritten as

(6.10)

$$x * h = y * ((h_2 * h_1') * h)$$

But h_2, h_1', and h are all in H, and because H is a subgroup, it must be closed under the operation $*$. Hence, $(h_2 * h_1') * h$ is an element of H. Thus, Equation (6.10) tells us that the element $x * h$ can be written as

$$x * h = y * (\text{some element of } H)$$

implying that $x * h$ is in $y * H$. Because this is true for *any* $x * h$ in the coset $x * H$, it follows that $x * H$ is a subset of $y * H$.

By an argument like the one we just did, except for the use of h_2' instead of h_1' in passing from Equation (6.7) to an equation like (6.8), we can show that $y = x * (h_1 * h_2')$ and then that $y * H$ is a subset of $x * H$. Because each of the cosets $x * H$ and $y * H$ is a subset of the other, they must be equal (by the definition of equal sets); so **Condition 1** has been verified. That is to say, every element of G is in *exactly one* of the different cosets of H in G. (*Whew!*)

EXAMPLE 6.20 (b)

As an illustration of the main argument in (the preceding) Part 1 of the proof, for the set S of Example 6.20, consider the cosets $3 \oplus S$ and $7 \oplus S$, both of which contain 5 because $5 = 3 \oplus 2$ and $5 = 7 \oplus 8$. Thus,

$$3 \oplus 2 = 7 \oplus 8$$

The inverse of 2, which is 8, is also in S, so

$$(3 \oplus 2) \oplus 8 = (7 \oplus 8) \oplus 8$$
$$3 \oplus (2 \oplus 8) = 7 \oplus (8 \oplus 8)$$
$$3 = 7 \oplus 6 \qquad (\text{Why?})$$

Now, any element in $3 \oplus S$ must be in $7 \oplus S$; we illustrate this by considering the element 1 of $3 \oplus S$. We have

$$1 = 3 \oplus 8 = (7 \oplus 6) \oplus 8 = 7 \oplus (6 \oplus 8) = 7 \oplus 4$$

which is also an element of $7 \oplus S$.

Proof of Lagrange's Theorem, *continued*

Condition 2. This part is easy. If we denote the identity element of G by z, then the coset $z * H$ is H itself (because $z * h = h$ for each element h in H). Therefore, H is one of the cosets, satisfying **Condition 2**.

EXAMPLE 6.20 (c)

Since 0 is the identity of D,

$$0 \oplus S = \{0 \oplus 0, 0 \oplus 2, 0 \oplus 4, 0 \oplus 6, 0 \oplus 8\}$$
$$= \{0, 2, 4, 6, 8\}$$
$$= S$$

Proof of Lagrange's Theorem, *continued*

Condition 3. To prove that all cosets have the same number of elements, we show that any coset of H in G contains the same number of elements as H itself. Consider any coset $g * H$. If we write out the elements of H and of $g * H$, there appears to be a natural matching between the two sets that shows they are the same size:

$$H = \{ h_1, \quad h_2, \quad h_3, \quad \ldots, \quad h_r \}$$
$$g * H = \{g * h_1, \ g * h_2, \ g * h_3, \ldots, \ g * h_r\}$$

One thing must be checked, however, before this argument is complete: We must be sure that different elements h are paired with *different* coset elements. That is, whenever we have two different elements of H, say $h_i \neq h_j$, then we must be sure that

$$g * h_i \neq g * h_j$$

But this is easy to verify: If $g * h_i = g * h_j$, then Cancellation Law 1 of Statement (6.2) would imply $h_i = h_j$, contrary to our hypothesis. Hence, each coset contains r elements, satisfying **Condition 3**.

EXAMPLE 6.20 (d)

Here is an example of the 1-1 correspondence between the subgroup S and one of its cosets in D.

$$S = \{ 0, \quad 2, \quad 4, \quad 6, \quad 8 \}$$
$$3 \oplus S = \{3 \oplus 0, 3 \oplus 2, 3 \oplus 4, 3 \oplus 6, 3 \oplus 8\}$$

This correspondence is a matching between distinct elements of the subgroup S and the coset $3 \oplus S$ because $3 \oplus x$ can equal $3 \oplus y$ if and only if $x = y$ (by cancellation).

Proof of Lagrange's Theorem, *continued*

At this point the hard work has been done. All that remains is to put the pieces together, as suggested earlier.

There are only a finite number of different cosets; say there are k of them. Because each coset contains r elements and no two cosets have any elements in common, $r \cdot k$ is the total number of elements in the various cosets (that is, in their union).[9] But each element of G must be in *some* coset, so $r \cdot k = n$. Therefore, r is a divisor of n, and Lagrange's Theorem is proved!

6.8 EXERCISES

Table 6.43 represents a six-element group with identity element 1. The sets $S = \{1, 3, 5\}$ and $T = \{1, 4\}$ are subgroups. (Can you verify this?) Use this table to find the cosets for Exercises 1–6.

*	1	2	3	4	5	6
1	1	2	3	4	5	6
2	2	3	4	5	6	1
3	3	4	5	6	1	2
4	4	5	6	1	2	3
5	5	6	1	2	3	4
6	6	1	2	3	4	5

Table 6.43 *The operation * on the group* $G = \{1, 2, 3, 4, 5, 6\}$.

1. $2 * S =$
2. $2 * T =$
3. $4 * S =$
4. $4 * T =$
5. $5 * S =$
6. $5 * T =$

For Exercises 7–14, recall the 12-element group $(C, \oplus)$ defined for Exercises 8–13 of Section 6.6:

$$C = \{0, 1, 2, 3, 4, 5, 6, 7, 8, 9, 10, 11\}$$

and $x \oplus y =$ the remainder when $x + y$ is divided by 12. Two of its subgroups are $S = \{0, 3, 6, 9\}$ and $T = \{0, 4, 8\}$. List the elements of each given coset in C.

7. $5 \oplus S$
8. $5 \oplus T$
9. $2 \oplus S$
10. $2 \oplus T$
11. $1 \oplus T$
12. $9 \oplus S$
13. $3 \oplus T$
14. $4 \oplus S$

15. This exercise refers to the group defined by *Table 6.9* in Section 6.2.

   (a) Write out all the distinct cosets of the subgroup $\{a, c\}$.

   (b) Write out all the distinct cosets of the subgroup $\{a, b\}$.

   (c) Based on your work for Parts (a) and (b), predict the distinct cosets of the subgroup $\{a, d\}$. Then check your prediction by actually writing them out.

16. Suppose G is a 50-element group.

   (a) What are the possible sizes for subgroups of G?

   (b) For each possible size, assume that G has a subgroup of that size, and specify how many distinct cosets that subgroup has.

17. Suppose G is a 55-element group.

   (a) What are the possible sizes for subgroups of G?

   (b) For each possible size, assume that G has a subgroup of that size, and specify how many distinct cosets that subgroup has.

18. Suppose that H is a subgroup of a group G and that some coset of H contains exactly 10 elements. Prove that G contains a 2-element subgroup.

9. The **union** of a collection of sets is the set of all elements that are in at least one of the sets.

19. These questions refer to a group of the form $(\mathbf{I}_n, \oplus)$, as defined for Exercises 14–19 of Section 6.7:

$$\mathbf{I}_n = \{0, 1, 2, \ldots, n - 1\}$$

and $x \oplus y =$ the remainder when $x + y$ is divided by n. Suppose that $\{4, 10, 16, 22, 28\}$ is one of the cosets of a subgroup H of this group.

(a) How many elements are in H?

(b) List the elements of H. Explain how you got them.

(c) List the elements of $1 \oplus H$.

(d) How many distinct cosets does H have?

(e) What is the number n? How do you know?

20. Let $(K, \circ)$ denote the system described by *Table 6.44*, and let $L = \{p, q, r\}$.

∘	p	q	r	s	t	u
p	p	q	r	s	t	u
q	q	p	s	s	u	t
r	r	s	p	s	t	q
s	s	s	s	p	q	u
t	t	u	t	q	p	r
u	u	t	q	u	r	p

Table 6.44 The operation ∘ on the set $K = \{p, q, r, s, t, u\}$.

(a) For each element x in K, write out the elements of the set

$$x \circ L = \{x \circ p, x \circ q, x \circ r\}$$

(b) How many different sets did you get in Part (a)? Do all these sets contain the same number of elements?

(c) Is $(K, \circ)$ a group? Justify your answer here in terms of your answer to Part (b).

21. Let $(G, *)$ be defined by *Table 6.45*.

(a) Construct a table describing this same operation $*$ on the subset $H = \{1, 4\}$.

(b) Is $(H, *)$ a group? Why?

(c) Write out the sets $1 * H$, $2 * H$, $3 * H$, $4 * H$, $5 * H$, and $6 * H$.

(d) What information do your results supply about whether or not $(G, *)$ is a group? Why?

*	1	2	3	4	5	6
1	1	2	3	4	5	6
2	2	5	4	3	6	1
3	3	4	5	6	1	2
4	4	3	6	1	2	5
5	5	6	1	2	3	4
6	6	1	2	5	4	3

Table 6.45 The operation $*$ on the set $G = \{1, 2, 3, 4, 5, 6\}$.

WRITING EXERCISES

1. Does the mathematical notation used to present the proof of Lagrange's Theorem in this section help or hinder your understanding? Would it be easier or more difficult to present this proof without so much (or any) notation? Discuss.

2. In Section 6.7 we checked quite a few groups for which Lagrange's Theorem works. Why is a proof of this theorem still necessary? What is the purpose of proof in mathematics?

3. Near the end of the proof of **Condition 1**, the text says,

   By an argument like the one we just did, except for the use of $h_2{}'$ instead of $h_1{}'$ in passing from Equation (6.7) to an equation like (6.8), we can show that $y = x * (h_1 * h_2{}')$ and then that $y * H$ is a subset of $x * H$.

   Write out this argument in detail.

GROUPS OF SYMMETRIES

This section provides a quick look at how groups can be used to describe the symmetries of some simple planar shapes. Many current applications of groups are merely more complicated versions of these elementary ideas.

Consider a straight split log lying flat side down in sand on a beach. If the log is picked up, there are exactly two ways to replace it so that its original and replacement imprints in the sand coincide (flat side down again, of course)—it can either be returned to its original position, or be reversed and then put down. See *Figure 6.1*. The rigidity of the log ensures that we can describe each of these movements by the initial and final positions of its ends. Thus, if the ends are numbered 1 and 2, the log movements may be symbolized by

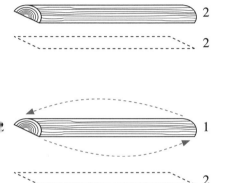

$$1 \longrightarrow 1 \qquad 1 \diagdown\!\!\!\!\diagup 1$$
$$2 \longrightarrow 2 \qquad 2 \diagup\!\!\!\!\diagdown 2$$

If we call these two movements of the log s (for *same*) and r (for *reverse*) and look at what happens when we follow one of these movements by another, we can see that

s followed by r is the same as r

r followed by s is the same as r

r followed by r is the same as s

s followed by s is the same as s

Figure 6.1 The two ways to replace a log so that its original and replacement imprints coincide.

This information is summarized in *Table 6.46*, where • denotes "followed by." Notice that this is a group table. The log movements s and r are examples of **rigid motions**—geometric figure movements that do not allow any distortion of the figure. With this understanding, we could easily replace our log model by a strictly geometric one using a line segment.

A more interesting group can be obtained by considering the rigid motions of an equilateral triangle that returns to its original "imprint" in the plane. Such motions must take each vertex to some original vertex position (not necessarily its own), and, because the triangle cannot be distorted, the positions of the three vertices determine the position of the entire triangle. There are exactly six of these rigid motions for an equilateral triangle. (See Exercise 1.) Three are the rotations of 120°, 240°, and 360° (or 0°); the other three are the reflections with respect to the three *medians*. A **median** of a triangle is a line through a vertex of the triangle and the midpoint of the opposite side;

•	s	r
s	s	r
r	r	s

Table 6.46 The rigid motions of a line segment.

a **reflection** of the triangle with respect to a median can be pictured as holding the median fixed and flipping the triangle over, as shown in *Figure 6.2*.

If the vertices are numbered 1, 2, and 3 (counterclockwise), then any rigid motion can be represented by a chart indicating *how each vertex is matched with the original vertex positions*. For example, if the triangle is rotated 120° counterclockwise, vertex 1 goes to the place originally occupied by vertex 2, vertex 2 goes to vertex 3's place, and vertex 3 goes to vertex 1's place. The six rigid motions of the triangle are as diagramed in *Figure 6.3*.

Now let us consider what happens when we follow one of these rigid motions by another. For example, look at the 120° rotation followed by the flip fixing the original median 3. The vertices are moved around as follows:

- Vertex 1 is moved to position 2 by the rotation, and the flip fixing the original median 3 takes that position 2 to position 1; so vertex 1 ends up back where it started, in position 1.

- The rotation takes vertex 2 to position 3, and the flip leaves position 3 alone; so vertex 2 ends up in position 3.

- Vertex 3 is rotated to position 1, and then this vertex in position 1 is flipped over to position 2; so vertex 3 ends up in position 2.

Looking at the result of all this, we see that the final position of the triangle is the same as if it had only been moved by the flip fixing median 1.

The process of following one motion by another is actually much easier than the foregoing description suggests, provided we adopt some convenient notation. Let us name the six rigid motions by the letters shown in *Figure 6.3* and denote the *followed by* process by "•." Then, using the diagrams of *Figure 6.3*, it is easy to observe the result of this process just by following arrows, as shown in *Figure 6.4*. Following the arrows from the left column to the right one, we see that $1 \to 1$, $2 \to 3$, and $3 \to 2$; that is, $i \cdot g = f$.

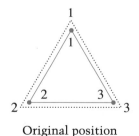

Original position

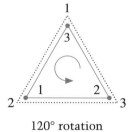

120° rotation

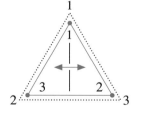

"Flip" fixing the median through vertex 1

Figure 6.2 Two rigid motions of an equilateral triangle. In each diagram, the *position* is indicated outside the triangle, and the *vertex* in that position is noted inside the triangle.

Figure 6.3 The six rigid motions of an equilateral triangle represented by vertex movement.

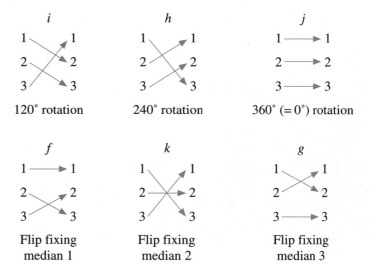

Because all possible rigid motions of the equilateral triangle are listed in *Figure 6.3*, any one of them followed by another must yield the same result as one of the original six motions; that is,

the set {*i, h, j, f, k, g*} of rigid motions is closed under the operation •

In fact, if we compute all possible combinations of rigid motions (using the method described) and arrange the results in an operation table, we actually get the group shown in *Table 6.37* on page 309. Thus:

×××

The set of all rigid motions of an equilateral triangle forms a six-element noncommutative group.

×××

Similar groups may be formed by considering the rigid motions of various geometric figures. For instance, there are eight rigid motions of a square—four rotations (90°, 180°, 270°, 360°) and four reflections

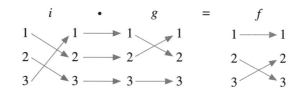

Figure 6.4 Successive rigid motions: i followed by g equals f.

(holding fixed each of the four dotted lines in *Figure 6.5*.) The resulting eight-element group is called the **octic group**. This group and the rigid-motion groups for several other planar figures are explored in the exercises.

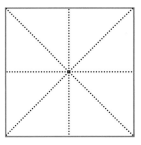

Figure 6.5 Four of the rigid motions of a square. The dotted lines are the axes for the four reflections of a square.

<table>
<tr><td>6.9</td><td>EXERCISES</td></tr>
</table>

1. Cut an equilateral triangle out of cardboard and mark the vertices 1, 2, 3 on both sides. (Make sure each vertex has the same number on both sides of the cardboard.) Trace the outline of your triangle on a piece of paper and number the vertices (outside the traced figure). Then, by moving your triangle around, verify that there are only six different ways to replace the triangle in its tracing. These replacements correspond to the six rigid motions listed in *Figure 6.3*.

2. **(a)** Compute the following combinations of rigid motions of an equilateral triangle according to the process illustrated by *Figure 6.4*:

   $f \bullet k, \ g \bullet i, \ i \bullet h, \ j \bullet f, \ g \bullet g, \ h \bullet h$

   (b) Compute the six combinations listed in Part **(a)** by actually moving around the triangle you constructed in Exercise 1.

3. Construct an operation table for the octic group (the group of rigid motions of a square), and verify that all the defining properties of a group are satisfied. (You may assume that associativity holds.) Is this group commutative?

4. **(a)** Find all the rigid motions of a rectangle that is not a square (there are four), and make an operation table for them.

   (b) Verify that this table satisfies all the defining properties of a group. Is this group commutative?

   (c) Relabel these four rigid motions using the letters a, b, c, and d so that your operation table matches *Table 6.9*.

   (d) Find a subgroup of the octic group (Exercise 3) that matches this rectangle group.

   (e) Repeat Parts **(a)**–**(d)** for a rhombus that is not a square.

5. **(a)** Find all the rigid motions of an isosceles right triangle and construct an operation table for them.

   (b) Verify that this table describes a group. Is this group commutative?

   (c) Relabel these rigid motions so that your operation table matches one of the tables in Section 6.2. (Which of those tables does it match?)

(d) Can you find a subgroup of the equilateral-triangle group that matches this isosceles-right-triangle group? Why or why not?

6. **(a)** Find all the rigid motions of a regular pentagon and construct an operation table for them.

(b) Assuming associativity, check the other properties needed to verify that this table describes a group. Is this group commutative?

(c) Find all the subgroups of this pentagon group.

(d) Is this group like any group we have seen earlier in the chapter? If so, which one, and how are they similar?

Exercises 7–10 use the following definition:

The line that stays fixed when a figure is reflected ("flipped over") is called a *line of symmetry* for that figure. Each line of symmetry corresponds to a reflection of the figure.

7. How many lines of symmetry does a regular pentagon have? Where are they located?

8. How many lines of symmetry does a regular hexagon have? Where are they located?

9. If a regular n-sided polygon has an odd number of sides, how many lines of symmetry does it have? Where are they located?

10. If a regular n-sided polygon has an even number of sides, how many lines of symmetry does it have? Where are they located?

11. Exercises 17–20 of Section 6.5 show that the sets of rotations of a square, an equilateral triangle, a regular pentagon, and a regular octagon all form groups. Generalize these results by proving that, for any positive integer n, the set of all rotations of a regular n-sided polygon forms a subgroup of the group of rigid motions of that figure. (Begin by asking yourself: How many rotations are there? How can they be described?)

12. If you have a single rotation of a regular n-gon, can you form all the rest of the rotations by repeated applications of the *followed by* process? Does the answer depend on which rotation you have? Does it depend on n? If so, how? Justify your answers.

13. Does the set of all reflections of a regular n-sided polygon form a subgroup of the rigid-motion group of the figure? What about the reflections together with the identity element? Justify your answer.

14. Prove that the rigid-motion group for a regular n-sided polygon contains (at least) n two-element subgroups if n is odd and $n + 1$ two-element subgroups if n is even.

The examples and exercises of this section suggest that the rigid motions of a regular n-sided polygon consist of *n* rotations and *n* reflections, so that the rigid-motion group contains 2*n* elements. For Exercises 15 and 16, assume that this is true.

15. **(a)** If you have the set of all rotations and a single reflection of a regular 20-sided polygon, can you make all the other reflections, using the *followed by* operation? If not, which ones cannot be made? Does the answer depend on which reflection you have? Justify your answers.

(b) Generalize your results for Part **(a)** to the case of a regular n-sided polygon. (Does it matter whether n is even or odd?)

16. **(a)** If you have the set of all reflections and a single rotation of a regular 20-sided polygon, can you make all the other rotations, using the *followed by* operation? If not, which ones cannot be made? Does the answer depend on which rotation you have? Justify your answers.

(b) Generalize your results for Part **(a)** to the case of a regular n-sided polygon. (Does it matter whether n is odd or even?)

WRITING EXERCISES

1. Look up the word "symmetry" in a dictionary and discuss how the definition(s) given there relate to the use of "symmetry" in this section.

2. Explain how the ideas of this section are related to the ideas presented in Section 6.1.

3. Does *every* polygon have a group of symmetries? Explain.

LINK: GROUPS IN MUSIC AND IN CHEMISTRY

6.10

We end this chapter with two elementary examples of groups as they appear in other subject areas. The first one is in music, and it is really just an observation of a particular pattern, rather than a serious application of the theory. Nevertheless, formalizing observed harmonic and melodic patterns both in the theory of harmony and in the analysis of classical compositions has led to the inclusion of deliberate mathematical patterns in the works of some twentieth-century composers, such as Schoenberg and Webern. Thus, this sort of link between mathematics and music is not as far-fetched as it may initially appear.

Our musical example uses the twelve-tone chromatic scale. Imagine yourself sitting at a piano. If you play middle C and then continue up the keyboard playing every key, black and white, until you reach high C, you will have played the notes of the chromatic scale in the order

C	C♯	D	D♯	E	F	F♯	G	G♯	A	A♯	B	(C)
0	1	2	3	4	5	6	7	8	9	10	11	(12)

If you were to continue playing notes beyond high C, you would just repeat the same scale notes an octave higher. Consider two notes to be the same if they have the same name, regardless of the octaves they are in. Thus, the high C in the scale shown here is in parentheses to indicate that it is to be considered as a copy of the first C in the scale. Now we define an operation ⊕ on the chromatic scale by using the numbers listed below the notes. To combine two notes, add

Picture Collection, The Branch Libraries, The New York Public Library

Patterns in Music.

their corresponding numbers, subtracting 12 if the sum is larger than 11; the result is the number of the "answer" note.[10] For example,

$$G \oplus D\sharp = A\sharp \quad \text{because} \quad 7 + 3 = 10$$

and

$$F \oplus A = D \text{ because } 5 + 9 = 14 \text{ and } 14 - 12 = 2$$

C is the identity note for this group, because 0 is the identity for addition. Each note has an inverse—the inverse of C$\sharp$ is B, the inverse of D is A$\sharp$, and so forth. In other words, the notes of the twelve-tone scale form a group. (This group is a disguised version of the group $(C, \oplus)$ defined for Exercises 8–13 of Section 6.6. You might find it interesting to compare the two.)

This formal way of looking at the scale allows us to identify mathematically some items of musical interest. For instance, two subgroups of this 12-element group are musically significant. The (only) four-element subgroup is

10. This is simply addition modulo 12.

$$\{C, D\sharp, F\sharp, A\}$$

(Why?), and these are precisely the notes of the "diminished seventh" chord in any of the keys represented by those four notes. The other two diminished-seventh chords,

$$\{C\sharp, E, G, A\sharp\} \qquad \text{and} \qquad \{D, F, G\sharp, B\}$$

can be obtained by "adding" any single note in the new chord to each note of the original one. For instance,

$$\{D, F, G\sharp, B\} = \{D \oplus C, \ D \oplus D\sharp, \ D \oplus F\sharp, \ D \oplus A\}$$

(If you have read Section 6.8, you should recognize this as a *coset*. Using this terminology, we are saying that the three distinct cosets of the subgroup

$$\{C, D\sharp, F\sharp, A\}$$

are the (only) three diminished-seventh chords.)

The (only) 3-element subgroup is

$$\{C, E, G\sharp\}$$

the notes of the C-augmented chord (and also the E-augmented chord and the G♯-augmented chord). Any of the other augmented chords—

$$\{C\sharp, F, A\}, \quad \{D, F\sharp, A\sharp\}, \quad \text{and} \quad \{D\sharp, G, B\}$$

—may be obtained by "adding" any single note of the desired chord to the subgroup chord. Thus,

$$\{C\sharp, F, A\} = \{C\sharp \oplus C, \ C\sharp \oplus E, \ C\sharp \oplus G\sharp\}$$

(In the terminology of Section 6.9, the augmented chords are the distinct cosets of the 3-element subgroup chord.) We observe in passing that the diminished-seventh and augmented chords are the only two common chord types that remain the same regardless of which of their notes is considered to be the root.

If you are interested in exploring group patterns in music, an excellent source of further information is Chapter 23 of Item 1 in the list *For Further Reading* at the end of this chapter. You will find in there a more detailed discussion of scales and harmony, and also an analysis of mathematical patterns in rounds, fugues, canons, and other musical forms.

Our second example of groups in other disciplines comes from the study of molecular structure, an important area of investigation in chemistry and other physical sciences. The shape of a molecule influences the way it behaves with respect to electromagnetic radiation. Because of this, a molecule's symmetric structure is closely related to its spectral properties. The symmetric structure of a molecule can be described as a group, known as the *point group* of the molecule. This name, which is also used in the study of crystalline symmetric structure, comes from the fact that any symmetry operation (rotation or reflection) that takes a molecule back into itself cannot move the point at its center of mass.

Let us look at a specific example of a point group. The methyl chloride molecule can be diagramed as in *Figure 6.6*, with three hydrogen atoms forming the equilateral base of a triangular pyramid that is topped by a carbon atom linked to a chlorine atom.

The requirement that any symmetric operation on this molecule must leave its center of mass fixed implies in this case that both the carbon and chlorine atoms must stay where they are. This means that all the symmetries of this molecule are determined by the rigid motions of the equilateral base triangle $H_1H_2H_3$. This group is easy to describe. In fact, we have already studied it in Section 6.9! It is the six-element group of rotations and reflections displayed in *Figure 6.3*.

If you are interested in finding out more about molecular symmetry, Chapter 2 of Item 7 in the list *For Further Reading* is an excellent source. It provides a wealth of detail about the use of groups and other abstract mathematical systems in the study of molecular structure.

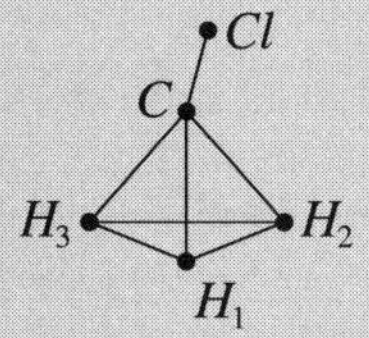

Figure 6.6 A methyl chloride molecule.

TOPICS FOR PAPERS

General Instructions

Consider the reader of your paper to be an anonymous student in your class—someone who knows the material of Chapter 6 fairly well, but does not know anything about your particular topic. Make sure the paper introduces your topic clearly and takes the reader step by step through an orderly development of your ideas to your desired conclusion.

1. Refer to the twelve problem-solving techniques described in Section 1.2. Write a paper that treats Chapter 6 as an extended example of how these problem-solving techniques are used. In particular, find and describe one or two instances of each of these techniques as they occur in Chapter 6,

and comment on how useful each instance seems to be. Conclude with some general opinions—and supporting arguments—about how effective and/or how distracting it is to focus on the problem-solving techniques when doing mathematics.

Suggestion: The most annoying part of writing this paper may well be deciding how to organize it. Take a little extra time to think it through, particularly from the viewpoint of your reader. You might find it helpful to write a brief, specific outline first.

2. Think of an operation-table pattern that you might like to investigate. Here are some suggestions to prod your imagination, but feel free to make up something different, if you wish.

   ■ The same element appears everywhere in the upper-left/lower-right diagonal.

   ■ The same element appears everywhere in the upper-right/lower-left diagonal.

   ■ Two rows (and/or columns) are identical.

   ■ Each row (or column) that appears also appears in reverse order.

   ■ A particular pair of elements always appears next to each other.

   ■ Only certain rearrangements of the order of the elements are allowable (as in the clock-arithmetic "cycles").

   Choose a specific pattern—one of those listed here or one of your own—and get your instructor's approval to use it as a topic for your paper. The type of structure (set and operation) described by your pattern will be yours to name and investigate, using as many of the problem-solving techniques from Chapter 1 as you find helpful. Pay particular attention to the concepts described in this chapter—associativity, commutativity, cancellation, identity element, inverses, substructures—but don't feel limited to them. Your paper is to be a description of this investigation, reporting what specific questions you asked yourself, describing in detail the particular problem-solving tactics you pursued, showing where they led, etc. Begin with a clear definition of your kind of structure, along with two or three examples of tables that fit the definition, and end it with some questions that seem to be promising avenues for further investigation.

3. This is an open-ended exploration assignment. Since it is difficult to observe associativity from an operation table, the construction of a group table is often done as follows:

   ■ Fill in the identity row and column.

   ■ Choose an inverse for each element.

   ■ Fill in the rest of the table so that there are no repetitions in any row or column. (That is, left and right cancellation hold.)

   ■ Hope that the resulting table is associative; check (somehow).

   For 2-, 3-, and 4-element sets, the first three of these steps actually yield groups; associativity follows automatically. However, for sets of 5 or more elements, an operation table constructed in this way may not be associative. What can be said about such tables? How are they like groups? How are they different? Investigate these questions and write a paper about your investigation. Begin with the following definition:

Definition: A **loop** is a set with an operation on it such that

    i. there is an identity element,

    ii. every element has an inverse, and

    iii. the left and right Cancellation Laws hold.[11]

Investigate loops in a way that parallels the development of groups in this chapter. Wherever possible, use the problem-solving techniques from Chapter 1, and describe how you are using them. Here are some particular questions you might consider:

(a) Can you construct some loops that are not groups?

(b) Is every loop commutative?

(c) Can you construct a loop of every size such that each element is its own inverse?

(d) How would you define "subloops"? What can be said about them?

(e) Does Lagrange's Theorem hold for loops?

Don't be limited by these questions; rather, use them to help you think of new ideas. If your explorations lead you in an interesting direction, follow it, even if you do not cover all the questions listed here.

Your paper is to be a description of *how you went about investigating loops*. Report what specific questions you asked, describe in detail the particular problem-solving techniques you pursued, show where they led, etc. (In this type of investigation, even a blind alley is something to report, provided you can describe the alley.) Be as creative as you like in finding examples, looking for patterns, and making conjectures, but try to prove your conjectures or at least supply some evidence or argument to support them.

FOR FURTHER READING

1. Budden, F. J. *The Fascination of Groups*. London: Cambridge University Press, 1972.

2. Davis, Philip J., and Reuben Hersh. *The Mathematical Experience*. Boston: Birkhäuser Boston, Inc., 1981, pp. 203–209.

11. This is a defined term in the mathematical literature. There were several articles about loops in the *American Mathematical Monthly* during the mid-1960s.

3. Devlin, Keith. *Mathematics: The New Golden Age*. London: Penguin Books, 1988, Chapter 5.

4. Grossman, Israel, and Wilhelm Magnus. *Groups and Their Graphs*. New York: Random House, Inc., 1964.

5. Kemeny, John G. *Random Essays on Mathematics, Education and Computers*. Englewood Cliffs, NJ: Prentice-Hall, Inc., 1964, Chapter 7.

6. Newman, James R., ed. *The World of Mathematics*, Vol. 3. New York: Simon and Schuster, Inc., 1956, Part IX.

7. Schonland, David S. *Molecular Symmetry*. London: D. Van Nostrand Company, Ltd., 1965, Chapter 2.

CHAPTER

7

MATHEMATICS OF SPACE AND TIME: FOUR-DIMENSIONAL GEOMETRY

WHAT IS FOUR-DIMENSIONAL GEOMETRY?

7.1

One of the most profound and far-reaching insights of twentieth-century science has been the realization that the world we live in is four-dimensional. More precisely, it is the realization that a proper understanding of physical laws depends on treating time as a dimension, just like length, width, and depth.

When Isaac Newton formulated his laws of motion in the late seventeenth century, he thought of the universe as existing in the stationary, uniform three-dimensional space described by Euclid's solid geometry. These laws of motion, which de-scribed the physical theory of gravitation, were based on the idea that time was an absolute concept, moving onward at a steady pace that is the same for all observers everywhere in the universe. Thus, for Newton, space was a fixed frame of reference within which the laws of gravity governed the relative motions of objects—from atoms to planets—as they lived out their allotted time span measured by the universal clock of the Creator. This view of the universe dominated physical science for two hundred years.

In 1881, however, a disturbing event occurred in Cleveland, Ohio. Two American physicists, A. A. Michelson and E. W. Morley, performed an experiment whose results shocked the scientific world. The specifics of that experiment need not concern us here. Suffice it to say that they developed an apparatus that could measure the speed of a

337

light beam to within a fraction of a mile per second. Newton's laws of motion as applied to light rays flashed from a point on our moving planet Earth assert that the speed of those light rays (186,284 miles per second) must vary some 20 miles per second one way or the other, depending on whether the light is flashed with or against the direction of the Earth's motion. The Michelson-Morley experiment showed that this variation does *not* occur! The results of this experiment, verified by later work of Morley and other scientists, called into question the theory of gravity and with it the entire structure of the Newtonian universe.

Albert Einstein was just a toddler when the Michelson-Morley experiment was performed; the problem it posed had to wait for a solution until he formulated his theory of relativity at the age of twenty-six. That theory scrapped not only Newton's laws of gravity, but also the very foundation upon which they rested. Einstein rejected the notions of absolute space and time, claiming, instead, that length, width, depth, *and* duration could only be measured *in relation to* some observer, and that these dimensions varied from observer to observer.

We are used to dealing with some aspects of this relativity of space. For example, if I am looking out a restaurant window and notice that on the adjacent sidewalk a lamppost is to the left of a fire hydrant, I know that you, standing across the street, see the lamppost as being to the right of the fire hydrant. With Einstein's relativity theory, this reversibility of order, based on the perspective of the observer, is extended to the before/after relation of time. Two events that appear to occur in one time order to me may appear to occur in the opposite order to someone in a distant galaxy, and *both* of us could be correct! Just as there is no absolute sense of left and right, there is no absolute sense of before and after.

To locate an object or an event in our universe accurately, then, we must be able to say *where* and *when* it is, relative to some chosen frame of reference. This idea did not originate with Einstein. The view that the mathematical science of mechanics can be regarded as the geometry of four dimensions—three spatial dimensions and a time dimension—first appeared in the eighteenth century, in the writings of the French mathematicians d'Alembert (in 1754) and Lagrange (in 1788). However, it was not until the emergence of time-space questions in early twentieth-century physics that this mathematical device was recognized as a valuable tool for describing reality. When Einstein solved the problem of motion by denying the absolute nature of time, scientists were suddenly faced with the need for four-dimensional coordinate systems to describe the where-when locations of objects in space-time.

The relative nature of time-and-space measurements with respect to these coordinate systems is a bit beyond the scope of this book. If you are interested in pursuing this topic further, there is a clear, simple explanation of relativity theory in Lincoln Barnett's book, *The Universe and Dr. Einstein*, Item 2 in the list *For Further Reading* at the end of this chapter; a more recent discussion of it appears in Item 8 of that list. We shall confine our attention to an explanation of the coordinate systems themselves, starting with the simple one-dimensional world of a single straight line and building step by step to the four dimensions of space and time. In the course of this development we shall see how a few elementary geometric and algebraic techniques and the insight provided by analogy combine to produce a mathematical device powerful enough to carry us past the three-dimensional limit of our spatial intuition to the real and fictional worlds of four dimensions and beyond.

7.1   EXERCISES

WRITING EXERCISES

1. Using complete sentences, write a half-page summary that identifies the main theme of this section and outlines how that theme is carried out.

2. Do you think that Einstein's theory of relativity has any effect at all on your daily life? If so, give an example; if not, why not?

3. Using the library as a resource, write a short biographical sketch of Isaac Newton or Albert Einstein. Be sure to list the source(s) you use and be careful that you do not simply copy or paraphrase the material from your source(s).

4. From a political and/or social standpoint, what was going on in the United States around 1881 (the time of the Michelson-Morley experiment)? What about in 1905 (about the time that Einstein's theory of relativity emerged)? Feel free to consult a history book, but be sure to list your source(s).

ONE-DIMENSIONAL SPACE

7.2

The word *dimension* comes from the Latin word *dimensus*, which literally means "measured out" or "measured separately." This meaning expresses the underlying idea of dimension very well. For instance, a line segment is considered one dimensional because it has only one measure, its length. A rectangle, on the other hand, requires two separate measurements, length and width, so it is considered two-dimensional.

A cardboard carton is three-dimensional because it requires three separate measurements—length, width, and depth (or height). Space-time objects are considered four-dimensional because four separate measurements are needed to describe them—length, width, depth, and duration. An object that requires five separate measurements (length, width, depth, duration, and temperature, for instance) is five dimensional, and so on. Thus, our study of dimensions must begin with the notion of measurement on a single straight line.

Note   Readers familiar with coordinates and distance on a plane can skim this material until the definition of "taxicab path" on page 349.

To measure length on a straight line, we must first choose a **unit length**; that is, we must decide what length will be assigned the number one. This choice can be made in any way we please. The measurement theory will be the same, regardless of our choice of unit, so long as we are consistent throughout the process. We could use feet or inches or meters or miles, or we could make up a unit, such as ————.

Different units are convenient for different purposes—inches or centimeters for cabinet makers; kilometers, miles, or light years for astronomers; angstroms or microns for nuclear engineers; and so on. A unit length is even used to represent time—a second, a minute, an hour, a day, etc. To emphasize that our theoretical work is independent of the choice of unit length, we shall often just use the word *unit* to denote the chosen basis for measurement.

| EXAMPLE 7.1 | The distance between points p and q on the line in *Figure 7.1* can be described (approximately) as: |

$\qquad$ 2, if the unit length is an inch;

$\qquad$ 5, if the unit length is a centimeter;

$\qquad$ $\frac{1}{6}$, if the unit length is a foot;

$\qquad$ 3, if the unit length is ————.

Coordinate geometry is based on the fact that, once a unit length is chosen, every point on a straight line can be labeled with a real number in such a way that the line can be used as a sort of "infinite ruler" to measure distances from some specific fixed point. This fixed point is called the **origin** and is assigned the number 0. Once the point of

p $\qquad\qquad\qquad\qquad\qquad$ q

Figure 7.1 The distance between p and q.

origin and the unit length have been chosen, the line is "coordinatized" by choosing a positive direction and marking off the integer points in successive unit lengths. If the line is horizontal, we usually regard *right* as the positive direction; if the line is vertical, we usually think of *up* as positive. However, those choices are simply a matter of custom and may be varied if it is more convenient to pick a different direction as the positive one. All the other points on the line have numerical labels, too. Each real number can be matched with exactly one point on the line, and vice versa.

Specific methods for matching numbers with points need not concern us here; it is sufficient for us to know that each point on the line corresponds to a unique real number. However, we note in passing that the Greeks had a geometric-construction method capable of locating the point for any rational number (fraction), and that irrational-number locations can be approximated to any desired degree of accuracy by using decimal expansions.[1] For the diagrams of this chapter we shall simply estimate the approximate locations of numbers relative to the unit length and the origin.

A line labeled with real numbers by specifying an origin, a unit length, and a positive direction is called a **number line** or **real line**, and is denoted by **R**. The points on that line are usually referred to by their numerical labels. A single straight line is a **one-dimensional space**, sometimes called **1-space**.

If each point in that space has been assigned a real number, then the distance between two points is easy to find. Recall that the **absolute value** of a number x is defined by

$$|x| = \begin{cases} x & \text{if } x \text{ is positive or zero} \\ -x & \text{if } x \text{ is negative} \end{cases}$$

DEFINITION   The **distance** between two points p and q on a real line is $|p - q|$, the absolute value of their difference.

EXAMPLE 7.2   The distance between 5 and 2 is 3 because $|5 - 2| = 3$. The absolute value is used so that the order in which the points are chosen does not affect the distance between them. In particular, the distance between 2 and 5 is the same as the distance between 5 and 2 because $|2 - 5| = |-3| = 3$.

EXAMPLE 7.3   The distance between -4 and 7 is

$$|-4 - 7| = |-11| = 11$$

1. For a discussion of the decimal expansions of rational and irrational numbers, see Section 5.4.

EXAMPLE 7.4

The absolute value of any point is its distance from the origin:

the distance between 2 and 0 is $|2 - 0| = 2$

the distance between $\frac{-3}{8}$ and 0 is $\left|\frac{-3}{8} - 0\right| = \left|\frac{-3}{8}\right| = \frac{3}{8}$

If we were to travel from a point p to a point q on a line, our path would cover the part of that line lying between p and q. More formally, if p and q are points in a one-dimensional space, with $p < q$, then the **path** between p and q is the line segment consisting of p and q themselves and all points between them. We symbolize this by

$$[p, q] = \{x \mid p \leq x \leq q\}$$

The **length** of this path is the distance between p and q.

Note   Recall that < means *less than* and ≤ means *less than or equal to*. If you are unfamiliar with this way of writing sets, look up "set-builder notation" in Section 4.2. We shall call such an expression a **point-set** description of the line segment.

EXAMPLE 7.5

The path between 1 and 4 is $[1, 4] = \{x \mid 1 \leq x \leq 4\}$. It can be pictured as in *Figure 7.2*. The path between 4 and 1 is exactly the same set of points. To be consistent with the use of the "≤" sign, we customarily write the smaller number first when describing paths.

EXAMPLE 7.6

The path between -2 and 7 is $[-2, 7] = \{x \mid -2 \leq x \leq 7\}$. Its length is $|-2 - 7| = |-9| = 9$.

The only basic geometric figures in a one-dimensional space are single points and line segments; all other figures are just collections of separate segments and points. Aesthetically speaking, this makes 1-space pretty dull. Moreover, travel in a one-dimensional world can be difficult. A single point provides a barrier that cannot be avoided. For instance, there is no way to travel from 1 to 4 without crossing the point 2. Two points completely enclose the segment between them; there is no way to get from a point in that segment to a point not in it without crossing one of the endpoints. Getting around a one-point barrier is like avoiding a waiting catcher on the way from third base

-3 -2 -1   0   1   2   3   4   5   6   7   *Figure 7.2 The path between 1 and 4.*

to home plate—unless you go outside the base path, you're bound for a collision. Like it or not, there is no way to go outside the base path in one-dimensional space. That option requires a space of at least two dimensions.

7.2   EXERCISES

For Exercises 1–8, measure with a ruler the distance between the points *p* and *q* on the line in *Figure 7.3*. Express this distance using each of the unit lengths given. (Your answers should be accurate to the nearest whole number.)

1. an inch
2. a centimeter
3. a quarter-inch
4. a millimeter
5. a foot
6. a meter
7. _____
8. _____

For Exercises 9–16, mark the approximate locations of the following points on the number line in *Figure 7.4*.

9. 3
10. −2
11. $\frac{1}{2}$
12. $\frac{-4}{5}$
13. $\frac{-10}{3}$
14. $\frac{39}{17}$
15. $\sqrt{2}$
16. $-\sqrt{3}$

For Exercises 17–26, find the distance between the given pair of points on a real (number) line.

17. 7 and 3
18. 15 and 2
19. −5 and 2
20. −5 and 5
21. −5 and −5
22. −5 and 0
23. $2\sqrt{2}$ and 0
24. $\sqrt{3}$ and $2\sqrt{3}$
25. 6 and x
26. x and x + 1

For Exercises 27–32, write in set notation the path between the given pair of points, find the length of that path, and sketch the path on a number line.

27. 2 and 3
28. 0 and 6
29. −2 and 1
30. $\sqrt{2}$ and $\sqrt{3}$
31. −5 and $\frac{-1}{2}$
32. −2.5 and .75

If a centimeter represents 10 minutes, mark the time intervals in Exercises 33–38 on the number line in *Figure 7.5*, assuming that each interval starts at 0 minutes.

33. 40 minutes
34. 5 minutes
35. one hour
36. half an hour
37. 37 minutes
38. 1 hour and 15 minutes

p   *q*

Figure 7.3 The distance between p and q in terms of various unit lengths.

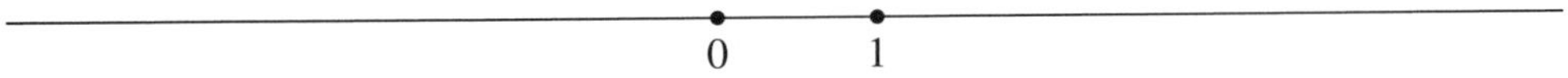

0   1

Figure 7.4 Points on a number line.

0 min. 10 min

Figure 7.5 Time intervals starting at 0 minutes.

39. In a *plane*, a **circle** is defined as:

> the set of all points that are a fixed distance (the radius) from a particular point (the center).

Using this definition, what does a circle in 1-space look like?

Exercises 40–42 are independent of each other, but they all depend on Exercise 39.

40. Specify the points that form the 1-space circle of radius 5 centered at 0. Write this set of points in set-builder notation and also list its specific points.

41. Specify the points that form the 1-space circle of radius 5 centered at 8. Write this set of points in set-builder notation, and also list its specific points.

42. If you remove the points of a circle from a plane, the rest of the plane is divided into two connected pieces, the *inside* and the *outside*. Is that true of 1-space, too? Explain.

WRITING EXERCISES

1. Name two units of length measure besides the ones mentioned in this section. Relate them to some familiar unit of measure, and describe their origins and most common uses. (If you can't think of any, you might try looking up *fathom*, *furlong*, or *rod*, but there are many others. Try to find at least one on your own.)

2. Explain *absolute value* as if you were talking to a ninth grader, without using any symbols at all.

TWO-DIMENSIONAL SPACE

7.3

When we go from the world of a single line to the world of a plane, we need a second measurement to fix the location of a point. That is, if we have a single real line in a plane, then instructions for reaching any point in that plane from the origin can be specified by *two* numbers. We need one number to tell us which way and how far to go along the line, and a second number to tell us which way and how far to go above or below the line. *Figure 7.6* shows two examples of locating points in this way. Starting at the origin of the line, we can reach the point *P* by traveling along the line 3 units in a positive direction, then moving up 2 units. Thus, assuming *up* is considered as the positive direction, we can give the "address" of *P* relative to the origin as the *ordered pair* of numbers 3, then 2. Similarly, the address of the point *Q* in *Figure 7.6* can be given as 4, then −1. Because the location of any point in the plane can be

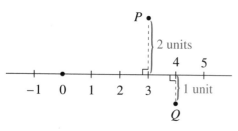

Figure 7.6 Locating points in two-dimensional space.

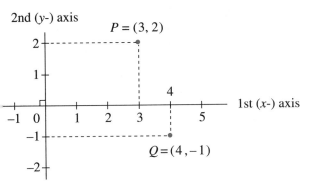

2nd (y-) axis

Figure 7.7 A Cartesian coordinate system.

determined by two numbers in this way, the plane is called **two-dimensional space**, or, simply, **2-space**.

The second numbers in the location process just described indicate distance and direction, just as the first numbers do; so we can also consider the second numbers as describing locations on a single real line, a line perpendicular to the first one at a common origin point, as shown in *Figure 7.7*. This arrangement of perpendicular copies of the real line crossing at their common origin is called a **Cartesian coordinate system**,[2] and each of the two number lines is called a **coordinate axis**.

The most common form of the Cartesian coordinate system uses a horizontal line as the first axis and a vertical line as the second. Then the location of any point in the plane is specified by an **ordered pair** (x, y) of real numbers, called the **coordinates** of the point. The first number denotes a position on the horizontal axis, which is usually called the **x-axis**; the second number denotes a position on the vertical axis, or **y-axis**. The point itself is the intersection of the two perpendicular lines through the axis points. *Figure 7.7* shows this for the points (3, 2) and (4, −1). In this way the set of all points in the plane can be represented by the *Cartesian product* **R** × **R**, the set of all ordered pairs of real numbers.

DEFINITION

The **Cartesian product** of any two sets A and B is the set of all ordered pairs with first elements from A and second elements from B. In symbols

$$A \times B = \{(a, b) \mid a \in A, b \in B\}$$

The use of perpendicular coordinate axes makes it easy to compute the straight-line distance between two points in the plane by using the Pythagorean Theorem.[3] For example, the straight-line distance d between (2, 1) and (6, 4) is the length of the hypotenuse of a right triangle whose right-angle vertex is (6, 1), as shown in *Figure 7.8* on page 346. The lengths of the other two sides of this triangle are 4 (the x-axis distance between the first coordinates) and 3 (the y-axis distance

2. Named for seventeenth-century French mathematician René Descartes, who used this type of system as the key to his development of analytic geometry.
3. "The square of the hypotenuse of a right triangle equals the sum of the squares of the other two sides." (See Figure B.1 of Appendix B.)

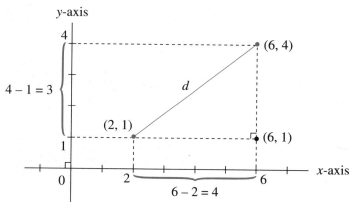

Figure 7.8
The distance between (2, 1) and (6, 4).

between the second coordinates). The Pythagorean Theorem then tells us that

$$d^2 = 4^2 + 3^2$$

and thus,

$$d = \sqrt{16 + 9} = \sqrt{25} = 5$$

DEFINITION

The **distance** between two points (x_1, y_1) and (x_2, y_2) in the plane is

$$\sqrt{(x_2 - x_1)^2 + (y_2 - y_1)^2}$$

EXAMPLE 7.7

The distance between $(3, 5)$ and $(8, -2)$ is

$$\sqrt{(8 - 3)^2 + (-2 - 5)^2} = \sqrt{5^2 + (-7)^2} = \sqrt{25 + 49} = \sqrt{74}$$

It does not matter which point is taken first. For example, reversing the order of the points here, we get

$$\sqrt{(3 - 8)^2 + (5 - (-2))^2} = \sqrt{(-5)^2 + 7^2} = \sqrt{25 + 49} = \sqrt{74}$$

EXAMPLE 7.8

The two-dimensional distance between two points on the same horizontal or vertical line (including the axes) is the same as the one-dimensional distance between them. For instance, the one-dimensional distance between -2 and 5 on the x-axis is $|-2 - 5| = 7$. In two-dimensional space, the distance between the same points, $(-2, 0)$ and $(5, 0)$, is

$$\sqrt{(5 - (-2))^2 + (0 - 0)^2} = \sqrt{49} = 7$$

Similarly, the distance between $(2, 3)$ and $(2, 7)$, which are on the same vertical line, is

$$\sqrt{(2 - 2)^2 + (7 - 3)^2} = \sqrt{4^2} = 4$$

which is the same as the one-dimensional distance between 3 and 7.

One of the simplest, and most useful, 2-dimensional figures can be described entirely in terms of the distance formula. A **circle** is the set of all 2-space points that are a given distance (its **radius**) from a particular point (its **center**). The coordinate description of a circle becomes particularly simple if the origin of the coordinate system is placed at its center.

In this chapter, unless you are specifically told otherwise, assume that all circles are centered at the origin.

A circle of radius 5, then, is the set of all points (x, y) whose distance from $(0, 0)$ is 5. Using the distance formula, this becomes

$$\{(x, y) \mid \sqrt{(x - 0)^2 + (y - 0)^2} = 5\}$$

which simplifies to

$$\{(x, y) \mid x^2 + y^2 = 25\}$$

In general, a circle of radius r can be described as:

$$\{(x, y) \mid x^2 + y^2 = r^2\}$$

EXAMPLE 7.9

The coordinate description of a circle of radius 3 (centered at the origin) is

$$\{(x, y) \mid x^2 + y^2 = 9\}$$

Determining whether a point (x, y) is inside, outside, or on this circle is a simple matter of observing whether $x^2 + y^2$ is less than, greater than, or equal to 9, respectively:

▪ The point $(2, 2)$ is inside this circle because

$$2^2 + 2^2 = 8 < 9$$

▪ The point $(-1.6, 2.6)$ is outside this circle because

$$(-1.6)^2 + 2.6^2 = 9.32 > 9$$

▪ To find a coordinate value y such that $(1.5, y)$ is on this circle, just solve the equation $1.5^2 + y^2 = 9$:

$$y^2 = 9 - 2.25 = 6.75$$

so y can be either

$$\sqrt{6.75} \text{ or } -\sqrt{6.75} \quad (\text{approximately}, \pm 2.6)$$

Figure 7.9 illustrates the locations of these points in relation to the circle.

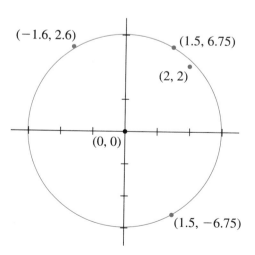

Figure 7.9 Points relative to the circle of radius 3 (centered at the origin).

The shortest road between two points is not always the easiest one to travel or describe. For instance, a taxicab in mid-Manhattan seldom takes its passengers from pick-up to drop-off in a straight path because it has to follow the rectangular grid of the streets and avenues. We, too, shall find it convenient to describe paths "taxicab fashion"—that is, by using segments of lines parallel to the coordinate axes. Although they are not always the shortest paths between points, we can visualize them easily, describe them in a simple symbolic form, and generalize them to higher-dimensional spaces in an obvious way.

The coordinates of all the points on a line parallel to an axis are easy to describe. For instance, the x-axis itself is the set of all points whose second coordinate is 0 because all those points are at the 0 level relative to the y-axis. Thus, in point-set notation, the x-axis is the set of points

$$\{(x, 0) \mid x \in \mathbf{R}\}$$

The segment of the x-axis between 2 and 5 is the set of points

$$\{(x, 0) \mid 2 \le x \le 5\}$$

Similarly, a horizontal line is parallel to the x-axis. Thus, the horizontal line 3 units above the x-axis is the set of all points with second coordinate 3; that is,

$$\{(x, 3) \mid x \in \mathbf{R}\}$$

It is usually called "the line $y = 3$." The segment of this line between $x = 2$ and $x = 5$ is the set

$$\{(x, 3) \mid 2 \le x \le 5\}$$

Figure 7.10(a) illustrates these lines and segments. Lines and segments parallel to the y-axis are described similarly, as shown in the following examples.

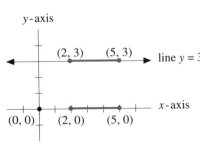

(a) Horizontal

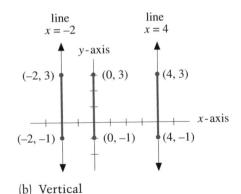

(b) Vertical

Figure 7.10 Lines and segments: (a) horizontal, (b) vertical.

EXAMPLE 7.10

(*Figure 7.10(b)* illustrates this example.) The *y*-axis is the set of all real-number pairs with first coordinate 0; that is,

$$y\text{-axis} = \{(0, y) \mid y \in \mathbf{R}\}$$

The segment on it between -1 and 3 is $\{(0, y) \mid -1 \leq y \leq 3\}$.

EXAMPLE 7.11

(*Figure 7.10(b)* illustrates this example.) The vertical line at $x = 4$ is

$$\{(4, y) \mid y \in \mathbf{R}\}$$

and the segment on it between -1 and 3 is

$$\{(4, y) \mid -1 \leq y \leq 3\}$$

Similarly, the line $x = -2$ is

$$\{(-2, y) \mid y \in \mathbf{R}\}$$

and the segment on it between -1 and 3 is

$$\{(-2, y) \mid -1 \leq y \leq 3\}$$

Now it is easy to give numerical descriptions of "taxicab paths" in the plane. Suppose, for instance, we want to go from the origin, $(0, 0)$, to the point $(5, 3)$. We could travel along the *x*-axis from 0 to 5, then travel up along the line perpendicular to the *x*-axis at 5 from the 0 level to the 3 level. (See *Figure 7.11*.) The two line segments of this path are

$$\{(x, 0) \mid 0 \leq x \leq 5\} \quad \text{and} \quad \{(5, y) \mid 0 \leq y \leq 3\}$$

(Note that the two segments have exactly one point in common—$(5, 0)$, the place where they are joined.) Thus, a path from $(0, 0)$ to $(5, 3)$ is the union of these two line segments:

$$\{(x, 0) \mid 0 \leq x \leq 5\} \cup \{(5, y) \mid 0 \leq y \leq 3\}$$

In general, we call two line segments **connected** if they have a common endpoint. A **taxicab path** is the union of connected line segments that are parallel to the coordinate axes, with successive segments parallel to different axes. Because we shall use only taxicab paths for the remainder of this chapter, we shall use the term **path** to mean a taxicab path. Now, all but two of the endpoints of line segments in a (taxicab) path are common to two (or more) line segments. The remaining two points are called the **endpoints** of the path. If a path has endpoints P and Q, we call it a **path between** P and Q.

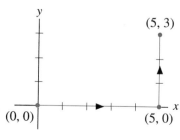

Figure 7.11   A taxicab path in the plane.

EXAMPLE 7.12

A path between $(-1, 3)$ and $(4, -2)$ is

$$\{(x, 3) \mid -1 \leq x \leq 4\} \cup \{(4, y) \mid -2 \leq y \leq 3\}$$

The intersection point of the two segments is $(4, 3)$. Another path between $(-1, 3)$ and $(4, -2)$ is

$$\{(-1, y) \mid -2 \leq y \leq 3\} \cup \{(x, -2) \mid -1 \leq x \leq 4\}$$

The intersection of these two segments is the point $(-1, -2)$. A third path between $(-1, 3)$ and $(4, -2)$ is

$$\{(x, 3) \mid -1 \leq x \leq 1\} \cup \{(1, y) \mid -2 \leq y \leq 3\} \cup \{(x, -2) \mid -1 \leq x \leq 4\}$$

The segments in this path are connected at $(1, 3)$ and $(1, -2)$. All three of these paths are pictured in *Figure 7.12*.

Another basic two-dimensional figure is the square. It can be built by using four copies of the basic one-dimensional figure, a line segment, in an obvious way:

Place one copy of the segment horizontally in the plane and attach another one at each end so that they are pointing vertically upward from the first segment; then "cap" the two vertical segments with the fourth one by attaching it to their upper endpoints.

This can also be done by attaching the four segments end-to-end on a single line, then folding them up, as in *Figure 7.13*. This diagram shows segment 2 held fixed while segments 1 and 3-4 are folded vertically upward. Then segment 4 is folded over until it meets segment 1, completing the square.

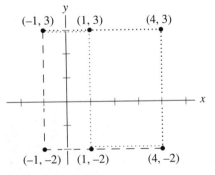

Figure 7.12 *Three paths between* $(-1, 3)$ *and* $(4, -2)$.

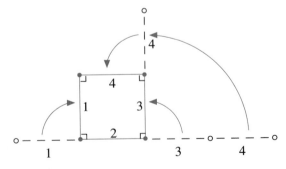

Figure 7.13 *Constructing a square from line segments.*

The region in the plane enclosed by a square can be thought of in another way that has interesting implications for us. If a line segment is moved in a direction perpendicular to itself for a distance as long as itself, then it "sweeps out" a square region in the plane. Now, if this perpendicular direction is taken to represent time, then the square region represents the existence of the line segment during a period of time.

For example, consider a two-dimensional coordinate system in which the vertical axis represents height (in inches) and the horizontal axis represents time (in minutes), as in *Figure 7.14.* Suppose that you have a very thin 3-inch candle that burns at the constant rate of 1 inch every 2 minutes. Set up the candle (but don't light it) and start your stopwatch. At the end of exactly 3 minutes the unlit candle will have "swept out" a 3-by-3 square in the height-time plane. After 3 minutes light the candle. As it burns, its height decreases by 1 inch every 2 minutes, so that it will be completely gone 6 minutes after you light it. The candle, set up for 9 minutes and burning from the end of the third minute until the end of the ninth minute, can be represented by a single two-dimensional region, as shown in *Figure 7.14.* In this way,

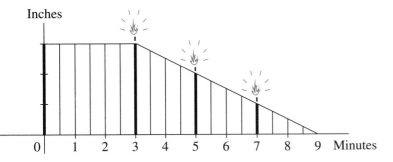

Figure 7.14   *The "life" of a candle in the height-time plane.*

> regions in two-dimensional space-time can be used to represent the existence in time of objects in one-dimensional space.

EXAMPLE 7.13

The outdoor temperature can be represented by a column of mercury calibrated in degrees Celsius, represented by the vertical axis. The time can be represented by the horizontal axis. During a 24-hour period the temperature rises from a midnight low of 10° to a high of 25° at 2 p.m., then drops back to 15° by the following midnight. All the changes in this column of mercury during that day are represented by the single height-time region in *Figure 7.15* on page 352.

Finally, let us consider the 1-space barrier problem described at the end of the preceding section. It is impossible to find a path between the point 4 on the number line and a point inside the interval [−2, 2]— say 1—that does not pass through the endpoint 2. However, if we consider that number line as the *x*-axis in a two-dimensional coordinate system, then the problem becomes:

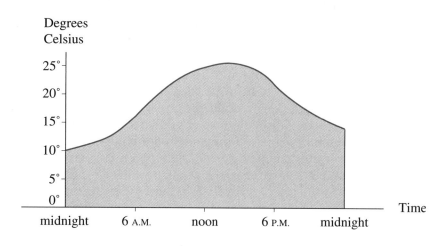

Figure 7.15 *Temperature change represented by a height-time region.*

Find a path from (4, 0) to (1, 0) that does not pass through (2, 0).

That's easy; we just go up, over, and down. *Figure 7.16* shows such a path; its point-set description is

$$\{(4, y) \mid 0 \leq y \leq 1\} \cup \{(x, 1) \mid 1 \leq x \leq 4\} \cup \{(1, y) \mid 0 \leq y \leq 1\}$$

Even without a picture, it is clear from this description that (2, 0) is not a point of this path because the only points of the path that have second coordinate 0 are the endpoints (4, 0) and (1, 0). The rest of the path lies outside the original one-dimensional space.

There is an analogous barrier problem for 2-space:

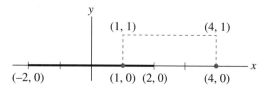

Figure 7.16 *A 2-space solution to the 1-space barrier problem.*

Between a point outside a bounded region (square, circular, etc.) of the plane and a point inside the region, find a path that does not pass through the boundary.

There is no two-dimensional solution of this problem; but a third dimension makes it easy, as we shall see in the next section.

7.3   **EXERCISES**

For Exercises 1–8, set up a single Cartesian co-ordinate system for the plane, and mark on it the location of the given points.

1. (2, 4)   2. (3, 3)   3. (−1, 3)   4. $\left(\frac{1}{2}, 4\right)$

5. (−1, −1)   6. (1, −2)   7. (−5, 0)   8. $\left(\frac{3}{4}, \frac{1}{4}\right)$

For Exercises 9–14, compute the distance between the given pair of points.

9. (2, 1) and (8, 9)   10. (2, 3) and (4, 5)

11. (0, 0) and (1, 1)   12. (−2, 7) and (7, −5)

13. (9, −1) and (9, 4)   14. (−1, −3) and (1, 3)

For Exercises 15–20, indicate whether the given point is *inside*, *outside*, or *on* the circle of radius 5 centered at the origin. Justify your answers.

15. $(2, 3)$ 16. $(-5, 0)$ 17. $(4, -3)$

18. $(7, -2)$ 19. $(2, \sqrt{21})$ 20. $(4, 4)$

21. Write a set-builder description of the circle of radius 12 centered at the origin. Then find the y-coordinate of a point $(7, y)$ inside the circle, the x-coordinate of a point $(x, 2.5)$ outside it, and the y-coordinate of a point $(6, y)$ on it.

22. Write a set-builder description of the circle of radius 10 centered at the origin. Then find the coordinates of three points not on either axis—a point inside the circle, a point outside it, and a point on it.

For Exercises 23–32, sketch the line, segment, or path in a Cartesian coordinate plane.

23. $\{(x, 5) \mid x \in \mathbf{R}\}$ 24. $\{(-1, y) \mid y \in \mathbf{R}\}$

25. $\{(x, 0) \mid 0 \le x \le 3\}$ 26. $\{(x, 4) \mid -1 \le x \le 2\}$

27. $\{(1, y) \mid -1 \le y \le 1\}$

28. $\{(-3, y) \mid -3 \le y \le 0\}$

29. $\left\{(x, 2) \;\middle|\; \dfrac{-1}{2} \le x \le \dfrac{1}{2}\right\}$

30. $\left\{(5, y) \;\middle|\; \dfrac{1}{2} \le y \le \dfrac{3}{4}\right\}$

31. $\{(x, -1) \mid -2 \le x \le 3\} \cup \{(3, y) \mid -1 \le y \le 4\}$

32. $\{(4, y) \mid 2 \le y \le 5\} \cup \{(x, 2) \mid 0 \le x \le 4\}$

For Exercises 33–40, describe in set-builder notation two taxicab paths between the two given points. Also, specify the intersection points that connect the segments in the paths. Sketch the paths.

33. $(0, 0)$ and $(3, 4)$ 34. $(2, 2)$ and $(0, 0)$

35. $(1, 2)$ and $(5, -1)$ 36. $(0, 1)$ and $(3, 0)$

37. $(2, 0)$ and $(-1, -1)$ 38. $(-1, 3)$ and $(4, 3)$

39. $\left(2, \dfrac{-5}{2}\right)$ and $\left(2, \dfrac{7}{3}\right)$ 40. $\left(\dfrac{1}{2}, 1\right)$ and $\left(5, \dfrac{2}{3}\right)$

For Exercises 41–44, draw a two-dimensional space-time figure to represent (approximately) each situation.

41. A 2-inch candle that burns at the constant rate of one inch every 5 minutes is lit and burns until nothing remains.

42. A 5-inch candle that burns at the constant rate of 1 inch every 2 minutes is lit, burns for 4 minutes, is blown out for 3 minutes, then is relit, and burns until nothing remains.

43. A stone is dropped from the top of a 140-foot cliff and hits the ground in 3 seconds.

44. A mercury thermometer measures the heat in a factory as it rises at a constant rate from 15° (Celsius) at 7 a.m. to 35° at 3 p.m.

45. Consider the point 3 and the interval $[0, 1]$ on the real line. By considering that line as part of 2-space, find a path between 3 and some point inside $[0, 1]$ such that this path does not pass through the endpoint 1. Sketch this path and write its set-builder description.

46. (a) Write a set-builder description of a square (line figure) with corners $(0, 0)$, $(2, 0)$, $(0, 2)$, $(2, 2)$. Sketch this square.

(b) Write a set-builder description of the planar region bounded by the square in Part (a).

(c) List the coordinates of three points that are inside the square region of Part (b) and of three points that are outside it.

47. A **disk** is the set of all points on or inside a circle.

(a) Describe in set-builder notation a disk of radius 5 centered at the origin.

(b) Describe in set-builder notation a disk of radius r centered at the origin.

The distance formula can be used to form equations for circles that are not centered at the origin. Exercises 48–50 explore this idea in stages.

48. (a) Write a formula for the distance between $(2, 7)$ and a point (x, y).

   (b) Use your answer to Part (a) to write a set-builder description of a circle of radius 5 centered at the point $(2, 7)$.

   (c) Is $(4, 4)$ inside, outside, or on this circle? How do you know?

49. (a) Describe in set-builder notation the circle centered at $(-1, 4)$ that passes through the point $(6, 8)$.

   (b) Is $(8, 0)$ inside, outside, or on this circle? What about $(0, 8)$? Justify your answers.

50. (a) Write a formula for the distance between a point (a, b) and a point (x, y).

   (b) Use your answer to Part (a) to describe in set-builder notation a circle of radius 5 centered at the point (a, b).

   (c) Generalize Part (b) by describing in set-builder notation a circle of radius r centered at the point (a, b).

51. You are located at $(1, 2)$, inside a circle of radius 3 centered at the origin. You want to get to the point $(-5, 2)$ by a straight path. At what point does your path intersect the circle?

52. You are located at $(2, 1)$, inside a circle of radius 3 centered at the origin. You want to get to the point $(5, 4)$ by a taxicab path. At what point does your path intersect the circle? Justify your answer. (There is more than one correct answer.)

WRITING EXERCISES

1. The construction of this section is analogous to that of Section 7.2. Write a comparative outline of the two sections that displays all points of the analogy.

2. Cartesian-product (ordered-pair) labeling is used to identify location in many situations outside of mathematics. For example, the location of a seat in a theater or stadium is often specified by an ordered pair consisting of a letter (or pair of letters) and a number, with the letter(s) specifying the row and the number specifying the seat within that row. Find and describe two more such examples.

3. Is the term "taxicab path" helpfully suggestive of its meaning, or do you find its imagery distracting? Explain. Can you think of a better term for this concept?

THREE-DIMENSIONAL SPACE

7.4

Now that we have seen how to generalize basic spatial concepts from one dimension to two dimensions, the step from 2-space to 3-space should be easy. Let us begin with the floor beneath your feet, extending it to form an unbounded plane. Any point in our spatial universe that is not on the plane itself is directly above or directly below some point of this plane. Now, because the floor plane is a two-dimensional space, each of its points can be specified by an ordered pair of real

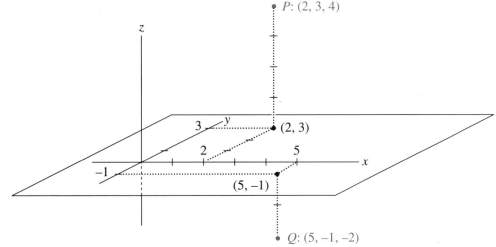

numbers in some Cartesian coordinate system. We can extend this coordinate system in order to label any point in the universe by using a third number to indicate exactly how far directly above or below the plane that point is.

Figure 7.17 shows this for two specific points—*P*, which is 4 units above the point (2, 3) of the plane, and *Q*, which is 2 units below the point (5, −1) of the plane. If at the origin we attach a third coordinate axis perpendicular to the two for the plane, this third axis can be used to measure the height above or below the plane. By this means, every location in our spatial universe can be specified by three separate coordinate measurements; thus, we speak of the universe as a **three-dimensional space**, or simply as **3-space**.

As in 2-space, this arrangement of perpendicular axes is called a *Cartesian coordinate system;* in fact, all the terminology carries over to higher-dimensional spaces in the obvious way. The third coordinate axis is usually called the **z-axis**. The plane determined by a pair of co-ordinate axes is named by the axis letters; we shall call such a plane a **basic plane**. (Sometimes a basic plane is also called a **coordinate plane**.) In 3-space there are three basic planes—the *xy*-plane, the *xz*-plane, and the *yz*-plane. The basic plane pictured in *Figure 7.17* is the *xy*-plane.

EXAMPLE **7.14**	The ordered triple (3, 1, 2) specifies locations on the *x*-, *y*-, and *z*-axes, respectively. As shown in *Figure 7.18* on page 356, the location of the point (3, 1, 2) can be found by moving 3 units along the *x*-axis from the origin to (3, 0, 0), then 1 unit parallel to the *y*-axis in the positive direction to (3, 1, 0), then 2 units parallel to the *z*-axis in the positive direction to (3, 1, 2).

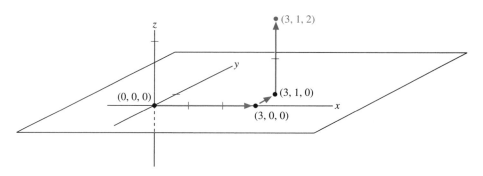

Figure 7.18 The point (3, 1, 2).

The formula for finding the distance between any two points in 3-space is analogous to the distance formula for 2-space. As a typical example, let us compute the distance between (1, 2, 3) and (5, 7, 9). These points in space can be visualized as the diagonally opposite corners of a rectangular box with sides parallel to the basic planes, as shown in *Figure 7.19*.

The length d of the diagonal between those points can be found by two applications of the Pythagorean Theorem. First, the length b of

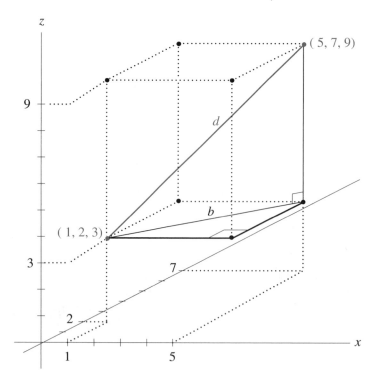

Figure 7.19 Points (1, 2, 3) and (5, 7, 9) at the diagonally opposite corners of a rectangular box.

the diagonal at the bottom of the box is found by using the x- and y-coordinates:

$$b = \sqrt{(5-1)^2 + (7-2)^2}$$

Then the length of d is found using b and the z-coordinates:

$$d = \sqrt{b^2 + (9-3)^2}$$
$$= \sqrt{(5-1)^2 + (7-2)^2 + (9-3)^2}$$
$$= \sqrt{4^2 + 5^2 + 6^2}$$
$$= \sqrt{77}$$

DEFINITION

The **distance** between two points (x_1, y_1, z_1) and (x_2, y_2, z_2) in 3-space is given by

$$\sqrt{(x_2 - x_1)^2 + (y_2 - y_1)^2 + (z_2 - z_1)^2}$$

The 3-space analogue of the circle is the sphere. A **sphere** is the set of all 3-space points that are a given distance (its **radius**) from a particular point (its **center**). As in 2-space, the coordinate description of a sphere is simplified by placing the origin of the coordinate system at its center. We shall assume this to be the case whenever possible. A sphere of radius 5 centered at the origin is the set of all points (x, y, z) whose distance from $(0, 0, 0)$ is 5. Using the definition of distance, this becomes

$$\{(x, y, z) \,|\, \sqrt{(x-0)^2 + (y-0)^2 + (z-0)^2} = 5\}$$

which simplifies to

$$\{(x, y, z) \,|\, x^2 + y^2 + z^2 = 25\}$$

In general, a sphere of radius r (centered at the origin) can be described as:

$$\{(x, y, z) \,|\, x^2 + y^2 + z^2 = r^2\}$$

EXAMPLE 7.15

The coordinate description of a sphere of radius 3, centered at the origin, is

$$\{(x, y, z) \,|\, x^2 + y^2 + z^2 = 9\}$$

Determining whether a point (x, y, z) is inside, outside, or on this circle is simply a matter of observing whether $x^2 + y^2 + z^2$ is less than, greater than, or equal to 9, respectively:

- ■ $(2, 2, 2)$ is outside this circle because $2^2 + 2^2 + 2^2 = 12 > 9$;
- ■ $(1, 2, 1)$ is inside this circle because $1^2 + 2^2 + 1^2 = 6 < 9$.
- ■ $(2, 1, 2)$ is on this circle because $2^2 + 1^2 + 2^2 = 9$

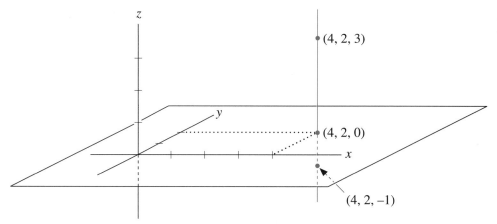

Figure 7.20 The line through (4, 2, 0) parallel to the z-axis.

As in 2-space, lines and segments parallel to the coordinate axes are relatively easy to visualize and to describe in point-set notation. For example, *Figure 7.20* shows the line parallel to the z-axis through the point (4, 2, 0) on the xy-plane; it is the set of points

$$\{(4, 2, z) \mid z \in \mathbf{R}\}$$

The segment on it between $z = -1$ and $z = 3$ is the set

$$\{(4, 2, z) \mid -1 \le z \le 3\}$$

The general method for describing such lines and segments should be clear from this example and the ones that follow.

EXAMPLE 7.16

The segment between $x = -7$ and $x = 25$ on the line parallel to the x-axis through the point (0, 9, −1) is

$$\{(x, 9, -1) \mid -7 \le x \le 25\}$$

EXAMPLE 7.17

The following three lines all go through the point (2, 8, .6):

$\{(x, 8, .6) \mid x \in \mathbf{R}\}$ is parallel to the x-axis

$\{(2, y, .6) \mid y \in \mathbf{R}\}$ is parallel to the y-axis

$\{(2, 8, z) \mid z \in \mathbf{R}\}$ is parallel to the z-axis

---

The descriptions for segments parallel to the coordinate axes make it convenient to use taxicab paths to connect points in 3-space. If we want to travel from one point to another, we can simply move in the axis directions, getting each coordinate of the "traveling point" to match those of our destination point, one by one. For instance, to go from $(1, 2, 3)$ to $(5, 7, 9)$, we can travel from $(1, 2, 3)$ along the segment

$$\{(x, 2, 3) \mid 1 \leq x \leq 5\}$$

to the point $(5, 2, 3)$, then along

$$\{(5, y, 3) \mid 2 \leq y \leq 7\}$$

to $(5, 7, 3)$, and finally along

$$\{(5, 7, z) \mid 3 \leq z \leq 9\}$$

to $(5, 7, 9)$. Thus, a path between $(1, 2, 3)$ and $(5, 7, 9)$ is

$$\{(x, 2, 3) \mid 1 \leq x \leq 5\} \cup \{(5, y, 3) \mid 2 \leq y \leq 7\} \cup \{(5, 7, z) \mid 3 \leq z \leq 9\}$$

The cube in 3-space is analogous to the square in 2-space. In Section 7.3 we saw how a square can be built from four copies of a line segment. A (hollow) cube can be built in a similar way, using six copies of a square region:

Consider one copy of the square in some plane to be a fixed "base" of the cube and attach another copy on each of the base edges. Attach a sixth square in the plane, as shown in *Figure 7.21*; this sixth square acts as the "cap" of the cube. Fold them up in 3-space as shown in *Figure 7.22*. The edges in *Figure 7.21* are numbered in such a way that edges with the same number are matched in the folding process.

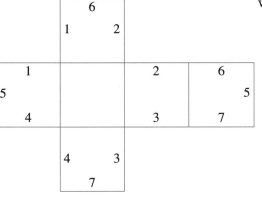

Figure 7.21 Six copies of a square, to be folded into a cube.

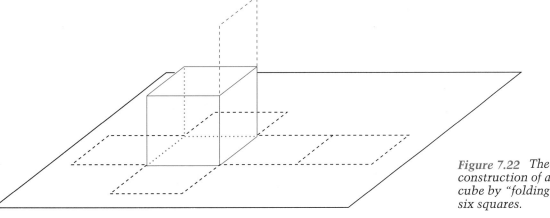

Figure 7.22 The construction of a cube by "folding up" six squares.

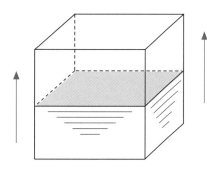

Figure 7.23 A cubic region as an infinite stack of square regions.

The three-dimensional solid cube can be thought of as being made from a two-dimensional square region in a way like that of making the square region from a line segment. If a square region is moved in a direction perpendicular to its original plane for a distance as long as one of its sides, then it sweeps out a cubic region in 3-space, as in *Figure 7.23*. If that perpendicular direction is taken to represent time, then the cubic solid represents the existence of the square region during a period of time. It is an infinite stack of square regions, one for each instant in the time interval.

For example, consider a square piece of paper, 6 centimeters on each side, lying on a desk during a 10-minute observation period. After 6 minutes, someone comes along with a razor, cuts the paper in half (top to bottom), and throws one half away. The "life" of the paper on the desk during the 10-minute observation period can be represented by a single solid region in 3-space, in which two coordinates measure length (in centimeters) and one measures time (in minutes). The first 6 minutes of the paper's 10-minute existence on the desk are represented by a cube and the last 4 minutes by a rectangular box, as shown in Figure 7.24. In this way, regions in 3-dimensional space-time can be used to represent the existence in time of objects in 2-dimensional space.

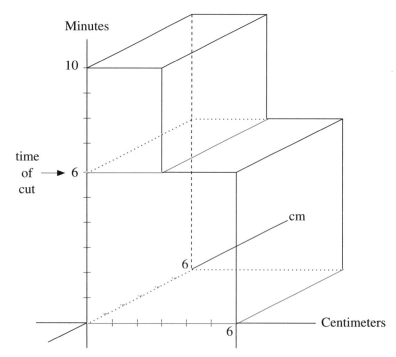

Figure 7.24 The existence in time of the cut piece of paper.

EXAMPLE 7.18

A thin pancake is to be cooked on a hot griddle. The batter is poured slowly and steadily, so that by the end of 5 seconds a circular pancake 20 centimeters in diameter is formed on the griddle. After it has cooked for 40 seconds, the pancake is flipped over (in 2 seconds) and cooked for the rest of one minute, then is removed from the griddle.

Figure 7.25 represents the cooking pancake. The *xy*-plane represents the griddle upon which the batter is about to be poured. The pancake, centered at the *xy*-origin, grows to a diameter of 20 centimeters in 5 seconds; this is represented by a downward-pointing space-time cone whose base radius is 10 centimeters. It then cooks at that size until 40 seconds have elapsed, as is shown by a space-time cylinder 35

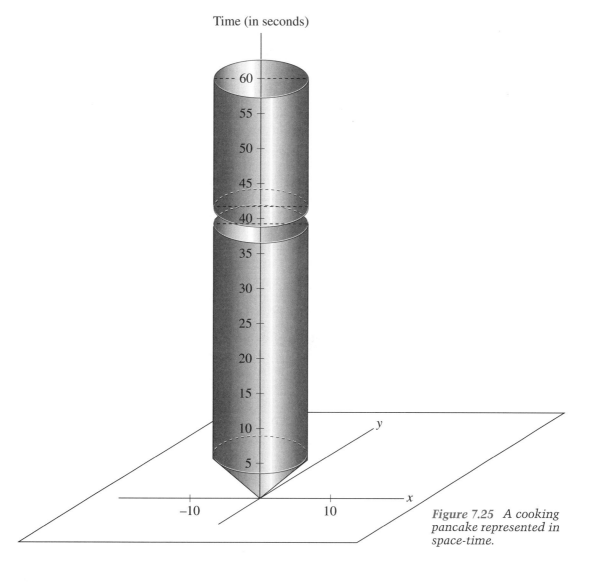

Figure 7.25 A cooking pancake represented in space-time.

seconds high. It is removed from its cooking plane for 2 seconds, then returned (upside down) to cook for the rest of the minute, as shown by another cylinder. Note that the gap between the cylinders at 40 seconds does not show the flipping of the pancake because that motion happens off the surface of the griddle; so it is outside the two-dimensional space being pictured in time.

Finally, let us look at the 2-space barrier problem described at the end of the preceding section. Consider, for example, a plane containing a square with vertices at $(0, 0)$, $(2, 0)$, $(0, 2)$, and $(2, 2)$. There is no way to make a 2-space path between the point $(1, 1)$ at the center of this square and the point $(3, 1)$ outside the square without intersecting the square itself. However, if we consider the plane as the xy-plane in a three-dimensional coordinate system, then the problem becomes:

> Find a path between $(1, 1, 0)$ and $(3, 1, 0)$ that does not intersect the given 2-by-2 square in the xy-plane.

That's easy; we just go up, over, and down. *Figure 7.26* shows such a path; its point-set description is

$$\{(1, 1, z) \mid 0 \leq z \leq 1\} \cup \{(x, 1, 1) \mid 1 \leq x \leq 3\} \cup \{(3, 1, z) \mid 0 \leq z \leq 1\}$$

Even without a picture, it is clear from the description that this path does not intersect the square. All points of the square have third coordinate 0 (why?); but the only points of the path set with third coordinate 0 are the endpoints $(3, 1, 0)$ and $(1, 1, 0)$. Thus, all the rest of the path lies outside the original two-dimensional space. (If we regard the third axis as representing time, this solution says: Move ahead one unit in time; then change your space location; then move back one time unit to the spatial world you started from!)

There is an analogous barrier problem for 3-space:

> Between a point outside a cubic region and a point inside this region, find a path that does not pass through any wall of the cube.

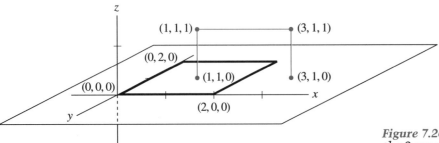

Figure 7.26 A 3-space solution to the 2-space barrier problem.

There is no three-dimensional solution of this problem, but a fourth dimension makes it easy. Before reading the next section, you might find it instructive to try constructing a solution analogous to our 3-space solution of the 2-space barrier problem.

7.4   EXERCISES

For Exercises 1–8, compute the distance between the two given points in 3-space.

1. $(0, 0, 0)$ and $(1, 1, 1)$

2. $(0, 0, 0)$ and $(3, 3, -3)$

3. $(2, 1, 1)$ and $(3, 8, 5)$

4. $(-4, 6, -1)$ and $(7, 7, -1)$

5. $(3, 2, 7)$ and $(0, 2, 3)$

6. $(7, 6, 4)$ and $(7, -5, 4)$

7. $(\sqrt{2}, \sqrt{3}, \sqrt{4})$ and $(0, 0, 0)$

8. $\left(\frac{1}{2}, \frac{1}{3}, \frac{1}{4}\right)$ and $(1, 1, 1)$

For Exercises 9–14, indicate whether the given point is *inside*, *outside*, or *on* the sphere of radius 7 centered at the origin. Justify your answers.

9. $(5, 1, 4)$ 10. $(2, 6, 3)$

11. $(4, 4, 4)$ 12. $(5, 0, -5)$

13. $(\sqrt{13}, 6, 1)$ 14. $(-4, 5, \sqrt{8})$

15. Write a set-builder description of the sphere of radius 12 centered at the origin. Then find the y-coordinate of a point $(7, y, 5)$ inside the sphere, the x-coordinate of a point $(x, 3.5, 2)$ outside it, and the z-coordinate of a point $(-8, 8, z)$ on it.

16. Write a set-builder description of the sphere of radius 10 centered at the origin. Then find the coordinates of three points *not* on any axis—a point inside the sphere, a point outside it, and a point on it.

For Exercises 17–19, describe each line and segment in set-builder notation.

17. The line through $(-8, 6, 1)$ parallel to the x-axis, and the segment of this line between $x = 4$ and $x = 7$.

18. The line through $(7, 0, -3)$ parallel to the y-axis, and the segment of this line between $y = -1$ and $y = 1$.

19. The line through $(-1, 2, -3)$ parallel to the z-axis, and the segment of this line between the xy-plane and $z = 2$.

For Exercises 20–24, write the set-builder description of a taxicab path between the two given points. Find the intersection points that connect the segments in your paths.

20. $(1, 2, 3)$ and $(4, 5, 6)$

21. $(0, 0, 0)$ and $(2, -5, -1)$

22. $(2, 1, 1)$ and $(2, 2, 1)$

23. $(-4, 3, 5)$ and $(7, -1, -9)$

24. $(\sqrt{2}, \sqrt{3}, k)$ and $(0, 0, 0)$

25. On a piece of cardboard, draw six copies of the same square, attached as in *Figure 7.21* on page 359. Then cut out the entire region in one piece and fold it up to form a cube, as shown in *Figure 7.22*.

26. Find the coordinates of each vertex of the 3-space solid in *Figure 7.24* on page 360.

27. A 30-centimeter square glass window is frosted with a thin layer of ice. The bottom edge of the glass is heated so that the ice melts evenly upward at a constant rate, and at the end of 5 minutes all the ice is gone.

   (a) Represent the ice in this process by a three-dimensional space-time solid.

(b) Determine the coordinates of each vertex of the solid figure of Part **(a)**.

28. A kitchen table, whose top is a square, 1 meter on a side, is standing with one edge against a wall. The table is constructed so that a 40-centimeter (by 1-meter) center leaf can be inserted with its longer sides parallel to the wall. Ten seconds into a 30-second observation period, the table is opened by pulling the edge farthest from the wall directly away from the wall without disturbing the opposite edge. The table is pulled open at a constant rate for 4 seconds, so that the gap is just wide enough to accept the leaf. Six seconds later, the leaf is put in place.

    (a) Represent the 30-second observation period for this table top by a three-dimensional space-time solid. (Draw a picture.)

    (b) Find the coordinates of each vertex of the solid figure of Part **(a)**. (*Hint*: There are 20 vertices in all.)

For Exercises 29–32 describe a path between the given point and the point (1, 1) at the center of a 2-by-2 square, constructed so that the path does not intersect any side of the square. Write a set-builder description of the path, and draw a sketch.

29. (1, 3)   30. (5, 4)   31. (0, 6)   32. (−3, 1)

33. The set of all points on or inside a sphere forms a solid ball.

    (a) Describe in set-builder notation a solid ball of radius 8 centered at the origin.

    (b) Describe in set-builder notation a solid ball of radius r centered at the origin.

The distance formula can be used to form equations for spheres that are not centered at the origin. Exercises 34–36 explore this idea in stages.

34. **(a)** Write a formula for the distance between (3, 5, 7) and a point (x, y, z).

    (b) Use your answer to Part **(a)** to write a set-builder description of a sphere of radius 8 centered at the point (3, 5, 7).

    (c) Is (−1, 0, 2) inside, outside, or on this sphere? How do you know?

35. **(a)** Describe in set-builder notation the sphere centered at (2, −1, 4) that passes through the point (5, 3, 6).

    (b) Is (−1, −1, 0) inside, outside, or on this sphere? What about (0, −1, −1)? What about (−1, 0, −1)? Justify your answers.

36. **(a)** Write a formula for the distance between a point (a, b, c) and a point (x, y, z).

    (b) Use your answer to Part **(a)** to describe in set-builder notation a sphere of radius 5 centered at the point (a, b, c).

    (c) Generalize Part **(b)** by describing a sphere of radius r centered at the point (a, b, c) in set-builder notation.

37. You are located at (1, 2, 3), inside a sphere of radius 5 centered at the origin. You want to get to the point (1, 2, 7) by a straight path. At what point does your path intersect the sphere?

38. You are located at (1, 2, 3), inside a sphere of radius 5 centered at the origin. You want to get to the point (5, 6, −7) by a taxicab path. At what point does your path intersect the sphere? Justify your answer. (There is more than one correct answer.)

WRITING EXERCISES

1. The development of this section is analogous to that of Section 7.3. Write a comparative outline of the two sections that displays all points of the analogy.

2. The parallel developments of Sections 7.2, 7.3, and 7.4 suggest that an analogous discussion of 4-dimensional space will appear in Section 7.5. Without looking ahead, try to

anticipate this discussion by writing analogous descriptions for 4 dimensions of each of these items:

(a) a point;

(b) the distance between two points;

(c) the construction of a sphere-like figure;

(d) the construction of a hollow cube-like figure;

(e) the construction of a solid cube-like figure;

(f) a solution to the 3-space barrier problem.

3. Choose a homework problem that is giving you difficulty and write three simpler questions that you think might provide helpful steppingstones for solving it. Be as specific as possible.

4. Describe how at least three of the dozen problem-solving tactics of Section 1.2 can be used to help in the solution of Exercise 28. (Even if you can't complete Exercise 28, you should be able to do this writing exercise.)

FOUR-DIMENSIONAL SPACE

In the light of Sections 7.3 and 7.4, we can now view the transition from three dimensions to four as a natural extension of the step from 2-space to 3-space. At this step the guidance of visual imagination must be replaced by a sense of analogy. **Four-dimensional space** (or just **4-space**) is a world where the address of every location requires four separate measurements. In other words, every point in 4-space is specified by four real numbers, so that four copies of the real line are needed as the axes for a Cartesian coordinate system.

In lower dimensions we have visualized the coordinate axes as perpendicular to each other, but the requirement of the word *dimension* is just that they be separate in an essential way. (The technical term is *independent*.) To replace our visual imagination, we might think of 4-space as infinitely many copies of 3-space, each one corresponding to a specific real number that is the fourth coordinate of every point in that copy of 3-space. This is analogous to thinking of 3-space as an infinite collection of planar "layers," each layer corresponding to a specific number on the z-axis, which is the third coordinate of every point at that level. (Compare this with the view of a cube as an infinite stack of squares, illustrated by *Figure 7.23* on page 360.)

If the fourth coordinate is considered to be time, then 4-space is the whole succession of copies of our spatial universe, one copy for each instant of its existence—much as motion in a cartoon world is drawn by an animator, one still frame at a time, for viewing in rapid succession. This is the space-time view of our physical space; we shall emphasize this view throughout the rest of the chapter by calling the fourth coordinate axis the **t-axis**.

EXAMPLE 7.19	If we choose three spatial coordinate axes (measured in meters, say) and a starting time for a fourth coordinate (measured in hours), then the space-time point (3, 5, 7, 2) represents the 3-space point (3, 5, 7) two hours after the starting time. The space-time point (3, 5, 7, 6) represents the same spatial point four hours later, and (3, 9, 7, 2) represents a point four meters away from the first point, but simultaneous with it.

The formula for finding straight-line distances in 4-space is exactly analogous to the 3-space distance formula.

DEFINITION	The **distance** between two points (x_1, y_1, z_1, t_1) and (x_2, y_2, z_2, t_2) in 4-space is given by

$$\sqrt{(x_2 - x_1)^2 + (y_2 - y_1)^2 + (z_2 - z_1)^2 + (t_2 - t_1)^2}$$

EXAMPLE 7.20	The distance between (1, 2, 3, 4) and (5, 7, 9, 11) is

$$\sqrt{(5 - 1)^2 + (7 - 2)^2 + (9 - 3)^2 + (11 - 4)^2} = \sqrt{16 + 25 + 36 + 49}$$
$$= \sqrt{126}$$

The 4-space analogue of the circle and the sphere is the **hypersphere**, the set of all 4-space points that are a given distance (its **radius**) from a particular point (its **center**). Just as the distance formulas for 2-space and 3-space provide easy ways to describe circles and spheres algebraically, so the distance formula for 4-space permits easy algebraic descriptions of hyperspheres. The hypersphere of radius r centered at (0, 0, 0, 0) is

$$\{(x, y, z, t) \mid x^2 + y^2 + z^2 + t^2 = r^2\}$$

EXAMPLE 7.21	The hypersphere of radius 4, centered at the origin, can be described as

$$\{(x, y, z, t) \mid x^2 + y^2 + z^2 + t^2 = 16\}$$

- (1, 2, 3, 1) is inside the hypersphere because

$$1^2 + 2^2 + 3^2 + 1^2 = 15 < 16$$

- (1, 2, 3, 2) is outside the hypersphere because

$$1^2 + 2^2 + 3^2 + 2^2 = 18 > 16$$

- (2, 2, 2, 2) is on the hypersphere because

$$2^2 + 2^2 + 2^2 + 2^2 = 16$$

As in 3-space, lines and segments parallel to the coordinate axes are easy to describe. For instance,

$$\{(2, y, -1, 4) \mid y \in \mathbf{R}\}$$

is a line through the point $(2, 0, -1, 4)$ parallel to the y-axis. In space-time this is a line parallel to the (spatial) y-axis at time 4. Similarly,

$$\{(3, -5, 8, t) \mid t \in \mathbf{R}\}$$

is a line through $(3, -5, 8, 0)$ parallel to the t-axis. In space-time this line represents the single spatial point $(3, -5, 8)$ existing for all eternity; the segment

$$\{(3, -5, 8, t) \mid 1 \leq t \leq 7\}$$

represents six time units in the life of that point. This ease of description means that, once again, "taxicab paths" provide a convenient way to connect points in 4-space.

EXAMPLE 7.22

A path between $(1, 2, 3, 4)$ and $(5, 7, 9, 11)$ is given by

$$\{(x, 2, 3, 4) \mid 1 \leq x \leq 5\} \cup \{(5, y, 3, 4) \mid 2 \leq y \leq 7\}$$
$$\cup \{(5, 7, z, 4) \mid 3 \leq z \leq 9\} \cup \{(5, 7, 9, t) \mid 4 \leq t \leq 11\}$$

The intersection points connecting the segments of this path are $(5, 2, 3, 4)$, $(5, 7, 3, 4)$, and $(5, 7, 9, 4)$. Notice that, without drawing a picture, a path such as this is easy to construct. We just change one coordinate at a time, connecting the points by line segments parallel to the axis of the coordinate being changed. In this case, the pattern of changes is:

$$(1, \; 2, \; 3, \; 4)$$
$$\downarrow$$
$$(5, \; 2, \; 3, \; 4)$$
$$\downarrow$$
$$(5, \; 7, \; 3, \; 4)$$
$$\downarrow$$
$$(5, \; 7, \; 9, \; 4)$$
$$\downarrow$$
$$(5, \; 7, \; 9, \; 11)$$

This path can be visualized in space-time measured, say, in meters and hours: From $(1, 2, 3, 4)$ the path is 4 meters parallel to the x-axis in the positive direction, followed by 5 meters parallel to the y-axis (positive direction), followed by 6 meters parallel to the z-axis (positive direction), followed by a 7-hour wait.

The four-dimensional analogue of a cube is called a **hypercube**, or a **tesseract**. We have seen (in *Figures 7.21 and 7.22* on page 359) how a cube can be constructed by connecting six square regions in a plane and then "folding them up" and attaching their edges in 3-space. A tesseract is made in much the same way, using eight copies of a cube: Consider one copy as the "base" cube and attach a copy to each of its six faces. Then attach the eighth cube to one of the outer six, as in *Figure 7.27*; this last cube, with faces 4, 5, and 6, is the "cap" that will close up the tesseract when it is folded up in 4-space. (The base cube is not visible in *Figure 7.27*; it is "in the middle" and has one face in common with each cube shown, except for the "cap" cube.)

The folding process itself is impossible to visualize because it requires more than three spatial dimensions. Nevertheless, the analogy with the cube construction gives us some helpful information. Recall that all the exposed edges of the unfolded cube in the plane must be sealed together in pairs when the cube is folded up in 3-space. In the same way, all the exposed square faces of the unfolded tesseract must be sealed together in pairs when it is folded up in 4-space. Some of the pairings are shown by the numbers in *Figure 7.27*; the rest are left for you in Exercise 21.

If we allow some distortion of the cubes in the folding process, we can get a rough idea of what must happen. Try to visualize the faces marked 1, 2, and 3 in *Figure 7.27* coming together in pairs. Of course, the cubes will have to be distorted to do this in 3-space; *Figure 7.28* illustrates this distortion for the two cubes that have a face numbered "2." (The faces are numbered as in *Figure 7.27*.) Notice that the two faces numbered "2" have been joined here. The partial cube in this figure is the "cap" cube, which would have to be cut open, turned inside out, and snapped around the rest of the cubes to close the tesseract in 3-space. None of these distortions are needed in 4-space; with the extra dimension to allow "folding," all of these matchings can be done without stretching, cutting, etc.

The analogous "folding up" of an unfolded cube, if the process were restricted to 2-space, would require similar types of distortions, as you can see from *Figure 7.29*. The final figure, (c), does not look very much like a cube if you look at it as a planar

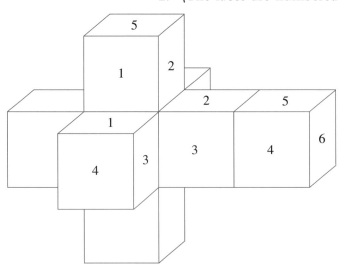

Figure 7.27 Eight copies of a cube to be "folded up" in 4-space.

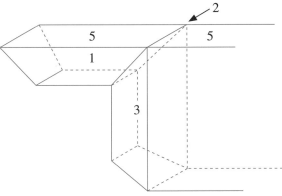

Figure 7.28 Faces of two cubes sealed together in the folding process.

figure. However, if you pretend to be looking down a mine shaft and think of all the angles you see as right angles, you may get a feeling for the 3-space object it represents. A similar reinterpretation of what we see is required to understand the folding up of the tesseract. Unfortunately, our visual intuition is restricted from making the leap into 4-space; so we must be content with reasoning from the cube analogy.

The *solid* hypercube can be thought of as the 4-space region swept out by a cube moving in the direction of the fourth axis for a distance as long as one of its sides. This is analogous to the construction of square areas and cubic solids. If the fourth axis is taken to represent time, then a tesseract is a cube whose "life span" in time units equals the length (in distance units) of one of its edges.

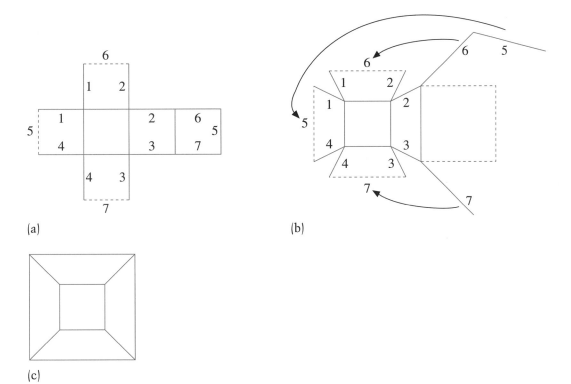

Figure 7.29 An analogy: "Folding up" a cube from six squares without leaving the plane requires distortion of angles and line segments.

EXAMPLE 7.23

In the space-time of meters and hours, a tesseract 2 units on a side is a 2-meter-by-2-meter-by-2-meter cube that exists for 2 hours.[4] It has 16 corners. (Why?) If one corner of this tesseract is at the origin and its edges are parallel to the coordinate axes (in their positive directions), then the coordinates of the corners are

(0, 0, 0, 0)	(0, 0, 0, 2)
(2, 0, 0, 0)	(2, 0, 0, 2)
(0, 2, 0, 0)	(0, 2, 0, 2)
(0, 0, 2, 0)	(0, 0, 2, 2)
(2, 2, 0, 0)	(2, 2, 0, 2)
(2, 0, 2, 0)	(2, 0, 2, 2)
(0, 2, 2, 0)	(0, 2, 2, 2)
(2, 2, 2, 0)	(2, 2, 2, 2)

In space-time, the eight corners listed in the left column are the corners of the cube at time 0, and the eight corners listed in the right column are the corresponding corners of the cube two hours later.

We close this section by solving the 3-space barrier problem posed at the end of Section 7.4. Consider a 2-by-2-by-2 cube in 3-space, situated as in *Figure 7.30*. The point (1, 1, 1) is at the center of the cube and the point (3, 1, 1) is outside it. Clearly, every path in 3-space between (1, 1, 1) and (3, 1, 1) intersects a wall of the cube. For instance, the straight line segment

$$\{(x, 1, 1) \mid 1 \leq x \leq 3\}$$

between the points intersects the cube wall at (2, 1, 1). However, if we consider this situation as part of a three-dimensional "slice" of a four-dimensional space, it is easy to find a path between the two points (1, 1, 1, 0) and (3, 1, 1, 0) that does not intersect the cube walls: Just go "out" in the fourth coordinate direction, move over, then come back "in" to this copy of 3-space. For instance, the path

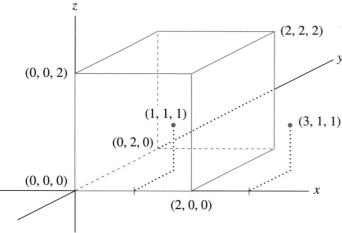

Figure 7.30 The 3-space barrier problem.

---

4. We will describe this cube as a "2-by-2-by-2-meter cube." Descriptions of 3- and 4-dimensional figures measured by coordinate units that are clear from the context will be abbreviated similarly from here on.

$$\{(1, 1, 1, t) \mid 0 \le t \le 1\} \; \cup \; \{(x, 1, 1, 1) \mid 1 \le x \le 3\} \; \cup \; \{(3, 1, 1, t) \mid 0 \le t \le 1\}$$

connects the point outside the cube with the point inside it. Moreover, this path does not intersect the cube walls anywhere—every point of the cube has fourth coordinate 0, but none of the path points (except the endpoints) have fourth coordinate 0.

7.5   EXERCISES

For Exercises 1–8, compute the distance between the two given points in 4-space.

1. $(0, 0, 0, 0)$ and $(1, 1, 1, 1)$

2. $(2, -1, 5, 8)$ and $(-2, 3, 0, 5)$

3. $(3, 3, 3, 3)$ and $(5, 5, 5, 5)$

4. $(6, 1, 5, -7)$ and $(-2, -4, 3, -1)$

5. $(0.5, 7, 8.4, -0.2)$ and $(1.8, -0.35, -6, 4.9)$

6. $(3.5, \sqrt{5}, -\sqrt{26}, -7)$ and $(-1.5, 3\sqrt{5}, 0, 3)$

7. $(0, 0, 0, 0)$ and $(\sqrt{2}, \sqrt{3}, \sqrt{5}, \sqrt{6})$

8. $\left(\frac{1}{2}, \frac{3}{4}, \frac{7}{8}, \frac{15}{16}\right)$ and $(1, 1, 1, 1)$

For Exercises 9–14, indicate whether the given point is *inside*, *outside*, or *on* the hypersphere of radius 6 centered at the origin. Justify your answers.

9. $(1, 2, 3, 4)$

10. $(2, 3, 4, 5)$

11. $(3, 3, 3, 3)$

12. $(6, 0, -6, 0)$

13. $(\sqrt{10}, 4, 2, 2)$

14. $(5, 0, 3, \sqrt{2})$

15. Write a set-builder description of the hypersphere of radius 12 centered at the origin. Then find the y-coordinate of a point $(5, y, 6, 7)$ inside the sphere, the z-coordinate of a point $(1, 3, z, 4)$ outside it, and the t-coordinate of a point $(2, -9, 6, t)$ on it.

16. Write a set-builder description of the hypersphere of radius 10 centered at the origin. Then find the coordinates of 3 points not on any axis—a point inside the hypersphere, a point outside it, and a point on it.

For Exercises 17–20, describe each line and segment in set-builder notation.

17. The line through $(8, 6, 4, 2)$ parallel to the x-axis, and the segment of this line between $x = -1$ and $x = 5$.

18. The line through $(1, 0, 1, 0)$ parallel to the y-axis, and the segment of this line between $y = 0$ and $y = 1$.

19. The line through $(-3, 7, -4, 2)$ parallel to the z-axis, and the segment of this line between $z = -8$ and $z = -4$.

20. The line through $\left(2, 3, 9, \frac{1}{2}\right)$ parallel to the t-axis, and the segment of this line between $t = -.5$ and $t = 2.6$.

For Exercises 21–26, write the set-builder description of a taxicab path between the two given 4-space points, and specify the intersection points that connect the segments in the path.

21. $(0, 0, 0, 0)$ and $(1, 1, 1, 1)$

22. $(2, 3, 4, 5)$ and $(6, 7, 8, 9)$

23. $(7, -3, 4, -1)$ and $(2, 5, 4, 0)$

24. $(1, 0, -1, 0)$ and $(0, 1, 0, -1)$

25. $(2, 7, 6, -1)$ and $(5, 7, 6, 3)$

26. $(8, 9, -6, -7)$ and $(4, -2, 1, 0)$

27–32. Propose a space-time interpretation of each of the paths in Exercises 21–26.

33. Find or construct eight cubes of the same size and glue them together to form an unfolded tesseract, as shown in *Figure 7.27* on page 368. Label each of the 34 exposed square faces with the numbers 1 through 17 to indicate how these faces must be paired together when the tesseract is folded up in 4-space.

34. This exercise refers to the tesseract described in Example 7.23.

    (a) How many edges does it have? (Think about how many edges meet at each corner; then adjust for the fact that each edge connects two corners.)

    (b) Write set-builder descriptions of five of its edges. For each edge you describe, specify the two corners it connects.

    (c) Write a taxicab path that connects (0, 0, 0, 0) and (2, 2, 2, 2).

    (d) The points (0, 0, 0, 0) and (2, 2, 2, 2) are diagonally opposite corners of the tesseract. Match the other fourteen corners in diagonally opposite pairs.

    (e) Describe a taxicab path that connects (2, 0, 0, 2) and its diagonally opposite corner.

35. This exercise refers to the cube of *Figure 7.30* on page 370, considered as existing in the 3-space "slice" of 4-space, for which $t = 0$.

    (a) List the coordinates of the eight corners of the cube.

    (b) Describe a path in 4-space between (1, 1, 1, 0) and (3, 1, 1, 0) that intersects the cube wall in the xy-plane.

    (c) Describe a path in 4-space between (1, 1, 1, 0) and (3, 3, 3, 0) that does not intersect any of the cube walls.

36. Prove that if each edge of a tesseract is n units long, then the length of its diagonal is $2n$ units.

37. The set of all points on or inside a hypersphere forms a solid "hyperball."

    (a) Describe in set-builder notation a solid hyperball of radius 9 centered at the origin.

    (b) Describe in set-builder notation a solid hyperball of radius r centered at the origin.

The distance formula can be used to form equations for hyperspheres that are not centered at the origin. Exercises 38–40 explore this idea in stages.

38. (a) Write a formula for the distance between (3, 5, 7, 9) and a point (x, y, z, t).

    (b) Use your answer to Part (a) to write a set-builder description of a hypersphere of radius 8 centered at the point (3, 5, 7, 9).

    (c) Is (−1, 0, 2, −3) inside, outside, or on this hypersphere? How do you know?

39. (a) Describe in set-builder notation the hypersphere centered at (2, −1, 4, 3) that passes through the point (5, 8, 6, 7).

    (b) Is (6, 1, −5, 0) inside, outside, or on this hypersphere? What about (−5, 6, 0, 1)? What about (0, −5, 6, 1)? Justify your answers.

40. (a) Write a formula for the distance between a point (a, b, c, d) and a point (x, y, z, t).

    (b) Use your answer to Part (a) to describe a hypersphere of radius 5 centered at the point (a, b, c, d) in set-builder notation.

    (c) Generalize Part (b) by describing a hypersphere of radius r centered at the point (a, b, c, d) in set-builder notation.

41. You are located at (1, 2, 3, 4), inside a hypersphere of radius 6 centered at the origin. You want to get to the point (1, 2, 3, −10) by a straight path. At what point does your path intersect the hypersphere?

42. You are located at (1, 2, 3, 4) , inside a hypersphere of radius 6 centered at the origin. You want to get to the point (−5, 6, −7, 8) by a taxicab path. At what point does your path intersect the hypersphere? Justify your answer. (There is more than one correct answer.)

WRITING EXERCISES

1. The development of this section is analogous to that of Section 7.4. Write a comparative outline of the two sections that displays all points of the analogy.

2. What is the etymology of the word "tesseract"? That is, what are its linguistic origin and root meaning?

3. As we have seen, one interpretation of 4-space treats time as if it were a spatial dimension, like length, width, or height. In what ways are time and these three spatial dimensions similar? In what ways are they different?

4. Write a brief description of a tesseract that would explain this "figure" to a curious high-school sophomore. Do not use any pictures.

5. The "folding-up" construction of a tesseract from eight cubes (described in this section in *Figures 7.27 and 7.28* on pages 368–69) yields a *hollow* tesseract, even though the eight cubes are solid. Write a convincing justification of this statement.

CROSS SECTIONS

7.6

A common way to represent and analyze 3-dimensional shapes is by using "cross sections." In 3-space, a *cross section* of a figure is its intersection with a plane—the 2-dimensional "face" made by "slicing" the figure with a plane. For example, a (hollow) sphere sliced through its center results in a cross section that is a circle with the same radius as the sphere. A solid ball sliced through its center yields a cross section that is a disk with the same radius as the ball. (See *Figure 7.31*.)

One of the most important tools in modern medical technology, the CAT scan, takes cross-sectional x-rays of a person and then uses a computer to assemble these images into a 3-dimensional picture of that person's inner organs. Cross sections are also used for making blueprints of buildings, contour maps of countryside, and planar representations of a wide variety of 3-dimensional figures.

When a coordinate system is used, it is particularly convenient to "slice" a figure with planes that are perpendicular to one of the axes. Such a plane is very easy to describe algebraically: It is the set of all points with one coordinate that is a fixed

Figure 7.31 A disk formed as the cross section of a solid ball sliced through its center.

number. In 3-space, for example, the set of all points for which $z = 0$ is the xy-plane; the set of all points for which $z = 5$ is the plane parallel to the xy-plane and 5 units above it (in the z direction). The z-axis is perpendicular to both of these planes.

EXAMPLE 7.24

Suppose you have a solid 5-by-2-by-3-inch rectangular block (of butter or wood or anything else that you can imagine being cut). *Figure 7.32* shows such a block in a coordinate system with the origin at the front lower right corner of the block. The figure also shows three cross sections—three rectangular regions—formed by intersecting the block with the planes $x = 1$, $x = 3$, and $z = 1.5$.

The block and its cross sections can be described algebraically as follows:

$$\text{Block: } \{(x, y, z) \mid 0 \le x \le 5, 0 \le y \le 2, 0 \le z \le 3\}$$

Cross section at $x = 1$: $\{(1, y, z) \mid 0 \le y \le 2, 0 \le z \le 3\}$
Cross section at $x = 3$: $\{(3, y, z) \mid 0 \le y \le 2, 0 \le z \le 3\}$
Cross section at $z = 1.5$: $\{(x, y, 1.5) \mid 0 \le x \le 5, 0 \le y \le 2\}$

Cross sections provide a way of reducing the number of dimensions by 1. This is particularly helpful in trying to analyze or describe 4-space figures because our visual experience is confined to 3 dimensions. To use cross sections in 4-space, we need to define them in a way that does not depend on our ability to "see" perpendicularity. The fixed-coordinate approach provides just the right tool.

DEFINITION

A **cross section** of a figure is the set of all points of the figure that have one coordinate fixed at a particular value.[5]

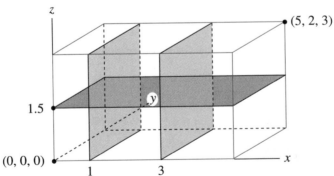

Figure 7.32 Some cross sections of a rectangular block.

5. These cross sections are all of a particularly simple kind, in that each one is "parallel" to a lower-dimensional coordinate space. However, since this is the only kind we shall work with, we use the term *cross section* without any qualifying adjectives.

Now let us take a second look at the space-time tesseract described in Example 7.23 on page 360—a 2-by-2-by-2-meter cube that exists for two hours. The eight points listed in the left column are the corners of the "beginning" cubic face (at 0 hours) and the points in the right column are the corners of the "ending" cubic face (at 2 hours). If we choose any particular instant of time and consider all points of the tesseract whose time coordinate is that instant, we get a 3-dimensional cross section of the tesseract and this cross section is itself a cube. For example, if we fix the time at exactly one hour, the cross section of the tesseract we get is the cube exactly halfway through its life span; its eight corners are:

$$(0, 0, 0, 1) \quad (2, 2, 0, 1)$$
$$(2, 0, 0, 1) \quad (2, 0, 2, 1)$$
$$(0, 2, 0, 1) \quad (0, 2, 2, 1)$$
$$(0, 0, 2, 1) \quad (2, 2, 2, 1)$$

Similarly, any cross section of the space-time tesseract obtained by specifying the time coordinate is a cube. But we have said that time in 4-space behaves just like a spatial coordinate. This suggests that a tesseract cross section obtained by fixing *any one* of the coordinate values should be a cube, and it is, as the following examples show.

EXAMPLE 7.25

If we take all the points of the 2-by-2-by-2-meter cube of Example 7.23 with x-coordinate 1, then we get a square cross section of that cube; we have "sliced it in half" parallel to the yz-plane. Now, if we consider that 2-by-2 square slice for its entire 2-hour life span, we get a 2-by-2-by-2 cube, which is the cross section of the tesseract at $x = 1$.

EXAMPLE 7.26

The six faces of the cube in the preceding example are just the square cross sections obtained by fixing the x-, y-, or z-coordinate at either 0 meters or 2 meters. With the time coordinate taken into consideration, these six cross sections of the tesseract are cubes. Each cube is formed by one of the squares "living out" its two-hour time span.

The fact that all the tesseract cross sections are cubes might tempt you to make some bad guesses about the behavior of 4-space; so let us consider another example, a little more complicated, but perhaps more enlightening. To keep the mathematics as simple as possible, we shall describe our example in a somewhat artificial way; but this should serve to make the point of the illustration clearer.

Consider a 2-by-2-by-2-meter block of ice enclosed in a porous box of the same size, and suppose the ice melts so that each edge of the block shrinks at the constant rate of 1 meter per hour. Tilt the box slightly toward one corner, so that as the ice shrinks, it always touches that corner of the box. Regard that corner as the origin of the spatial coordinate system. (See *Figure 7.33*.) The ice block takes two hours to melt; its shrinking existence during those two hours can be represented by a single four-dimensional space-time figure.

Our imagination is not capable of visualizing this 4-space figure all at once. However, we can get an idea of it by looking at some of its cross sections, just as a builder might learn about a structure by looking at its blueprints.

If we specify a particular time t, the cross section we get is a cube because the ice block has shrunk uniformly during the time from 0 hours to t hours. In fact, because of our simplifying assumption about the rate of shrinkage, the cross section at a fixed time t is a cube $2 - t$ meters on a side.

EXAMPLE 7.27

Figure 7.33 shows four cross sections of the melting ice block for fixed times:

When $t = 0$, the ice block is 2 meters on a side;

when $t = 1$, the ice block is 1 meter on a side;

when $t = \frac{3}{2}$, the ice block is $\frac{1}{2}$ meter on a side;

when $t = 2$, the ice block is 0 meters on a side, and hence it is just the point $(0, 0, 0, 2)$.

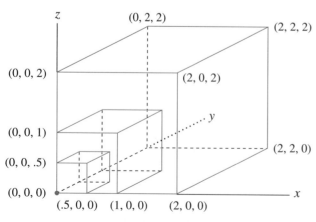

Figure 7.33 *Four cross sections of the melting ice block.*

On the other hand, if we specify a particular *spatial* coordinate value, the resulting cross section is not a cube at all. For instance, if we choose $y = 0$, the cross section of the cube in *Figure 7.33* is the front face, on the xz-plane. Now, as time progresses from 0 hours to 2 hours, this ice face shrinks uniformly from a 2-by-2 square to a point. Thus, the three-dimensional space-time cross section representing that melting face over time is the off-center pyramid shown in *Figure 7.34(a)*.

If we choose $x = 1$, the cross section of the cube is again a 2-by-2 square. Think of cutting the box that holds the ice block in half at $x = 1$. As time passes from 0 hours to 1 hour, the ice square at that cut in the box shrinks from 2 meters by 2 meters to 1 meter by 1 meter. But after 1 hour has elapsed, there is no ice at all at that cut in the box; hence, the space-time cross section at $x = 1$ is a truncated pyramid, as shown in *Figure 7.34(b)*.

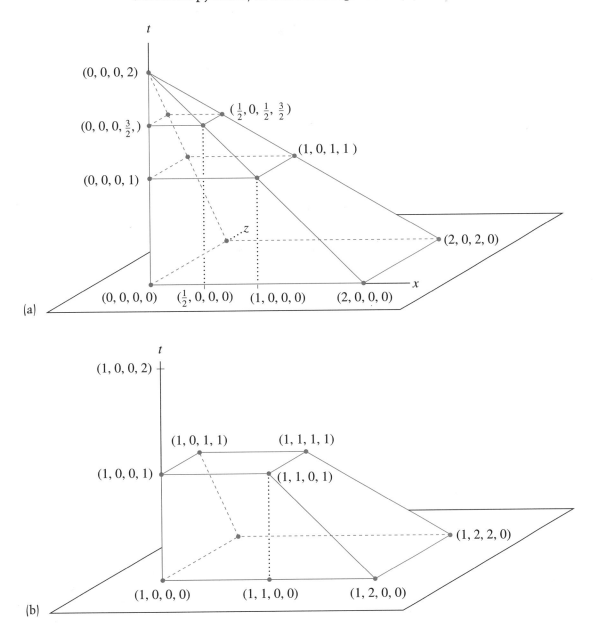

Figure 7.34 *Cross sections of the melting block* (a) *at y = 0, the xz-plane and* (b) *at x = 1, a plane parallel to the yz-plane.*

Another elementary 4-space figure that defies our visual imagination is the hypersphere, the set of all 4-space points that are a given distance from a particular point. Nevertheless, its cross sections are easy to describe, and they provide us with another confirmation of the close analogy between 4-space and 3-space.

Consider first the cross sections of a sphere in 3-space. If we slice the sphere directly through the center, the resulting cross section is a circle of the same radius. If we slice it elsewhere, the cross section is still a circle but of smaller radius. The farther from the center we slice it, the smaller the radius of the circular cross section is.

Now let us look at a hypersphere of radius 5 centered at the origin, which can be described algebraically as

$$\{(x, y, z, t) \mid x^2 + y^2 + z^2 + t^2 = 5^2\}$$

■ The cross section at $t = 0$ is

$$\{(x, y, z, 0) \mid x^2 + y^2 + z^2 + 0^2 = 5^2\}$$

which is just a sphere of radius 5 in the 3-space formed by the x-, y-, and z-coordinate axes.

■ The cross section at $t = 2$ is

$$\{(x, y, z, 2) \mid x^2 + y^2 + z^2 + 2^2 = 5^2\}$$

which is just a copy of the set

$$\{(x, y, z) \mid x^2 + y^2 + z^2 = 5^2 - 2^2\}$$

a sphere of radius $\sqrt{5^2 - 2^2} = \sqrt{21}$.

■ Similarly, the cross section at $t = 4$ is

$$\{(x, y, z, 4) \mid x^2 + y^2 + z^2 = 5^2 - 4^2\}$$

a copy of the sphere

$$\{(x, y, z) \mid x^2 + y^2 + z^2 = 9\}$$

whose radius is 3.

Thus, the cross sections of the hypersphere are spheres, and the farther from the center the cross section is taken, the smaller the radius of the spherical cross section is.

This last observation about cross sections provides a way of visualizing a hypersphere in 4-dimensional space-time. A hypersphere of radius 5 in the space-time of meters and hours, for instance, is a sphere that starts its life as a point, grows like a balloon for 5 hours until it reaches its maximum radius of 5 meters, then shrinks back to a point during the next 5 hours. Moreover, the growing and shrinking of the radius is determined by

$$5^2 - t^2$$

where the time t varies from -5 hours to 5 hours. Thus, a hypersphere in space-time is a growing and shrinking sphere whose life span in hours is equal to its maximum diameter in meters!

Beyond 4-Space

The extension of the concept of space need not stop with four dimensions. As we have seen, the key to higher-dimensional spaces is the concept of a coordinate system. We can go beyond four dimensions to five or six or however many we want simply by using more coordinate axes. Thus,

5-space is the set of all ordered quintuples of real numbers

6-space is the set of all ordered sextuples of real numbers

and so forth. The formulas and figures we have developed in 4-space have analogues in these higher-dimensional spaces, and so do cross sections. By fixing one coordinate value, we reduce by 1 the number of dimensions required to analyze or visualize a figure. By fixing another coordinate value (taking a cross section of a cross section), we reduce the number of dimensions by 2; and so on.

Spaces of large dimension have become quite useful in many areas of science. For example, because the location of a point in space can be specified by three numbers, the location of two points can be specified by six numbers. Thus, just as the movement over time of a particle in 3-space can be represented by a single figure in 4-space, so the movement over time of a two-particle system can be represented by a single figure in 7-space. The geometric properties of these figures often provide insights about the dynamic properties of the systems they represent. In this way, higher-dimensional geometry is a basis for the mathematical treatment of dynamical systems in physics.

7.6   EXERCISES

Exercises 1–8 refer to the rectangular block of Example 7.24 and *Figure 7.32* on page 374. In that example, the cross section at $x = 1$ was described algebraically (in set-builder notation). It can also be described geometrically, as a 2-inch-by-3-inch rectangular region. For Exercises 1–6, give similar algebraic and geometric descriptions of each of the following cross sections of that rectangular block. Also, try to describe where the cross section is located.

1. $x = 4$ 2. $y = 1$ 3. $z = 0$

4. $x = 2.7$ 5. $y = 2$ 6. $z = 3$

7. Describe the intersection of the $x = 4$ and $y = 1$ cross sections of this block.

8. Describe the intersection of the $x = 2.7$ and $z = 3$ cross sections of this block.

Exercises 9–20 begin with a 4-dimensional analogue of the rectangular block of Example 7.24, a 4-by-6-by-8-by-10 block in 4-space, described algebraically as

$$\{(x, y, z, t) \mid 0 \le x \le 4, 0 \le y \le 6,$$
$$0 \le z \le 8, 0 \le t \le 10\}$$

Exercises 9–16 specify cross sections of this figure. Write an algebraic description of each one and also describe its size and shape geometrically. If the cross section is a 3-dimensional "face" of the figure, say so.

9. $x = 2$ 10. $y = 5$ 11. $z = 8$

12. $t = 8$ 13. $x = 1.2$ 14. $y = 0$

15. $z = 3\sqrt{2}$ 16. $t = 10$

17. Describe the intersection of the $x = 2$ and $y = 5$ cross sections of this figure.

18. Describe the intersection of the $z = 8$ and $t = 8$ cross sections of this figure.

19. Write a space-time interpretation of this 4-dimensional figure, using t as the time axis.

20. Write a space-time interpretation of the four cross sections of Exercises 9–12.

21. Specify the coordinates of all of the corners of each of the first three cross sections given in Example 7.27, that is, the cross sections determined by $t = 0$, $t = 1$, and $t = \frac{3}{2}$.

For Exercises 22–29, describe the cross section of the melting ice block of *Figure 7.33* determined by each given coordinate value. In your description be sure to specify the coordinates of all corners of the cross section.

22. $t = \frac{1}{2}$ hour   23. $t = 45$ minutes

24. $x = 0$ meters   25. $z = 0$ meters

26. $x = \frac{2}{3}$ meter   27. $y = 1$ meter

28. $z = 1.5$ meters   29. $y = 2$ meters

Exercises 30–34 refer to a sphere of radius 5 centered at the origin of 3-space.

30. Use set-builder notation to write an algebraic description of this sphere.

31. Describe algebraically and geometrically the cross sections of this sphere at $x = 0$ and at $y = 0$.

32. Describe algebraically and geometrically the cross sections of this sphere at $z = -1$, at $z = -2$, at $z = -3$, at $z = -4$, and at $z = -5$.

33. (a) How are the cross sections at $x = 3$ and at $x = -3$ related? Be specific.

   (b) In general, for any number a, how are the cross sections at $x = a$ and at $x = -a$ related?

34. (a) How are the cross sections at $x = 3$ and at $z = 3$ related? Be specific.

   (b) In general, for any number a, how are the cross sections $x = a$ and $z = a$ related?

Exercises 35–39 refer to a solid ball of radius 5 centered at the origin of 3-space.

35. Use set-builder notation to write an algebraic description of this solid ball.

36. Describe algebraically and geometrically the cross sections of this ball at $y = 0$ and at $z = 0$. How are they related?

37. Describe algebraically and geometrically the cross sections of this ball at $x = -1$, at $x = -2$, at $x = -3$, at $x = -4$, and at $x = -5$.

38. Sometimes people confuse cross sections of a figure with halves or other pieces of it. Write an algebraic description of the half of this solid ball that is on one side of its cross section at $z = 0$.

39. Describe algebraically and geometrically the $x = 0$ cross section of the half of the solid ball described in Exercise 38.

For Exercises 40–43, consider a hypersphere of radius 5 centered at the origin of 4-space. Describe the cross section determined by each given coordinate value, and specify a point on it.

40. $t = 3$ 41. $x = 2$ 42. $y = 1$ 43. $z = 0$

Exercises 44–49 refer to a solid "hyperball" of radius 7 centered at the origin of 4-space.

44. Use set-builder notation to write an algebraic description of this solid hyperball.

45. Describe algebraically and geometrically the $y = 0$ and $z = 0$ cross sections of this hyperball. What, if anything, is their intersection?

46. Describe algebraically and geometrically the cross sections of this hyperball at $x = 1$, at $x = 3$, at $x = 5$, and at $x = 7$. Do any of them intersect?

47. What is the intersection of the $x = 3$ and $t = 2$ cross sections? Justify your answer.

48. What is the intersection of the $y = 3$ and $y = 2$ cross sections? Justify your answer.

49. What is the intersection of the $z = 5$ and $t = 5$ cross sections? Justify your answer.

For Exercises 50–53, describe, as explicitly as you can, the 5-space analogue of the given 4-space concept.

50. the distance between two points

51. a taxicab path

52. a tesseract

53. a hypersphere

WRITING EXERCISES

1. What part of the ice-block example do you find most confusing or difficult to understand? Write three specific questions that you think might help to clarify it for you.

2. Describe a plane in 3-space. How can you generalize your answer to describe a "hyperplane" in 4-space? (If this question seems too difficult to answer in general, start with a simpler case: First describe coordinate planes and their 4-space analogues; then describe planes parallel to coordinate planes and their 4-space analogues; finally, try the general case.)

3. What is the difference between a solid 4-dimensional figure and a "hollow" one? Is the 4-dimensional ice-block figure described in this section solid or hollow? What about the hypersphere? When we construct a 4-space figure from a 3-space figure, what determines whether the result is solid or hollow?

4. Picking up on the idea explained in the last paragraph of this section, explain how two balls on a billiard table can be used to represent 4-space. Test your explanation by applying it to the following questions.

   (a) What is a *point* in this example?

   (b) What are the four coordinate axes?

   (c) Can you describe a line segment? A taxicab path? How?

   (d) How can you describe a tesseract or a hypersphere in this setting? (These concepts may be difficult. You might try just "thinking with your pen" about this question—just writing down what comes to mind without worrying about paragraph structure or organization.)

CYLINDERS AND CONES [OPTIONAL]

7.7

In this section we investigate the 4-dimensional analogues of two other 3-dimensional shapes—the cylinder and the cone. As in the case of the descriptions in Section 7.6, we shall see how the interplay of geometry and algebra lets us use algebraic methods to "see" the 3-dimensional cross sections of a figure, thereby getting some sense of the shape of the 4-dimensional figure as a whole. That is, for each of these shapes, we shall do the following:

- describe the 3-space figure geometrically, and look at its 2-dimensional cross sections;
- translate these geometric descriptions into algebraic ones;
- generalize the algebraic description of the figure to 4-space by the analogous addition of a fourth coordinate;
- look at the algebraic descriptions of 3-dimensional cross sections of the 4-space figure by fixing points on different axes;
- interpret these algebraic descriptions geometrically to get a visual image of the cross sections in 3-space.

Don't let this apparently complicated menu discourage you. It is merely a guide to help you sort out the main steps in the material to come. If you refer to it as you work through this section, you should find it useful in breaking the discussion into manageable pieces.

Cylinders

Let us begin with the **cylinder**, a hollow tube with a (very thin!) wall perpendicular to a circular base. If we place coordinate axes so that the cylinder can be thought of as sitting on the xy-plane with the center of its base at the origin, as in *Figure 7.35(a)*, then it is easy to give an algebraic description of this figure. Suppose the cylinder is h units high and the radius of the base is r units. Then the base, which is a circle of radius r centered at $(0,0,0)$, is just "copied" all the way up the cylinder; that is, we can think of the cylinder as an infinite stack of circles, from height 0 to height h (in the positive z direction). Algebraically, we have

(7.1)
$$\{(x, y, z) \mid x^2 + y^2 = r^2 \text{ and } 0 \leq z \leq h\}$$

If the cylinder is cut horizontally (parallel to the xy-plane) at any height k between 0 and h, its cross section on the plane $z = k$, then, is just a circle of radius r, as in *Figure 7.35(b)*; that is, replace z by k in Statement (7.1) to obtain

$$\{(x, y, k) \mid x^2 + y^2 = r^2\}$$

Figure 7.35 A cylinder
and some of its cross
sections.

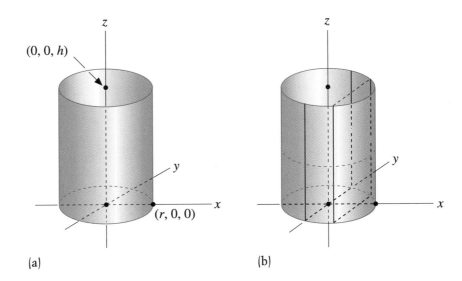

(a)

(b)

If, on the other hand, the cylinder is cut vertically, then its cross section is just a pair of parallel lines. *Figure 7.35(b)* shows two such cross sections—one on the *yz*-plane and one parallel to the *yz*-plane but over a little in the positive *x*-direction. Notice that the parallel lines get closer as the cut moves further out. Example 7.28 shows how the algebra reflects what we see in the figure.

EXAMPLE 7.28	Consider a cylinder of radius 5 inches and height 12 inches, situated as in *Figure 7.35*. Its algebraic description is

$$\{(x, y, z) \mid x^2 + y^2 = 25 \text{ and } 0 \le z \le 12\}$$

At height $z = 6$, its horizontal cross section is the circle

$$\{(x, y, 6) \mid x^2 + y^2 = 25\}$$

Since the *yz*-plane is the plane where $x = 0$, its (vertical) cross section on this plane is

$$\{(0, y, z) \mid 0^2 + y^2 = 25 \text{ and } 0 \le z \le 12\}$$

But $y^2 = 25$ means that y must equal either 5 or -5, so this cross section is just

$$\{(0, 5, z) \mid 0 \le z \le 12\} \cup \{(0, -5, z) \mid 0 \le z \le 12\}$$

that is, two 12-inch line segments that are 10 inches apart on opposite sides of the *z*-axis. If we take the vertical on the plane $x = 3$, we get

$$\{(3, y, z) \mid 3^2 + y^2 = 25 \text{ and } 0 \le z \le 12\}$$

Now, $3^2 + y^2 = 25$ simplifies to $y^2 = 25 - 9 = 16$, implying that y must equal 4 or -4 in this case. Thus, this cross section is

$$\{(3, 4, z) \mid 0 \le z \le 12\} \cup \{(3, -4, z) \mid 0 \le z \le 12\}$$

two 12-inch line segments that are 8 units apart on the plane $x = 3$.

Now, what is a natural 4-space analogue of the cylinder? One way to proceed is to start with the 3-dimensional analogue of the circular base, which is a sphere of radius r, and extend that for some fixed "height" h in the t direction.[6] We shall call this 4-dimensional object a **hypercylinder**. If we again agree to put the origin of our coordinate system at the center of the base, this hypercylinder has an easy algebraic description:

(7.2) $$\{(x, y, z, t) \mid x^2 + y^2 + z^2 = r^2 \text{ and } 0 \le t \le h\}$$

To examine 3-dimensional cross sections of this object, we can:

- ▪ fix one of the four coordinates in Statement (7.2) at 0 or at some other number;
- ▪ work out the resulting algebraic expression; then
- ▪ interpret this expression as a geometric figure in 3-space.

Rather than carrying out these steps for the general case of a hypercylinder of base radius r and height h, let us look at the following specific, but typical, example, which is the direct analogue of Example 7.28—a hypercylinder of base radius 5 inches and height 12 inches.

EXAMPLE 7.29

Consider the hypercylinder

$$\{(x, y, z, t) \mid x^2 + y^2 + z^2 = 25 \text{ and } 0 \le t \le 12\}$$

As you might expect from the analogy used in constructing the hypercylinder, its cross section for any fixed t—for instance, 6—is just a sphere of radius 5 inches in the appropriate hyperplane (at $t = 6$):

$$\{(x, y, z, 6) \mid x^2 + y^2 + z^2 = 25$$

The analogous cross section obtained from fixing one of the other variables is not so obvious geometrically, but the algebra is easy. If we let $x = 0$, we get

---

6. This is not the only way to proceed. Another, slightly more cumbersome approach is to consider a figure with a *cylindrical* base that extends for some fixed length in the t direction. In time-space this would represent a 3-dimensional cylinder existing for some length of time. We shall not explore this alternative; you might like to consider it on your own.

(7.3) $\{(0, y, z, t) \mid 0^2 + y^2 + z^2 = 25 \text{ and } 0 \le t \le 12\}$

The equation in this set description, which reduces immediately to $y^2 + z^2 = 25$, is a circle of radius 5 inches in the yz-plane. Therefore, the cross section at $x = 0$, described by Statement (7.3), is a (3-dimensional) cylinder of base radius 5 inches and height 12 inches in yzt-space!

Continuing the analogy with Example 7.28, let us see what happens when we fix x at some other number. For $x = 3$, we get

$$\{(3, y, z, t) \mid 3^2 + y^2 + z^2 = 25 \text{ and } 0 \le t \le 12\}$$

But $3^2 + y^2 + z^2 = 25$ reduces to

$$y^2 + z^2 = 25 - 9 = 16$$

which is a circle of radius 4 in the yz-plane. Therefore, the cross section at $x = 3$ is a cylinder of base radius 4 inches and height 12 inches. Similarly, if we choose $x = 4$, we get

$$\{(4, y, z, t) \mid 4^2 + y^2 + z^2 = 25 \text{ and } 0 \le t \le 12\}$$

By simplifying the equation to $y^2 + z^2 = 25 - 16 = 9$, we can see that the cross section at $x = 4$ is again a cylinder of height 12 inches; but this one has a base radius of 3 inches. Thus, the cross sections for values of x chosen farther and farther away from 0 are cylinders that get thinner and thinner.

Here is an illustration of how the algebraic formulas can be used to obtain more precise information than our visual intuition provides. Observing that the thinness of a cylindrical cross section depends on its distance from the origin, we can formulate the exact radius of the cylindrical cross section in terms of its distance from the origin. For instance, if we think of x as being some fixed number k, then the distance of the cross section from the origin at $x = k$ is just $|k|$, and the formula for the radius of this cylinder is

$$k^2 + y^2 + z^2 = 25$$

or

$$y^2 + z^2 = 25 - k^2$$

Thus, the radius of the circular base of this cylinder is $\sqrt{25 - k^2}$, a number that shrinks to 0 as k grows from 0 to 5.

The coordinates x, y, and z in the algebraic description of the hypercylinder are interchangeable in some sense; that is, each of these three coordinates is squared and they are all added together in the same equation. Thus, cross sections obtained by picking a particular

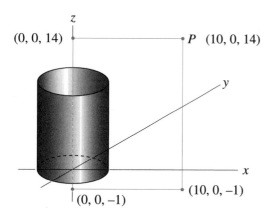

Figure 7.36 A "taxicab loop" through a cylinder.

value for y or z will look like those we got by picking a value for x, as in Example 7.29. Thus, a hypercylinder "looks" something like a bunch of spheres stacked up in the t direction, and cross sections parallel to the t-axis in *any* of the other directions are cylinders that shrink in diameter as we get farther away from the origin!

In case you think that you are beginning to get a picture of what a hypercylinder "really" looks like in 4-space, let us offer one more observation about cylinders and hypercylinders to test your intuition. Unlike a sphere, a cylinder does *not* separate 3-space into two separate pieces, an inside and an outside. Its hollow "interior" is actually just part of all the rest of 3-space. As you can see from *Figure 7.36*, it is possible to put a taxicab path right through the middle of the cylinder from any point P not on the cylinder, regardless of where P is, without intersecting the cylinder itself. That's not surprising; a piece of pipe wouldn't be very useful if water couldn't get through it! The surprise is that, despite the fact that a hypercylinder has a spherical base, it is hollow in exactly the same way in 4-space as a cylinder is in 3-space! That is:

A hypercylinder does *not* separate 4-space into two separate pieces.

It is possible to put a path right through the middle of the hypercylinder from any point P not on the hypercylinder, regardless of where P is, without intersecting the hypercylinder itself. To verify this (perhaps surprising) claim, let us look at some specific taxicab paths, using the cylinder and the hypercylinder of Examples 7.28 and 7.29 as typical figures.

Referring first to Example 7.28, consider a point P somewhere fairly far away from the cylinder—at $(10, 0, 14)$, for instance. Since the "core" of the cylinder centers around the z-axis, a taxicab path passing through the cylinder is easy to describe. Here is one possible path:

Go down in the z direction to a level below the base of the cylinder ($z = -1$); move over to the z-axis; travel up the z-axis to the level of P, beyond the top of the cylinder ($z = 14$); move back out to point P.

The point-set description of this path is

$$\{(10, 0, z) \mid -1 \le z \le 14\} \cup \{(x, 0, -1) \mid 0 \le x \le 10\}$$
$$\cup \{(0, 0, z) \mid -1 \le z \le 14\} \cup \{(x, 0, 14) \mid 0 \le x \le 10\}$$

This path is pictured in *Figure 7.36*. Of course, if P were less conveniently placed, the path would be a little more complicated, but the main idea would be exactly the same: Go down to a level below the base of the cylinder, travel up the z-axis to a point above the top, then go back to your starting point. Notice that the point-set descriptions of this path and the cylinder make it quite clear that the path doesn't intersect the cylinder anywhere:

- For all points (x, y, z) of the cylinder, $0 \le z \le 12$.

- For no point of the cylinder do *both* x and y equal 0 (because $x^2 + y^2 = 25$).

- For no point of the cylinder does $x = 10$ (because $x^2 + y^2 = 25$, so that $-5 \le x \le 5$).

- For the only points of the path for which $0 \le z \le 12$, either both x and y equal 0 or else $x = 10$.

Now consider the point $Q = (10, 0, 0, 14)$ in 4-space, which is fairly far away from the hypercylinder in Example 7.29. We can construct a taxicab "loop" from Q through the hypercylinder and back to Q, just as we did for the cylinder in 3-space:

Go "down" in the t direction to a level below the base of the cylinder ($t = -1$); move over to the t-axis; travel "up" the t-axis to the level of Q, beyond the "top" of the cylinder ($t = 14$); move back out to point Q.

The point-set description of this path is

$$\{(10, 0, 0, t) \mid -1 \le t \le 14\} \cup \{(x, 0, 0, -1) \mid 0 \le x \le 10\}$$
$$\cup \ \{(0, 0, 0, t) \mid -1 \le t \le 14\} \cup \{(x, 0, 0, 14) \mid 0 \le x \le 10\}$$

Again, the point-set descriptions of this path and the hypercylinder make it quite clear that the path doesn't intersect the hypercylinder anywhere:

- For all points of the cylinder, $0 \le t \le 12$.

- For no point of the cylinder do the x-, y-, and z-coordinates *all* equal 0 (because $x^2 + y^2 + z^2 = 25$).

- For no point of the cylinder does $x = 10$ (because $x^2 + y^2 + z^2 = 25$, so that $x^2 \le 25$, and consequently, $-5 \le x \le 5$).

- For the only points of the path for which $0 \le t \le 12$ either x, y, and z *all* equal 0 or else $x = 10$.

Thus, a hypercylinder does not enclose any part of 4-space. ("Hyperwater" can flow through a hypercylindrical pipe!)

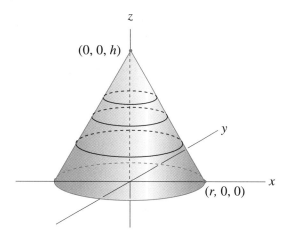

Figure 7.37 A cone in 3-space.

Cones

In order to give a simple coordinate description of a cone in 3-space, we shall think of it as "upside down" (relative to its use for holding ice cream), so that it has a circular base and tapers upward to a point. (See *Figure 7.37*.) We can think of a **cone**, in fact, as an infinite stack of circles whose radii shrink uniformly from the radius of the base down to 0, the radius of the "circle" that is the top point of the cone. That is, the cross section of this cone parallel to its base at any height from bottom to top is just a circle, and the closer to the top it is, the smaller the circle is.

The algebra in the coordinate description of such a cone is a little more complicated than that for a cylinder, but this stack-of-circles viewpoint makes it a little easier to understand. Let r stand for the base radius of the cone and h stand for its height. To get the radii of the circles in the stack to shrink from r to 0 as z goes from 0 to h, we ought to subtract more and more from the initial radius r as the height z increases—but how much more? Let us look first at a typical example:[7]

EXAMPLE 7.30

Consider a cone with base radius 5 inches and height 10 inches, situated as in *Figure 7.37*. To find the radius of its circular cross section at a height z between 0 and 10, first observe that the height of the cone is exactly twice the radius of its base. Thus, we would expect to shrink the base radius by half as much as the height of the cross section. This means that the circular cross section at height z should have radius $5 - \frac{1}{2}z$; so its equation is

$$x^2 + y^2 = \left(5 - \frac{1}{2}z\right)^2$$

Thus, the point-set description of this cone is

$$\{(x, y, z) \mid x^2 + y^2 = \left(5 - \frac{1}{2}z\right)^2 \text{ and } 0 \le z \le 10\}$$

The cross section at $z = 4$ is

$$\left\{(x, y, 4) \mid x^2 + y^2 = \left(5 - \frac{1}{2} \cdot 4\right)^2\right\}$$

a circle in the plane $z = 4$ with radius 3 inches. Similarly, the cross section at $z = 7$ is

---

7. Parts of the remaining material presuppose some familiarity with slopes and equations of straight lines.

$$\left\{(x, y, 7) \mid x^2 + y^2 = \left(5 - \frac{1}{2} \cdot 7\right)^2\right\}$$

which is a circle with radius $1\frac{1}{2}$ inches in the plane $z = 7$.

As Example 7.30 illustrates, the algebraic description of a cone depends on the "shrinkage ratio" for the circles. The general form of that ratio, in terms of base radius r and height h, is probably not obvious from this one example, but it is not too hard to figure out with some experimentation and/or a little algebra. For every unit we go up from the base, we want to subtract $\frac{r}{h}$ units; that is, we want the radius at height z to be $r - \frac{r}{h}z$.[8] At the base, $z = 0$, so that

$$r - \frac{r}{h} \cdot z = r - \frac{r}{h} \cdot 0 = r - 0 = r$$

while at the top, $z = h$, so that

$$r - \frac{r}{h} \cdot z = r - \frac{r}{h} \cdot h = r - r = 0$$

as required. The general algebraic description of such a cone, then, is

(7.4)
$$\{(x, y, z) \mid x^2 + y^2 = \left(r - \frac{r}{h} \cdot z\right)^2 \text{ and } 0 \le z \le h\}$$

Cross sections of the cone by planes other than horizontal ones have been of particular interest ever since early Greek times. The so-called **conic sections** were first thoroughly studied by Apollonius in the third century B.C. They are the four essentially different kinds of curves that result from the intersection of a plane and a cone—the *circle*, the *ellipse*, the *parabola*, and the *hyperbola*. Each of these figures can be represented by a quadratic equation in two variables![9] Since we have been considering only cross sections that are parallel to coordinate axes, our treatment just covers two of the four kinds of conic sections. One, the circle, we have already seen; the other results from cross sections perpendicular to the base of the cone, as Example 7.31 shows.

EXAMPLE 7.31

The cross section $x = 0$ of the cone described in Example 7.30 is an inverted V-shaped figure on the yz-plane, as shown in *Figure 7.38*.

8. This can also be found by observing that the slope of the line through $(0, 0, h)$ and $(r, 0, 0)$ in the xz-plane is $\frac{-h}{r}$ and then by solving the equation $z = \frac{-h}{r} \cdot x + h$ for x. The radius at height z is the x-value of the point $(x, 0, z)$ on the cone.

9. A *quadratic equation* is an equation of degree 2.

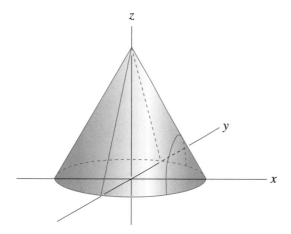

Figure 7.38
Vertical cross
sections of a cone.

Algebraically, it is found by setting $x = 0$ in the defining equation for the cone:

$$0^2 + y^2 = \left(5 - \frac{1}{2}z\right)^2$$

$$y = \pm\left(5 - \frac{1}{2}z\right)$$

The "$\pm$" sign indicates that there are really two equations represented here:

$$y = -\frac{1}{2}z + 5 \qquad \text{and} \qquad y = \frac{1}{2}z - 5$$

Each of these equations represents a straight line in the yz-plane, and they cross when $y = 0$ and $z = 10$ (at the tip of the cone). The vertical cross sections away from the z-axis are somewhat different, however. At $x = 3$, for example, we get

$$3^2 + y^2 = \left(5 - \frac{1}{2}z\right)^2$$

$$y^2 = \left(25 - 5z + \frac{1}{4}z\right)^2 - 9$$

$$y^2 - \frac{1}{4}z^2 = -5z + 16$$

which is the equation of the hyperbola that is partly shown in *Figure 7.38*.

If we generalize the idea of *cone* from 3-space to 4-space using the same approach as we did for the cylinder, we think of a **hypercone** as an infinite stack of spheres whose radii shrink uniformly from the radius of the base down to 0.[10] If we again put the origin of our coordinate system at the center of the base, the algebraic description of the cone, given by Statement (7.4), generalizes easily to 4-space:

$$\{(x, y, z, t) \mid x^2 + y^2 + z^2 = \left(r - \frac{r}{h} \cdot t\right)^2 \text{ and } 0 \le t \le h\}$$

For each value of t between 0 and h, the 3-dimensional cross section is a sphere of radius $r - \frac{r}{h} \cdot z$; this radius value decreases from r to 0 as t increases from 0 to h.

---

10. Again, this is not the only way to proceed. Another approach is to consider a figure with a *conical* base that extends for some fixed length in the t direction. In timespace, this would represent a 3-dimensional cone existing for some length of time. We shall not explore this alternative, but you might like to do so.

EXAMPLE 7.32

The point-set description of a hypercone with base radius 5 inches and height 10 inches is

$$\{(x, y, z, t) \mid x^2 + y^2 + z^2 = \left(5 - \frac{1}{2}t\right)^2 \text{ and } 0 \le t \le 10\}$$

The cross section at $t = 4$ is

$$\left\{(x, y, z, 4) \mid x^2 + y^2 + z^2 = \left(5 - \frac{1}{2} \cdot 4\right)^2\right\}$$

a sphere with radius 3 inches in the space determined by $t = 4$. Similarly, the cross section at $z = 7$ is

$$\left\{(x, y, z, 7) \mid x^2 + y^2 + z^2 = \left(5 - \frac{1}{2} \cdot 7\right)^2\right\}$$

a sphere with radius $\frac{3}{2}$ inches in the space determined by $t = 7$. As in the 3-dimensional case, cross sections perpendicular to the base are less obvious. For $x = 0$, we get

$$\{(0, y, z, t) \mid y^2 + z^2 = \left(5 - \frac{1}{2}t\right)^2 \text{ and } 0 \le t \le 10\}$$

Comparing this with the algebraic description in Statement (7.4), it is not hard to see that this figure is a 3-dimensional cone in yzt-space! For $x = 3$, however, we obtain

$$\{(3, y, z, t) \mid 3^2 + y^2 + z^2 = \left(5 - \frac{1}{2}t\right)^2 \text{ and } 0 \le t \le 10\}$$

The defining equation here is not something we have seen before; but in some ways it is exactly the analogous result you might expect. Just as the vertical cross sections of the (3-dimensional) cone change from an inverted "V" at $x = 0$ to a hyperbola at $x = 3$, the corresponding cross sections of the hypercone change from a cone at $x = 0$ to an inverted cup with a hyperbolically curved top at $x = 3$.

7.7 EXERCISES

1. Give the coordinates of two points that are on the cylinder of Example 7.28 and that have no coordinate equal to zero.

2. Give the coordinates of two points that are on the hypercylinder of Example 7.29 and that have no coordinate equal to zero.

3. Give the coordinates of two points that are on the cone of Example 7.30 and that have no coordinate equal to zero.

4. Give the coordinates of two points that are on the hypercone of Example 7.32 and that have no coordinate equal to zero.

For Exercises 5–8, write a point-set description of each figure.

5. A cylinder with base radius 3 and height 7.

6. A hypercylinder with base radius 12 and height 6.

7. A cone with base radius 4 and height 20.

8. A hypercone with base radius 7 and height 15.

In Exercises 9–20, fill in the blank by a coordinate so that the resulting point is on the figure described in the example that is noted.

9. $(3, \underline{\quad}, 2)$ on the cylinder of Example 7.28.

10. $(\underline{\quad}, 1, 9)$ on the cylinder of Example 7.28.

11. $(4, \underline{\quad}, 0, 3)$ on the hypercylinder of Example 7.29.

12. $(1, 2, \underline{\quad}, 4)$ on the hypercylinder of Example 7.29.

13. $(1, \underline{\quad}, 8)$ on the cone of Example 7.30.

14. $(\underline{\quad}, 1, 6)$ on the cone of Example 7.30.

15. $(3, 4, \underline{\quad})$ on the cone of Example 7.30.

16. $(0, 4, \underline{\quad})$ on the cone of Example 7.30.

17. $(0, 0, 0, \underline{\quad})$ on the hypercone of Example 7.32.

18. $(2, 1, 2, \underline{\quad})$ on the hypercone of Example 7.32.

19. $(3, \underline{\quad}, \sqrt{3}, 2)$ on the hypercone of Example 7.32.

20. $(\underline{\quad}, 1, 1, 6)$ on the hypercone of Example 7.32.

Exercises 21–24 refer to Example 7.28. Describe the cross section of this cylinder determined by each value.

21. $z = 1$ 22. $y = 2$

23. $x = -4$ 24. $y = 5$

Exercises 25–32 refer to Example 7.29. Describe the cross section of this hypercylinder determined by each value.

25. $t = 3$ 26. $t = 12$

27. $y = 0$ 28. $z = 0$

29. $x = 1$ 30. $x = 2$

31. $x = 5$ 32. $x = 6$

Exercises 33–36 refer to Example 7.30. Write an equation of the circular cross section of this cone determined by each value.

33. $z = 0$ 34. $z = 5$

35. $z = 8$ 36. $z = 10$

For Exercises 37–40, write the point-set descriptions of *two* cross sections of the given figure that contain the specified point.

37. $(-3, 4, 9)$ on the cylinder of Example 7.28.

38. $(4, \sqrt{5}, 2, 7)$ on the hypercylinder of Example 7.29.

39. $(3, \sqrt{7}, 2)$ on the cone of Example 7.30.

40. $(1, 2, -2, 4)$ on the hypercone of Example 7.32.

41. This exercise refers Example 7.28. Write the point-set description of a taxicab path that starts at $(7, 8, 9)$, passes through the "inside" of the cylinder without touching the cylinder itself or the z-axis, and ends up back at $(7, 8, 9)$.

42. This exercise refers to Example 7.29. Write the point-set description of a taxicab path that starts at $(6, 7, 8, 9)$, passes through the "inside" of the hypercylinder without touching the cylinder itself or the t-axis, and ends up back at $(6, 7, 8, 9)$.

WRITING EXERCISES

1. Explain how a cylinder in 3-space is the appropriate analogue of two parallel line segments in a plane. (*Hint*: What is a "circle" in 1-space?)

2. Extending the analogy used in this section by one more dimension, describe a "hyperhypercylinder" and a "hyperhypercone."

3. Use the point-set descriptions of a cone and a hypercone to describe how a cone-shaped cup in 4-space can hold "hyperwater" without leaking.

LINK: 4-SPACE IN FICTION AND IN ART

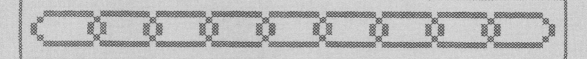

Cubism. An awareness of the notion of four-dimensional space in the first decade of the twentieth century in Paris led to this important movement in art.

7.8 About 100 years ago, a Shakespearian scholar and theologian named Edwin Abbott Abbott (whose initials might be abbreviated as EA^2) wrote a short novel of mathematical fiction entitled *Flatland*. Its narrator, A. Square, is an inhabitant of a 2-dimensional world visited by a 3-dimensional being, a Sphere. The Sphere tries to explain the existence of a third, independent direction, "Upward, not Northward," with little initial success. The Sphere can only show himself to the Square as a circle, his cross section formed by intersecting with the planar world of Flatland. A. Square can only observe the Sphere's third dimension as growth of the circle in time, as the Sphere slowly descends through the plane. The Sphere finally convinces A. Square of the reality of the third dimension by "lifting" him into it. As the Square ponders this new world taught to him first by analogy, he is led to a startling conclusion:

> . . . just as there was close at hand, and touching my frame, the land of Three Dimensions, though I, blind senseless wretch, had no power to touch it, no eye in my interior to discern it, so of a surety there is a Fourth Dimension. . . .[11]

Later in the narrative, the skepticism of the Sphere about a space of more dimensions than his own is at once an insightful comment on human nature and an invitation to the further exploration of the analogies established earlier in the book.

As fiction, *Flatland* is light and enjoyable; as science, it was prophetic in that it preceded Einstein's relativity theory and the resulting interest in space-time by some thirty years or so; as pedagogy, *Flatland* remains to this day one of the clearest, simplest explanations of the idea of 4-space that has been written. But the very fact that *Flatland* was deliberately written to explain 4-space makes it a somewhat artificial example of 4-space in fiction. Our next example is more in keeping with the spirit of this section. It is a novel in which 4-space occurs as a natural part of the story, not as a vehicle for teaching the reader about mathematics.

Robert A. Heinlein is one of the acknowledged masters of twentieth-century science fiction, and one of his best-known novels is *Stranger in a Strange Land*, first published in 1961. The stranger in the story is Valentine Michael Smith, a human raised as a Martian. Among his many unusual abilities, Smith had a peculiar way of disposing of "wrong things":

> He stepped toward Berquist; the gun swung to cover him. He reached out— and Berquist was no longer there.[12]

In an attempt to understand this strange power, his friends had him demonstrate it while they photographed the process from two different angles (on solid-sight-sound 4 mm. film, projected in a 3-dimensional tank). An empty brandy case was thrown into the air, and, one witness begins to describe the scene:

> "The box did not simply vanish. The process lasted some fraction of a second. From where I am sitting it appeared to shrink, as if it were disap-

---

11. Edwin A. Abbott, page 94 of Item 1 in the list *For Further Reading* at the end of this chapter.
12. Robert A. Heinlein, *Stranger in a Strange Land*. New York: G. P. Putnam's Sons, 1961. (New York: Berkley Books, 1981, p. 67.) © 1961 Robert A. Heinlein; used by permission.

pearing into the distance. But it did not go outside the room; I could see it up to the instant it disappeared."[13]

In slow motion, the film from the first camera showed the box shrinking, smaller and smaller, until it was no longer there. . . . When they ran the second film, Duke cursed.

> "Something fouled the second camera, too. . . . It was shooting from the side so the box should have gone out of the frame to one side. Instead it went straight away from us again. You saw it."
> "Yes," agreed Jubal. "Straight away from us."
> "But it can't—not from both angles."
> "What do you mean, 'it can't'? It did. . . . Duke, the cameras are okay. What is ninety degrees from everything else?"
> "I'm no good at riddles."
> "It's not a riddle. I could refer you to Mr. A. Square from Flatland, but I'll answer it. What is perpendicular to everything else? Answer: two bodies, one pistol, and an empty case."[14]

As the excerpts quoted above suggest, both *Flatland* and *Stranger in a Strange Land* treat the fourth dimension primarily as another spatial direction. Time as the fourth dimension has also been a popular theme in the fiction of the last hundred years. One of the best-known early examples of this is H. G. Wells' *The Time Machine*, published in 1895.[15] Early in the book, Wells' "Time Traveller" makes the case for the credibility of time travel by explicitly citing the similarity between time and the spatial dimensions, saying (in part):

> There are really four dimensions, three which we call the three planes of Space, and a fourth, Time. There is, however, a tendency to draw an unreal distinction between the former three dimensions and the latter, because it happens that our consciousness moves intermittently in one direction along the latter from the beginning to the end of our lives.

The fourth dimension became a popular topic in European and American fiction late in the nineteenth century. A comprehensive catalog of nineteenth- and twentieth-century fiction in which it appears as a theme would run to many pages and would include the names of many well-known authors. We list here just a few of the most prominent early writers and their works to provide some sense of how pervasive this topic has been. The fourth dimension is a major theme in H. G. Wells' *The Wonderful Visit* (1895) and *The Invisible Man* (1897), in Oscar Wilde's *The Canterville Ghost* (1891), in George Macdonald's *Lilith* (1895), and in Joseph Conrad and Ford Madox Hueffer's *The Inheritors* (1901). It is mentioned in Fyodor Dostoevsky's *The Brothers Karamazov* (1879),

13. Ibid., p. 111.
14. Ibid., pp. 124–25.
15. H. G. Wells, *The Time Machine: An Invention.* London: W. Heinemann, 1895.

in Marcel Proust's *Duc Sté de Chez Swann* (1913), in P. G. Wodehouse's *The Amazing Hat Mystery* (1922), and (at least implicitly) in Gertrude Stein's writings of the 1910s.

The late nineteenth-century surge of popular interest in the fourth dimension can be traced back to several major mathematical events that had occurred in the middle of the century. In the 1840s, Arthur Cayley in England and Hermann Grassmann in Germany published major works describing the theory of n-dimensional space. At the same time, Bernhard Riemann was developing a type of non-Euclidean geometry[16]; a decade later (in 1854), he gave a landmark speech describing this geometry for a space of n dimensions. From the next several decades of mathematical research and the continuing philosophical debate engendered by the conflict between the theory of non-Euclidean geometry and the philosophy of Immanuel Kant, there gradually evolved increasingly concrete representations of the idea of higher dimensions than the third. These representations caught the popular interest, aided in part by the deliberate efforts of mathematicians such as Henri Poincaré in France and James Joseph Sylvester in England, who sought to make the mathematical theories more intelligible to interested nonmathematicians.

By the early 1900s, "the fourth dimension" was discussed widely in the popular press. In the United States, for example, there were frequent articles in magazines such as *Science, Harper's Weekly*, and *Current Literature*. In 1909, *Scientific American* offered a $500 prize for "the best popular explanation of the Fourth Dimension." It is worth noting that all of the twenty or so essays published in this contest described the fourth dimension in *spatial* terms. Within the following decade, the announcement of Einstein's Theory of Relativity abruptly changed the interpretation of the fourth dimension to *time* (and spawned another *Scientific American* essay contest).

The last quarter of the nineteenth century also saw the emergence of a popular philosophy based on the acceptance of the reality of four spatial dimensions. The most influential early figures in this movement were J. C. F. Zöllner, of England, and Charles Howard Hinton, originally of England. In the early twentieth century, P. D. Ouspensky, of Russia, and Claude Bragdon, of the United States, were widely read proponents of this philosophy. Zöllner was also an active supporter of Henry Slade, an American medium who was a controversial figure in London in the late 1870s. The publicity surrounding this link between spiritualism and the fourth dimension added a mysterious, mystical flavor to the mathematics of 4-space.

In France, the popular writings of Henri Poincaré, a highly respected mathematician and philosopher, brought an awareness of four-dimensional space to Paris in the first decade of the twentieth century. It was here that cubist painters became intrigued by "new measures of space which, in the language of the

16. Non-Euclidean geometries are described in Section 3.6.

modern studios, are designated by the term *fourth dimension*."[17] They were working at a time when H. G. Wells' novels were enjoying widespread popularity, as were the 4-space science-fiction writings of French authors (especially Gaston de Pawlowski and Alfred Jarry). Several of the cubist artists in Paris studied the writings of Poincaré and other mathematicians on the geometry of four-dimensional space. Maurice Princet, a friend of the artist Jean Metzinger and an actuary with an interest in modern painting, often joined in the artists' discussions and helped work through the connections between geometry and cubist art.

Perhaps the most famous of the cubist artists involved in those Paris discussions in the 1910s was Marcel Duchamp. During the years 1912–14, Duchamp began working on *The Bride Stripped Bare by Her Bachelors, Even (The Large Glass)*. The notes he made explicitly identify aspects of the mathematical writings of Poincaré and others in his design of "The Bride. . . as if it were the projection of a four-dimensional object."[18] Other Duchamp works depicting aspects of four-dimensionality are *Portrait of Chess Players* (1911) and *Nude Descending a Staircase, No. 2* (1912). The latter painting, shown on page 398, is noteworthy because it is one of the few cubist works that views the fourth dimension as time. (All three of these works can be seen at the Philadelphia Museum of Art.)

Cubist art arrived in the United States with the opening of the 1913 Armory Show in New York City. To say it got mixed reviews would be to err on the charitable side of accuracy. Typical of the reaction in the popular press was a page-one story in the March 20, 1913, *Chicago Record-Herald*:

CUBIST ART IS HERE/AS CLEAR AS MUD . . .

CUBIST ART— . . . a "woosy" attempt to express the fourth dimension.

FUTURIST ART—Same as the former, only more so, with primeval instincts thrown in; Cubism carried to the extreme or fifth dimension.

Negative "popular" reaction notwithstanding, the cubist art in the show was clearly recognized as related somehow to four-dimensional space. This relationship covered a wide range of work, however, including not only Duchamp's analytical *Nude Descending a Staircase*, but also the completely abstract work of Francis Picabia, who used the cubists' appeal to 4-space to justify the total rejection of perspective. In this regard, Picabia is more representative than Duchamp of the modern artist's view of the fourth dimension. In the words of one art historian:

17. Guillaume Apollinaire, "La Peinture nouvelle: Notes d'art." *Les Soirés de Paris*, no. 3 (Apr. 1912), p. 90.

18. As quoted in Henderson, *The Fourth Dimension . . .* , [5], from Pierre Cabanne, *Dialogues with Marcel Duchamp* (1967) (trans. Ron Padgett). New York: Viking Press, 1971.

Philadelphia Museum of Art: Louise and Walter Arensberg Collection

Nude Descending a Staircase, No. 2, by Marcel Duchamp

. . . the fourth dimension was primarily a symbol of liberation for artists. . . . Specifically, belief in a fourth dimension encouraged artists to depart from visual reality and to reject completely the one-point perspective system that for centuries had portrayed the world as three-dimensional.[19]

The 1913 Armory Show provided the impetus for American artists to explore the theory of four-dimensional space. Both Picabia and Duchamp were key figures in the development of American cubism. However, the popularization of Einstein's theory of relativity a few years later shifted the focus of interest in the fourth dimension away from the spatial interpretation of the early cubists to a time-space interpretation. That, along with an increasing tendency to reject perspective of any kind, led, by about 1930, to a fading of interest in 4-space among most artistic movements.

The one notable exception to this trend was the surrealist movement. The surrealists adopted the spatial interpretation of four-dimensional space, along with many of its mystical overtones. One of the most striking examples of 4-space imagery in surrealist art hangs in the Metropolitan Museum of Art in New York City. It is a painting titled *Crucifixion*, by Salvador Dali. It was painted in 1954, a period during which Dali was deeply involved with both mathematical and religious themes, and he originally titled it *Corpus Hypercubus* (*Hypercubic Body*). The "cross" in the scene is, in fact, a hypercube, or tesseract, unfolded in three dimensions. Its "shadow" on the planar surface below is an unfolded cube, and a checkerboard pattern of squares stretches out toward the horizon.

As we approach the present, there are signs of recurring artistic interest in the visual representation of 4-space. Some of this interest has been sparked by the development of powerful techniques in computer graphics, techniques that represent accurately on a screen the various cross sections and projections of four-dimensional objects. Other interest stems from a more speculative urge, similar to the kind of curiosity that drives a research scientist or mathematician. In the words of one modern artist:

Artists who are interested in four dimensional space . . . are motivated by a desire to complete our subjective experience by inventing new aesthetic and conceptual capabilities. . . . Our reading of the history of culture has shown us that in the development of new metaphors for space artists, physicists, and mathematicians are usually in step.[20]

The foregoing examples by no means exhaust the occurrences of 4-space in fiction and art. In fact, they are intended merely as illustrations to whet your

19. Henderson, *The Fourth Dimension . . .* , [5], p. 340.
20. Tony Robbin, "The New Art of 4-Dimensional Space: Spatial Complexity in Recent New York Work." *Artscribe* (London), no. 9: 20, 1977.

appetite for finding more examples on your own. Now that you have had a glimpse of the world of four dimensions, you should have little trouble finding other instances of it in the literature you read and the art you see.

Acknowledgement

Much of the material for this section was drawn from Linda Dalrymple Henderson's superb book, *The Fourth Dimension and Non-Euclidean Geometry in Modern Art*, [5] in the list *For Further Reading* at the end of the chapter. If you are interested in learning more about the connection between modern geometry and modern art, this is a richly detailed, very readable source.

TOPICS FOR PAPERS

1. In this chapter we have explored some 4-dimensional figures that are natural analogues of 3-dimensional figures—the tesseract, hypersphere, hypercylinder, and hypercone (analogues of the cube, sphere, cylinder, and cone, respectively). These figures have been described in several different ways, starting with an extension of their geometric construction in 3-space, then by considering the fourth dimension as time, then by looking at cross sections, etc.

    This paper is to be a thorough description of a 4-dimensional figure of your choosing. You may choose to describe the 4-dimensional analogue of any one of the following figures—

    rectangular box   cylinder   cone   half-cylinder ("quonset hut")
    hemisphere   prism   pyramid   torus ("doughnut")

    —or you may choose something else (at your own risk. See your instructor if you have any doubts about your choice). Your paper should include the following ingredients:

    ■ A preliminary note indicating the typical readers you have in mind and the mathematical background they would need.

    ■ A general geometric description of your 3-dimensional figure and also a coordinate description of a particular example of it.

    ■ A description of the 4-dimensional analogue of your figure, both in general geometric terms (how to build or visualize it) and by using specific coordinates in 4-space.

- A time-space interpretation of your 4-dimensional figure.

- Some discussion of its 3-dimensional cross sections.

- Anything else you find interesting about your 4-dimensional object.

- Any problem-solving strategies you used in attacking this project.

A word of advice: As in many exhibition sports, originality and degree of difficulty undoubtedly will be factors considered in grading your performance. You might take this into account in choosing your figure and in deciding on your overall approach to the topic. (In particular, if you choose a figure that is described in detail in the chapter, you must do more with it than just repeat what is in the book.) If you are whimsically inclined, an allegorical or fictional approach to this topic would be appropriate, *provided that it contains substantial mathematical ingredients.*

2. Any 2-dimensional figure can be generalized to 3 dimensions in two standard ways: it can be "pushed" perpendicularly in the third coordinate direction or it can be shrunk to a point located somewhere off the original plane. A circular region that is pushed perpendicularly is a cylinder; one that is shrunk to a point (directly "above" or "below" its center) is a cone. A square that is pushed perpendicularly is a cube or a rectangular box (depending on how far it is pushed); one that is shrunk to a point (directly above its center) is a pyramid. In addition, the way some planar figures are constructed can be generalized to 3-space, as we have seen in this chapter. The circle and square constructions generalize to yield a sphere and a cube, respectively.

These three ways of generalizing figures can be extended from 3-space to 4-space. A solid cube "pushed" perpendicularly in the fourth coordinate direction yields a solid tesseract; one that is shrunk to a point becomes a figure somewhat like the ice block in Section 7.6; the figure obtained by generalizing the construction of the cube (as in Section 7.5) is a hollow tesseract. A sphere pushed perpendicularly in the fourth coordinate direction yields a hypercylinder; one that is shrunk to a point becomes a hypercone; the generalization of the construction of the sphere yields a hypersphere.

Describe the three 3-space figures that are obtained when each of these three generalization methods is applied to *one* of the following planar figures:

- A right triangular region.

- An equilateral triangular region.

- A nonsquare rectangular region.

- A half-disk.

Then generalize from 3-space to 4-space using these three methods on *each* of the three 3-dimensional figures you obtained in the first part. (At this point, you should have nine 4-dimensional figures, but they may not all be different.) Compare and contrast these nine figures using whatever techniques from this chapter you find useful for clarifying your ideas—

algebraic and/or geometric descriptions, time-space, cross sections, etc. Consider the reader of your paper to be an anonymous student in your class—someone who knows the material of Chapter 7 fairly well, but does not know anything about your particular topic. Introduce your topic clearly and take your reader step by step through an orderly development of your ideas to your desired conclusion.

FOR FURTHER READING

1. Abbott, Edwin A. *Flatland*, 1884. Reprint, 5th Ed., Rev. New York: Barnes & Noble, Inc., 1969.

2. Barnett, Lincoln. *The Universe and Dr. Einstein*, Rev. Ed. New York: William Sloane Associates, 1957.

3. Dirac, P. A. M. "The Evolution of the Physicist's Picture of Nature," *Mathematics in the Modern World*. San Francisco: W. H. Freeman and Company, 1968.

4. Hawking, Stephen. *A Brief History of Time*. New York: Bantam Books, 1988.

5. Henderson, Linda Dalrymple. *The Fourth Dimension and Non-Euclidean Geometry in Modern Art*. Princeton: Princeton University Press, 1983.

6. Manning, Henry P. *Geometry of Four Dimensions*. New York: Dover Publications, Inc., 1956.

7. Marr, Richard F. *4-Dimensional Geometry*. Boston: Houghton Mifflin Company, 1970.

8. Rucker, Rudolf v. B. *Geometry, Relativity and the Fourth Dimension*. New York: Dover Publications, Inc., 1977.

9. Weeks, Jeffrey R. *The Shape of Space*. New York: Marcel Dekker, Inc., 1985.

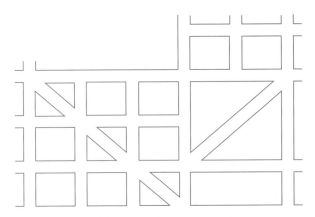

Figure 8.3 *Problem: Is there a route that goes through each street on this map exactly once?*

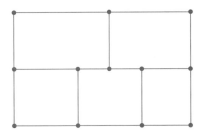

Figure 8.4 *Problem: Can a continuous curve in the plane cross each segment exactly once?*

EXAMPLE 8.5	**Problem:** Referring to *Figure 8.4*, is it possible to draw a continuous curve in the plane of this page that crosses each line segment exactly once?

Before moving on to the next section, try your hand at the puzzles described in these problems. For each one you solve successfully, give yourself a well-deserved pat on the back. For those you cannot solve, consider the possibility that there may be *no* solution of the type called for. (Can you prove that there is no solution? If so, this would also "solve" the problem, although in a different sense.)

8.1	**EXERCISES**

WRITING EXERCISES

1. Choose one of the problems presented in Examples 8.1–8.5 and describe how at least four of the dozen problem-solving techniques of Section 1.2 can be used to help solve it. (Even if you can't completely solve the problem, you should be able to do this writing exercise.)

2. Examine the dozen problem-solving techniques of Section 1.2 and decide which two of them are least likely to be useful in solving the problems described in Examples 8.1–8.5. Justify your choices.

3. Look up Leonhard Euler in a book on the history of mathematics or the history of graph theory and write a short biographical sketch of him. (You should be able to find a considerable amount of material, so be selective. Be sure to list the source(s) you used in writing the paper. Do not simply copy or paraphrase the material from your sources.

SOME BASIC TERMS

8.2

We begin our study of graph theory by introducing some basic terminology. We shall see that all the problems posed in Section 8.1 can be rephrased in standardized terms using the language of graph theory.

DEFINITION

A **graph** is a nonempty, finite set of points together with a finite (possibly empty) set of line segments or arcs. The points are called **vertices** (singular, **vertex**) and the line segments or arcs are called **edges**. Each edge must have a vertex at each of its two ends.

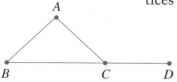

Figure 8.5 *A graph.*

A graph, then, is a bare-bones geometric figure, made up of vertices (endpoints) and edges (line segments or arcs). You should note that the definition requires that a graph have at least one vertex; however, it either may or may not have edges. For the sake of identification, we shall label the vertices of a graph with capital letters. Edges can normally be identified by the labels of their endpoints. Thus, in *Figure 8.5*, the vertices are *A*, *B*, *C*, and *D*; the edges are *AB*, *AC*, *BC*, and *CD* (or *BA*, *CA*, and so on).

The definition of a graph requires that each edge have a vertex at each of its ends; but note that it allows for the possibility of more than one edge joining a given pair of vertices or of a single vertex serving as "both" endpoints of an edge. A graph with these features is shown in *Figure 8.6*. Sometimes it is useful to assign numbers to the multiple edges joining a pair of vertices as has been done there with the edges connecting *E* and *F* and with those connecting *G* and *H*. The edge *HH* in *Figure 8.6* is an example of a **loop**, an edge whose end-points coincide at a single vertex. It is important that you think of a loop not as an edge without an end or with only one end, but rather as an edge with two ends that coincide. Every edge *must* have two endpoints, both of which are vertices; however, these vertices need not be distinct.

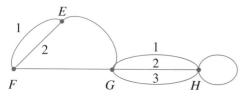

Figure 8.6 *A graph with multiple edges between vertices and with a loop.*

EXAMPLE 8.6

Figure 8.7 as labeled is *not* an example of a graph; the line segment with *J* at one end has no vertex at its other end, and thus does not represent an edge. *Figure 8.8*, on the other hand, *is* a graph because every edge has a vertex at either end. Note that vertex *M* is not an

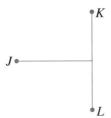

Figure 8.7 Not a graph as labeled.

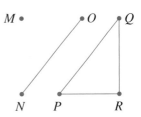

Figure 8.8 A graph with several "pieces."

endpoint of any edge. Also, this graph is in several "pieces," but is nevertheless a single graph according to the definition.

It may seem odd to allow a graph to have multiple "pieces," as in *Figure 8.8*, or to have such features as loops and "extra" edges, as in *Figure 8.6*. If graphs were no more than "connect-the-dots" pictures, this might be a valid criticism; but the wide variety of physical and mathematical situations that can usefully be represented by graphs, some of which call for such features, argue for the definition of a graph as given.

Although both *Figure 8.6* and *Figure 8.8* are examples of graphs, you can see that there are differences between them. In the former, one can "travel" from any vertex to another by following appropriate edges; this is not the case in the latter. The formal terminology for this is given in the following definition.

DEFINITION

A **path** (**from** A_0 **to** A_n) in a graph is a finite sequence of (one or more) edges $A_0A_1, A_1A_2, \ldots, A_{n-1}A_n$, such that each pair of consecutive edges in the sequence shares a common vertex. The path **begins** at A_0 and **ends** at A_n. An **open path** is one that begins at one vertex and ends at a different vertex; a **closed path** is one that begins and ends at the same vertex.

EXAMPLE 8.7

In *Figure 8.5*, the sequence of edges *AB*, *BC*, *CD* form a path from *A* to *D*. For brevity, we shall denote this path by *ABCD*. (Notice that with this notation, a three-edge path is denoted by four letters.) Another path from *A* to *D* is given by *ACD*. Thus, *ABCBA* is a closed path from *A* to *A*. (A path may use some edges more than once.) Note that *ABD* does not designate a path from *A* to *D* because there is no edge from *B* to *D*; a path is a feature of a given graph, not just a string of letters.

Where there are multiple edges between vertices, we can distinguish their use in paths by numbering, as indicated previously. Thus, in

Figure 8.6, one path from *E* to *H* is given by E_2FG_1H. If we are not concerned with which edge is used, we may omit the numerical subscripts.

Note   You may have observed that our notation seems to allow some ambiguity: Does *XY* name the edge joining *X* and *Y* or a one-edge path from *X* to *Y*? The answer is "Yes" to both. In other words, every edge is a path. We do not really have two *different* ideas represented by an ambiguous label; rather, one idea incorporates the other, and this is represented by the notation.

Returning to the graph in *Figure 8.8,* we can use the terminology of paths to distinguish its "piecemeal" character from those in *Figures 8.5 and 8.6.*

DEFINITION   If every pair of distinct vertices of a graph is joined by a path, the graph is **connected**; otherwise, it is **disconnected**.

One more clarification is in order here. In *Figure 8.9,* notice that the two edges *SU* and *TV* have no vertex in common, although, geometrically, they do intersect. Nevertheless, *Figure 8.9* is a graph because every edge does, in fact, connect two vertices. However, it is a disconnected graph because there are distinct vertices, such as *S* and *V,* that are not joined by any path. Connectedness depends on the existence of a path, not on the geometric configuration. The graph of *Figure 8.9* is distinguished from earlier graphs by the following terminology:

A Graph Pattern on an Urban Road.

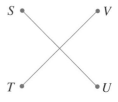

Figure 8.9 A disconnected graph.

DEFINITION

A **flat** graph is one in which two edges, or two parts of the same edge, meet only at a vertex.

Note

The use of the term "flat" is nonstandard in graph theory, but is analogous to "planar" in the standard terminology. Modern graph theory conceptualizes the notion of a graph more abstractly than is the case here, with the result that "planar" is a property of how the graph *can be drawn*, whereas "flat" is a property of how it *is drawn*. For the type and level of problems we encounter in this chapter, the more abstract treatment is not warranted.

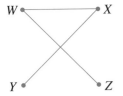

Figure 8.10
A nonflat,
connected graph.

Thus, *Figures 8.5 and 8.6* show connected flat graphs; *Figure 8.8* is flat and disconnected; *Figure 8.9* is nonflat and disconnected; and, to round out the possibilities, *Figure 8.10* is nonflat and connected. Note that what makes a graph flat is, in a sense, a matter of labelling; *Figure 8.9* is, of course, a figure in a plane, and it is a graph. Specifically, it is a graph with four vertices and two edges. Geometrically, the line segments *SU* and *TV* intersect; but that point of intersection is not a vertex because it is not so labelled. If that point were labelled, say as vertex *R*, the resulting figure would be a graph with *five* vertices and *four* edges; this graph would be flat and connected, and it would be a *different* graph than the given one, despite the similarity in appearance of the geometric figures.

Let us now return to the puzzles of Section 8.1 and rephrase them using the language of graph theory. The figure to be traced in Example 8.1 (*Figure 8.1*) is essentially a graph already; we need only label the vertices. This is done in *Figure 8.11*. (Note that we opt to think of the figure as a flat graph by identifying a vertex at *D*.) The problem of tracing translates into one of finding a path with each edge in the graph used exactly once. For future reference, we define such a path as an **edge path**.

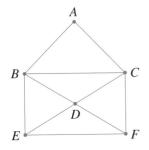

Figure 8.11 Figure 8.1
with vertex labels.

Turning to the Königsburg Bridge Problem, we can translate *Figure 8.2* into a graph containing all of the significant features of the original picture. Since walking around on one side or the other of the river or on one of the islands has no effect on the problem (the challenge is finding a way to cross the bridges appropriately), we "shrink" each of these land areas down to a single point, represent it as a vertex, and then represent each bridge as an edge. This is done in *Figure 8.12*. The problem of crossing each bridge exactly once then translates into finding an edge path in this graph—the same kind of problem as the tracing puzzle!

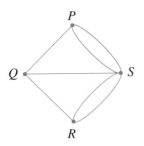

Figure 8.12 The
Seven Bridges of
Königsburg.

The street map of Examples 8.3 and 8.4 translates in an obvious way into a graph, with each intersection becoming a vertex and each street an edge. This transformation is left as an exercise. The problem for the pothole-repair crew, to travel each street exactly once, is again the edge-path problem. The task for the street-sign crew is different, but analogous. Visiting each intersection exactly once translates to finding a path that uses each vertex exactly once. Quite naturally, we define such a path as a **vertex path**.

As is the case with the first puzzle, *Figure 8.4* is essentially already a graph, except for the labeling of the vertices. The problem of Example 8.5, though, does not translate into finding a path of any sort. Here, we are challenged to find some continuous curve that *crosses* edges and *avoids* vertices, instead of following them as in a path. We coin the term **crossing curve** to denote a continuous curve that crosses each edge in a graph exactly once and passes through no vertex. Notice that a crossing curve may have neither, one, or both of its ends inside one of the regions (if any) enclosed by edges of the graph. Notice also that a crossing curve may cross itself one or more times; however, it must cross each edge of the graph once and only once.

8.2   EXERCISES

1. Which of the drawings as labeled in *Figure 8.13* represent graphs?

2. Which of the graphs in *Figure 8.13* are connected?

3. Which of the graphs in *Figure 8.13* are flat?

4. Which of the graphs in *Figure 8.13* contain a loop?

5. Referring to *Figure 8.13(g)*:

   (a) Identify a path from Z to X that uses no edge more than once.

   (b) Identify a path from W to X that uses every edge at least once.

   (c) Identify a path from W to Z that is an edge path.

   (d) How many different edge paths are there from W to Z?

   (e) How many different edge paths are there from X to Y?

   (f) How many different paths are there from X to Y?

6. Referring to *Figure 8.13(i)*:

   (a) Identify a path from E to G that uses no edge more than once.

   (b) How many paths are there from E to G that use no edge more than once?

   (c) Identify a path from E to G that uses every edge at least once.

   (d) Is there an edge path from E to G?

7. If we measure the *distance* between two vertices as the number of edges in the shortest path joining them:

   (a) In *Figure 8.13(f)*, what is the distance from P to T?

   (b) In *Figure 8.13(f)*, what is the distance from P to S?

   (c) In *Figure 8.13(g)*, what is the distance from W to Y?

Figure 8.13 Graphs
and nongraphs for
Exercises 1–7.

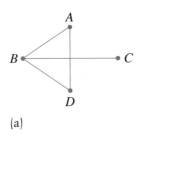

(a)

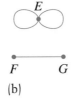

(b)

(c)

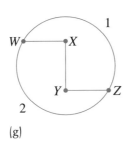

(d)

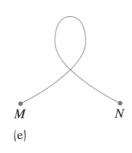

(e)

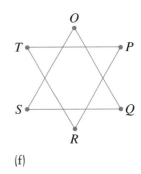

(f)

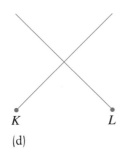

(g)

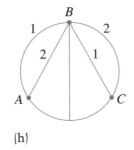

(h)

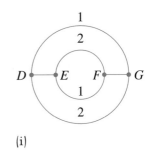

(i)

(d) In *Figure 8.13(g)*, what is the distance from *W* to *Z*?

(e) In Figure 8.13(i), which vertex is closer to *E* (in the sense of a shorter path): *F* or *G*?

(f) In *Figure 8.13(i)*, which vertex is closer to *D*: *F* or *G*?

8. In a certain graph, *AFEBBCEAGCB* is a path. Determine whether the following statements are *true*, *false*, or *cannot be determined* without additional information.

(a) This is an open path.

(b) This path contains 11 edges.

(c) The graph contains at least 10 edges.

(d) The graph contains at least 8 edges.

(e) The graph contains at least 7 vertices.

(f) The graph contains a loop.

(g) There is no edge joining *A* and *B*.

(h) There is a two-edge path joining *A* and *B*.

(i) There is a closed path that uses only vertices *A*, *B*, and *C*.

(j) The graph contains multiple edges joining some pairs of vertices.

(k) The graph is connected.

(l) The graph is flat.

9. Draw a graph that represents the street map given in *Figure 8.3*.

10. Answer Exercise 8 with the additional assumption that the given path is an edge path.

11. The Town of Mud Flats is built on the banks of a river, much like Königsburg. (See *Figure 8.14*.) Many years ago the river followed a different channel (indicated by dashed lines) and one of the old bridges was left standing for its scenic beauty. Draw a graph representing the current land masses as vertices and the bridges as edges. Can you find an edge path for this graph?

12. For each of the following sets of conditions, give an example of a graph with four vertices that satisfies all of the given conditions. If this is not possible, explain why not.

   (a) Disconnected, nonflat.

   (b) Disconnected, with an edge path.

   (c) Disconnected, with a vertex path.

   (d) Connected, with no vertex path.

   (e) With an edge path, but no vertex path.

   (f) With a vertex path, but no edge path.

   (g) With an edge path and a vertex path.

13. Repeat Exercise 12 by giving examples of graphs with three vertices, if possible, rather than with four.

WRITING EXERCISES

1. As an alternative to the graph of *Figure 8.11* on page 409, the puzzle of Example 8.1 could also be represented by a graph that *looks* the same, but has no vertex D. Write a one-page paper that explains the difference between the two graphs and either supports or contradicts the position that one graph is as good as the other as a representation of the puzzle.

2. Consider a graph that represents the Königsburg Bridge Problem by treating each bridge as a vertex and streets connecting the bridges as edges. Write a one-page paper on whether or not finding a vertex path in this graph is essentially the same problem as finding an edge path in the graph of *Figure 8.12* on page 409. If they are different problems in graph theory, which one represents the original puzzle? If they are essentially the same, is there any reason to prefer one over the other?

3. Select one part of Exercise 8 whose answer is different from the corresponding part of Exercise 9. Write a paragraph explaining why the assumption that the path is an edge path makes a difference.

4. Select one part of Exercise 12 whose answer is different from the corresponding part of Exercise 13. Write a paragraph explaining why the number of vertices makes a difference.

5. Two different closed paths in a given graph might contain the same edges. (For example, in *Figure 8.13(a)*, consider *ABDA* and *BDAB*.) How could the definition of path be rephrased so that this distinction would be eliminated?

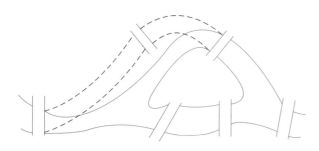

Figure 8.14 The bridges of Mud Flats.

EDGE PATHS

8.3

Let us thoroughly examine the first type of puzzle identified in the preceding sections—that of finding an edge path. By trial and error, you may have found an edge path in some of the puzzles of Section 8.1. Maybe you were lucky and found an edge path with little effort; maybe you found one only after many trials. But you surely did not find an edge path in all of them, for in some of the puzzles posed, none exists.

Unfortunately, failure to find an edge path does not solve the puzzle. For even moderately complex graphs, trial and error alone will not guarantee that there is *no* edge path; we may have merely overlooked one. In principle, we could consider each pair of vertices in the graph, record every path joining them that uses no edge more than once, stopping if an edge path is found. Then if no edge path is found between any two vertices, we could conclude that no edge path exists. However, for any but the simplest graphs (where the existence or nonexistence of edge paths is already obvious), the huge number of possibilities makes this approach thoroughly impractical. For example, in the relatively simple "envelope" graph of *Figure 8.11* on page 409, there are twenty-one pairs of vertices (including such pairs as *C*, *C*) and between each pair there are over fifty different paths that use no edge more than once—over a thousand paths! We must find a more efficient strategy.

The Front of a Building in the Form of a Graph.

Even when an edge path is found and it solves the particular puzzle at hand, there are still unsettled questions. Have we found the *only* edge path or are there others? Is there an edge path between *any* two vertices or only between the pair we happened upon? Why? Why do some graphs have an edge path while others do not? Is there some property or combination of properties that will allow us to distinguish between these two classes of graphs without the immense effort required by exhaustive trial and error?

The distinction between graphs that have edge paths and graphs that do not can be made by using a common mathematical technique. Without constructing a particular example, we assume that we are dealing with a "typical" or "generic" graph that happens to have an edge path, and we try to deduce whatever we can about this graph. Effectively, then, we are observing properties possessed by *every* graph that has an edge path. With luck, we may find that some of these properties will apply *only* to graphs with edge paths, and thus we can identify such graphs just by finding these critical properties.

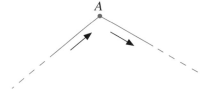

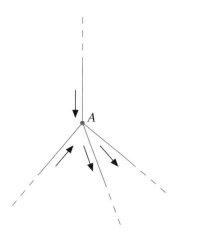

Figure 8.15 *A path that does not return to A.*

Taking this approach, then, suppose that A is a vertex of our hypothetical graph-with-edge-path; and suppose, further, that A is neither the beginning nor the end of the path, but rather is some vertex encountered along the way. Clearly, there must be at least two edges at A, one by which the path first comes to A and one by which it leaves. If the edge path does not return to A, then there are exactly two edges at A (*Figure 8.15*). On the other hand, if the path does return to A, it cannot do so by one of the two edges previously referred to because it is an edge path; thus, there must be a third edge to return by *and a fourth edge* to leave by (*Figure 8.16*). If the path does not return to A again, then there are exactly four edges at A. If the path does return a second time, there must be a fifth and a sixth edge at A, to allow the path to come and then leave. In general, although we cannot say exactly how many edges there are at A, we can conclude that *there must be an even number* because the edge path neither begins nor ends at A.

Figure 8.16 *A path that returns to A.*

There is one slight flaw in this analysis. We have assumed that when the path leaves from A, it goes on to some *other* vertex (and comes back to A from some other vertex). If there is a loop at A, the path leaves from A and returns to A on the same edge, so that only one edge is involved in this leaving and returning.[3] From the localized viewpoint, however, we can incorporate this possibility into our earlier

3. For this reason, among others, some authors exclude loops in defining a graph.

analysis by noting that there are still an even number of *ends* of edges at *A*. This discussion motivates the next definition.

DEFINITION

The **degree** of a vertex is the number of ends of edges incident with it. An **even vertex** is one of even degree; an **odd vertex** is one of odd degree. An **isolated vertex** is one of degree zero.[4]

EXAMPLE 8.8

In *Figure 8.8* on page 407, vertices *P*, *Q*, and *R* each have degree 2 and thus are even vertices; *N* and *O* each have degree 1 and are odd vertices; and *M* is an isolated vertex, having degree zero. (Of course, *M* is also an even vertex because zero is an even number.)

As we have seen, in a graph with an edge path, a vertex that neither begins nor ends the path must be an even vertex. By a similar analysis, a vertex that is either a beginning or an ending endpoint of an open edge path must be an odd vertex. If the edge path is closed, then any vertex can be used as the beginning and ending point and it must be even. Thus, we have proved the following general property:

xxxxxxxxxxxxxxxxxxxxxxxxxxxxxxxxxxxxxxxxxxxxxxxxxxx

(8.1)

If a graph has an edge path, then either all the vertices are even and the path is closed, or there are exactly two odd vertices and the edge path is open, beginning at one odd vertex and ending at the other.

xxxxxxxxxxxxxxxxxxxxxxxxxxxxxxxxxxxxxxxxxxxxxxxxxxx

EXAMPLE 8.9

In *Figure 8.17*, $PRTSRQ_1S_2QP$ is a closed edge path beginning and ending at vertex *P*. (There are closed edge paths beginning and ending at each vertex.) Observe that each vertex in the graph is even. In *Figure 8.18*, *ACDAB* is an open edge path beginning at one odd vertex, *A*, and ending at a second, *B*; all the other vertices are even.

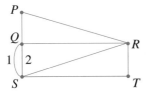

Figure 8.17 *A graph in which every vertex is even.*

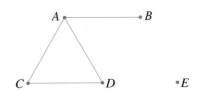

Figure 8.18 *A graph with two odd vertices, A and B.*

4. Some authors exclude isolated vertices in defining a graph.

Because a graph with an edge path can have at most two odd vertices, we have a relatively easy way to rule out a large number of graphs as having no possible edge paths. We need merely look at each vertex to see if it is odd; if we find three such vertices, there is no need to proceed any further. This is formalized in the following statement.

xxxxxxxxxxxxxxxxxxxxxxxxxxxxxxxxxxxxxxxxxxxxxxxxxxxxxxxxxxxxxxxx

(8.2)   If a graph has more than two odd vertices, then it has no edge path.

xxxxxxxxxxxxxxxxxxxxxxxxxxxxxxxxxxxxxxxxxxxxxxxxxxxxxxxxxxxxxxxx

Does the converse of Statement (8.2) hold true? In other words, does it follow that a graph with at most two odd vertices must have an edge path? Clearly not, as *Figure 8.19* shows. The obvious problem with this graph, of course, is its disconnectedness.

Returning to our examination of our hypothetical graph-with-edge-path, we can see that the existence of an edge path guarantees our ability to get from one vertex to another—with one exception. That exception is the possible presence of isolated vertices in the graph. Since no edges are incident with an isolated vertex, the presence of such a vertex has no effect on the existence of an edge path, although it does cause the graph to be disconnected. To coin a phrase for the sort of graph we are now considering, we will describe a graph that is connected, or connected except for one or more isolated vertices, as a **nearly connected** graph.

Figure 8.19   A graph with only two odd vertices, but no edge path.

EXAMPLE 8.10	*Figure 8.17* is connected, and thus is nearly connected. Neither *Figure 8.18* nor *Figure 8.19* is connected. *Figure 8.18* is nearly connected, but *Figure 8.19* is not.

With this terminology, the following statement becomes obvious.

xxxxxxxxxxxxxxxxxxxxxxxxxxxxxxxxxxxxxxxxxxxxxxxxxxxxxxxxxxxxxxxx

(8.3)   A graph that has an edge path is nearly connected.

xxxxxxxxxxxxxxxxxxxxxxxxxxxxxxxxxxxxxxxxxxxxxxxxxxxxxxxxxxxxxxxx

Now, if we combine the two properties we have identified, we obtain a characteristic way to distinguish graphs with edge paths:

xxxxxxxxxxxxxxxxxxxxxxxxxxxxxxxxxxxxxxxxxxxxxxxxxxxxxxxxxxxxxxxx

(8.4)   If a graph is nearly connected and has at most two odd vertices, then it has an edge path.

xxxxxxxxxxxxxxxxxxxxxxxxxxxxxxxxxxxxxxxxxxxxxxxxxxxxxxxxxxxxxxxx

Since near-connectedness is obvious at a glance, this theorem effectively reduces the problem of determining the existence of an edge path to that of counting the degrees of the vertices. (Actually, since only the odd or even parity need be determined, it is literally as simple as performing the numerical analog of "She loves me, she loves me not," by "counting": "odd, even, odd, even, . . .")

Although our analysis of edge-path graphs should make Statement (8.4) plausible, *we have not yet proved that it is true*. The proof is somewhat lengthy, but it is not difficult. In studying it, proceed slowly, making sure you understand each step before going on to the next one.

Proof of Statement (8.4)

We shall prove the case of a nearly connected graph with two odd vertices; the remainder is left to the exercises.

Step 1. Let us assume that the two odd vertices are labeled P and Q. We start at one, say P, and *randomly* construct a path, using no edge more than once, continuing as long as there are unused edges available when the path comes to a vertex.

Each time this path comes to a vertex other than P or Q, the path uses one of an even number of available edges, so there must be another edge at this vertex that is not yet used, by which the path can continue. Coming to and leaving such a vertex uses two of an even number of edges (or ends of the same edge in the case of a loop), leaving an even number yet to be used. Thus, the path cannot end at such a vertex.

At P, one of an odd number of edges is used (the very first one of the path), leaving an even number. Thus, if the path happens to return to P, the situation is exactly like coming to an even vertex; that is, the path cannot end at P. Eventually, the path must end since there are only a finite number of edges. Our analysis has shown that the only place it can end is at Q. By the time this random path ends, *all* the edges at Q must have been used up—otherwise, the path could continue.

To recapitulate, at this stage an odd number of edges at P have been used, all the edges at Q have been used, and an even number of edges (possibly zero) at every other vertex have been used. There may well be edges in the graph not used in this path, but if there are any, there are an even number remaining at any vertex where they exist.

Step 2. If the path constructed in *Step 1* has used all the edges, then it is an edge path and we are finished. On the other hand, if there are unused edges, then there must be a vertex having *some* edges that were used in the path and *some* that were not. (Otherwise, the graph would not be nearly connected.) We choose one such vertex, labeling it X for reference.

Step 3. Starting at X, we again randomly construct a second path, using no edge that has been used previously (in *Step 1* or in this step), and continuing as long as there are available edges. As was the case in *Step 1*, this path must end; but it cannot end at a vertex having an even number of edges available. Thus, it must stop at X, where the use of one edge to start left an odd number of edges unused.

Step 4. We now "cut and paste" the two paths constructed in *Steps 1 and 3* at the place where we know they share a common vertex, X. That is, we follow the path constructed in *Step 1* from P until we come to X; we then follow the path constructed in *Step 3* to its end, back at X, where we resume the original path, following it to its end at Q. The result is an expanded path from P to Q, using no edge more than once.

Step 5. If all edges have now been used, the expanded path is an edge path and we are finished. If not, the analysis of *Step 2* applies; we choose an appropriate vertex—one with both "used" and "unused" edges—and repeat *Steps 3 and 4*. Each time we do this, we use up more edges. Eventually, all edges in the graph must be included in the expanded path, and we have an edge path.

Statements (8.1), (8.3), and (8.4) completely characterize graphs that have edge paths. Note that there are effectively four slight variations possible among graphs with edge paths: the graph may be connected or not (but it must at least be nearly connected), and the edge path may be open or closed. We will not create special terminology for each of these cases, but we would be remiss not to identify the classic label for the "nicest" of these possibilities:

DEFINITION   An **Euler graph** is a connected graph with a closed edge path.

8.3   EXERCISES

1. For each of the seven graphs in *Figure 8.20*, determine which vertices are even and which are odd.

2. In *Figure 8.20(c)*, find the degree of each vertex.

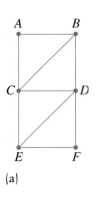

(a)

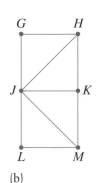

(b)

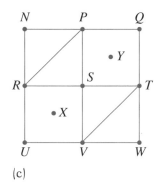

(c)

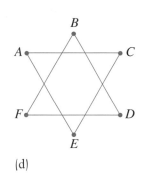

(d)

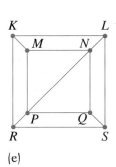

(e)

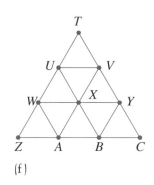

(f)

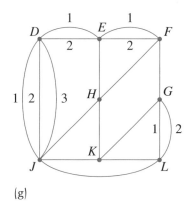

(g)

Figure 8.20 *Graphs for Exercises 1–5, 16, 18, and 20.*

3. In *Figure 8.20*(g), find the degree of each vertex.

4. For each of the seven graphs in *Figure 8.20*, how many odd vertices are there?

5. **(a)** Which of the seven graphs in *Figure 8.20* have no edge path?
   (b) Which have an open edge path?
   (c) Which have a closed edge path?
   (d) Which are Euler graphs?

6. Consider the twenty-six capital letters of the alphabet, each written in a simple block font, *sans serif*, as illustrated below:

   ABCDEFGHIJKLMNOPQRSTUVWXYZ

   (a) Considering each letter as just a geometric figure, convert it to a graph, with vertices at each "end," "corner," and "joint."
   (b) Which letter(s) have the smallest number of vertices?

   (c) Which letter(s) have the largest number of edges?
   (d) Which letters correspond to Euler graphs?
   (e) Which correspond to graphs with an open edge path?
   (f) Which correspond to graphs with no edge path?

7. In the text only one case of Statement (8.4) was proved—that of two odd vertices. Modify this proof to cover the case of no odd vertices.

8. Can you construct a graph with exactly one odd vertex? (*Hint:* Study the reasoning of *Step 1* in the proof of Statement (8.4).)

9. Generalize the reasoning of *Step 1* in the proof of Statement (8.4) to show that *every* graph (connected or not) has an even number of odd vertices.

10. Apply the methods of this section to Example 8.1 to answer the question posed there.

11. Apply the methods of this section to Example 8.2 to answer the question posed there.

12. Apply the methods of this section to Example 8.3 to answer the question posed there.

13. Show that if the people of Königsburg built one additional bridge from any land mass to another, they could find a path that crosses each of the bridges exactly once. (See Example 8.2 and *Figures 8.2* on page 404 and *8.12* on page 409.)

14. Show that if the people of Königsburg tore down one bridge connecting any land mass to another one, they could find a path that crosses each of the bridges exactly once. (See the preceding exercise.)

15. How many new bridges should be built in Königsburg if the townspeople wish to be able to cross every bridge exactly once and return to their starting place? (See Exercise 13.) Where should the new bridge(s) be built?

16. **(a)** If a connected graph has exactly six odd vertices, how many new edges must be added to change it into an Euler graph? (See Exercises 13 and 15.)

    (b) Where should these edges be added?

    (c) Apply this idea to the graph of *Figure 8.20(e)*. Which vertices can be joined by edges to convert the graph into an Euler graph?

17. If a connected graph has exactly n odd vertices (where n is some natural number), how many new edges must be added to change it into an Euler graph? (See Exercise 16.)

18. **(a)** Show that you can convert the graph of *Figure 8.20(d)* to one that has an edge path by adding appropriate edges.

    (b) What is the smallest number of edges that can be added to convert that graph to an Euler graph?

19. Show that an Euler graph must have at least as many edges as it has vertices.

20. **(a)** Show that by deleting certain edges from the graph of *Figure 8.20(e)* you can convert it into an Euler graph. (See Exercise 15.)

    (b) What is the smallest number of edges that can be deleted to accomplish this?

    (c) What is the largest number? (See Exercise 19.)

21. Give an example of a connected graph that is not an Euler graph and that cannot be converted into one by the deletion of any number of edges. (See Exercise 20.)

22. A **null graph** is the name given to a graph with no edges (that is, to a nonempty finite set of vertices). Suppose that a certain graph is not a null graph and is not an Euler graph. What is the smallest number of vertices this graph can have?

23. Suppose that a certain graph is not a null graph and has no edge path. What is the smallest number of vertices this graph can have? (See Exercise 22.)

WRITING EXERCISES

1. Define the term *graph* so as to exclude loops and isolated vertices.

2. Write a one-page paper arguing the merits of the definition described in Exercise 1 when one is studying the question of which graphs have edge paths.

3. The text uses visual aids (*Figures 8.15, 8.16*) to support the argument that a vertex that

neither starts nor ends an edge path must be even. Write a paper making the same argument, but based on a consideration of the "verbal" representation of a path as a string of letters. (Don't forget to consider the possibility of repetitions such as . . . *AA* . . . appearing somewhere in the string.)

4. Carefully consider the appropriate definitions to write a short paper that answers the following questions. (See Exercise 22.) Explain the reasoning behind each of your answers.

(a) Is a null graph connected?

(b) Is a null graph nearly connected?

(c) Is there a path in a null graph?

(d) Is a null graph an Euler graph?

VERTEX PATHS

We now turn to the problem suggested by Example 8.4 of Section 8.1 —finding a path that passes exactly once through each vertex of a graph. Noting the close parallel in form with the problem of Section 8.3, we shall approach our present problem with the same method.

Specifically, let us imagine that we have some arbitrary "typical" graph that has a vertex path. What can we logically assert about such a graph? Perhaps the most obvious fact is that the graph must be connected; because the vertex path passes (exactly once) through every vertex of the graph, some part of the vertex path must be a path between any two vertices we choose. Thus, we have quickly and easily proved our first result:

xxxxxxxxxxxxxxxxxxxxxxxxxxxxxxxxxxxxxxxxxxxxxxxxxxxxxxxxxxxxxxx

(8.5)   If a graph has a vertex path, it is connected.

xxxxxxxxxxxxxxxxxxxxxxxxxxxxxxxxxxxxxxxxxxxxxxxxxxxxxxxxxxxxxxx

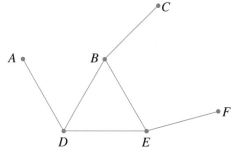

Figure 8.21   A connected graph with no vertex path.

Although connectedness is clearly a necessary condition, it is not sufficient to guarantee that a graph has a vertex path, as *Figure 8.21* shows. Although this graph is connected, it cannot have a vertex path. Vertex *A* has degree 1, and thus any vertex path must either begin or end at *A*. But the same is true of vertices *C* and *F*, and a path cannot have three ends. For brevity, we will call a vertex of degree 1 a **terminal vertex**, and the one edge at such a vertex, a **terminal edge**. With this terminology, we make the following observation:

xx

(8.6)   If a graph has a vertex path, it has at most two terminal vertices.

xxxxxxxxxxxxxxxxxxxxxxxxxxxxxxxxxxxxxxxxxxxxxxxxxxxxxx

A little thought should make it clear that a vertex path in a graph with one or two terminal vertices must be an open vertex path. In other words:

xxxxxxxxxxxxxxxxxxxxxxxxxxxxxxxxxxxxxxxxxxxxxxxxxxxxxx

(8.7)   If a graph has a closed vertex path, it has no terminal vertices.

xxxxxxxxxxxxxxxxxxxxxxxxxxxxxxxxxxxxxxxxxxxxxxxxxxxxxx

We could continue to pursue conditions pertaining to open versus closed vertex paths, stating appropriate pairs of theorems, such as Statements (8.6) and (8.7); but this would quickly become tiresome. Instead, let us restrict ourselves to considering one type of vertex path, leaving the similar observations about the other type for the exercises. For simplicity and historical interest, we shall consider only closed vertex paths for the rest of the section. This analogy to an Euler graph is formalized in the following definition:

DEFINITION   A **Hamilton graph** is a graph with a closed vertex path.[5]

Suppose we have a connected graph with no terminal vertices. Is it necessarily a Hamilton graph? *Figure 8.22* shows us that the answer is *no*. Here, we see a graph that has no terminal edges and that is connected, but that is not a Hamilton graph. The problem is that it has two "halves," with a single vertex, T, connecting them. A closed vertex path must pass through vertices on either side of T and form a closed path. There is no way to do this without passing through T twice, contradicting the definition of a vertex path; therefore, no closed vertex path exists. In order to characterize Hamilton graphs, we must describe the feature exemplified by

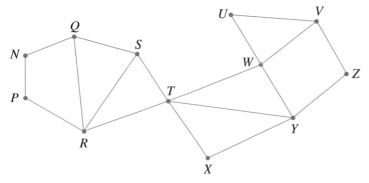

Figure 8.22   Not a Hamilton graph.

---

5. This concept is named after William Rowan Hamilton; see Section B.8 of the Appendix.

A Church Window Frame Forming a Graph.

vertices such as *T*. This we do in Statement (8.8), using the following definition:

DEFINITION   A vertex in a connected graph is a **critical vertex** if its removal, together with all edges incident with it, would leave a disconnected graph.

xxxxxxxxxxxxxxxxxxxxxxxxxxxxxxxxxxxxxxxxxxxxxxxxxxx

(8.8)   A Hamilton graph contains no critical vertices.

xxxxxxxxxxxxxxxxxxxxxxxxxxxxxxxxxxxxxxxxxxxxxxxxxxx

Do we now have a characteristic set of properties for a Hamilton graph? If a graph is connected and has neither terminal vertices nor critical vertices, is it necessarily a Hamilton graph? Again, the answer is *no*, as is shown in *Figure 8.23*.

In the graph of *Figure 8.23*, which meets all the conditions specified in the preceding paragraph, a closed vertex path must go through edges *SP* and *PQ* because this is the only way to reach (and leave) vertex *P*. But the same is true of edges *SR* and *RQ*, relative to vertex *R*, and of edges *ST* and *TQ*, relative to vertex *T*. Thus, three edges at *Q* (and also at *S*) must be included in any possible closed vertex path in order to reach every vertex; but this implies passing through *Q* (and *S*)

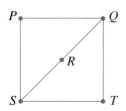

Figure 8.23 *Not a Hamilton graph, either.*

more than once. Again, we have a new necessary condition that prompts us to define a new term:

DEFINITION An edge incident with a vertex of degree two is called a **Hamilton edge**. (We abbreviate this as **_H_-edge**.)

xxxxxxxxxxxxxxxxxxxxxxxxxxxxxxxxxxxxxxxxxxxxxxxxxxxxx

(8.9) A closed vertex path of a graph must contain every _H_-edge.

(8.10) In a Hamilton graph, no vertex is incident with more than two _H_-edges.

xxxxxxxxxxxxxxxxxxxxxxxxxxxxxxxxxxxxxxxxxxxxxxxxxxxxx

Even with this added condition, though, we still do not have a characterization of Hamilton graphs. _Figure 8.24_ suggests two alternative ways to modify the graph of _Figure 8.23_ so as to satisfy the condition of Statement (8.10) without really eliminating the underlying barrier to having a Hamilton graph. In Graph (a) of _Figure 8.24_, multiple edges join the vertices; in Graph (b), loops have been added. The net effect in either case is to increase the degree of the vertices, thus technically eliminating _H_-edges; but the graphs still have the same basic drawback as _Figure 8.23_ in failing to be Hamilton graphs. As this example suggests, multiple edges and loops do not provide _useful_ additional edges to be used in a closed vertex path. They only serve to obscure our search for a characteristic of Hamilton graphs. So we shall exclude them for the remainder of this section, restricting ourselves to what we shall call **simple graphs**—graphs with no loops and no multiple edges.

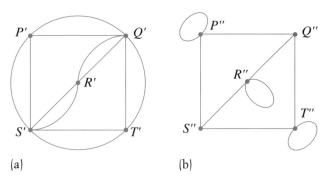

(a)   (b)

Figure 8.24 _Figure 8.23 with (a) added edges and (b) added loops._

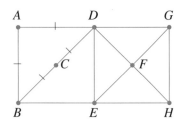

Figure 8.25 _Still not a Hamilton graph._

Collecting all of our observations so far, we are ready to pose the characterization question again. If a simple, connected graph has no terminal vertices, no critical vertices, and no vertex incident with more than two _H_-edges, is it necessarily a Hamilton graph? You guessed it. The answer is still _no. Figure 8.25_ shows a graph with all the conditions called for; for clarity, each _H_-edge is marked with a short cross-mark. We can identify our new problem in terms of these edges. However, it is not that there are too many edges at one vertex. Rather, the _H_-edges at A, B, C, and D form a closed path. As was observed in Statement (8.9), each of these edges must be used in a closed vertex path; by definition, the other vertices, E, F, G, and H, must also be used. But there is no way to do both of these things. Therefore, we have another exclusion to add to our growing list.

xxxxxxxxxxxxxxxxxxxxxxxxxxxxxxxxxxxxxxxxxxxxxxxxxxxx

(8.11)   If the H-edges of a graph include a closed path among some, but not all, vertices, then the graph is not a Hamilton graph.

xxxxxxxxxxxxxxxxxxxxxxxxxxxxxxxxxxxxxxxxxxxxxxxxxxxx

Are we finally finished? If a simple graph passes all the tests implied in Statements (8.5), (8.7), (8.8), (8.10), and (8.11), is it necessarily a Hamilton graph? Unfortunately, *no*. *Figure 8.26* is a counterexample. As before, H-edges are marked with a single cross-mark. Thus, any closed vertex path must contain $QRSTU$ (or, of course, $UTSRQ$). Clearly, this cannot be accommodated in a vertex path that must reach both P and V and also be closed. One way to describe and classify this problem is to note that since RS and ST are H-edges, PS and SV cannot be used. Since these edges are unavailable as ways to pass through P or V, then, effectively, PQ, QV, VU, and UP are "secondary" H-edges because they are the only edges actually available to reach P and V after the H-edges have been identified. (These are marked with double cross-marks.) Then the configuration of these H-edges and "H_2-edges" together rules out a closed vertex path because it violates the conditions of both Statements (8.10) and (8.11).

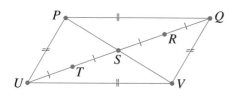

Figure 8.26   Still not a Hamilton graph.

We could formally define "H_2-edges," "H_3-edges," and so on, and generalize Statements (8.10) and (8.11) accordingly. But even then we would fail to cover every property that characterizes Hamilton graphs. *Figure 8.27* shows a simple, connected graph with no isolated or critical vertices and no H-edges. Yet it takes very little study to conclude that there is no closed vertex path for this graph.

What we have experienced in this section is a taste of the frustration met by every individual who has tried to characterize Hamilton graphs. Although the problem is over a century old, no one has yet found any combination of properties whose presence in a graph will guarantee the existence of a closed vertex

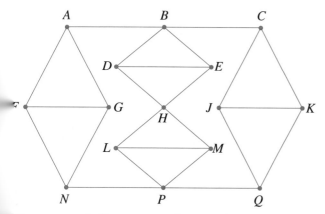

Figure 8.27   Still not a Hamilton graph.

path. This is not to say that our efforts have been wasted. We have identified a number of properties that are *necessary* for a Hamilton graph. The absence of any of them, then, allows us easily to identify when a given graph is *not* Hamilton. And our observations about *H*-edges can help us find a closed vertex path if one exists, even though we have no theorems to *guarantee* the existence of the desired path. Similarly, although we cannot characterize whether or not a graph has an open edge path, we can modify the statements about closed edge paths to provide useful information.

In all, determining whether or not a graph has a vertex path remains largely a matter of trial and error, but the necessary conditions we have stated can greatly reduce the number of trials and provide an organized guide as to what to try.

8.4 EXERCISES

Exercises 1–8 refer to *Figure 8.28*.

1. Which graphs have terminal vertices?

2. Identify all the terminal edges.

3. Which graphs have critical vertices?

4. Which graphs are not connected?

5. Which graphs are not simple?

6. Identify the *H*-edges in Graphs (b), (e), and (h).

7. Which graphs are Hamilton graphs?

8. Which graphs have open vertex paths?

9. Can the street-sign crew of Example 8.3 do their job of visiting each intersection exactly once? (See *Figure 8.3* on page 405.)

10. (a) With reference to Exercise 6 of Section 8.3, which capital letters of the alphabet correspond to graphs with critical vertices?

   (b) Which correspond to graphs that are not simple?

   (c) Which correspond to Hamilton graphs?

   (d) Which correspond to graphs with open vertex paths?

11. (a) Which of the graphs in *Figures 8.22 through 8.27* have open vertex paths?

   (b) For those that do, identify a path.

12. (a) Show that every Hamilton graph with at least two vertices has an open vertex path.

   (b) Why is a Hamilton graph with only one vertex different?

   (c) Give an example to show that the converse of Part (a) is not true; that is, a graph with an open edge path need not be a Hamilton graph.

13. Prove that an open vertex path may fail to contain at most two *H*-edges in a graph. (*Hint:* See Statement (8.9).)

14. Prove that if a graph has a vertex incident with five or more *H*-edges, it contains no vertex path. (*Hint:* See Statement (8.10).)

15. Prove that a connected graph with more than two vertices and with a terminal vertex must have a critical vertex.

16. Show that if a closed path is both a vertex path and an edge path for a graph, the number of vertices must equal the number of edges.

Figure 8.28
Graphs for
Exercises 1–8.

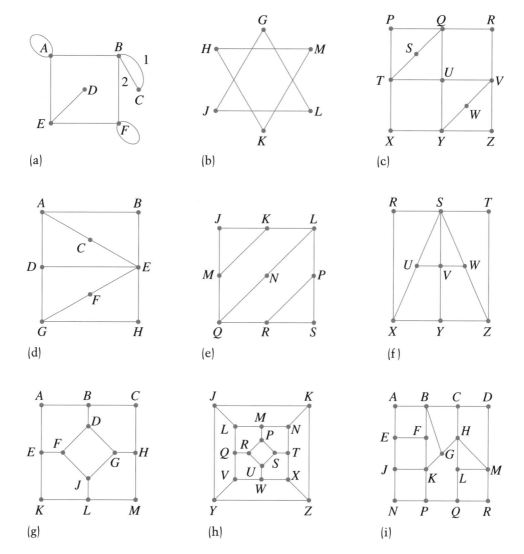

(a)   (b)   (c)

(d)   (e)   (f)

(g)   (h)   (i)

WRITING EXERCISES

1. Give a formal definition of an H_2-*edge*.

2. Using the definitions of an *H*-edge as well as that of an H_2-edge, which was the preceding exercise, formulate the generalizations of Statements (8.10) and (8.11), and present some argument for their validity.

3. Write a paper identifying the reason that *Figure 8.27* fails to be a Hamilton graph. If possible, formulate that identification into a definition, and state the appropriate theorem linking that concept to Hamilton graphs.

4. Identify as many distinctions as you can between edges and vertices, and discuss how these distinctions might explain the difference between the ease of characterizing Euler graphs and the continuing failure to do so for Hamilton graphs.

CROSSING CURVES

Now we consider the type of problem posed in the puzzle of Example 8.5—finding a continuous curve crossing each line segment in a figure exactly once. In that puzzle, we asserted that the figure could be labeled in an obvious and unambiguous way to become a graph. This was a correct assertion for that graph and many others, but it is not always the case; and the choice of how to represent a figure as a graph may dramatically alter the answer to the crossing-curve question.

Figure 8.29(a) shows a simple figure which, with appropriate identification of vertices, is a graph. Two different ways of making this identification are given in *Figure 8.29(b) and (c)*. Graph (b) is nonflat, it has six edges, and it has a crossing curve, as indicated. But this technically correct crossing curve *for the graph* seems inadequate to represent the problem posed by the original picture, almost as if we were somehow cheating. For this reason, we shall restrict ourselves to flat graphs in our consideration of crossing curves. Graph (c) is flat (and has eight edges) and appears to be a more honest graph representation of *Figure 8.29(a)*; but no crossing curve is possible. We will also limit ourselves to connected graphs because for the purpose of finding a crossing curve, a disconnected graph can be treated as several different connected graphs.

DEFINITION

A **polygonal graph** is a connected flat graph. Each region of the plane determined by the edges of a polygonal graph is called a **face** of the graph.

Note

The choice of the phrase "determined by," rather than "contained within," is deliberate because it turns out that it is useful to consider the portion of the plane "outside" of a polygonal graph to be a face. For example, in *Figure 8.29(c)*, there are *five* faces—the four triangular regions that share vertex *R*, and the fifth face, the region of the plane outside of the square *PQST*.

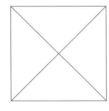

(a)

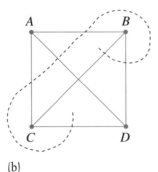

(b)

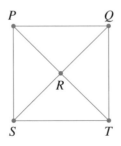

(c)

Figure 8.29 Two different graphs for a figure.

A crossing curve, then, is a curve that crosses from one face of a polygonal graph to another, that intersects each edge exactly once, and that does not pass through any vertex. An exception occurs if a graph has a terminal edge. In such an example, the crossing curve remains within a given face when crossing such an edge. In other words, terminal edges pose no barrier to the existence of a crossing curve; as long as we are able to reach every edge, a crossing curve exists. A crossing curve will be impossible if all the edges surrounding some face are used up (crossed once already) and the curve is "trapped" within that face while some edges, elsewhere, are not yet crossed. Thus, the existence of a crossing curve is determined by the number of times a possible curve must enter or leave a face. We define a term to identify this number.

DEFINITION

The **order** of a face is the number of edges adjacent to it. A terminal edge is counted twice.

The peculiarity of this definition with regard to terminal edges can be explained by observing that a crossing curve, in crossing a terminal edge, leaves and enters the face simultaneously. Alternatively, we could think of each edge as having two "sides," and the process of determining the order of a face as that of counting these sides. For most edges, one side is in one face and the other is in a distinct neighboring face; but both sides of a terminal edge lie in the same face.[6]

EXAMPLE **8.11**

Problem: Determine the order of each face in the graphs of *Figure 8.30*.

Solution: In *Figure 8.30(a)*, there are three faces; one of order three, one of order four, and one (the "outside" face) of order five. In *Figure 8.30(b)*, there are four faces: two of order three, one (contained by the loop) of order one, and one of order seven.

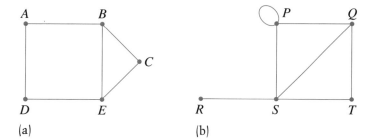

Figure 8.30 Problem: Determine the order of each face in these graphs.

(a)   (b)

---

6. This peculiarity could be avoided by excluding terminal edges in the definition of polygonal graphs. However, that option also has its problems.

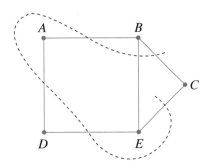

Figure 8.31 A crossing curve for the graph of Figure 8.30(a).

Let us look for a crossing curve for *Figure 8.30(a)*. Since the face enclosed by the path *BCEB* has order three, a crossing curve can enter this face twice and leave once, or leave twice and enter once. In other words, either the start or finish of the curve must lie in this face. The same is true of the "outside" face of order five. Fortunately, the face of order four must be entered twice and exited twice, so that the curve does not have to begin or end in this face. Therefore, a crossing curve is possible—one such curve is indicated in *Figure 8.31*.

In *Figure 8.30(b)*, each of the four faces has an odd order, thereby requiring one more entry than exit, or vice versa. Thus, a crossing curve must have one of its ends in each of four faces—an impossibility. There is no crossing curve for this graph.

If this "even versus odd" analysis is reminiscent of the discussion of edge paths we went through in Section 8.3, there is good reason. The problem of finding a crossing curve from face to face across edges is quite similar to that of finding an edge path from vertex to vertex along edges. In the current instance, the existence of a crossing curve depends on the (odd or even) order of the faces; in the earlier case, the existence of an edge path depended, in a completely analogous way, on the degree of the vertices. This suggests that there should be some way to translate from a given graph (in which we are seeking a crossing curve) to some related graph (in which we could determine whether or not there is an edge path). We make the connection formally by using the following definition and Statement (8.12).

DEFINITION

The **dual** of a given polygonal graph is a graph in which each face of the given graph has a corresponding vertex in the dual, and each edge in the given graph has a corresponding edge in the dual. The face-to-vertex and edge-to-edge correspondences must be such that an edge in the given graph that separates two faces corresponds to an edge in the dual that connects the vertices corresponding to those faces.

EXAMPLE 8.12

The graph of *Figure 8.32(a)*, with its faces labeled by lowercase letters, has as its dual the graph of *Figure 8.32(b)*. The face-to-vertex and edge-to-edge correspondences are given in *Table 8.1*. For example, because in Graph (a) (the given graph) face q is separated from face r by two edges, BC and CE, in Graph (b) (the dual graph) vertex Q is joined to vertex R by two edges, Q_1R and Q_2R.

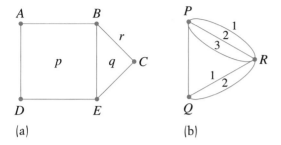

GRAPH (a) FACE	DUAL (b) VERTEX	GRAPH (a) EDGE	DUAL (b) EDGE
p	P	AB	P_1R
q	Q	AD	P_2R
r	R	BC	Q_1R
		BE	PQ
		CE	Q_2R
		DE	P_3R

Figure 8.32 A graph and its dual.

Table 8.1 The face–vertex and edge–edge correspondences for the graphs of Figure 8.32.

xxxxxxxxxxxxxxxxxxxxxxxxxxxxxxxxxxxxxxxxxxxxxxxxxxx

(8.12) A polygonal graph has a crossing curve if and only if its dual graph has an edge path.

xxxxxxxxxxxxxxxxxxxxxxxxxxxxxxxxxxxxxxxxxxxxxxxxxxx

The proof of Statement (8.12) is essentially contained in the previous discussion. We need only observe that in the correspondence from a polygonal graph to its dual, the order of a face corresponds to the degree of the corresponding vertex, so that the discussion and proofs in Section 8.3 apply exactly to crossing curves by means of the dual-to-graph correspondence. For example, the crossing curve shown in *Figure 8.31* corresponds to the edge path $Q_2R_3P_2R_1PQ_1R$ in the dual graph (*Figure 8.32(b)*). Conversely, any edge path in the dual can be translated to a crossing curve in the original graph.

In order to determine whether or not a crossing curve exists, then, we can construct the dual graph and apply the results of Section 8.3. More simply, we can just translate the statements of Section 8.3 to corresponding statements about crossing curves by replacing *vertex* with *face*, *degree of vertex* with *order of face*, and so on. In particular, we obtain Statements (8.13) and (8.14):

xxxxxxxxxxxxxxxxxxxxxxxxxxxxxxxxxxxxxxxxxxxxxxxxxxx

(8.13) If a polygonal graph has a crossing curve, then either all the faces have even order, or there are exactly two faces of odd order and the curve begins in one odd face and ends in the other.

(8.14) If a polygonal graph has more than two odd faces, then it has no crossing curve.

xxxxxxxxxxxxxxxxxxxxxxxxxxxxxxxxxxxxxxxxxxxxxxxxxxx

EXAMPLE **8.13**	**Problem:** Is it possible to draw a crossing curve in the graph of *Figure 8.33*?

Solution: We first determine the order of each face; the results are given in *Table 8.2*. Because there are only two odd faces (*a* and *f*), there is a crossing curve, with one end in face *a* and the other in face *f*. The drawing of a crossing curve is left as an exercise.

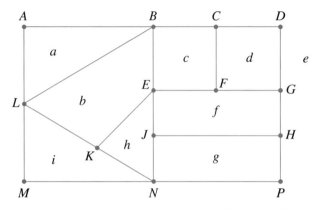

Figure 8.33 *Can a crossing curve for this graph be drawn?*

FACE	ORDER	FACE	ORDER
a	3	*f*	5
b	4	*g*	4
c	4	*h*	4
d	4	*j*	4
e	10		

Table 8.2 *The orders of the faces in the graph of Figure 8.33.*

8.5   EXERCISES

1. Find a crossing curve for the graph of *Figure 8.33*.

2. Does the graph in Example 8.5 have a crossing curve? If so, draw one; if not, why not?

3. For each graph in Figure *8.34*, identify the number of odd faces.

4. Which of the graphs in *Figure 8.34* have dual graphs that are Euler graphs?

5. Which of the graphs in *Figure 8.34* have crossing curves? For each one that does, draw a crossing curve.

6. Construct a graph that is the dual of *Figure 8.34(c)*.

7. Construct a graph that is the dual of *Figure 8.34(e)*.

8. In the graph of *Figure 8.32(b)*, label the faces with the letters *a, b, c, d,* and *e* in a way that shows that *Figure 8.32(a)* is the dual graph with its vertices, *A, B, C, D,* and *E,* respectively, corresponding to these faces.

Exercises 9–12 refer to a self-dual graph, that is, one which, with the appropriate correspondence of faces and edges, is dual to itself.

9. In the graph of *Figure 8.34(a)*, label the faces with the letters *a, b, c, d,* and *e* in a way that shows that it is self-dual.

10. Show that *Figure 8.34 (b)* is self-dual.

11. If possible, construct a self-dual graph with exactly three vertices. If this is impossible, explain why.

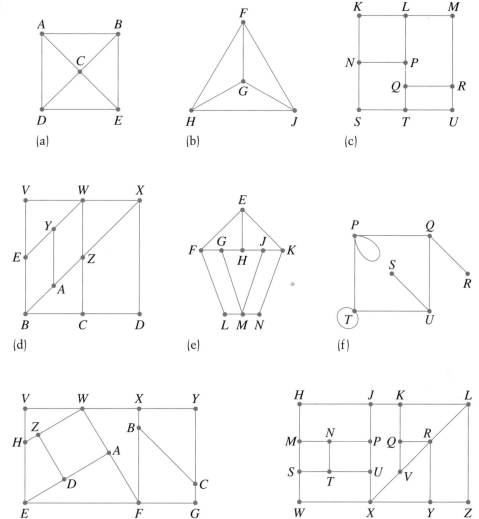

Figure 8.34 Graphs for Exercises 3–7, 9, 10, and 12.

12. If possible, construct a self-dual graph with exactly four vertices, which is different from the graph of *Figure 8.34(b)* (that is, not with four faces each of order three). If this is impossible, explain why.

13. In the process of forming dual graphs, the concepts *face* and *vertex* are duals of each other, *crossing curve* and *edge path* are duals, and so on. Identify the dual of each of the following:

(a) Terminal edge.

(b) Terminal vertex.

(c) Isolated vertex.

14. Construct a graph with exactly one odd face. If this is impossible, explain why.

WRITING EXERCISES

1. The definition of the order of a face somewhat arbitrarily counts a terminal edge twice. Discuss the pros and cons of counting a terminal edge once.

2. Discuss the pros and cons of ignoring terminal edges in determining the order of a face.

3. Give definitions of "open crossing curve" and "closed crossing curve" that are dual to "open edge path" and "closed edge path." (See Exercise 13.)

4. Explain why the dual of the dual of a graph is the original graph. (See Exercise 8.)

EULER'S FORMULA

In our search for edge paths and crossing curves, we have noted the significance of numbers in graph theory. Thus, the even or odd degree of vertices (or order of faces) largely determines the possibility of the existence of an edge path (or crossing curve). There are many other numerical properties of graphs. For example, if we add up the degrees of all of the vertices in a graph, the result must be even because we have effectively counted all the ends of all the edges, which must come in pairs. This fact allows us to prove that there must be an even number of odd vertices (if any), and it provides an alternative method of proof from that suggested in Exercise 9 of Section 8.3.

A more significant and more useful relation exists among the numbers of vertices, edges, and faces in a polygonal graph. For a source of data, let us turn to the graphs of *Figure 8.34* in the exercises at the end of Section 8.5. In *Table 8.3*, we have recorded the numbers of vertices, edges, and faces in each of these graphs; they are denoted by V, E, and F, respectively.

At first glance, there may not seem to be much of a pattern. For instance, graphs with the same number of vertices do not have the same number of edges, and vice versa. Those with more vertices tend to have more edges and faces, but in apparently random fashion. However, we can see that in every instance there are more edges than either vertices or faces. In fact, the number of edges is nearly as large as the total number of vertices and edges combined. Moreover, if we expand the data to include the *sum* of the number of vertices and the number of faces, a striking pattern emerges. (See *Table 8.4*.)

GRAPH	V	E	F
(a)	5	8	5
(b)	4	6	4
(c)	10	13	5
(d)	10	15	7
(e)	9	13	6
(f)	6	8	4
(g)	12	18	8
(h)	17	25	10

Table 8.3
The numbers of vertices, edges, and faces in the graphs of Figure 8.34.

Table 8.4 *A pattern in the data of Table 8.3.*

GRAPH	V	E	F	V + F
(a)	5	8	5	10
(b)	4	6	4	8
(c)	10	13	5	15
(d)	10	15	7	17
(e)	9	13	6	15
(f)	6	8	4	10
(g)	12	18	8	20
(h)	17	25	10	27

Comparing the last column of *Table 8.4* with the middle column, we see that $V + F$ is exactly 2 more than E in every instance. Let us state this observation in the form of an algebraic equation.

(8.15)
$$V + F = E + 2$$

Equation (8.15) can be rearranged algebraically in many ways; regardless of the form, the relation is called *Euler's Formula*.[7]

xxxxxxxxxxxxxxxxxxxxxxxxxxxxxxxxxxxxxxxxxxxxxxxxxxxxxx

(8.16)

Euler's Formula: In any polygonal graph,
$$V - E + F = 2$$
where V, E, and F are the number of vertices, edges, and faces, respectively.

xxxxxxxxxxxxxxxxxxxxxxxxxxxxxxxxxxxxxxxxxxxxxxxxxxxxxx

Of course, the fact that Formula (8.16) is correct for eight examples is not a proof that it is always correct. The general argument that verifies the formula is not too difficult, as we shall see.

Proof of Euler's Formula

Any polygonal graph can be constructed by starting with one vertex and expanding step by step with one of two types of procedures:

Type 1. Add a new vertex and join it by an edge to some vertex already present.

Type 2. Add a new edge (possibly a loop) joining two (not necessarily distinct) vertices already present.

---

7. This formula, reportedly, was known to mathematicians prior to Euler; but his proof is the earliest recorded one—hence the attribution.

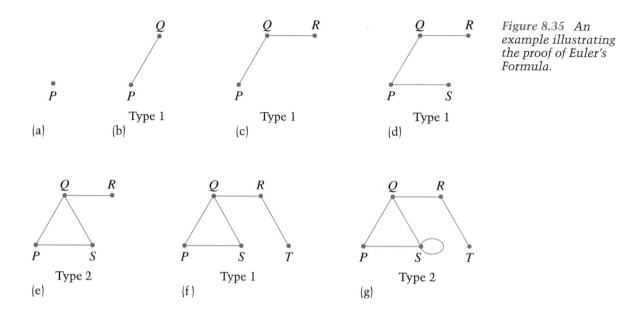

Figure 8.35 An example illustrating the proof of Euler's Formula.

For example, we can construct the polygonal graph of *Figure 8.35(g)* by starting with a single vertex in Figure 8.35(a) and adding vertices and edges according to these two types of steps. Each intermediate step is shown in *Figure 8.35*, labeled *Type 1* or *Type 2*. This is not the only sequence of steps you can follow, but this polygonal graph or any other one can be constructed using only these two types of procedures.

Now, let us keep track of the quantity $V - E + F$ in Euler's Formula at each stage of construction. Starting with a single isolated vertex, we have $V = 1$, $E = 0$, and $F = 1$ since the plane is the one face; thus, Euler's Formula is satisfied. Each time a step of *Type 1* is used, we add a vertex and an edge, so V and E each increase by 1. Since the graph we are building is polygonal, it is flat. Thus, the new edge must lie entirely within a face that already exists, and no new faces are created by this construction (that is, F remains unchanged). Consequently, the value of $V - E + F$ does not change in the course of the step. Each time a *Type 2* step is used, E increases by 1 and V remains fixed. Since the added edge in a flat graph lies within some face and since the graph is connected at each step because new vertices are always connected by a new edge using *Type 1* steps, this added edge must "cut off" part of the "old" face, creating a new, additional face. That is, F increases by 1, and again, the value of $V - E + F$ does not change. In this way, starting with $V - E + F = 2$, we build up the graph by a series of construction steps that leave the quantity $V - E + F$ unchanged at each step. Therefore, we must end as we began, with a graph satisfying Euler's Formula.

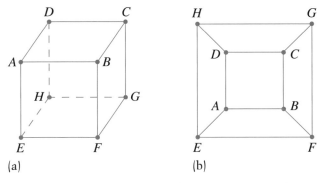

(a)   (b)

Figure 8.36 *A polygonal graph representing a cube.*

There are a number of applications and generalizations of Euler's Formula that are of importance in graph theory, geometry, and topology.[8] One classic case involves solid geometric figures, such as cubes, prisms, and pyramids. Specifically, let us look at any solid whose edges are straight-line segments, whose faces are polygons (triangles, quadrilaterals, and so on), and which have no "holes" (no square doughnuts allowed, for instance). First, envision such a solid as reduced to an open-latticework figure, like a toothpick model, as in *Figure 8.36(a)*. Now, think of those edges as being elastic, so that we can "flatten" the solid onto a plane, forming a polygonal graph.[9] For example, in *Figure 8.36(b)*, we have a "graphic" representation of the cube indicated in *Figure 8.36(a)*.

Notice that, although the graph distorts the geometric shape of most of the faces (from square to trapezoid), it accurately represents the number of vertices, edges, and faces of the solid, as well as the number of edges meeting at each vertex (the degree) and the number of edges surrounding each face (the order). Note that one face of the cube (*EFGH*) corresponds to the "outside" face of the graph. Since the numbers of vertices, edges, and faces translate unchanged from geometric solid to graph and vice versa, Euler's Formula must hold for such solids. This allows us to establish several important facts about geometric solids.

As one example, we focus on **regular solids**—that is, solids with each face an equilateral, equiangular polygon, and with the same number of faces at each vertex. The cube is probably the most familiar example of a regular solid, with its six square faces, three meeting at each vertex. Are there other such solids? If so, how many kinds are there, and what do they look like? Euler's Formula together with a little numerical reasoning can answer these questions.

Consider a typical regular solid. If we let D be the degree of each vertex (the degree must be the same for each one if the solid is regular), then $D \cdot V$ counts the sum of all the degrees of all the vertices. This must equal twice the number of edges, as noted earlier. Similarly, letting R be the order of each face, $R \cdot F$ must also equal $2 \cdot E$; this follows either by duality or by noting that each edge has two faces adjacent to it. Thus, we know that

8. Topology is the study of the properties of geometric figures or solids that are unaffected by changes in size or shape.

9. It is precisely this connection that lends the name "polygonal" to polygonal graphs.

(8.17)
$$D \cdot V = 2 \cdot E \quad \text{or} \quad V = \frac{2 \cdot E}{D}$$

and

(8.18)
$$R \cdot F = 2 \cdot E \quad \text{or} \quad F = \frac{2 \cdot E}{R}$$

Substituting these into Euler's Formula, we obtain

$$\frac{2 \cdot E}{D} - E + \frac{2 \cdot E}{R} = 2$$

and dividing this expression by E produces the following result, which *must hold for every regular solid*, since we made no special assumptions about the ones under discussion.

xxxxxxxxxxxxxxxxxxxxxxxxxxxxxxxxxxxxxxxxxxxxxxxxxxxxxxxxx

(8.19)
> In any regular solid,
> $$\frac{2}{D} - 1 + \frac{2}{R} = \frac{2}{E}$$
> where D is the degree of each vertex, R is the order of each face, and E is the number of edges.

xxxxxxxxxxxxxxxxxxxxxxxxxxxxxxxxxxxxxxxxxxxxxxxxxxxxxxxxx

We can now try some particular values of D and R, and determine the corresponding values of E from Statement (8.20). The order of each face, R, must be at least 3 (a triangle is the simplest regular polygon) and D must be at least 3 (each vertex of a solid must join at least three faces). If we let $D = R = 3$, the left side of the formula in Statement (8.19) becomes $\frac{2}{3} - 1 + \frac{2}{3}$, which equals $\frac{1}{3}$. Then E must equal 6 to balance the equation. Statements (8.17) and (8.18) tell us that V and F are each 4 in this case, and we have described a solid with four vertices and four triangular faces, three meeting at each vertex. This figure is called a (regular) **tetrahedron**.

As a consequence of Statement (8.19), the number of allowable choices for D and R turns out to be quite limited. Because E, and thus $\frac{2}{E}$, must be positive, the left side of the equation must also be positive. If either D or R is too large, $\frac{2}{D}$ and $\frac{2}{R}$ become so small that their sum is less than 1, and the left side is negative or zero. Moreover, as noted previously, D and R must each be at least 3. By trial, there are only five possible pairs of values for D and R that fit all these constraints. These values are given in *Table 8.5*, together with the resulting values of E, V, and F required by Statements (8.19), (8.17), and (8.18), respectively.

D	R	E	V	F	Name of Solid
3	3	6	4	4	Tetrahedron
3	4	12	8	6	Hexahedron (cube)
3	5	30	20	12	Dodecahedron
4	3	12	6	8	Octahedron
5	3	30	12	20	Icosahedron

Table 8.5 The five regular solids.

These five regular solids, called the **Platonic solids**, were known to the ancient Greek mathematicians, who could find no other examples of regular solids, but could not *prove* that there were no others. It was not until the advent of graph theory that the relatively simple numerical relationships, which were "hidden" in the original geometric forms of these figures, became clear from their graph-theoretic forms and then could easily be stated and proved.

8.6   EXERCISES

1. For each of the graphs of *Figure 8.28* on page 427, except Graph (b), count the number of vertices, edges, and faces, and verify that Euler's Formula is satisfied.

2. In Exercise 1, why is the graph of *Figure 8.28(b)* excluded?

3. The graph of *Figure 8.37* is flat, but not connected. If we stretch the definition of *face* a little, we can identify three faces. Count the vertices and edges and compute

$$V - E + F$$

4. Construct several flat graphs that are disconnected into two parts. (See Exercise 3.) For each graph, compute

$$V - E + F$$

Construct a formula for such graphs. Prove your conjecture if you can.

5. Repeat Exercise 4 for flat graphs disconnected into three parts.

6. Let P be the number of parts in a flat graph. (For a connected graph, $P = 1$; for the graph of *Figure 8.37*, $P = 2$; etc.) Determine a formula for flat graphs that relates V, E, F, and P. (See Exercises 4 and 5.)

7. In a polygonal graph, does the degree of a vertex always equal the number of faces meeting at that vertex? If not, why not?

8. In Statement (8.19), if $R = 3$ and $D = 6$, the left side of the formula becomes zero. The right side, $\frac{2}{E}$, can be thought of as equaling zero if E is infinite. Because $D = 6$ and $R = 3$, as a polygonal graph, this equation corresponds to a figure with six equilateral triangles at each vertex, which extends infinitely throughout the plane (so that there are an infinite number of edges, vertices, and faces). This formation is called a **regular tesselation** of the plane with triangles (see *Figure 8.38*). Find any other values of R and D that have this property. Describe the corresponding tesselation.

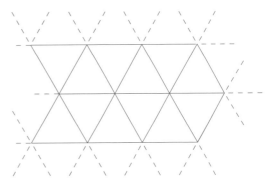

Figure 8.38 *A regular tesselation of the plane with triangles.*

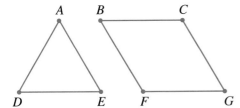

Figure 8.37 *A flat graph that is not connected.*

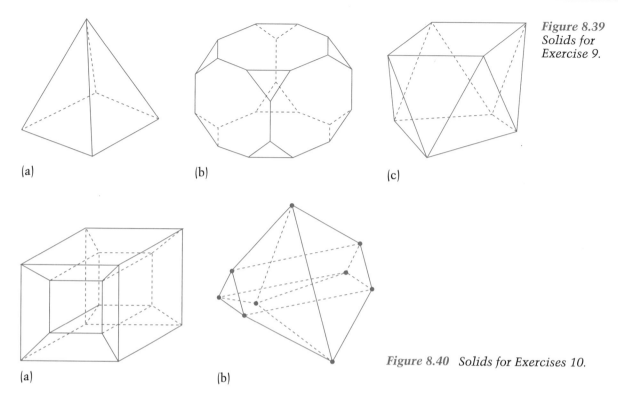

Figure 8.39
Solids for
Exercise 9.

(a)   (b)   (c)

(a)   (b)

Figure 8.40   Solids for Exercises 10.

9. Verify Euler's Formula for the solids pictured in *Figure 8.39*.

10. Calculate
$$V - E + F$$
for the solids of *Figure 8.40*. These solids, with polygonal faces, have one "hole" through them. Note that no single face has a hole; rather, the hole is surrounded by several polygons.

11. Let H be the number of holes in a solid. (For the solids of *Figure 8.39*, $H = 0$; for those of *Figure 8.40*, $H = 1$; and so on.) Determine a formula for solids that relates V, E, F, and H. (See Exercises 9 and 10.)

12. Which of the graphs representing the five regular (Platonic) solids are Euler graphs?

13. Which of the graphs representing the five regular solids are Hamilton graphs?

WRITING EXERCISES

1. Explain the etymology of the names for the five regular solids. (What are the root meanings of the parts of these words? What language(s) do they come from?)

2. In the proof of Euler's Formula, there are two types of steps that, in combination, can describe the construction of any polygonal graph. Suppose, instead, that we describe the construction of a polygonal graph by first locating all vertices, then drawing in all the edges. Discuss the pros and cons of this procedure as a method of describing the construction, then as a tool for proving Euler's Formula.

3. In the discussion about regular solids, the text asserts that the formula
$$R \cdot F = 2 \cdot E$$
follows by duality. Write a paragraph justifying this assertion.

4. Write a short paper in which you:

 (a) Determine a formula that relates the degree of a vertex to the number of faces at that vertex.

 (b) Identify and precisely define the other quantities that need to be counted for your formula.

 (c) Give at least two examples to demonstrate your formula.

 (d) Try to prove that your formula is always correct. (See Exercise 7.)

5. Examine the relationships among the graphs of the five regular solids relative to dualism. (*Hint*: study the data in *Table 8.5*.) For each regular solid, describe the *inscribed solid* formed when the points at the center of each face are used as vertices of a new solid. How does this relate to the dual graph relationships?

Looking Back

8.7

In reflecting on what you have studied in this chapter, it is important to observe that some aspects of our exploration of graph theory typify common general approaches and procedures in the exploration of mathematical ideas. Here we summarize a few parts of our discussion that are significant in this regard.

The method by which graphs with edge paths were identified in Section 8.3 is a common one in mathematics. Recall that we were looking for characteristic properties of graphs so that we could find an edge path. What we did was to assume that we had a solution (an edge path) and then investigate the "hypothetical" graph with that path. In a sense, we assumed the "answer" and worked backwards to examine the "question."

In Sections 8.3 and 8.4 we examined the problem of finding edge paths and vertex paths, respectively. The two problems are closely analogous, each dealing in exactly the same way with one of two fundamental components of a graph. But the results were anything but analogous. Edge paths were completely characterized in terms of connectedness and the degrees of vertices, but vertex paths defied characterization in terms of any combination of properties we could find. We were left with only partial results about vertex paths, and hence with a problem that remains unsolved, as it has been for some two centuries. Such disparity of solution or theory for apparently similar problems is not unusual in mathematics.

On the other hand, the problem of crossing curves, which at first seemed quite different from the path problems, turned out to be virtually identical with that of edge paths. This, too, is not uncommon.

In fact, a fundamental mathematical process is the attempt to abstract the essential structure from a given situation. By stripping away misleading or irrelevant details, mathematicians can often identify the true nature of a problem and then (hopefully) use known facts from previously solved problems which are essentially the same as the ones in the problem at hand. Even when solutions are not known, the identification of abstract similarities in two or more problems paves the way for the crossover of techniques from one problem to another, leading to a possible solution of *all* the problems involved. In many such instances, such as the one we encountered in this chapter, the similarities between the problems are so pervasive that each is the *dual* of the other; that is, each statement can be formed from the other simply by exchanging one or two key terms (and their related phrases) in one statement for corresponding key terms in the other.[10] Typically, the basic connection between the duals hinges on just one or two terms, although a wealth of related details may carry over from one to the other. In several areas of mathematics, the transfer of ideas via duality is a major source of new ideas and theorems.

Finally, the historical development of graph theory, from puzzles and curious diversions to a variety of applications, is also typical of much of mathematics. At various times in history, new concepts (such as negative numbers or complex numbers) or even an entire branch of mathematics (such as non-Euclidean geometry) were considered merely as curiosities, of value only when they clarified the understanding of "real" mathematics. Only later did these concepts take on profound practical significance as valuable tools for describing and analyzing real situations. So it is with graph theory, which is now beginning to flourish in the mathematical garden of powerful, useful ideas.

8.7 EXERCISES

WRITING EXERCISES

1. Explain how the use of equations in solving word problems in elementary algebra is an example of assuming an "answer."

2. Write a short paper defending the assertion that the analogy between the edge-path problem and the vertex-path problem is illusory. (Are there only two components to a graph? Are they analogous? What does Section 8.5 suggest?)

---

10. For instance, "Any two distinct points determine a unique line" and "Any two distinct lines determine a unique point" are dual statements, using *point* and *line* as the corresponding terms. In Euclidean geometry, only one of these statements is true; in projective geometry, however, they are both true. (See Sections 3.2 and 3.4.)

3. Write a short paper describing a pair of concepts or situations in some field other than mathematics which are, in some sense, duals of each other in a way analogous to that described in this chapter. Describe the pair clearly, and explain the sense in which they are duals of each other. In what ways is the duality similar to that in mathematics? In what ways is it dissimilar?

LINK: Digraphs and Project Management

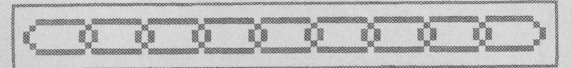

8.8

We mentioned at the outset that graph theory has been applied in a number of areas. Some of these areas are within mathematics. For example, as we saw in Exercise 11 of Section 8.6, there is a formula relating the number of vertices, edges, faces, and holes of various solids. There are many complicated mathematical figures for which it is difficult to determine whether or not something that appears to be a hole should be considered as such. The known formula becomes a way of deciding this and thereby classifying the solid. A similar technique helps topologists to classify the dimension and shape of various figures. Our discussion of edge paths and vertex paths can be applied in a variety of situations. Sometimes, such problems can be figured out (or avoided) by common sense, without appealing to graph theory; however, by abstracting the problem to its essential features, a solution that may be hard to see in its "real" setting can often be made much clearer in its graphic representation.

One area of application involves the construction of flat graphs with specific conditions as to which vertices are to be joined by edges. A classic puzzle illustrates this type of problem: Three eccentric recluses live in three houses. Each has a grudge against the other two and refuses even to walk on the same ground that the others do; however, each one needs to do business at each of three local shops. Is it possible to find paths from each house to each of the three shops in such a way that no recluse crosses another's path? In graph-theoretic terms, this is equivalent to constructing a flat graph with an edge connecting each of the vertices A, B, and C to each of the vertices X, Y, and Z. (We leave the solution of this problem to you.) Such problems find practical

High-rise Construction. Digraphs can be used to manage major projects, such as those arising in the construction industry.

application, for instance, in the design of printed circuits. The general theoretical form of this problem—that is, which sets of required properties can be satisfied by a graph and which cannot be—has been completely solved, but practical difficulties often arise in specific examples. In other words, the theory may guarantee that a certain flat graph exists, but the actual drawing of such a graph can be very difficult.

Many applications of graph theory require the added notion of restricting the *direction* of allowable paths by assigning a direction to some or all edges. (This is like making some streets one-way.) A **directed edge** is indicated by an arrowhead. A graph with *all* edges directed is called a **directed graph**, or simply a **digraph**. A **directed path**, as you would expect, is a path in a digraph that follows the prescribed directions of the edges.

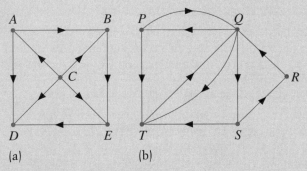

Figure 8.41 *Two digraphs.*

Figure 8.41 shows two examples of digraphs. In Graph (a), *ABED* is a directed path from *A* to *D*. There is no directed path from *D* to *A*. In Graph (b), there are exactly two odd vertices (*P* and *S*), but there is no directed edge path, since *P* and *S* each have two exiting edges and only one entering edge.

As the preceding demonstration suggests, the statements about paths in Section 8.3 invite revision for directed paths, but pursuing these revisions in detail would distract us from our purpose in introducing digraphs. The great utility of the digraph results from its applicability in so many areas. The vertices of a digraph may represent individuals or groups and the directed edges could signify their financial, social, or political influence on one another. The corresponding digraph would allow one to represent and then analyze the interrelations among the individuals or the groups. Or the vertices may represent locations in a computer's memory, and the directed edges may represent the ways in which information can be transferred from one location to another. Such a digraph can be extremely useful in data processing, where efficient manipulation of, and access to, stored information is essential. Digraphs can represent the interlocking food chains in an ecosystem. They can model the results of athletic contests; they can represent the organizational structure of a business. There are applications in virtually every field that involves any kind of relationship.

Graphs, directed or not, can be coupled with arithmetic to meet other needs. An obvious application of this kind uses the assignment of numbers to represent lengths of edges. The numbers might represent distance (as on a road map) or they might measure time, seniority, and so on. Alternatively, numbers can be assigned to vertices. (For historic reasons, this process is known as **coloring** a graph.[11]) For example, each vertex of a digraph might represent a portion of a computer program, with a number that specifies the time required to execute that portion; directed edges between some of the vertices might correspond to

---

11. "Coloring" derives its name from the famous **Four-Color Problem**, which resisted solution for over a century. An 1852 conjecture claimed that four colors are sufficient to color any map, drawn on a plane, in such a way that regions sharing a common border receive different colors. The direct approach of representing such maps as polygonal graphs and assigning colors (or numbers) to the faces failed to lead to a proof; in the 1930s, the problem was transformed via the concept of a dual graph to that of assigning appropriate "colors" to the vertices of a polygonal graph, a change of perspective that ultimately led to its solution in 1976. See Item 8 in the list *For Further Reading* at the end of the chapter.

the fact that certain things must be done before others, as when the results of one computation are necessary for another. Analysis of such a graph can predict, within limits, how long the program will take to run in various circumstances. This is critical, for example, in designing computer programs for a missile defense system.

In another important application of colored digraphs, each vertex represents a task to be accomplished in some project. Numbers represent the time required for each task, and the directed edges represent the required order of precedence among the tasks, where this is appropriate. The digraph is then a planning tool for determining the most efficient means of allocating resources and personnel. This is the heart of the management technique known as PERT (Program Evaluation and Review Technique). We shall examine this application by means of an extended, somewhat contrived example. Although the story may seem a bit facetious, the process it portrays can be of serious value in planning and managing large, complex projects.

Lester Leezure is a wealthy eccentric who lives on a remote private island in the Pacific Ocean. He contacts Manuel Werker, a contractor in California, to obtain an estimate for turning his unfinished basement into a recreation room. Lester wants a paneled, carpeted entertainment space, including a wet bar. Manuel faxes Lester an itemized estimate for the cost of materials and for the cost of labor based on the time needed for each task; this itemization is listed in *Table 8.6.*

Lester is not bothered by the cost of the materials, but he expresses some concern about the projected time span for completing the project. Manuel explains that the labor estimate is based on person-days of work. For reasons of safety and availability, he always uses two-person crews, and he plans to have the job finished in ten days. Lester still has a problem, though; he would like to minimize the inconvenience to his home and habits and get the work done as quickly as possible. "If one person can do the job in twenty days, or two can do it in ten days, why not employ ten workers and get the job done in two days?" he asks.

Manuel points out that increasing the number of workers does not automatically reduce the time proportionately—too many people often just get in each other's way. Through years of experience, he has found that putting more than two men to work on any task becomes counterproductive.

"Even so," Lester retorts, "the most

Task	Days
A. Frame walls	3
B. Panel walls	4
C. Hang acoustic ceiling	3
D. Install wiring	2
E. Install electrical fixtures	1
F. Do plumbing	3
G. Install wet bar	2
H. Lay carpet	2
Total	20

Table 8.6 Tasks for renovating Lester's basement.

time-consuming task is the paneling, which should take two workers only two days; the rest of the work could be done by other workers at the same time."

Manuel (by now contemplating physical violence) points out that it would not be efficient to complete all of the tasks using different people simultaneously. For example, the paneling can only be done after the framing, and the carpeting should not be laid while other work is still being done. To put it in black and white for Lester, the contractor produces another list, which is shown in *Table 8.7*.

Before Starting Task	We Must Complete Task
B	A, D, F
D	A
E	A, B, C, D
G	F
H	A, B, C, D, E, F, G

Table 8.7 *Task ordering for the basement project.*

"Well, how quickly can you do the job?" asks Lester.

"I've never worked that way before," replies Manuel in exasperation, "so I don't know. If you want to pay for them, I'll put on a larger crew and we'll find out how long it takes. But since you live so far from civilization, once a crew is assigned to your job, they'll have to stay with it for the whole time; I can't afford to ferry them back and forth."

Impatient but uncertain, Lester decides to compromise. He agrees to pay for three two-person crews—a total of six workers—in the hope that the job can be done in a little more than three days. Manuel, who is honest and competent, studies the job carefully and puts his people to work in the most efficient ways he can find, and none of the individual tasks takes longer than its estimated time. Nevertheless, the job takes six days! Needless to say, Lester is upset about having to pay for 36 person-days of labor for a job estimated at 20 person-days, and he complains about the amount of time some of the workers sat around doing nothing. Manuel is upset because the extra labor time seems to cast doubt on his ability and his integrity. If only they had consulted a graph theorist, they would have avoided all this wasted money, bad temper, and distress!

You see, a digraph and PERT can be used to come up with an answer to Lester's question. Using the contractor's two charts, we blend the information there into a single digraph, shown in *Figure 8.42* on page 448. The vertices represent the tasks, and the numerical "coloring" records the time each task will take; the directed edges convey the required precedence relations from *Table 8.7*.

Determining the time to do the construction job translates into adding up the numbers in the graph. The various paths in the graph correspond to *possible* different workers (or pairs of workers, in half the time) accomplishing different sequences of tasks. Several distinct, intertwining sequences of tasks must be carried out in order to accomplish the job—specifically, *CEH*, *ADBEH*, *FBEH*, and *FGH*. The total lengths (in days) of these paths are 6, 12, 10, and 7, respec-

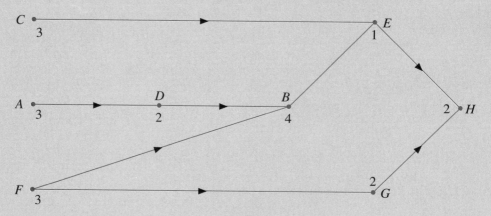

Figure 8.42 *The digraph for Lester's basement renovations.*

tively. The longest path, *ADBEH*, is called a **critical path**; its length sets a lower limit for the time in which the total job can be finished.

This graph shows Lester that the shortest work-time he can hope for is six days (with a pair of workers performing, in order, the tasks on the critical path while others work to do the remaining tasks). In fact, Manuel had done the most efficient scheduling job possible for three pairs of workers. He began with each pair of workers simultaneously doing tasks *A*, *C*, and *F*. However, when they finished their first tasks (after a day and a half), while one pair proceeded to *D* and another pair to *G*, the third pair had to sit around idle. Overall, one possible optimal schedule for a three-pair crew is *ADBEH* for one pair, *FG* for another, and *C* for the third; the total job takes six days, with a lot of wasted time for most of the workers. Moreover, because there are only three separate paths at any stage of the graph, adding a fourth pair of workers would have made no time difference whatsoever!

This little bit of graph theory could have saved Lester a lot of money by convincing him that he would lose no time using a two-pair crew: One pair would follow the tasks on the critical path while the other pair would do task *F* first, then *C* and *G*, in either order. Again the total job would take the minimum time of six days (there is no way to escape the maximal length of a critical path), but Lester would have cut his labor cost to 24 person-days. The critical-path analysis shows that to minimize time, Lester could do no better than six days, with some workers necessarily idle some of the time. To minimize cost, the workers would have to be fully employed at all times. This can only be done by using two workers for the entire job, doing the tasks in any order that does not violate the precedence requirements; as Manuel had said, this would take ten days.

One final observation about this case is instructive: If the second pair of workers (in the minimum-time solution described) made the mistake of doing task C before F, they would force a delay for the other pair because the latter, after two and a half days on tasks A and D, could not begin task B. Even with the assistance of a digraph, there is more to efficient planning than just identifying the critical path.

Generalizing from this example, it is not really surprising that putting more workers on a job does not necessarily translate in a simple, proportional way into getting the job done faster. The example also makes clear that there are times when putting more workers on a job has absolutely no effect on the total length of time needed to do the job. This particular example does not include some other counterintuitive features, such as the fact that assigning available workers successively to tasks in the longest critical path not yet begun is sometimes a very inefficient way of scheduling, or that efficiencies in performing individual tasks (such as reducing the time to do the paneling, lay the carpet, etc.) can sometimes actually increase the total time needed to do the job. Nevertheless, it does give us some idea of how digraphs are used to represent information in a scheduling problem, of the limits set by critical paths, and of the fact that, even with a digraph, determining a schedule that optimizes time or cost can still be a challenge. And ours was a trivial example. Imagine, for example, the problem of scheduling the more than 20,000 tasks in the Apollo space mission!

TOPICS FOR PAPERS

1. Investigate the potential use of dual graphs for pursuing the problem of characterizing those graphs that contain vertex paths. How does the problem carry over to its dual? How do the concepts defined in Section 8.4 transfer to their dual forms? Create appropriate terminology and give precise definitions. Look for insights into solving the dual problem; if you identify any, how do they translate back to the original problem via the dual process? Give your assessment of the value, real or potential, of the use of dual graphs to solve the original problem.

2. Generalize Euler's Formula for two-dimensional polygonal graphs to three-dimensional "polyhedral graphs." Note that there are four quantities to count: vertices, edges, faces, and three-dimensional regions separated by faces. Make any necessary definitions to clarify your discussion; consider any appropriate restrictions (for example, what would correspond to the restriction "polygonal"?); generate data from several examples; formulate a conjecture; if possible, generalize the proof of Euler's Formula to prove your conjecture. There may be different answers, depending on how you establish restrictions and define quantities; don't try to pursue all possibilities, but carefully explain the generalization you see.

FOR FURTHER READING

1. Behzad, Mehdi, Gary Chartrand, and Linda Lesniak. *Graphs & Digraphs*, 2nd ed. Monterey, CA: Wadsworth & Brooks-Cole, 1986.

2. Biggs, N. L., E. K. Lloyd, and Robin J. Wilson. *Graph Theory 1736–1936*. Oxford: Clarendon Press, 1976.

3. Chartrand, Gary. *Introductory Graph Theory*. New York: Dover Publications, Inc., 1985.

4. Devlin, Keith. Mathematics: *The New Golden Age*. London: Penguin Books, 1988, Chapter 7.

5. Flores, Ivan. *Data Structures and Management*. Englewood Cliffs, NJ: Prentice Hall, Inc., 1970, Chapter 2.

6. Harary, Frank. *Graph Theory*. Reading, MA: Addison-Wesley Publishing Company, Inc., 1969.

7. Maldevitch, Joseph, and Walter Meyer. *Graphs, Models, and Finite Mathematics*. Englewood Cliffs, NJ: Prentice-Hall, Inc., 1974.

8. Steen, Lynn Arthur, ed. *Mathematics Today, Twelve Informal Essays*. New York: Springer-Verlag, 1978.

9. Tutte, W. T. *Graph Theory*. Menlo Park, CA: Addison-Wesley Publishing Company, Inc., 1984.

MATHEMATICS OF MACHINES: COMPUTER ALGORITHMS

WHAT IS A COMPUTER?

A computer is a machine that can accept, store, and manipulate data, perform arithmetic and logical operations without human intervention, and report the results of these operations. In today's world we are, literally, met at every turn by computers. When we flip on a light switch, we are tapping into a power system that is constantly monitored and controlled by computers. The utility company maintains records of its thousands of customers by computer and sends out computer-generated bills. Every phone call we make engages a computer-controlled switching system. For many of us, the car we drive has computer components that report to us on fuel, battery, headlights, and other functions. We buy our groceries at supermarkets with electronic sensors that "read" a code on each package, consult a computer for the correct price and for inventory control, and print and total our bill. We use hand calculators; we play video games; we use credit cards; we receive letters that have been "personalized" by word processors. In these and dozens of other instances, we are interacting with computer technology, either directly or indirectly.

We tend to think of computers as a very recent phenomenon, and certainly ours is the age of computers. A more accurate perspective, though, recognizes that the history of computing machines is extremely old although the technology has mushroomed in the last twenty to

Abacus. This ancient computation device is still used in parts of the world today.

Credit: Smithsonian Institution

thirty years. Man's oldest aids in counting and computing are his fingers, which are almost certainly the motivation for our decimal system. And even if we discount this example as not being an external or mechanical aid to computation, we must allow the *abacus* as a bona fide instance of a computing machine, one that dates from approximately 3000 B.C. and is still in use in parts of the world today.

There are hints throughout early recorded history of attempts to devise machines that would calculate, though there is little evidence of success. Contact between European and Oriental cultures in the Middle Ages introduced not only the Hindu-Arabic numeration system to the Western world, but also the abacus. In the thirteenth century, Ramon Lull borrowed ideas from the *zairja,* an Arab "thinking machine," to design the *Ars Magna,* a partly mechanical, partly logical scheme that attempted to arrive at truth in a "machine-like" way.

In 1617, Lord Napier, the originator of logarithms, devised a calculating device now known as *Napier's Bones,* which utilized a set of marked sticks made from bone. This device was the precursor of the *slide rule,* which was developed by William Oughtred in 1630. The first mechanical adding machine was built in 1642 by mathematician and philosopher Blaise Pascal, and was improved and extended to handle multiplication and division later in the seventeenth century by Gottfried Leibniz, one of the discoverers of the calculus. (See Section B.6. of the Appendix.)

Certainly one of the most illustrious niches in the computer hall of fame must go to Charles Babbage (1792–1871). An outstanding mathematician, Babbage was fascinated by machines. In 1822, he built a small working model of his *Difference Engine,* a mechanical device used to compute and print tables of logarithms.

Napier's Bones. *At left:* Napier's Reckoning Board. *At right:* Napier's Reckoning Rod.

Slide Rule.

The British government was very interested in this machine because of the importance of precise tables for accurate navigation, and it supported Babbage's attempts to produce a larger machine. But Babbage became interested in an even grander scheme, his *Analytical Engine.* This machine was a general-purpose computing machine, remarkably like today's computers in its basic design. It had a section devoted to control, another section for logical and computational processes, and one for memory (of fifty-digit numbers). The design called for *punched cards* to be used in the computing section to control the choice of the operations to be performed, and other cards were to control the transfer of data to and from the memory. Babbage's idea of using punched cards

The Pascaline. An early mechanical calculator invented by Blaise Pascal.

Credit: IBM Archives

Leibniz Calculator.

Credit: IBM Archives

Key-punch. This is now an obsolete means of input.

Jacquard Loom.

Credit: IBM Archives

was borrowed from Joseph Jacquard, who had used them successfully to control weaving machines at the turn of the century (1801).[1]

For about ten years, Babbage was assisted in his work by Augusta Ada Lovelace (1815–52), the only child of the marriage between the English poet George Gordon, the sixth Lord Byron, and Anna Isabella Milbanke. While translating a French paper about Babbage's Analytical Engine, Lovelace expanded the work with copious notes which explained and extended the ideas of the original paper. In her notes, she developed in detail the idea of "programming" the machine. Because of this work, Ada Lovelace is considered to be the inventor of computer programming.[2]

1. Jacquard's punched-card idea was also the inspiration for Herman Hollerith, who revised the idea in 1887 and adapted it for use in counting machines for the 1890 U. S. census. Hollerith's machine was the prototype for others in his Tabulating Machine Company, which later became International Business Machines. Hollerith's cards were used by many IBM machines until the early 1980s.

2. As a tribute to her pioneering work, the programming language Ada, developed in the 1980s, is named after her.

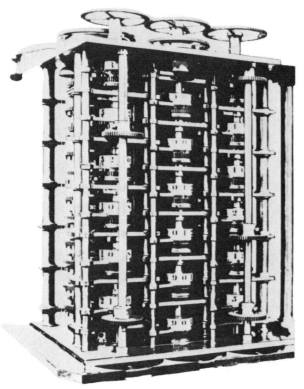

Credit: IBM Archives

Charles Babbage.

Babbage's Difference Engine.

Credit: IBM Archives

The difference between the Analytical Engine and twentieth century computers is not in any essential component, but, rather, in their forms. The Babbage machines were mechanical rather than electronic, and relied on a monstrous collection of gears, cams, etc.—all driven by steam power. This proved to be an insurmountable stumbling block, as the state of the art in the production of standardized-precision machine parts was insufficient to match Babbage's vision. Consequently, the Analytical Engine was never built.

Many of the advances in the precision and sophistication of machines that ultimately made computing machines a reality were made in the nineteenth and twentieth centuries. The first practical typewriter was patented by Samuel Soule in 1868. Ten years later, W. T. Odhner devised the "Odhner wheel," which played a role in the development of the adding machine and the cash register of the following decade. Important, too, was the theoretical work of Sir William Hamilton, Augustus DeMorgan, and George Boole. Boole's "An Investigation of the Laws of Thought on Which are Founded the Mathematical Theories of Logic and Probabilities," published in 1854, laid the mathematical groundwork for computer logic. Claude Shannon's 1937 master's thesis at M. I. T. applied Boole's ideas, now known as *Boolean algebra*, to describe the behavior of relay and switching circuits, and showed that electronic devices could mimic mathematical logic.

The Mark I.

Credit: IBM Archives

The first general-purpose computer, the IBM Mark I, developed by IBM and Harvard scientists in 1944, was more mechanical than electronic. Two years later, John Mauchly and J. Presper Eckert of the University of Pennsylvania completed construction of the first electronic computer, the ENIAC (Electronic Numerical Integrator And Calculator). And in 1949, John von Neumann's proposal that a computer could be controlled by a "stored program" was implemented at Cambridge University on the EDSAC (Electronic Delayed Storage Automatic Computer); thus, the century-old plan of Babbage was finally fulfilled. The first commercial computer was the Univac I, designed by Mauchly and Eckert and sold to the U. S. Census Bureau in 1951.

These "first generation" machines depended on vacuum tubes and relays to store and process data. They were large, they consumed sizable amounts of energy, and they were subject to frequent component failures. The transistor, invented at Bell Laboratories in the early 1950s, was successfully used in computer design by the end of that decade, thus marking the "second generation" of computers. The "third generation" machines, first introduced in the mid-sixties and still in use today, are those that employ integrated circuitry. The past thirty years have seen extensive refinements in solid-state circuitry, especially microminiaturization of integrated circuits. This has led from the

Credit: Smithsonian Institution

The ENIAC.

Courtesy of AT&T

From Vacuum Tubes to Transistors to Integrated Circuits.

minicomputers of the sixties and the microcomputers of the seventies to the work stations, desktops, laptops, and palmtops of the eighties and nineties.

Computers have experienced continued growth in speed, power, and storage capacity. For example, in 1986, a typical personal computer had magnetic (hard disk) memory that could store 5–10 million **characters** (letters of the alphabet, punctuation, or digits). By 1995, it was becoming more common to have personal computers with a storage capacity of 500–1000 million characters. Another storage medium, CD-ROM (the same compact disc that is used by the music industry), which can store 300 million or more characters, is becoming a standard way of storing and shipping large amounts of data. But while computer technology has been expanding, the size and cost of computers have shrunk by similar factors. One recent book summarizes this explosive growth by saying:

> There has never been a technology in the history of the world that has progressed as fast as computer technology. We could, for instance, compare advances in computer technology with those in the automotive industry: If automotive technology had progressed as fast as computer technology between 1960 and today, the car of today would have an engine less than one tenth of an inch across; the car would get 120,000 miles to a gallon of gas, have a top speed of 240,000 miles per hour, and would cost $4.[3]

This trend is not expected to change in the foreseeable future. In fact, instead of slowing down, this same tremendous growth is now expanding to other areas, such as intercomputer communication. Initially, most computers were isolated machines that did their work independent of one another. However, in the early 1970s, the Advanced Research Project Agency (ARPA), which was run by the U. S. Department of Defense, started a project in which computers would be linked (or "networked") together, allowing them to send messages to each other and to share data. This project led to the development of the ARPANET, a network connecting several dozen computer sites nationwide. Since then, through the support of the National Science Foundation (NSF), this network has evolved and joined with other networks to form the Internet of today, a worldwide network of computing sites.

The Internet has grown explosively in the last few years. For example, the number of users at the end of 1994 was estimated to be 35 million; by the year 2000, it is estimated that there will be 180 million

3. R. Decker and S. Hirschfield, *The Analytic Engine*, page 17 of Item 1 in the list *For Further Reading* at the end of this chapter.

users. The White House is now connected to the Internet, and from mid-1993 to mid-1994, it received over 200 thousand messages over the Internet.[4] In June, 1994, about 15 trillion characters of information, or an average of 3 million characters per second, 24 hours a day, was transmitted along the NSF backbone of the Internet. A year earlier, the NSF backbone carried less than 7 trillion characters per month. These figures do not include "local traffic," that is, traffic between computers on the network which are close enough so that they do not need the backbone in order to communicate.

In summary, we are truly in the midst of a "computer revolution," and, as is common with other revolutions, the growth and change are so rapid that it is almost impossible to predict what is in store for even the immediate future. However, mathematics can provide some insight into what may be possible and what may not be possible in certain areas of computing. For example, there are some problems that seem tailor-made for computer solution, but which, ironically, appear to be unsolvable by any computer in a period of time briefer than millions of years. One purpose of this chapter is to examine such problems, explaining what they have in common and why they appear to be so intractable.

9.1 EXERCISES

WRITING EXERCISES

1. When you hear the word "computer," what are the first ideas or images that come into your mind? Write a paragraph exploring this theme.

2. What is the literal meaning of the word *computer*? Look up the word in a dictionary that was published before 1940. How does that meaning relate to the way the word is used in this electronic age?

3. Are *compute* and *calculate* synonyms? Are *computer* and *calculator* synonyms? Explain your answers; then comment on any aspect of this question that you find interesting.

4. Find out whether or not your school has a connection to the Internet. If so, find out how to use it to send a message to one of the authors of this text, and then do so. (The authors will try to acknowledge as many messages as they can.) The author's e-mail addresses are:

   wpberlin@colby.edu
   grant@scsud.ctstateu.edu
   djskrien@colby.edu.

5. Compare the hard-drive capacity, price, and speed of several computers purchased at different times by your school to see whether the capacities and speeds have increased while the costs have gone down. Based on your findings, write a paragraph in which you discuss your estimate of the capacity of hard drives and the cost and speed of computers two years from now.

---

4. The first Internet mail message by a U. S. President was sent by Bill Clinton on March 2, 1993.

THE TRAVELING SALESMAN PROBLEM

The *Traveling Salesman Problem* is a mathematical problem with roots more than a hundred years old, although it was not until the 1930s that mathematicians gave it this name. *Figure 9.1* illustrates a simple, but typical, instance of this problem. You are given five cities, *A, B, C, D, E* (represented by circles), and roads between the cities (represented by lines connecting the circles). Assume that the label on each line indicates the time (in hours) it takes to travel between the two cities along that road. For example, the road from city *A* to city *E* takes 3 hours to travel. (The actual lengths of the lines are irrelevant, as are the positions of the cities and the fact that some of the roads cross one another.) Now pretend that you are a salesman who lives and works in city *A* and that you need to travel to the other four cities to sell your wares before returning home. Furthermore, you are allowed to visit each city only once. Any *path* that starts at one city and visits all the other cities exactly once before returning to the starting city is called a **tour**.

EXAMPLE **9.1**

You could visit the cities shown in *Figure 9.1* in the order *A–B–C–D–E–A*, in which case the total time of your traveling would be $6 + 7 + 4 + 6 + 3 = 26$ hours. However, to minimize your travel time, you might be interested in finding a tour with shortest total time. For this collection of cities, a shortest tour has a total time of 24 hours. Exercise 1 asks you to find such a tour.

When we talk of the Traveling Salesman Problem in general, we do not mean just this one situation involving the five cities shown in *Figure 9.1*. The Traveling Salesman Problem refers to the problem of finding a shortest tour through any number of cities with roads of any length connecting them. Therefore, to solve this problem, we need a general method for finding a shortest tour, one that will work regardless of the number of cities and the lengths of the roads. With this idea in mind, the Traveling Salesman Problem can be stated more formally as follows:

Traveling Salesman Problem (TSP)

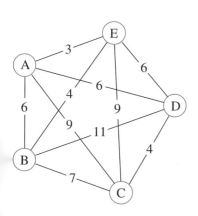

Figure 9.1 *An instance of the Traveling Salesman Problem.*

Input: A list of (any number of) cities and a table giving the travel times (in hours) between each pair of cities.

Problem: Of all the tours (that visit each city exactly once and return to the starting point), which one has shortest total time?[5]

Of course, traveling salesmen are not the only people interested in the TSP. For example, your mail carrier may want a solution to this problem when delivering the mail. That is, the carrier might like to know a shortest tour from the post office to all the homes on her route, ending with a return to the post office.

We can also interpret the circles to be other kinds of objects rather than cities and the lines to be other means of connecting the objects instead of roads. Thus, the circles could represent goods that need to be produced on one assembly line in a factory. The lines and numbers could represent the cost of switching the assembly line from the production of one type of good to that of another. The factory manager is then faced with the task of deciding in which order the goods are to be produced in order to minimize the total cost of conversion.

The TSP has many other applications as well, including the following:

1. The wiring of modules in a computer sometimes involves finding a shortest path for the wires connecting various modules.

2. In x-ray crystallography, a crystal is mounted on a diffractometer, which has a detector that is placed successively in thousands of positions around the crystal to take readings of x-rays scattered by the crystal. The positioning of the detector takes a significant amount of time; the operators want the optimal ordering of the positions so as to minimize the total time involved in repositioning the detector.

3. Machines that punch or drill holes in boards or metal sheets need to reposition the drill after completing each hole, and this can take a significant amount of time. The preferred ordering of the punching or drilling is the one that takes the minimum amount of time.

4. The TSP even shows up in problems involving wallpaper cutting and the design of dartboards. (The Link section at the end of this chapter describes these two applications.)

Items 11, 12, and 13 in the list *For Further Reading* at the end of this chapter include more details about these four applications and several others.

---

5. If you have studied Chapter 8, then you should notice that we are really asking a question about a graph. In fact, what we are calling a *tour* is called a *closed vertex path* in that chapter.

Let us now consider a general method for solving the TSP. The method that systematically checks all possible tours to see which one is shortest is certainly conceptually simple. This method is called **exhaustive search**. When the number of cities and roads is small this can easily be done with pencil and paper. However, with 100 objects and 4950 connections between them, it is no longer practical to attempt the problem in this way. That is when a computer comes in handy.

At first glance, the TSP seems like an ideal type of problem for a computer to solve. After all, numbers have to be repeatedly added (to find the total time of each possible tour) and these various sums have to be compared (to find a tour with the shortest total time). In fact, it is not too difficult for beginning programmers to write a program that will solve the TSP using exhaustive search. Such programs work well with small numbers of data, but they tend to bog down with larger volumes of data—taking days, weeks, or even years of computation to solve a problem involving only 20 cities. In short, although exhaustive search programs solve the problem correctly, they take far too much time, as we shall see in later sections of this chapter. (To get a sense of why this might be the case, do Exercises 12, 15, and 17.) Therefore, a program that uses a better method than exhaustive search is necessary for solving the problem. An *ideal* program would come up with a shortest tour quickly, that is, after only a few hours or at most one or two days of computation, even for problems with 1000 or more cities. A program that can solve the traveling salesman problem quickly is what we mean by an "efficient" program. However, *no one has yet been able to create such a program for the TSP*. In fact, the fastest programs in existence for solving the TSP may take many millions of centuries to solve problems involving only 100 cities! You will see why this situation exists by the time we have finished the chapter.

9.2   EXERCISES

1. Find a tour of the five cities represented in *Figure 9.1* that has the shortest total time, namely, 24 hours.

2. In the text it is mentioned that the tour *A–B–C–D–E–A* in *Figure 9.1* has a length of 26 hours. What happens if you reverse this tour? That is, how does the length of the tour *A–E–D–C–B–A* compare with the length of the tour *A–B–C–D–E–A*? When you solve the TSP, is there only one tour with the shortest total length?

3. Suppose there are four cities and the length of every road connecting a pair of the cities is 10. Find a shortest tour. What if there are five cities and the length of every road is 10? How about six cities?

4. Suppose there are four cities *A*, *B*, *C*, and *D*, and all the roads have length 10 except the road from *A* to *B*, which has length 5. Find a shortest tour of these cities. Find a different shortest tour of these cities. What do all shortest tours have in common?

5. Redo Exercise 4 assuming that the road from A to B has length 15.

6. Suppose there are five cities A, B, C, D, and E, and all the roads have length 10 except the roads from A to E and from B to C, each of which has length 5. Find a shortest tour of these cities. Find a different shortest tour of these cities. What do all shortest tours have in common?

7. Redo Exercise 6 assuming that the roads from A to E and from B to C have length 15.

8. Suppose there are five cities A, B, C, D, and E, and all the roads have length 10 except those to city E, all of which have length 5. Find a shortest tour.

9. Redo Exercise 8 assuming that the roads to city E all have length 15.

10. Suppose there are five cities A, B, C, D, and E, and all the roads have length 10 except for the roads to city E. The road from A to E has length 1, the road from B to E has length 2, the road from C to E has length 3, and the road from D to E has length 4. Find a shortest tour.

11. Suppose there are four cities A, B, C, and D. Create roads between them so that these six roads have lengths 1, 2, 3, 4, 5, and 6, and so that a shortest tour is A–B–C–D–A.

Exercises 12–14 refer to *Figure 9.2*.

12. Find all the tours of the four cities in which you start each tour at city A. How many such tours are there?

13. Solve the TSP. That is, find a shortest tour that starts at city A, visits each city exactly once, and returns to A.

14. Start and end your tours at city B instead of A. What is a shortest such tour? How does this relate to a shortest tour you get by starting and ending at A? Describe in general the effect of starting tours at different cities.

Exercises 15 and 16 refer to *Figure 9.3*.

15. Find all the tours of the five cities starting at A. How many tours are there?

16. Solve the TSP. That is, find a shortest tour that starts at city A, visits each city exactly once, and returns to A.

Exercises 17 and 18 refer to *Figure 9.4*.

17. Find all the tours of the six cities starting at A. How many tours are there?

18. Solve the TSP. That is, find a shortest tour that starts at city A, visits each city exactly once, and returns to A.

19. In *Figure 9.2*, change the length of time of the road from A to B from 10 to 8. Will that create a new shortest tour?

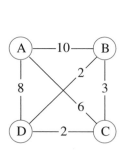

Figure 9.2 *Cities and roads for Exercises 12–14.*

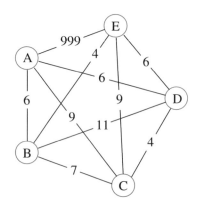

Figure 9.3 *Cities and roads for Exercises 15 and 16.*

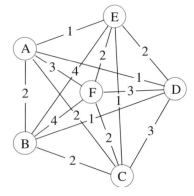

Figure 9.4 *Cities and roads for Exercises 17 and 18.*

20. In *Figure 9.3*, change the length of time of the road from *A* to *B* from 6 to 1. Will that create a new shortest tour?

21. Suppose the cities are *A*, *B*, *C*, *D*, and *E*. Create roads between them so that the shortest tour (and its reversal) has length 10 but no other tour has length less than 20.

22. Suppose the cities are *A*, *B*, *C*, *D*, and *E*. Create roads between them so that there are exactly two shortest tours (in addition to their reversals).

23. In the TSP, there must be a road between every two cities. What happens if at least one such road does not exist? For example, suppose that there is no way to get directly from city *A* to city *B*. What effect does this restriction have on the problem? What can you do to get around the restriction?

24. Suppose you are not interested in finding a shortest tour, but rather are interested in finding a tour in which no road takes "too long" to travel. For example, in *Figure 9.2*, is there a tour that uses no road for which the time is 7 hours or more? In *Figure 9.4*, is there a tour that uses no road for which the time is 3 hours or more? (If you have already studied Chapter 8, describe how this problem relates to Hamilton graphs, which are discussed in that chapter.)

25. (This problem requires knowledge of some of the material in Chapter 8.) What is the difference between the TSP and the problem of determining whether a graph is a Hamilton graph? (*Hint*: Do the edges of Hamilton graphs have lengths of time? Are there edges between every pair of vertices in Hamilton graphs?)

26. The travel time along roads has always been given in hours. Would it make any difference in determining the shortest tours if the time were given in any other unit, such as minutes or years?

WRITING EXERCISES

1. The method of exhaustive search systematically checks all tours to find a shortest one. As part of this process, the method requires knowing when all possible tours have been found and checked. Try to come up with and describe a systematic way of finding all tours. Your method should work with any number of cities. Do not worry about how much time your method takes, that is, about how efficient your method is. Use some of the problem-solving techniques from Chapter 1. Also, the material in Section 4.4 might be helpful.

2. Try to come up with and describe a systematic way of solving the TSP in general (not just for problems with four or five cities) other than by exhaustive search. Do not worry about how much time your method takes; that is, about how efficient your method is. Use some of the problem-solving techniques from Chapter 1.

3. Choose one of the applications of the TSP (other than touring cities) mentioned in this section, and write a paper describing in detail how the TSP applies to it.

4. The TSP is useful in contexts other than the ones mentioned in this section. Think of another application and describe how it is modeled by the TSP. What features in your application correspond to cities, roads, and distance in the TSP?

THE SPEED OF A COMPUTER

9.3

Computers are very good at executing repetitive, simple instructions. A typical computer instruction does one job, such as adding, subtracting, multiplying, or dividing two numbers, moving a number from one place in the computer's memory to another, or storing or loading data to or from memory.

EXAMPLE 9.2

In the programming language BASIC, an acronym for Beginner's All-purpose Symbolic Instruction Code, you can give the computer the instructions

```
LET A = 3
C = A + B
IF A < B THEN LET C = 3
GOTO 600
```

In these sample instructions, the letters A, B, and C refer to particular locations in the computer's memory. The command "LET A = 3" tells the computer to store the number 3 in location A. The command "C = A + B" tells the computer to find the numbers in locations A and B, add them together, and store the result in location C. The command "IF A < B THEN LET C = 3" tells the computer to check locations A and B to see whether the value in A is smaller than the value in B. If so, the computer should put the value 3 into location C. Finally, the command "GOTO 600" tells the computer to jump to the command numbered 600 and execute it next. As you can see, each instruction does little by itself; but a sequence of such instructions, called a **program**,[6] can do virtually every kind of computation that humans can do.

The speed at which these computations are performed is set by the computer's internal clock. Each cycle of this clock typically sets in motion the electronic signals that allow one computation to be done. Therefore, a computer that operates at a clock rate of 50 MHz (megahertz or million cycles per second), a typical speed for desktop computers in the mid-1990s, can execute up to 50 million instructions per second.[7] This incredible speed at which computers operate is what enables them to be used in the mind-boggling array of applications

---

6. See also the definition given on page 476.

7. Actually, it should be noted that each of the BASIC instructions mentioned in Example 9.2 may take more than one clock cycle because in order for the computer

mentioned at the beginning of Section 9.1. It is useful to remember that such applications were unthinkable before the advent of small, cheap, dependable, fast computers.

However, despite all of the advances in computer technology—in particular, despite the continued increase in the speed at which computations can be done—computers are still notoriously bad at doing certain things, such as understanding spoken and written commands. In fact, aural and visual pattern recognition, things that humans can do almost instinctively from birth, are among the hardest things for modern computers to do. Other problems that computers are hard-pressed to solve include the analysis of complex systems, such as weather conditions. Thus, in many ways, weather prediction is not significantly better than it was 40 years ago, before computers began helping to gather and analyze climatic data.

What is the problem with computers? A human brain, unlike a computer, has billions of neurons that are able to be fired in parallel. Consequently, millions of the neurons can receive and send signals at the same time, and so the human brain can process thousands or millions of signals (through the eyes, ears, and other organs) simultaneously. In contrast, most computers have only one "brain cell," called the **central processing unit** or **CPU**, which can only process a small amount of data at a time. Computers must then do computations sequentially. This difference between the human brain and computers has led computer scientists to study "neural networks," in which computers imitate the neurons of the human brain. With rapid advances in neural-network computers and in massively parallel computers that use thousands of CPUs instead of just one, the computer may soon be able to do successfully the kinds of things humans now do instinctively.

It is also useful to keep in mind that since the mid-1970s, computer hardware has doubled in power roughly every year. This yearly doubling is called **Moore's Law**. Some of the factors that have contributed to this growth include miniaturization of computer components, increased speeds of computer clocks, more efficient execution of instructions by

The Motherboard of a Power Macintosh 8100. The CPU and main memory reside on this board.

to understand and execute the instruction, it must execute several—often ten or more—low-level instructions.

the CPU, and increased use of parallel computation. If Moore's Law continues to hold in the years ahead, then the power of the computer of the future will be truly awesome. In particular, it would be easy to believe that nothing will stand in the way of the computer as a problem solver. For example, if present-day computers are too slow to solve some problems because they take ten times as long as you would like, then, according to Moore's Law, you only have to deal with this slowness for three to four more years (why?), at which point computers will be able to solve your problems in the desired amount of time.

However, despite all this promise, there are some fundamental limitations on the abilities of computers. In some situations, it is not enough to have computers that are two or four or even 4 million times more powerful than present-day computers. We will explore these limitations in the upcoming sections. When we are done, we hope you will have a feeling for why computers have not been efficient and probably will never be so in solving some problems, such as the Traveling Salesman Problem, for which they seem perfectly suited.

9.3	**EXERCISES**

1. Specify one aspect of your life that is not affected in any way by computers and find ten aspects of your life that are affected dramatically by computers.

2. If computer power doubles every year, how many years will it take for computers to be 32 times more powerful than they are now? 100 times more powerful? 1000 times more powerful?

For Exercises 3–5, which require the use of a calculator, suppose you have a computer that executes 25 million instructions per second. How long would it take to execute programs consisting of the indicated numbers of instructions? Which, if any, of these programs would be practical? Why?

3. (a) 10^5 instructions?

   (b) 20^5 instructions?

   (c) 100^5 instructions?

4. (a) 2^{10} instructions?

   (b) 2^{20} instructions?

   (c) 2^{100} instructions?

5. (a) 10! instructions?

   (b) 20! instructions?

   (c) 100! instructions?

   (For positive integers n, the number $n!$ is defined by
   $$n! = n \cdot (n - 1) \cdot (n - 2) \cdot \ldots \cdot 3 \cdot 2 \cdot 1$$
   It is discussed in more detail in Section 4.4.)

Miniaturization of computer components is important in speeding up computers because this allows electrical signals to travel shorter distances, and therefore take less time. In answering Exercises 6–10, use the fact that the speed of electrical signals is nearly the same as the speed of light, which is approximately 186,000 miles per second.

6. If a computer executes with a clock rate of 25 MHz, then each low-level instruction has only $\frac{1}{25,000,000}$ second in which to execute. How far can an electrical signal travel in $\frac{1}{25,000,000}$ second?

7. If a computer executes with a clock rate of 66 MHz, then each low-level instruction has only $\frac{1}{66,000,000}$ second in which to execute. How far can an electrical signal travel in that time?

8. If a computer executes with a clock rate of 110 MHz:

   (a) How long does it take for each low-level instruction to execute?

(b) How far can an electrical signal travel in that time?

9. If a signal needs to travel only one inch to execute one instruction, potentially how many instructions per second can be executed?

10. If a signal needs to travel only one thousandth of an inch to execute one instruction, potentially how many instructions per second can be executed?

11. Parallel computers contain two or more CPUs, and so can execute several instructions at once. Indicate a problem in which executing two or more instructions at a time would be useful and also one in which this would not be useful.

WRITING EXERCISES

1. What is a computer program? Can you give a precise definition? (What makes a definition precise?)

2. Do some research on the speed of the computers at your school. What are their clock speeds? How many instructions per second can be executed? Write a brief description of your research.

3. Do some research on how much better the weather service is at predicting the weather now that modern supercomputers are available to help. Write a one-page summary of your research.

4. Do some research on how well computers can perform tasks like speech recognition or visual pattern recognition. Write a one-page summary of your research.

ALGORITHMS AND SORTING

9.4

An **algorithm** is a formal step-by-step method for solving problems. An example of an algorithm is the grade-school method of performing long division with pencil and paper. To understand why computers are limited in what they can do, we need a method of analyzing computer algorithms that will allow us to compare the speed of one algorithm with that of another. In this section, we will develop two sorting algorithms to be used as examples. In later sections of this chapter, we will show how these algorithms can be analyzed and compared.

One of the most common operations performed by computers is sorting. For example, computers are continually called upon to sort lists of names alphabetically or lists of numbers in numerical order. There are many algorithms that computers can use for sorting. Some of these are fast and efficient, while others are slow and inefficient, and hence are not actually used in serious applications. Let us analyze two such algorithms, neither of which is the most efficient method of sorting; but both are simple to understand and to analyze, and are typical of the algorithms that we will study in this chapter.

In the examples that follow, we assume that the computer has been given a list of numbers that is to be sorted in numerical order from smallest to largest. We do not assume that the list is of a fixed length because we want our algorithm to be capable of sorting lists of numbers of any length. Thus, we assume that our list contains n numbers, where n is any positive integer. The *Sorting Problem* can be formally stated as follows:

Sorting Problem

> ***Input:*** A list of numbers which may be of any length.
>
> ***Problem:*** Rearrange the list so that the numbers are sorted from smallest to largest.

Before reading the next few paragraphs, try to think up an algorithm—a systematic procedure—by means of which you or a computer could sort a long list of numbers. Humans are so proficient at sorting a short list of numbers that it might be hard to pin down the series of steps we actually follow. It may help to apply the techniques you learned in Chapter 1. For example, you might first try to sort a list of only two numbers, then a list of three numbers, then a list of four numbers, and look for a pattern in these sorting methods. Then imagine using your method on a list of 1 million numbers to determine whether it would still work on such a long list. It might also help to think of three generic numbers, x, y, and z, and describe how you would prescribe a series of steps that, if followed, would invariably identify which of these is the smallest, which is the second smallest, and which is the largest.

One of the most obvious sorting procedures for lists of numbers is called *Selection Sort*. In this algorithm, a computer first searches the list for the largest value. Once this number is found, it is moved to the end of the list. Then the computer searches the rest of the list for the next largest number. When this number is found, it is moved into the second-to-last position, that is, just before the largest number. The computer continues this process of finding the largest remaining number and putting it in its proper place at the end of the list until all of the numbers have been sorted. Here is a summary of the algorithm:

Selection Sort

> ***Input:*** An unsorted list L of n numbers.
>
> ***Output:*** The list L with the numbers arranged in sorted order.
>
> ***Algorithm:***
>> Let $k = n$.
>> Repeat the following instructions n times:
>>> Find the largest of the first k numbers.
>>> Swap this number with the number in position k.
>>> Let $k = k - 1$.

Figure 9.5 shows how this algorithm might be written in the programming language Pascal. (It is not necessary to know Pascal for the remainder of this discussion.)

The series of instructions beginning with "Repeat the following instructions n times" forms a **loop**. The computer goes back to the

```
{ Procedure SelectionSort sorts the list of numbers A[1],}
{ A[2], A[3], ..., A[n].}
Procedure SelectionSort (n: integer; var A:
ListofNumbers);
var k,j,max,temp: integer;
begin
   for k := n downto 1 do
      begin
         {find the largest value from the first k numbers}
         max := 1;
         for j := 2 to k do
            if A[j] > A[max] then
               max := j;
         {swap this number with the number in position k}
         temp := A[k];
         A[k] := A[max];
         A[max] := temp
      end
```

Figure 9.5 The Selection Sort procedure written in Pascal.

beginning of the instructions repeatedly to reexecute them. In Selection Sort, this loop is executed n times.

The notation "Let $k = n$" means that k is a variable which is given the same value as n. Also, "$k = k - 1$" means that the value of k is to be decreased by one. Remember that n denotes the length of the list of numbers. The value of n does not change while the algorithm is being executed. The variable k indicates the length of the part of the list that still needs to be sorted. So, initially we set $k = n$ because all n numbers in the list still need to be sorted. After each execution of the loop, k is decreased by one because another number has been put in its proper place, leaving one less number to be sorted.

| EXAMPLE 9.3 | To make this algorithm clearer, let us write it out explicitly in the case in which our list consists of the three numbers: |

$$8, \ 27, \ 2$$

In this case, $n = 3$. Our method requires that the following be done:

Let $k = 3$.
Find the largest of the first 3 numbers.
   (27 is the largest of the 3 numbers.)
Swap this number with the number in position 3.
   (After swapping, our list becomes 8, 2, 27.)

Let $k = 2$.
Find the larger of the first 2 numbers.
   (8 is the larger of the 2 numbers.)
Swap this number with the number in position 2.
   (After swapping, our list becomes 2, 8, 27.)

Let $k = 1$.
Find the "largest" of the "first 1 numbers."
   (2 is the only remaining number and so is automatically the largest.)
Swap this number with the number in position 1.
   (2 is already in position 1; so our list remains 2, 8, 27.)

At this point the list is sorted and we are done. You will note that the last three steps (when $k = 1$) were unnecessary. You can easily eliminate them by a slight modification of the algorithm. (See Exercise 1 at the end of this section.)

The second sorting algorithm is called *Slow Sort*, for reasons you will understand shortly. Here is a summary of the algorithm:

Slow Sort

Input: An unsorted list L of n numbers.

Output: The list L with the numbers arranged in sorted order.

Algorithm:

Repeat the following instructions until a sorted list is found:
Arrange the numbers in a new order.
Check to see whether this ordering of the numbers is properly sorted, and if so, stop.

When we say, "Arrange the numbers in a new order," we do not mean that we want the computer to choose an arrangement *randomly*. Random selection would make the algorithm much harder to analyze and would also make it slower because the computer might randomly choose some arrangements two or more times and other arrangements not at all. Instead, we mean that the computer systematically checks all possible arrangements of the numbers in the list.

EXAMPLE 9.4

Let us see how this algorithm would behave using the sample list, 8, 27, 2, of the preceding example. These numbers can be rearranged in six possible ways:

(8, 27, 2), (27, 2, 8), (27, 8, 2), (2, 27, 8), (2, 8, 27), (8, 2, 27)

(Section 4.4 discusses rearrangements of numbers in more detail.) Let us assume that the arrangements are checked in this order. Here is a sample run of the algorithm:

Arrange the numbers in a new order.
(8, 27, 2 is the first order.)
Check this order, and if it is sorted, stop.
(It is not sorted properly; so we continue.)
Arrange the numbers in a new order.
(27, 2, 8 is the next order.)
Check this order, and if it is sorted, stop.
(It is not sorted properly; so we continue.)
Arrange the numbers in a new order.
(27, 8, 2 is the next order.)
Check this order, and if it is sorted, stop.
(It is not sorted properly; so we continue.)
Arrange the numbers in a new order.
(2, 27, 8 is the next order.)
Check this order, and if it is sorted, stop.
(It is not sorted properly; so we continue.)

Arrange the numbers in a new order.
(2, 8, 27 is the next order.)
Check this order, and if it is sorted, stop.
(This order is sorted properly; so we stop here.)

Note that we did not check the last arrangement of the numbers—namely, 8, 2, 27 because we happened to hit on the sorted arrangement before that final arrangement appeared.

There are two details of Slow Sort that deserve further comment. First of all, how does a computer check a list of numbers to see whether it is sorted? This question is left as Exercise 10. Secondly, how does the computer systematically come up with all possible arrangements of the numbers in the list? Unfortunately, this is a much harder question and one we cannot explore here. The interested student is encouraged to look in advanced programming books or in reference 9 in the list *For Further Reading* at the end of this chapter for more information. Luckily, for our purposes, you will see in the next few sections that this detail is actually of minor importance.

The important questions for us are the following: Which of these algorithms is better in the sense that it sorts lists faster? Is the difference in the speeds of the algorithms significant? To answer these questions, we need to develop a method of analyzing algorithms; that will be the subject of the next section.

A Macintosh Powerbook 520.
This is a modern notebook computer.

EXERCISES

1. Make one minor modification of the Selection Sort algorithm that eliminates the last three (unnecessary) steps, as seen in Example 9.3.

2. Think of four real-life situations in which computers are called upon to sort lists.

3. Suppose that a list of numbers that is already sorted is to be processed by each of the sorting algorithms discussed. Will either of these algorithms run faster in this case? Which one will and why?

4. Execute the steps of Selection Sort for this list of four numbers:

   13, 11, 19, 12

5. Execute the steps of Selection Sort for this list of five numbers:

   46, 2, 23, 1, 18

6. The letters A, B, and C can be arranged in six ways:

   ABC, ACB, BAC, BCA, CAB, CBA

   Suppose that A = 13, B = 2, and C = 6. Also, suppose that a computer checks the arrangements of these three numbers in the order given here when performing Slow Sort; that is, the computer checks the order ABC first, then ACB, then BAC, etc. Of these six different arrangements, how many does the computer check before stopping?

7. In how many ways can the letters A, B, C, and D be arranged? List them all.

8. In how many ways can the letters A, B, C, D, and E be arranged? You need not list them all.

9. In how many ways can the letters A, B, C, D, E, and F be arranged? You need not list them all.

10. Describe an algorithm that will take a list of numbers and check to see whether it is sorted. *Hint*: Think about how you would check by hand to see whether a list is sorted; then break your process into small steps (instructions) for a computer. One way to do it is to use a statement of the following form repeatedly:

    IF something < something else
    THEN do something

11. Suppose that just two of the three or more numbers on a list are equal. What impact does this have on each of the two sorting algorithms? Do the sorts still work? Does this affect the speed of either algorithm?

12. Suppose that all the numbers on a list are equal. What impact does this have on each of the two sorting algorithms? Do the sorts still work? Does this affect the speed of either algorithm?

13. Rewrite the Selection Sort algorithm so that it repeatedly finds the smallest number and moves it to the beginning of the list, instead of finding the largest number and moving it to the end. Is one version faster than the other?

14. Find two sorting algorithms other than Selection Sort and Slow Sort. Show how they work on a short list of three or four numbers.

15. On the lowest level, computers can actually deal only with numbers. In this case, how do you think computers sort lists of names alphabetically?

16. Suppose you and your roommate are given one long list of numbers to sort manually. Describe several ways that you could divide the labor. Assuming you are both equally fast at sorting, will the two of you finish in half the time, or would one of you have to wait for the other at any time during the sorting process? What if you had four equally fast people to work on the sorting? Would the list get sorted in one-fourth of the original time?

WRITING EXERCISES

1. Find out how computers *actually* compare two numbers. How does a computer know which number is bigger? Summarize your findings in a one-page paper.

2. Choose one of the exercises in this section that you found difficult and explain how the problem-solving tactics of Section 1.2 can be used to help in finding the solution.

COMPARING ALGORITHMS

9.5

Now let us compare the two sorting algorithms to see which is faster. An obvious way to do this is to convert the algorithms into programs and run both programs on a computer to see which one sorts a given list first. Here, a **program** means a specific list of steps, written in the language of a particular computer that implements an algorithm to run on that computer. (Think of a program as a translation of an algorithm into a language that a computer can understand.) A given algorithm can be implemented in many different ways, depending on the computer used, the choice of computer language (BASIC, Pascal, FORTRAN, etc.), and the programmer's methods for implementing various parts of the algorithm.

However, converting algorithms into programs which are to be run is not the best way to compare algorithms. To see why not, suppose we were to convert both sorting algorithms into programs and then run them on the same computer using a given list of numbers. Now, suppose that the program that implemented Selection Sort finished first. Can we conclude that Selection Sort is faster than Slow Sort? Unfortunately, *no.* The problem is that the speed of the programs depends on a number of factors besides the algorithms themselves:

1. There is more than one way to convert an algorithm into a computer program, just as there are many ways of expressing an idea in the English language. Some of these conversions yield programs faster than others. Therefore, the specific implementations of the algorithms that we use in the tests can affect the results. In other words, one algorithm may win the test run, not because the algorithm is better, but because the program into which it was translated is more efficient than the program for the other algorithm.

2. The particular order in which the numbers in the list are initially arranged can affect the test results. In particular, if the

list is already sorted before we start running the algorithm, certain algorithms will sort the list faster than others. For example, if Slow Sort is given an already sorted list, it need not rearrange the numbers at all; thus, it will finish much sooner than it would had it been given a list of numbers in random order.

3. One of these sorting algorithms may be faster than the other for short lists of numbers but slower for long lists. In that case, we can't say that one algorithm is *always* faster than the other.

It is hard to eliminate the first objection. Ideally, we should use the fastest implementation of each algorithm in our test. However, it is impossible to prove that among all possible (existing or future) implementations of an algorithm any one program is the fastest.

It is also hard to negate the second objection. We could insist that the list initially be in random order; but even then, some algorithms may arrive at the solution faster than others because they happen to run more efficiently on the particular random ordering chosen. Another way to deal with this particular objection would be to run both algorithms many times using various lists of numbers, and then to determine which one was faster *on average*. However, to obtain a truly representative answer, we would need to test the algorithms on so many lists that it would take a huge amount of time just to perform the tests. This is both inefficient and impractical.

The final objection is also an important one. To resolve it, we could run the algorithms on lists of different lengths, keeping a table that indicates which algorithm is faster at various lengths. Even this seems unsatisfactory.

As it turns out, there is a better way of comparing algorithms that avoids all of these problems by simply looking at the algorithms themselves and not worrying about the machine or program code used. In our case, we only want to know which is the faster of two algorithms. We do not care exactly how much faster it is or how many millionths of a second each takes to execute as a program. Therefore, we will simply count the number of "steps" necessary for execution of the algorithm. We can think of each step as one instruction, as discussed in Section 9.3.[8] The algorithm that takes fewer steps is the faster. This measure may at first appear somewhat sloppy; but it is still precise enough to allow for comparison of the general behavior of the two algorithms.

---

8. A more precise definition of "step" can be found using a *Turing machine*, a hypothetical simple machine that is sophisticated enough to be able to do any computation an existing computer can do. Item 3 in the list *For Further Reading* at the end of this chapter discusses Turing machines and their role in analyzing algorithms.

Let us analyze the Selection Sort procedure in order to count the number of steps.

- There is one step that initializes the variable k in order to give it the value n.
- There is a series of instructions beginning with "repeat the following instructions n times." These instructions form a loop that is executed n times.
- Each time through the loop, the algorithm needs to find the largest of the first k values. This process takes k steps.
- Then the algorithm makes a swap, which takes three steps. (Why three steps for the swap and not one or two? See Exercise 14.)
- Finally, there is one step to decrease the value of k by 1.

The first time through the loop there are n steps needed to find the largest value, 3 steps for the swap, and 1 step needed for decreasing the value of k. The second time through the loop there are $n - 1$ steps needed to find the next largest value, 3 steps for the swap, and 1 step for decreasing k. The third time there are $n - 2$ steps for finding the next largest value, 3 swap steps, and 1 step for decreasing k. The nth (or last) time through the loop there is only 1 step needed to find the largest value since there is only one number left, 3 swap steps in which the number is (wastefully) swapped with itself, and 1 step for decreasing k. The pattern should now be clear. The total number of steps is:

1	(initializing k to the value of n)
$n + 3 + 1$	(the first time through the loop)
$n - 1 + 3 + 1$	(the second time through the loop)
$n - 2 + 3 + 1$	(the third time through the loop)
. . .	
$1 + 3 + 1$	(the last time through the loop).

This total comes to:

$$1 + (n + 3 + 1) + (n - 1 + 3 + 1) + (n - 2 + 3 + 1) + \ldots + (1 + 3 + 1)$$

We can regroup by combining the 3 and 1 at the end of each parenthesis to obtain:

$$1 + (n + 4) + (n - 1 + 4) + (n - 2 + 4) + \ldots + (1 + 4)$$

There are n sets of parentheses in this expression and so there are n 4s in this sum. We can remove the 4s from the parentheses and put them all at the end of the expression to get:

(9.1)
$$1 + n + (n - 1) + (n - 2) + (n - 3) + \ldots + 1 + 4n$$

All of the terms of Sum (9.1) except for the first and last terms can be simplified using the formula for the sum of an arithmetic series:

$$1 + 2 + 3 + \ldots + n = \tfrac{1}{2}n \cdot (n + 1)$$

as demonstrated in Formula (1.6) of Chapter 1. For example, if $n = 4$, then Formula (1.6) says that

$$1 + 2 + 3 + 4 = \tfrac{1}{2} \cdot 4 \cdot (4 + 1) = \tfrac{1}{2} \cdot 4 \cdot 5 = 10$$

When we use Formula (1.6) to replace all but the first and last terms (with their order reversed) in our Sum (9.1), we see that the total number of steps is given by:

$$1 + \tfrac{1}{2}n \cdot (n + 1) + 4n = 1 + \tfrac{1}{2} \cdot (n^2 + n) + 4n$$

$$= 1 + \tfrac{1}{2}n^2 + \tfrac{1}{2}n + 4n$$

$$= 1 + \tfrac{1}{2}n^2 + \tfrac{9}{2}n$$

We restate this final result:

xxxxxxxxxxxxxxxxxxxxxxxxxxxxxxxxxxxxxxxxxxxxxxxxxxx

(9.2) The total number of steps required by Selection Sort to sort a list of n numbers is $1 + \tfrac{1}{2}n^2 + \tfrac{9}{2}n$.

xxxxxxxxxxxxxxxxxxxxxxxxxxxxxxxxxxxxxxxxxxxxxxxxxxx

EXAMPLE 9.5

To sort a list of $n = 3$ numbers, Selection Sort requires

$$1 + \tfrac{1}{2} \cdot 3^2 + \tfrac{9}{2} \cdot 3 = 1 + \tfrac{9}{2} + \tfrac{27}{2}$$

$$= 19 \text{ steps}$$

Now let us do the same computation for Slow Sort. Like Selection Sort, Slow Sort has a loop containing a series of instructions that are executed repeatedly. Each time through the loop, the Slow Sort algorithm generates a new arrangement of the numbers and then checks all the numbers in this arrangement to see whether they are sorted correctly. We shall ignore the time it takes the algorithm to come up with a new arrangement of the numbers. It is not trivial to generate such arrangements, but we will see shortly that it is irrelevant to our discussion. However, checking a list of n numbers to see whether they are sorted takes approximately n steps; this is left as Exercise 25. Therefore, we will assume that n steps are needed each time through the loop.

How many times is the loop executed? It might be executed only once if the first arrangement happens to be in sorted order, since Slow Sort stops searching as soon as it finds a sorted list. However, in the worst possible case, the initial list will not be sorted and the sorted list will not appear until the very end. In that case, the algorithm will take much longer. Therefore, since the number of steps required by Slow Sort can vary, depending on the original ordering of the list, how can we count the number of steps in any general sense?

This discussion is related to the second factor for determining the speed of our algorithm, described on page 477—namely, the initial order of the numbers in the list. Should we count the number of steps for our algorithm in the best possible case, in the worst possible case, or in some kind of average case? It turns out that analyses of all three cases are sometimes done when algorithms are being compared. However, the most commonly measured of the three cases is the worst-case analysis because people using an algorithm typically want to know the worst possible scenario. The average-case analysis may seem just as useful, but it is often not calculated because it is very hard to determine what the "average" case is. In our worst-case scenario, when we consider all possible input lists of n numbers, we want to identify the one that causes the algorithm to take the greatest number of steps.

To compute the worst-case execution time of Slow Sort, we need to know how many **permutations** (or arrangements) of a list of n numbers there are, since, in the worst case, that is the number of times Slow Sort will execute its loop. To figure out how many ways a list of n numbers can be arranged, we first consider a simple case.

| EXAMPLE 9.6 | We construct all arrangements of 4 numbers: 1, 2, 3, 4. There are 4 possible choices for the first number. For each of these choices there are only 3 possibilities for the second number since a number may not be used more than once, making a total of 12 different ways the first two numbers may be chosen. For each of these 12 situations there are 2 possible choices for the third number. Once this choice has been made, there is only one remaining number, which must go in the fourth position. Hence there are $4 \cdot 3 \cdot 2 \cdot 1 = 24$ possible arrangements of the 4 numbers. (*Figure 4.5*, on page 148, illustrates these 24 possibilities, using A, B, C, D, instead of 1, 2, 3, 4.) |

In general, if we have n numbers from which to choose, we then have n choices for the first number. For each of these n choices, we have $n - 1$ choices for the next number. Therefore, there are $n \cdot (n - 1)$ choices for the first two numbers. For each of these $n \cdot (n - 1)$ choices there are

$n - 2$ choices for the third number, and so a total of $n \cdot (n - 1) \cdot (n - 2)$ choices for the first 3 numbers. Continuing this way, we see that there are $n \cdot (n - 1) \cdot (n - 2) \cdot \ldots \cdot 3 \cdot 2 \cdot 1$ choices for arranging all n numbers. The number $n \cdot (n - 1) \cdot (n - 2) \cdot \ldots \cdot 3 \cdot 2 \cdot 1$ is computed by multiplying together all the integers from n down to 1; it is denoted by $n!$ and is read "n factorial."[9] For example,

$$5! = 5 \cdot 4 \cdot 3 \cdot 2 \cdot 1 = 120$$

and so there are 120 ways of arranging 5 numbers in a list.

Therefore, there are $n!$ ways to arrange n numbers in a list and so, in the worst case, Slow Sort's loop is executed $n!$ times. Since the computer needs to make n computations within each execution of the loop, we have the following result.

XXXXXXXXXXXXXXXXXXXXXXXXXXXXXXXXXXXXXXXXXXXXXXXXXXX

(9.3)   The number of steps that Slow Sort executes is $n(n!)$.

XXXXXXXXXXXXXXXXXXXXXXXXXXXXXXXXXXXXXXXXXXXXXXXXXXX

EXAMPLE 9.7   To sort a list of 3 numbers, Slow Sort requires

$$3 \cdot (3!) = 3 \cdot (3 \cdot 2 \cdot 1) = 3 \cdot 6 = 18 \text{ steps}$$

in the worst case.

INPUT	SELECTION SORT	SLOW SORT
n	$1 + \frac{9}{2}n + \frac{1}{2}n^2$	$n(n!)$
1	6	1
2	12	4
3	19	18
4	27	96
5	36	600
10	96	36,288,000
50	1476	1.52×10^{66}*

*Recall that "1.52×10^{66}" is in scientific notation. Here, 10^{66} is the number 1 followed by 66 zeros. Therefore, 1.52×10^{66} is the same as 152 followed by 64 zeros.

Table 9.1 A comparison of Selection Sort with Slow Sort.

Now that we have the formulas of our two algorithms, we should be able to determine which algorithm is faster. What we need to do is compare the two formulas to see which one yields smaller numbers of steps for various values of n (the number of items to be sorted). See Table 9.1 for comparisons using various values of n. Note that for small values of n, Slow Sort appears to be better; but as n grows, Selection Sort is better. In most commercial applications, such as when a company wants to sort a database of all of its customers or products, there are thousands or even millions of items to be sorted. Therefore, Selection Sort is the algorithm of choice between these two.

In fact, although Slow Sort appears to be better for lists of length 3 or less, that is not the case in actual practice. We have been ignoring the number

9. This subject is also discussed in more detail in Section 4.4.

of steps it takes to rearrange the numbers in the list in Slow Sort. When those steps are included, Selection Sort becomes the winner for *all* values of *n*. Furthermore, even if Slow Sort were better on lists of length 3 or less, such lists are so short that *any* sorting algorithm will perform acceptably with them. Therefore, Selection Sort is the algorithm of choice for all lists. (There are other sorting algorithms even faster than Selection Sort. These faster algorithms are not discussed in this section because they are somewhat complex. See Item 9 in the list *For Further Reading* for a thorough discussion of sorting algorithms.)

This type of analysis can be repeated for any algorithm. The idea is to construct a formula giving the number of steps required to execute the algorithm (in the worst case) as a function of *n*, the size of the input. This formula is called the (**worst-case**) **time complexity** of the algorithm. In the exercises, you are asked to find the time complexities of other algorithms, including one that solves the Traveling Salesman Problem.

Are the differences in the time complexities of these algorithms significant? For example, even though Slow Sort takes more time than Selection Sort for most (or all) lists of numbers, won't both algorithms work well in actual practice because computers are so fast? If that is the case, then why bother with the analysis that we have just completed? These questions will be answered in the next section.

9.5   EXERCISES

1. Add another row to *Table 9.1* that uses the value of 6 for *n*.

2. Add another row to *Table 9.1* that uses the value of 100 for *n*. (A calculator is useful here.)

In comparing the numbers in *Table 9.1*, it is worthwhile to determine how long it would take a computer to execute that many calculations. (Here we are thinking of one step as a calculation.) In Exercises 3–6, pretend that you have a computer that can perform 1 million calculations per second. Determine how long it will take your computer to do the following numbers of calculations.

3. 96   4. 1476   5. 36,288,000

6. 1.52×10^{66} (A calculator is useful here.)

7–10. Redo Exercises 3–6 assuming that your computer can do 1 billion calculations per second.

11. When sorting a list of 2 numbers, Selection Sort takes 12 steps. What is done in the first step? in the second step? Describe what is done in each of the remaining 10 steps.

12. When sorting a list of 3 numbers, Selection Sort takes 19 steps. What is done in the first step? in the second step? Describe what is done in each of the remaining 17 steps.

13. When sorting a list of 4 numbers, Selection Sort takes 27 steps. What is done in the first step? in the second step? Describe what is done in each of the remaining 25 steps.

14. Why does swapping two numbers normally take three steps instead of just one or two? (*Hint*: Each number is stored in a certain location in the computer's memory. Only one number at a time can be copied from one location to another. When a number is copied into a new location, the old number stored there is irretrievably lost. In the swap we want the two numbers to end up in each other's memory location.)

15. To test Slow Sort and Selection Sort, you could convert them into programs, run them on the same computer using the same input, and see which one finishes first. What is wrong with such a test?

16. When we calculated the time complexity of Slow Sort, we ignored the steps necessary to come up with each new arrangement of the numbers in the list. Suppose that it takes n steps to create each new arrangement. What happens to the time complexity of Slow Sort if we include these extra steps?

Exercises 17–20 concern another sorting procedure, called *Insertion Sort*, which can be described as follows:

The computer starts with the first two numbers in the list and arranges them in the proper order. Then it considers the third number in the list and inserts it in its proper place relative to the first two numbers. Now the first three numbers of the list are sorted. The computer repeatedly takes the next unsorted number and inserts it in its proper place relative to the already sorted numbers. Here, then, is the algorithm:

Insertion Sort

Input: An unsorted list L of n numbers.

Output: The list L with the numbers arranged in sorted order.

Algorithm:

Let $k = 1$.

Do the following instructions n times:

Extract the kth number from the list.

Insert this number into its proper place among the already sorted numbers in the first $k - 1$ positions in the list.

Let $k = k + 1$.

Note that finding the proper place to insert the kth number is not a one-step operation. The computer needs to search through the sorted numbers in the first $k - 1$ positions to find the proper place. Therefore, finding the proper place for the kth number can take up to $k - 1$ steps.

17. Try Insertion Sort on the list of numbers

$$9, 13, 7, 6$$

How many steps does it take? Describe how you calculated the number of steps.

18. Try Insertion Sort on the list of numbers

$$4, 1, 7, 2, 5$$

How many steps does it take? Describe how you calculated the number of steps.

19. What is the worst-case time complexity of Insertion Sort?

20. Would you prefer Insertion Sort to Selection Sort or Slow Sort? Why or why not?

21. In the best case, Slow Sort would take only n steps, so it has a best-case time complexity of n. What is the best-case time complexity of Selection Sort? of Insertion Sort?

22. The following algorithm solves the TSP:

Slow TSP

Input: A list of n cities and a table giving the time it takes to travel between each pair of cities.

Output: A tour of the n cities with the shortest total time.

Algorithm:

> Repeat the following instructions until all possible tours have been checked:
>> Arrange the cities in a new order to form a tour.
>> Check to see whether this tour has total length less than the previously shortest tour. If so, record it as the new shortest tour.
> Output the shortest tour found.

Slow TSP will solve the problem, but not necessarily in the fastest possible way. In fact, this algorithm is very similar to Slow Sort, which is why we named it "Slow TSP." The problem with it is that it tries all possibilities. Find the time-complexity of Slow TSP and explain how you arrived at your answer. (*Hint*: How many tours of n cities are there?)

23. Statements (9.2) and (9.3) indicate the total number of steps for Selection Sort and Slow Sort, respectively, in a *worst*-case scenario. If we count the number of steps of Selection Sort and Slow Sort in the *average* case instead of the worst case, the complexity of Selection Sort does not change, but the complexity of Slow Sort becomes $n(n!)/2$. Add a new column to *Table 9.1* for Slow Sort using the expression $n(n!)/2$. Does this make Slow Sort better than Selection Sort?

24. There are some situations in which we want to compare programs instead of algorithms. For example, if we are given two programs that both implement the same algorithm, we might want to compare them to see which is faster. Describe at least one more situation in which it is desirable to compare programs, instead of algorithms.

25. Show that checking a list of n numbers to see whether it is sorted takes approximately n steps. See Exercise 10 of the preceding section.

WRITING EXERCISES

1. In Sections 9.3 as well as in this section, the word "program" was defined. Compare the two definitions with each other and with the everyday use of the word "program." What are the differences? Does one of these definitions given in the text provide more information than the other? Explain.

2. Which idea in this section do you find most confusing or difficult to understand? After rereading the section, if you still do not understand the idea, write three specific questions that you think might help clarify the problem. If you now understand the idea, write a brief explanation that you think might help a classmate who is having trouble with it.

COMPLEXITY ANALYSIS

9.6

We now want to analyze a larger variety of algorithms than the sorting algorithms discussed in the preceding section. However, some algorithms have quite complicated time-complexity formulas, making them hard to compare with one another. Therefore, one of the purposes of this section will be to determine which parts of these formulas are essential for our analyses and which parts are irrelevant. This will allow us to simplify the formulas by eliminating inessential parts.

Simplifying a formula in this way doesn't necessarily make it less useful, just as discarding insignificant digits of a number may not make the number less meaningful. For example, when comparing the population of the United States with that of Canada, it is much easier to deal with figures rounded to the nearest million rather than with exact figures. Similarly, in analyzing algorithms we do not lose anything by simplifying formulas because we are interested only in which values grow faster as n gets larger, and not in the exact differences between these values.

To see how to simplify time-complexity formulas, let us return to the one for Selection Sort. Recall that Statement (9.2) indicated that Selection Sort requires $1 + \frac{1}{2}n^2 + \frac{9}{2}n$ steps to sort a list of n numbers. Which of the three terms—1, $\frac{9}{2}n$, or $\frac{1}{2}n^2$—is the most important term of the formula?

n	1	$\frac{9}{2}n$	$\frac{1}{2}n^2$
10	1	45	50
20	1	90	200
50	1	225	1250
100	1	450	5000
500	1	2250	125,000
5000	1	22,500	12,500,000

Table 9.2 A comparison of the terms 1, $\frac{9}{2}n$, and $\frac{1}{2}n^2$.

To answer this question, look at the values in *Table 9.2*. The column on the left contains various values of n. The other three columns give the corresponding values of the three terms involved in the formula. Keep in mind that the exact values are not that significant; it is the relative sizes and the changes in the sizes of the numbers that are important. As n increases, the term 1 does not change; it is a constant. The term $\frac{9}{2}n$ grows steadily, but the term $\frac{1}{2}n^2$ grows much faster. For example, when the value of n is doubled (say from 10 to 20), we get twice as big a value for the term $\frac{9}{2}n$, but we get *four* times as big a value for the term $\frac{1}{2}n^2$. Thus, for very large values of n, the terms other than $\frac{1}{2}n^2$ shrink in relative importance. Consequently, for long lists of numbers, in which case the value of n is large, the number of steps needed by Selection Sort is "dominated" by $\frac{1}{2}n^2$.

In summary, when comparing the complexities of algorithms, by far the most important term is the fastest growing one:

> The fastest growing terms are the ones that have the largest effect on how many steps the algorithm requires when being run with large numbers of input values (that is, with large n).

Therefore, when we simplify our formulas, we will ignore all but the fastest growing terms.

To make things even easier, consider the numerical coefficients: $\frac{9}{2}$ (of the term $\frac{9}{2}n$) and $\frac{1}{2}$ (of the term $\frac{1}{2}n^2$). How important for our analysis

n	$50n$	$500n$	$\frac{1}{10}n^2$
10	500	5000	10
20	1000	10,000	40
50	2500	25,000	250
100	5000	50,000	1000
500	25,000	250,000	25,000
5000	250,000	2,500,000	2,500,000
10,000	500,000	5,000,000	10,000,000

Table 9.3 Changing the coefficients of n and n².

are these two numbers? In *Table 9.3* we have changed the coefficients in these terms to see what effect this has. From the table, we can see that the term containing n^2 *still* grows faster than any of the terms containing n. The term n^2 outgrows the term n regardless of the coefficients of the two terms. Thus, since we are interested only in which characteristics of a formula cause rapid growth, we can ignore all numerical coefficients.

In summary:

> the dominant feature of a time-complexity formula is the fastest growing term, regardless of its numerical coefficient. Thus, we will ignore anything other than the fastest growing term, and will even ignore the numerical coefficient of that term.

What remains is still precise enough for the type of analysis we desire.

From now on, when we refer to the "complexity" of an algorithm, we will mean this simplified (approximate) formula rather than the complete (more precise) formula initially calculated. For example, we will treat the Selection Sort algorithm as having the complexity of n^2 (quadratic complexity) and the Slow Sort algorithm as having the complexity of $n!$ (factorial complexity), rather than use their more precise formulas involving $1 + \frac{1}{2}n^2 + \frac{9}{2}n$ and $n \cdot (n!)$, respectively.[10]

Algorithms come with a wide variety of complexity formulas. Two complexities that commonly occur but haven't yet been mentioned in this chapter are 2^n (exponential) and log n (logarithmic). You may have seen "log x" before as meaning $\log_{10}x$ (the logarithm of x to the base 10) since 10 is the base of our number system. But 2 is a more natural base for computers and so it is used here. Therefore,

> when we write log n, we mean the logarithm of n to the base 2.

Beware that many pocket calculators only calculate logarithms to the base 10 or to the base e, and not to the base 2.[11]

The logarithmic formula can be thought of as the "inverse" formula for 2^n. Thus, beginning with the computation

$$2^3 = 8 \qquad \text{(2 to the 3rd power is 8)}$$

10. Technically the number n in the expression $n \cdot (n!)$ is not a coefficient; but it contributes so little to the growth of the values compared with $n!$ that it actually can be ignored. See Exercise 31.

11. The number e, which is approximately 2.718, is the base of the so-called "natural" logarithms. For various technical reasons, it is often the base of choice when working with logarithms.

the log formula reverses this process by writing

$$\log_2 8 = 3 \qquad \text{(log 8 to the base 2 is 3)}$$

or, since base 2 is understood, simply

$$\log 8 = 3$$

More generally, for any natural number n, if

$$2^x = n$$

then

$$\log n = x$$

On the left side of *Table 9.4*, values of 2^n are given in the second column for the sample values of n in the first column. On the right side of this table, in the third column, the values that were obtained in the second column are now treated as the values of n. The fourth column shows the corresponding values of $\log n$. The graphs of 2^n and $\log n$ are shown in *Figure 9.6*.

For our purposes, the most important aspect of *Table 9.4* is showing that as n gets larger, 2^n grows explosively, whereas $\log n$ grows *extremely* slowly. In particular, an algorithm of logarithmic complexity is much more desirable than one of exponential complexity because, for large values of n—that is, for input consisting of a large number of values—the logarithmic algorithm takes far fewer steps.

The procedure for comparing algorithms is to find the worst-case time complexities of these algorithms and then to compare these complexities for large values of n. Remember that the complexity can be expressed by a simplified formula indicating the approximate number of steps required by the algorithm to solve a problem of size n. Therefore, whichever algorithm has the slowest growing complexity takes the fewest steps to solve large problems and is therefore the fastest algorithm.

In the rest of this chapter we will be discussing a variety of complexity formulas. Some of the formulas, such as those for $n!$, 2^n, and $\log n$, have already been introduced. The other formulas are ones that you have probably seen before, such as those for $\sqrt{n}$ and for n to a power. In particular, $\sqrt{n}$ denotes the nonnegative number which, when

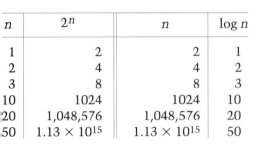

n	2^n	n	$\log n$
1	2	2	1
2	4	4	2
3	8	8	3
10	1024	1024	10
20	1,048,576	1,048,576	20
50	1.13×10^{15}	1.13×10^{15}	50

Table 9.4 Sample values for 2^n and $\log n$.

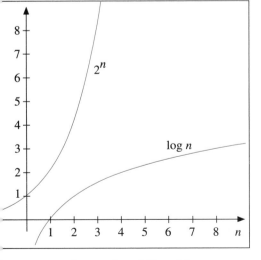

Figure 9.6 The graphs of 2^n and $\log n$.

n	$\log n$	$\sqrt{n}$	$n(\log n)$	n^2	n^3	n^5	2^n	$n!$
5	2.3	2.2	11.5	25	125	3125	32	120
10	3.3	3.2	33	100	1000	10^5	1024	3×10^6
20	4.3	4.5	86	400	8000	3×10^5	10^6	10^{18}
40	5.3	6.3	212	1600	64,000	10^8	10^{12}	10^{48}
50	5.6	7.1	280	2500	10^5	3×10^8	10^{15}	10^{64}
100	6.6	10.0	660	10,000	10^6	10^{10}	10^{30}	10^{158}
150	7.2	12.3	1080	22,500	3×10^6	8×10^{10}	10^{45}	10^{262}

Table 9.5
Common complexities of algorithms.

multiplied by itself, gives n. For example, if $n = 9$, then $\sqrt{n} = \sqrt{9} = 3$. Also, n raised to a power, such as n^4, denotes the number we get by multiplying 4 factors of the same number n. For example, if $n = 3$ then

$$n^4 = 3^4 = 3 \cdot 3 \cdot 3 \cdot 3 = 81$$

The more complicated formulas we will encounter are composed of combinations of these simple formulas. Thus, the formula for $\sqrt{n} \cdot \log(n^3)$ denotes the number we get when we compute n^3, take the logarithm (to the base 2) of the result, and then multiply that calculation by $\sqrt{n}$. For instance, if $n = 4$, then

$$\sqrt{n} \cdot \log(n^3) = \sqrt{4} \log(4^3) = 2 \log(64) = 2(6) = 12$$

Table 9.5 is useful for comparing various algorithms; it contains some of the most commonly appearing complexities:

$$\log n, \sqrt{n}, n(\log n), n^2, n^3, n^5, 2^n, n!$$

Each column shows the number of steps required to solve problems for various values of n. More complexities are included in the exercises. Looking down any column at the numbers that appear, we see that, as one would expect, they grow larger when larger values of n are used. However, notice the variation in how fast the numbers in the different columns grow. The numbers in the last few columns grow much faster than the numbers in the first few columns. In fact, the complexities in the table are listed in order; those with the slowest growing numbers (therefore indicating the fastest algorithms) on the left side, and those with the fastest growing numbers (the slowest algorithms) on the right side. From this table, we can see, for example, that an algorithm with complexity $\sqrt{n}$ is going to be faster than one with complexity n^2 because the former algorithm requires significantly fewer steps to complete its calculations.

Some of the numbers in this table are so large that it is hard to comprehend their significance. For example, the table says that an algorithm of complexity 2^n requires 10^{15} steps to solve a problem when the size of n is 50. (Recall that 10^{15} indicates the number 1 followed by 15 zeros.) Let us determine how long it would take a computer to execute this many steps. For convenience, we will assume that the computer can execute 1 million (10^6) steps per second, a reasonable speed for modern computer work stations. Then we can compute the total time as follows:

$$(10^{15} \text{ steps}) \times \left(\frac{1 \text{ sec}}{10^6 \text{ steps}}\right) \times \left(\frac{1 \text{ hour}}{3600 \text{ secs}}\right) \times \left(\frac{1 \text{ day}}{24 \text{ hours}}\right) \times \left(\frac{1 \text{ year}}{365 \text{ days}}\right)$$

$$= \frac{10^{15}}{(10^6 \times 3600 \times 24 \times 365)} \text{years}$$

$$= 32 \text{ years}$$

(On the first line, notice that all the units divide out, except for years.) Thus, it takes an algorithm of complexity 2^n about 32 years to solve a problem of size 50. Think of waiting that long for your computer to solve your problem! Furthermore, in this problem the size of n is only 50, whereas typical industrial problems use a value of n in the thousands or millions. Thus, for all practical purposes, a 2^n-complexity is worthless.

All the other numbers in *Table 9.5* can be converted into times in the same way; the result is shown in *Table 9.6*. Here "msec" means milliseconds (thousandths of a second) and "μsec" means microseconds (millionths of a second).[12]

Notice how the times grow in each column. In particular, the times in the first six columns grow at a rate that is *significantly* slower than the times in the last two columns. For instance, the last row of *Table 9.6* shows that algorithms with the first six complexities can solve problems of size $n = 150$ in less than a day, whereas algorithms of the last two complexities require billions of centuries to solve a problem of the same size.

This difference is clearly of major importance in the theory of complexity of algorithms. Algorithms with the first six complexities

n	$\log n$	$\sqrt{n}$	$n(\log n)$	n^2	n^3	n^5	2^n	$n!$
5	2.3 μsec	2.2 μsec	11.5 μsec	25 μsec	125 μsec	3.125 msec	32 μsec	120 μsec
10	3.3 μsec	3.2 μsec	33 μsec	100 μsec	1 msec	100 msec	1.0 msec	3 sec
20	4.3 μsec	4.5 μsec	86 μsec	400 μsec	8 msec	3.2 sec	1 sec	31,000 yrs
40	5.3 μsec	6.3 μsec	212 μsec	1.6 msec	64 msec	100 sec	11.6 days	10^{34} yrs
50	5.6 μsec	7.1 μsec	280 μsec	2.5 msec	100 msec	5 min	32 yrs	10^{50} yrs
100	6.6 μsec	10.0 μsec	660 μsec	10 msec	1 sec	2.8 hrs	10^{16} yrs	10^{144} yrs
150	7.2 μsec	12.3 μsec	1080 μsec	22.5 msec	3 sec	22.2 hrs	10^{31} yrs	10^{248} yrs

Table 9.6 Common complexities of algorithms measured in time units.

12. The symbol μ is the Greek letter *mu*, which is pronounced "mew."

are called **polynomial-time** algorithms since either their complexities are just n to a fixed power (which is a simple polynomial) or their complexities involve formulas that grow even slower than n to a fixed power. Any other algorithms, such as those with complexities 2^n and $n!$, are called **exponential-time** algorithms.[13] Algorithms that exhaustively test all possibilities in order to find a solution, such as Slow Sort, are usually exponential-time algorithms because of the huge number of possibilities that need to be checked. In contrast to that, polynomial-time algorithms usually employ some feature of the data being analyzed to solve the problem in a more reasonable way.

As you can see from the tables, exponential-time algorithms are really quite useless except when the input consists of only a very small number of values. Polynomial-time algorithms, on the other hand, are very useful in that they will solve problems in a reasonable amount of time even for a large value of n. Therefore, it is desirable to find a polynomial-time algorithm for every problem, whether it be a sorting problem or the Traveling Salesman Problem.

The difference between polynomial-time and exponential-time algorithms is more significant than we might initially believe. Moore's Law, noted in Section 9.3, states that computer hardware is doubling in power roughly every year. Therefore, we might be tempted to think that in the future, exponential algorithms will be executable in a reasonable amount of time because much faster computers will be around by then. In other words, we might be tempted to believe that the problem with exponential-time algorithms is just that we have not yet developed fast enough computers to run them.

Unfortunately, this is not the case. Suppose, for example, that we want to wait until computers are fast enough to solve problems twice as large as the largest problems solvable in a reasonable amount of time today. The ratios in *Table 9.7* indicate, for various complexities, how much faster a computer would have to be to solve a problem of size $n = 200$ in the same amount of time currently used to solve a problem of size $n = 100$.

n	$\log n$	$\sqrt{n}$	n	$n(\log n)$	n^2	n^3	n^5	2^n	$n!$
100	6.6	10	100	660	10,000	10^6	10^{10}	10^{30}	10^{158}
200	7.6	14	200	1529	40,000	8×10^6	32×10^{10}	10^{60}	10^{375}
Ratio	1.15	1.4	2	2.3	4	8	32	10^{30}	10^{217}

Table 9.7 Comparison of size n = 100 with n = 200 for various complexities.

13. A more precise name would be **superpolynomial-time** algorithms.

For instance, if we are using an n^2 algorithm, the ratio tells us that we will need a computer 4 times faster in order to solve a problem of size 200 in the same amount of time as we currently use to solve a problem of size 100. By Moore's law, we would have to wait approximately 2 years before such a computer would be available. In some cases ($\log n$, $\sqrt{n}$, n), we would need to speed up computers by only a factor of 2 or less, which means waiting one year or less. However, in the case of the 2^n or $n!$ algorithm, we would need to have a computer 10^{30} (one thousand billion billion billion) or 10^{217} times faster, respectively, than a current computer in order to solve such a problem in the same amount of time. We would have to wait about 100 years or 721 years, respectively, (assuming Moore's law holds that far into the future) before we would find computers that are 10^{30} or 10^{217} times more powerful than today's computers. (See Exercises 37 and 38.) Therefore, for exponential-time algorithms, waiting until we have a fast enough computer is not a viable option.

There is yet another way to compare these algorithms, as is illustrated in Example 9.8.

EXAMPLE 9.8

If a problem of size 100 takes only 1 millionth of a second to compute using an n^2-complexity algorithm, then a problem of size 200 would take only 4 millionths of a second. However, if a problem of size 100 takes only 1 millionth of a second (1 μsec) to compute using a 2^n-complexity algorithm, then a problem of size 200 would take 10^{30} millionths of a second, which can be converted into centuries as follows:

$$(10^{30} \text{ μsec}) \cdot \left(\frac{1 \text{ sec}}{10^6 \text{ μsec}}\right) \cdot \left(\frac{1 \text{ hour}}{3600 \text{ secs}}\right) \cdot \left(\frac{1 \text{ day}}{24 \text{ hours}}\right) \cdot \left(\frac{1 \text{ year}}{365 \text{ days}}\right) \cdot \left(\frac{1 \text{ century}}{100 \text{ years}}\right)$$

$$= \frac{10^{30}}{(10^6 \times 3600 \times 24 \times 365 \times 100)} \text{ centuries}$$

$$= 3 \times 10^{14} \text{ centuries}$$

$$= 300 \text{ trillion centuries}$$

Example 9.8 indicates that exponential-time algorithms are *inherently* slow, and faster computers will not have any significant effect on them in the foreseeable future. Therefore,

> when programs are written to solve problems, they must use polynomial-time algorithms and not exponential-time algorithms;

otherwise, they may take eons to run.

Note that it is not up to the computer to decide which algorithm to use. Rather, it is the responsibility of the programmer to use the appropriate (polynomial-time) algorithm when writing a program. Unfortunately, as you might guess, it is not always easy to find such algorithms. In fact, there are many problems for which no one has yet found a polynomial-time algorithm.[14] The Traveling Salesman Problem is one of them; that is, the only known algorithms for solving the TSP are exponential-time algorithms, which explains why we said (in Section 9.2) that current TSP algorithms may take many centuries to solve problems involving only 100 cities. Many computer scientists wonder why only exponential-time algorithms have been found for the TSP and other problems. Is the absence of polynomial-time algorithms for these problems the fault of insufficiently creative human programmers, or is there something inherent in the nature of such problems that prohibits the existence of such algorithms? This question leads to one of the major theoretical discoveries in the field of computer science, which is the topic of the next section.

9.6   EXERCISES

1. What is the value of log 32 (using base 2)?

2. What is the value of log 128 (using base 2)?

In Exercises 3–5, add extra rows to *Table 9.4,* **using the given values for** *n* **in the column furthest to the left.**

3. $n = 4$ 4. $n = 8$ 5. $n = 100$

In Exercises 6–9, add extra rows to *Table 9.5,* **using the given values for** *n* **in the column furthest to the left.**

6. $n = 8$ 7. $n = 15$

8. $n = 30$ 9. $n = 75$

In Exercises 10–13, add extra rows to *Table 9.6,* **using the given values for** *n* **in the column furthest to the left.**

10. $n = 8$ 11. $n = 15$

12. $n = 30$ 13. $n = 75$

Simplify each of the complexity formulas in Exercises 14–21 by indicating the fastest growing term and eliminating the coefficients.

14. $300 + 4n^2$ 15. $\log n + n!$

16. $n^5 + 89n^3$ 17. $1000 + n^3 + 1000n$

18. $46n^{100} + 2^n$ 19. $3n + \log n + 45{,}000$

20. $n! + 2^n + 3^n$ 21. $\sqrt[4]{n} + \sqrt{n} + n^2$

14. There are also problems for which *no* algorithm (polynomial-time or exponential-time) exists. An example is the Halting Problem, for which there is an elementary explanation on pp. 289–92 of Item 1 in the list *For Further Reading* at the end of this chapter.

22. Compare the complexities obtained in Exercises 14–21 and list them in order from slowest growing to fastest growing.

23. Which of the complexities obtained in Exercises 14–21 are polynomial formulas and which are exponential?

24. Use *Table 9.6* to estimate how long it will take Selection Sort and Slow Sort to sort a list of 10 numbers.

25. Use *Table 9.6* to estimate how long it will take Selection Sort and Slow Sort to sort a list of 50 numbers.

The numbers in *Table 9.6* were calculated under the assumption that there is a single computer which has one CPU and which can execute 1 million (10^6) instructions per second. Re-calculate the values in the last row of *Table 9.6* under the assumptions of Exercises 26–29.

26. Assume that the computer is 1000 times faster. That is, assume that the computer can execute 1 *billion* (10^9) instructions per second.

27. Assume that the computer is 1 million times faster. That is, assume that the computer can execute 1 *trillion* (10^{12}) instructions per second.

28. Instead of a single computer there is a parallel computer network with 10 CPUs to share the work.

29. Instead of a single computer there is a massively parallel computer with 1000 CPUs, or a massively parallel computer network of 1000 computers, each with a single CPU, to share the work.

30. Suppose you wish to run an $n(\log n)$ algorithm on a computer. If a problem of size 100 takes only 1 millionth of a second to compute using your algorithm, how long would a problem of size 200 take? A problem of size 400? (*Hint*: Table 9.7 can help you.)

31. Between which two columns in *Table 9.5* should each of the following complexities

be placed? That is, which of the complexities in *Table 9.5* are slower growing and which are faster growing than each of these new complexities? Then determine which of these complexities are polynomial and which are exponential.

(a) $\sqrt[4]{n}$ (b) n^n (c) $n^2 \cdot (\log n)$

(d) $n \cdot (n!)$

(*Hint*: Use the techniques you learned in Chapter 1.)

32. Suppose that an algorithm has worst-case time complexity 2^n. How large can the input be (that is, how large can n be) if the algorithm must solve the problem in an hour and if you use a computer that can execute one step in one-millionth of a second?

33. Suppose that an algorithm has worst-case time complexity $n!$. How large can the input be (that is, how large can n be) if the algorithm must solve the problem in an hour and if you use a computer that can execute one step in one-millionth of a second?

34. Suppose that an algorithm has worst-case time complexity n^2. How large can the input be (that is, how large can n be) if the algorithm must solve the problem in an hour and if you use a computer that can execute one step in one-millionth of a second?

35. Suppose that algorithm A has worst-case time complexity $1,000,000n^2$ and an algorithm B has worst-case time complexity $.00001n^3$.

(a) In the theoretical sense discussed in this section, which algorithm is faster?

(b) For which values of n is algorithm A faster than algorithm B? (A calculator would help here.)

(c) For which values of n is algorithm A slower than algorithm B? (A calculator would help here.)

(d) For which values of n, if any, do both algorithms have the same speed? (A calculator would help here.)

36. Another sorting procedure is called "Merge Sort." In Merge Sort, the computer first splits the unsorted list into halves over and over again. Thus, on the second split, each "half" consists of (approximately) one-quarter of the original list; on the third split, one-eighth of the original list, etc., until, finally, each half consists of one or two items. Then, in reverse order the halves are merged into longer lists by comparing the smallest items in each half to find the smallest one in the merged list, then the next smallest one, etc., until finally the original list has been sorted. What is the worst-case time complexity of Merge Sort?

37. Assuming that computer power doubles every year, show that we would need to wait 100 years before we will find a computer that is 10^{30} times more powerful than today's computers.

38. Assuming that computer power doubles every year, show that we would need to wait 721 years before we will find a computer that is 10^{217} times more powerful than today's computers.

WRITING EXERCISES

1. Our definition of a "step" has been somewhat vague. In most algorithms, each step actually takes up to 10 or 20 clock cycles to execute. Write at most one page describing why we can be vague about the definition of a step without destroying the validity of our analysis.

2. Some of the complexities, such as log n, are those of very fast algorithms, whereas others, such as $n!$, are those of slow algorithms. What if an algorithm has complexity $\log(n!)$? Is this the complexity of a fast or slow algorithm? What about $(\log n)!$? Use problem-

solving techniques from Chapter 1 and explain your results.

3. Come up with several examples of situations, like the one in the second paragraph of this section, where a precise numerical value is no more useful than an approximate value.

4. Defend the statement: "An algorithm of complexity $10,000n$ is better than an algorithm of complexity n^3." Assume that your audience has read Sections 9.1–9.5, but hasn't read this section.

NP-COMPLETENESS

9.7

As was mentioned in the preceding section, there are many problems, including the Traveling Salesman Problem, for which no computer scientist has yet been able to find a polynomial-time algorithm. This has not been for lack of trying. Many of the world's foremost computer scientists have attempted, but failed, to find such algorithms. There seems to be something inherently difficult about these problems. Nevertheless, in 1971, the computer scientist Stephen Cook shed some light on this situation by showing that many of these problems are actually related.[15] The purpose of this section is to discuss the nature of their relationship.

In order to understand this relationship, we first need to categorize our problems as follows: problems that require an optimal solution for an answer, such as the TSP, which requires a shortest tour, are called **optimization problems**, whereas **decision problems** are those that can be answered *yes* or *no*.

Decision problems arise naturally in many settings. For example, in some states, truck drivers are allowed to drive only 10 hours a day. An inspector reviewing a driver's logbook is interested in knowing only whether he or she has driven at most 10 hours each day (a *yes*-or-*no* problem), and not the exact number of hours driven. Similarly, when a student takes a course "pass/fail," the registrar's office is only interested in whether or not the grade is above the passing level.

As another example, consider the following way of rephrasing the Traveling Salesman Problem:

Traveling Salesman Decision Problem (TSDP)

> **Input:** A list of cities, a table giving the times of transversal between each pair of cities, and a positive number M.

> **Problem:** Determine whether there exists a tour of all the cities that visits each city exactly once, returns to the starting point, and has a total length of time less than M.

This is a decision problem. An algorithm that solves this problem need not produce a shortest tour. Such an algorithm need only yield an answer of *yes* or *no*, depending on whether or not there is a tour of length less than M.

EXAMPLE 9.9

Look back at the five cities in *Figure 9.1* on page 461. Suppose we want to know whether a tour of those cities with length less than 30 exists. An algorithm that solves this example of the TSDP, in which $M = 30$, should yield *yes* because there does exist a tour (namely, *A–B–C–D–E–A*) with total length less than 30. However, if we want to solve the TSDP with the same cities but using $M = 20$, our algorithm should produce *no* because no tour of those five cities has length less than 20. In fact, the shortest tours have length 24.

---

15. S. A. Cook, "The Complexity of Theorem-Proving Procedures," *Proceedings of the Third Annual ACM Symposium on the Theory of Computing.* New York: Association for Computing Machinery, 1971 pp. 151–58.

A Sun SPARCStation 5 Computer Workstation.

Unfortunately, decision versions of problems are not necessarily any easier to solve than optimization versions. For example, no one has yet come up with a polynomial-time algorithm for either the TSDP or the TSP. In fact, it has been proved that someone who comes up with a polynomial-time algorithm for the TSDP could use that algorithm to create a polynomial-time algorithm for the TSP, and vice versa. The main reason for studying the decision versions of problems instead of the optimization versions is that the former are easier to work with. In particular, it is easier to prove results about how hard it is to solve such problems in polynomial time.

Now, let us consider a particular property of some decision problems, such as the TSDP. Suppose we wanted to know whether a tour of length less than M exists for a particular collection of cities and roads. Suppose also that with this data, an algorithm for the TSDP produces the answer *yes*. One way the algorithm could have determined that answer was by finding a tour of length less than M. The tour found is evidence which justifies the claim that a tour of length less than M exists. We can easily check the validity of this evidence by checking whether the tour really visits every city exactly once, returns to the starting city, and has total length of time less than M.

EXAMPLE 9.10

Suppose that we have an algorithm for the TSDP and that the data consists of four cities A, B, C, and D, the lengths of the roads between each pair of them, and the number $M = 10$. Suppose that the execution of the algorithm results in *yes* and, as evidence, indicates the tour *A–C–B–D–A*. Then, to verify that *A–C–B–D–A* is a qualifying tour, we need to check only that each of the four cities is visited exactly once (which is indeed the case with this tour) and that the lengths of the roads from *A* to *C*, *C* to *B*, *B* to *D*, and *D* to *A* add up to less than 10.

In summary, for some decision problems, a *yes* answer can be justified with evidence, that is, with an example that shows that the desired situation exists. In the TSDP, the evidence would be a tour of length less than *M*. Moreover, in some cases, the evidence can be checked quite easily to see whether it is, in fact, valid for the problem. For example, with the TSDP, we could write a little algorithm of time complexity n that would verify the evidence for us. (You are asked to write such an algorithm in Exercise 9.)

It turns out that the ease with which evidence can be checked is an important property of decision problems. The set of all decision problems for which there is a polynomial-time algorithm that verifies evidence is called **NP**.[16] The TSDP is a problem in NP since there is an algorithm of complexity n that verifies evidence. For another example of a problem in NP, consider the Bin-Packing Problem discussed in Exercises 11–14. For an example of a problem that is believed *not* to be in NP, look at the problem in Writing Exercise 1. Note that, for problems in NP, we do not require that a polynomial-time algorithm *find* the evidence. There must only be such an algorithm to *verify* the evidence. For example, regardless of whether or not there is an algorithm that solves the Traveling Salesman Decision Problem in polynomial time, there is an algorithm that *verifies* in polynomial time whether a given tour is valid evidence.

This distinction between solving decision problems (that is, answering *yes* or *no*) and verifying that any given evidence is legitimate is an important one, and is not unique to the TSDP. Many of the decision problems that occur in actual settings are of the type for which evidence can be verified by a polynomial-time algorithm, even though the problem itself may not be solvable by one. In other words, the hardest part of many decision problems is finding a solution. Once a solution is found, it is easy to check any evidence for validity.

The set NP turns out to be very important in the theoretical analysis of problems, as we will soon see. In particular, it is of fundamental importance for Cook's theorem.

---

16. NP stands for "nondeterministic polynomial time." It refers to the fact that problems in NP can be solved in polynomial time using nondeterministic Turing machines, which are beyond the scope of our coverage.

This extremely important theorem involves another problem in NP, known as the *Satisfiability Problem*, which concerns so-called *Boolean variables*. For our purpose it is not necessary to understand even the statement of the Satisfiability Problem in order to appreciate the significance of Cook's Theorem. (An optional discussion of the Satisfiability Problem is included at the end of this section.[17]) Cook's result is that the Satisfiability Problem has the following surprising property:

xxxxxxxxxxxxxxxxxxxxxxxxxxxxxxxxxxxxxxxxxxxxxxxxxxxxxx

If we can solve the Satisfiability Problem in polynomial time, then we can solve *all* problems in NP in polynomial time.

xxxxxxxxxxxxxxxxxxxxxxxxxxxxxxxxxxxxxxxxxxxxxxxxxxxxxx

In fact, if a polynomial-time algorithm that solves the Satisfiability Problem can be found, then this algorithm can be modified to solve any other problem in NP in polynomial time. This is an amazing result! It says that the Satisfiability Problem is as hard or harder to solve than any other problem in NP because once we have solved the Satisfiability Problem in polynomial time, we can solve all other problems in NP in polynomial time, including the TSDP.

To prove his remarkable theorem, Cook showed that the input (let us call it I_H) for any problem H in NP can be rewritten—in polynomial time—as an input (call it I_S) for the Satisfiability Problem in such a way that the I_H results in a *yes* outcome using an algorithm that solves problem H if and only if I_S results in a *yes* outcome using an algorithm that solves the Satisfiability Problem. Therefore, if we were to have a polynomial-time algorithm for the Satisfiability Problem, we could construct a polynomial-time algorithm for problem H that works as follows: Our algorithm would take any input to H and convert it (in polynomial time) into input for the Satisfiability Problem. Then our algorithm would use the polynomial-time algorithm for the Satisfiability Problem to obtain a *yes* or *no*. Finally, our algorithm would output that answer. (Understanding how the input to H is rewritten into input for the Satisfiability Problem requires knowledge of the computer model called a *Turing machine* and requires more precise definitions of "polynomial time" and of the set NP. These matters will not be covered here.)

Very soon afterwards, many other problems in NP were found to have the same property as the Satisfiability Problem; that is, if we

17. Section 9.8 and the Link that conclude this chapter do *not* depend on the discussion of the Satisfiability Problem.

could find a polynomial-time algorithm for the problem at hand, then we could find polynomial-time algorithms for *all* problems in NP. Problems with this property are called **NP-complete** problems. There are now several hundred types of problems known to be in the NP-complete set; the Traveling Salesman Decision Problem is one such problem. The discovery of NP-complete problems has had a profound effect on research in theoretical computer science, as will be discussed in the next section.

In the event that you are curious about the Satisfiability Problem, we now give a brief description. To this end, we let $a, b, c, \ldots$ denote variables that can have one of only two values: "true" or "false."[18] Such variables are called **Boolean variables**. If we like, we can think of the variables as sentences that are true ("$1 + 1 = 2$") or false ("The authors of this text were born on Mars"). But we need not concern ourselves with the meanings of "true" and "false" here; we just think of them as the possible values of the variables. We can **negate** these variables: a negated variable has the opposite value of the original variable. We denote the negation of a by "$\sim a$." If $a = $ true, then $\sim a = $ false, and vice versa. Using the connecting word "or," we can put together some of these variables and negations of variables into a **disjunctive clause**, which we will enclose by parentheses. For example, (a or $\sim b$ or c) is such a clause. The variables or negations of variables in a clause are called the **terms** of the clause. We will say that a disjunctive clause is true if at least one of its terms is true.

EXAMPLE 9.11

If $a = $ true, $b = $ false, and $c = $ false, then the clause (a or $\sim b$ or c) is true because one of the terms, namely a, is true. Even if $a = $ false, $b = $ false, and $c = $ false, the clause (a or $\sim b$ or c) is still true, because one of the terms, namely $\sim b$, is true (since b is false). However, if $a = $ false, $b = $ true, and $c = $ false, then the clause (a or $\sim b$ or c) is false, because all three of the terms of the clause are false.

Now, suppose there are several clauses, such as $C_1 = $ (a or b), $C_2 = $ (b or c), $C_3 = $ ($\sim a$ or $\sim c$), and $C_4 = $ (a or $\sim b$ or c), and we want to know whether there are values for the variables so that all four clauses are true at the same time. (We are not allowed to let $a = $ true in the first clause and $a = $ false in the second. In other words, each particular variable must have the same value in all the clauses.) In this example, we could let $a = $ true, $b = $ true, and $c = $ false. Because $a = $ true,

---

18. It is suggested that students who lack a background in elementary logic should read Appendix A before reading the remainder of this section.

C_1 and C_4 are true. Because $b =$ true, C_2 is true. Because $c =$ false, C_3 is true. Therefore, for these values of a, b, and c, all four clauses are true. This example serves as an introduction to the Satisfiability Problem.

Satisfiability Problem

> ***Input:*** A set of clauses containing Boolean variables.
>
> ***Problem:*** Is it possible to assign values to each of the variables in such a way that all the clauses are true at once?

An algorithm that solves this problem could take as input any set of clauses, and, if possible, find values for the variables so that all the clauses are true at the same time. If it is able to find such values, then the result is *yes*. For evidence, this algorithm could provide the values of the variables. If it is not able to find such values for the variables, the result is *no*. In the exercises that follow, you will have the chance to try your hand at solving this problem in a variety of cases and at analyzing an algorithm called *Slow SAT* that solves the Satisfiability Problem.

9.7   EXERCISES

In Exercises 1–3, consider the set of cities in Figure 9.2 on page 464. For each value of M, is there a tour of length less than M? If so, give such a tour.

1. $M = 10$ 2. $M = 18$ 3. $M = 20$

In Exercises 4–6, consider the set of cities in Figure 9.3 on page 464. For each value of M, is there a tour of length less than M?

4. $M = 20$ 5. $M = 28$ 6. $M = 38$

7. What is the difference between the TSP and the TSDP? Why would someone prefer to study the TSDP rather than the TSP?

8. An algorithm that solves the TSP can be used to construct one to solve the TSDP. Show how this can be done. That is, assume we have an algorithm A for the TSP. We need to construct an algorithm B that takes as input a set of cities, the distances between any pair of them, and a positive integer M, and determines whether there is a tour of length less that M. Our algorithm B can make use of algorithm A.

9. Construct a verification algorithm for the TSDP with time complexity n. Your algorithm should take any tour and a positive integer M as its input and determine whether the length of the tour is less than M. (*Hint*: Create a loop in which the length of each road in the tour is added to the sum of the lengths of the previous roads in the tour. Then compare the final sum with M.)

10. In Section 9.4, we introduced the Sorting Problem. The decision version of the problem is:

    Sorting Decision Problem

    Input: A list of numbers.

    Problem: Does there exist an ordering of the numbers in which they are sorted?

    Is this problem in NP? Is it NP-complete?

Exercises 11–14 concern another problem that is difficult to solve in polynomial time:

Bin-Packing Problem (BPP)

We are given a list of n numbers, each representing the weight of an object, and we are given one more number representing the capacity of a bin (all bins have the same capacity). We are supposed to put the n objects into several bins such that the sum of the weights of the objects in each bin does not exceed the capacity of the bin. Our goal is to put the objects in the bins in such a way as to use the fewest bins.

For example, if there are 5 objects with weights 1, 2, 4, 4, 5 and the bins all have capacity 6, then we can fit the 5 objects into 4 bins by putting the first two objects into one bin and the last three objects each into their own bin. But that is not the smallest number of bins that could be used. If we put the object with weight 1 and an object with weight 4 together into a bin, the object with weight 2 and an object with weight 4 together into another bin, and then the object with weight 5 into its own bin, we need use only 3 bins; this is the smallest number of bins that can be used for these 5 objects.

11. Suppose there are 5 objects with weights 1, 3, 3, 4, and 5, and suppose that the bins have a capacity of 6. What is the smallest number of bins that we can use?

12. Suppose there are 9 objects with weights 1, 3, 3, 4, 5, 5, 5, 6, and 7, and suppose that the bins have a capacity of 9. What is the smallest number of bins that we can use?

13. Restate the Bin-Packing Problem as a Bin-Packing Decision Problem (BPDP), just as was done with the TSP.[19]

14. Show that if we are given a solution to an instance of BPDP in which there are n objects to be placed in bins, then we can verify that a solution is correct in time n. That is,

we can come up with an algorithm with time complexity n that takes a solution to an instance of the BPDP as its input and verifies that the solution is correct.

15. (This problem assumes a knowledge of Chapter 8.)

Hamilton Graph Problem (HGP)

Input: A graph.

Problem: Does there exist a closed vertex path in the graph?

Suppose we have an algorithm that solves the TSDP. Show how to use that algorithm to construct an algorithm that solves the HGP. Our new algorithm should take a graph as input, somehow convert it into input for the TSDP, and then use the TSDP algorithm to obtain a solution.

16. There are 4 ways to assign truth values to two variables a and b: [a = true and b = true], [a = true and b = false], [a = false and b = true], and [a = false and b = false]. In how many different ways can we assign truth values to three variables a, b, and c?

17. In how many different ways can we assign truth values to four variables a, b, c, and d?

For each of the sets of clauses in Exercises 18–23, if possible, find values of the variables so that *all* the clauses are true.

18. (a or $\sim a$)

19. (a or b), (a or c), (a or d), (a or e), (a or f)

20. (a or b), ($\sim a$ or $\sim b$)

21. (a or b), ($\sim a$ or $\sim b$), ($\sim a$ or b), (a or $\sim b$)

22. (a or b or $\sim c$), ($\sim a$ or c), (b or $\sim c$), ($\sim a$ or $\sim b$ or $\sim c$)

23. (a or b or $\sim c$ or $\sim d$), (a or $\sim b$ or c), (b or c or d), ($\sim a$ or b or $\sim d$), (a or d), (b or $\sim c$), ($\sim a$ or c or $\sim d$), ($\sim b$ or $\sim d$)

19. It has been proven that the Bin-Packing Decision Problem is NP-complete. (See Item 3 in the list *For Further Reading* at the end of this chapter.)

24. If we have a set of clauses, all of which contain the variable a and none of which contain $\sim a$, find a set of values that makes all the clauses true.

25. If we have a set of clauses, all of which contain the variables a or b and none of which contain $\sim a$ and $\sim b$, find a set of values that makes all the clauses true.

26. The following algorithm solves the Satisfiability Problem:

    Slow SAT

    Input: A set of m clauses using n variables.

    Output: *Yes* or *no*, which indicates whether or not there exists values for the n variables that make all the clauses true.

    Algorithm:

    Repeat the following two instructions until all the clauses are true or until all possible values for the variables have been checked:

    Find a new set of values for the variables.

    Check to see whether these values cause all the clauses to be true. If so, stop and output *yes*.

    If no values for the variables were found that make all the clauses true, stop and output *no*.

    This algorithm will solve the problem, but not necessarily in the fastest possible way. It is very similar to Slow Sort, which is why we gave it the name "Slow SAT." Find the time-complexity of the Slow SAT algorithm. (*Hint*: In the worst case, there will not be values for the variables that make all the clauses true. In that case, all possible values for the n variables have to be checked. How many ways are there to assign true and false values to n variables?)

WRITING EXERCISES

1. If a decision problem is not in NP, then not all solutions of the problem can be verified in polynomial time. One problem that is not known to be in NP, and that is, in fact, believed *not* to be in NP, is the following:

   Number-of-Tours Problem

   Input: A list of n cities, a table of lengths of time between pairs of them, and a positive number M.

   Problem: Are there at least 2^n tours in which the length of time is greater than M?

   To verify a solution consisting of 2^n tours, we would have to check all of the tours to see that they are legitimate and that their lengths of time are greater than M. This verification would take at least 2^n steps and so would take exponential time. In spite of that, the Number of Tours Problem may be in NP because there may be some other evidence that could be checked in polynomial time.

   However, it is commonly believed that this problem is not in NP. Think of another decision problem that is probably not in NP and explain why you think it is not.

2. Using complete sentences, write a half-page summary that identifies the main theme of this section and outlines how that theme is carried out.

3. Describe several situations, as was done in the third paragraph of this section, in which a decision problem is more appropriate than an optimization problem. Describe several situations in which an optimization problem is more appropriate.

4. Why would anyone care about finding truth values for Boolean variables that make a set of clauses true? Find and describe one application of the Satisfiability Problem in the real world.

IMPLICATIONS OF NP-COMPLETENESS

The developments in the theory of NP-completeness have had a major influence on attempts to find polynomial-time algorithms for problems. To understand why, let us return to our old friend, the Traveling Salesman Decision Problem, for which there is no known polynomial-time algorithm. What can be done about this particular problem? We cannot just ignore the TSDP, because there are many applications in which the TSP or TSDP plays an important role. Should we spend a lot of time trying to find a polynomial-time algorithm for it? Before the problem was proved to be NP-complete, there might have been a valid reason to do so; but now we recognize that it would most likely be a waste of time. The reason for this is that if we *were* able to find a polynomial-time algorithm for the TSDP, then, because the TSDP is NP-complete, our algorithm could be adapted to construct polynomial-time algorithms for *all* problems in NP. In other words, if we were able to find such an algorithm, our algorithm would be extremely useful and would make us very famous. Unfortunately, the NP-completeness of TSDP also means that it is very unlikely that there is a polynomial-time algorithm for it. After all, many brilliant people have looked for polynomial-time algorithms for various NP problems and have failed to find them.

For the same reasons, it is unlikely that there are polynomial-time algorithms for any NP-complete problem. However, no one has been able either to prove or disprove this conjecture. It is still an open question as to whether any (and therefore all) NP-complete problems have polynomial-time algorithms; but it is commonly believed that no polynomial-time algorithms for NP-complete problems exist.

There is another way to phrase this conjecture. Let P denote the set of decision problems that do have polynomial-time algorithms. The sorting problem is an example of a problem in P. Then we have four classes of problems, which are related as follows: The set of all problems contains the subset NP, which in turn contains the subset P and the subset consisting of the NP-complete problems. The conjecture that NP-complete problems have no polynomial-time algorithms can be restated as

(9.4)

$$P \neq NP$$

Figure 9.7 on page 504 shows the relationship among these sets under the assumption that P ≠ NP. If P = NP, then the oval indicating the set P would coincide with the oval indicating the set NP, and the NP-complete problems would form a subclass, as shown in *Figure 9.8*.

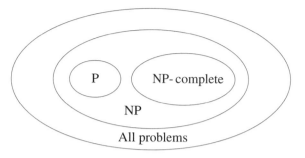

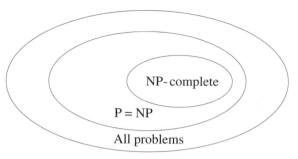

Figure 9.7 *The relationship among various sets of problems assuming that* P ≠ NP.

Figure 9.8 *The relationship among various sets of problems assuming that* P = NP.

Conjecture (9.4) remains one of the most important unproved results in theoretical computer science. What remains to be done is for someone either to prove this conjecture or else to find a polynomial-time algorithm for some NP-complete problem, which will then imply that P = NP. Until either of these is done, the problem remains open. In the meanwhile, people who need a solution to a problem that is NP-complete have to shift the focus of their work.

Suppose you work as programmer for a company and your boss tells you to come up with an algorithm for an NP-complete problem, such as the TSDP. What should you do?

You should first explain the implications of the NP-completeness of the problem to your boss so that he or she will understand that you cannot easily create an efficient algorithm for the problem. Then you should consider how to get around the NP-completeness as best you can.

For example, it would be possible to work around the NP-completeness if the problem that your boss is interested in uses only a small number of input values. In that case, an exponential-time algorithm may be fast enough because there is very little data to be processed. In particular, if the algorithm needs to be run only once, instead of repeatedly, it might be acceptable to let the algorithm

An IBM Power Workstation and a Thinkpad Power Laptop Computer.

The Shortest Tour of 4461 Cities in the Former East Germany. It was computed in 1992 by D. Applegate, V. Chvátal, R. Bixby, and W. Cook, using several Sparc 10/30 workstations in parallel. The computation would have taken 1.9 years if they had used only one workstation.

take several days or even weeks to run before arriving at a solution. (Of course, before you run the algorithm, you still want to make it as efficient as possible so that it will take as little time as possible.)

Another way to attack the problem is to accept the fact that there may not be any algorithms that guarantee a solution in a reasonable amount of time, and instead, try to find some exponential-time algorithms that work well in special cases. In particular, you may be able to find an exponential-time algorithm that solves your particular problem and that is solvable in a very small amount of time. For example, one algorithm was able to solve a TSP instance involving 500,000 cities in only 3.5 hours on a Cray 2 supercomputer because it was a particular case that the algorithm could easily handle. Unfortunately, not all instances fall into that category. Another instance involving only 60 cities was never solved because the algorithm was taking too much time. (See Item 15 in the list *For Further Reading* at the end of this chapter.)

As another alternative, you might want to inspect your boss's problem carefully to see whether it really is equivalent to an NP-complete problem or whether it has different conditions. If the latter is the case, then you could try to see whether you can use these conditions to get around NP-completeness. That is, there may exist an efficient solution for your case because your problem is not really NP-complete. As a trivial example, suppose you are asked to solve the TSP in the case in which all roads have the same length. This problem is easy to solve since *all* tours along those roads have the same total length, which means that your algorithm need choose only an arbitrary tour as its solution.

Finally, if all these methods fail, you could accept the fact that you will not find an exact solution efficiently, and instead, try to find an *approximate* solution efficiently. Of course, finding an approximating algorithm is helpful only if such a solution is acceptable to your boss.

For example, consider the TSP with the extra restriction that the road lengths satisfy the *triangle inequality*. The **triangle inequality** says that for any three cities X, Y, and Z, the length from X to Z is less than or equal to the sum of the lengths from X to Y and Y to Z. One

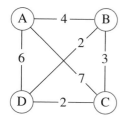

Figure 9.9 A situation
in which road lengths
satisfy the triangle
inequality.

way to think of this inequality is that going from X to Z by way of Y can never be shorter than going directly from X to Z.

Figure 9.9 shows one situation in which the road lengths satisfy the triangle inequality. For example, the length from city A to city C is 7, which is *less than* the sum of the lengths of the roads from A to D and from D to C, which is $6 + 2 = 8$. On the other hand, the sum of the lengths from A to B and from B to C is $4 + 3 = 7$, which *equals* the length from A to C. You can check yourself that the triangle inequality also holds true for the road lengths among all other sets of three cities in this figure. Thus, this is an instance of the TSP in which the triangle inequality is satisfied.

Figure 9.2 on page 464 shows a situation in which the road lengths do *not* satisfy the triangle inequality. For example, the length of the road from A to B is 10, which is *greater than* the length from A to C plus the length from C to B, which is $6 + 3 = 9$. In this case, going *directly* from A to B is *longer* than going from A to B by way of C.

It might appear at first that the problem of finding a shortest tour among roads whose lengths satisfy the triangle inequality is easier to solve than the general TSP because of the extra restriction on the lengths of the roads. However, it can be shown that the TSDP with the triangle inequality is still an NP-complete problem, and so there are no known fast algorithms that give you an exact answer. But there do exist fast approximation algorithms that will give you tours that are at most a few percent longer than the optimal tour. Such approximations may be sufficient for your application.[20]

Other approximation techniques are continually being developed. One such technique is **simulated annealing**, in which the algorithms imitate the process of cooling liquid metals so that they solidify into an optimal structure. The cooling must be done at a very gradual and specific rate to ensure a good resulting structure. A simulated annealing algorithm for the TSP would start with a random tour and use it repeatedly to find new tours with lower total lengths by

Courtesy of International Business Machines

A 4,000,000-bit Memory Chip. It can store over 500,000 characters.

---

20. The triangle inequality helps here because it prohibits roads from being arbitrarily long. That is, if the roads from A to B and B to C are both of length 10, then the road from A to C can be of length at most 20. This fact allows the programmer first to

gradually reducing one of the parameters used in the creation of new tours. (See Item 8 in the list *For Further Reading* at the end of this chapter.) Such techniques are providing useful ways around the problem of NP-completeness. Nevertheless, it cannot be denied that NP-completeness has had a dramatic effect on our understanding of the ability of computers to solve large problems, even ones that initially seem quite simple.

This chapter has used the Traveling Salesman Problem to illustrate the power and importance of mathematics in determining the efficiency of computer algorithms. Without the mathematics discussed here, computer programmers would have to rely on tests or guesswork, neither of which are sufficient for the types of problems they must face on a regular basis. Mathematics has provided them with a tool that can prevent fruitless searches for nonexistent algorithms or the time-wasting use of inefficient algorithms.

9.8 EXERCISES

1. In *Figure 9.7*, the ovals representing the set P and the set of NP-complete problems do not overlap. What would happen if they did overlap? That is, what would result from knowing that there is a problem that is in P and that is NP-complete?

2. Consider a variation on the TSDP that is only concerned with situations in which all roads between pairs of cities have the same length. The problem, as before, is to find a tour of length less than M. Give a polynomial-time algorithm that solves the problem.

3. Do the 5 cities and 10 roads in *Figure 9.3* on page 464 satisfy the triangle inequality?

4. Do the 6 cities and 15 roads in *Figure 9.4* on page 464 satisfy the triangle inequality?

In Exercises 5–7, suppose there are 5 cities and all the roads except one have length 10. Furthermore, suppose that the triangle inequality is satisfied by all the roads. Which of the following lengths are possible for the remaining road?

5. length 1   6. length 10   7. length 30

8. Do the 5 cities and 10 roads in *Figure 9.10* satisfy the triangle inequality? If not, find three cities for which the triangle inequality fails.

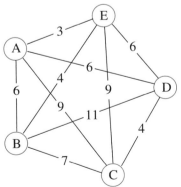

Figure 9.10 A longest-tour problem.

create approximations that are not necessarily tours (due to the fact that they might repeat a road or in some other way not form a complete tour), but which can then be modified so as to become tours without drastically increasing the total length.

9. Suppose your goal is to find a *longest* tour instead of a shortest.

   (a) Find a longest tour using the cities in *Figure 9.10*. Does it seem easier to find a longest tour rather than a shortest tour?

   (b) Now modify the lengths of all the roads by subtracting each length from 12 and using these differences as the new lengths. For example, change the length of the road from A to E so that it equals $12 - 3 = 9$. Now find a *shortest* tour using the new lengths. How does this answer relate to the answer you found in Part (a)?

   (c) If you know a method for finding a longest tour, how can you use that method to find a shortest tour, or vice versa?

10. One suggestion for a polynomial-time algorithm for the TSP is the following "greedy" algorithm:

Starting with city A, find the shortest of all the roads from A to other cities and follow that road (or any such shortest road). Then from the new city find the shortest of all the roads from this new city to cities other than A. Repeat this process of choosing the shortest road from the current city to another unvisited city until all cities have been visited. Then return directly to A.

   (a) What is the worst-case time complexity of this greedy algorithm?

   (b) Find a set of cities and roads for which the greedy algorithm does not give a shortest tour. (*Hint*: You need only 4 cities.)

   (c) Find a set of cities and roads for which the tour found by the greedy algorithm is more than 100 times as long as a shortest tour. (*Hint*: You need only 4 cities.)

WRITING EXERCISES

1. Research "greedy" algorithms such as the one described in Exercise 10, and explain several of them. Determine why they are called by that name.

2. Using complete sentences, write a half-page summary that identifies the main theme of this section and outlines how that theme is carried out.

3. Defend the statement: "The average person should not attempt to find a polynomial-time algorithm for the Traveling Salesman Problem."

4. For which of the applications of the TSP mentioned in Section 9.2 is it essential that the best tour be found? For which of the applications would it be sufficient to find a tour whose length is within 10% of that of the best tour? Explain your reasoning as well as the implications of your answers.

5. Which of the applications of the TSP mentioned in Section 9.2 involve the triangle inequality? Explain your reasoning.

6. Small problems are solvable even with very slow algorithms. That is, even an exponential-time algorithm can solve problems in a reasonable amount of time if the problems are "small enough." Therefore, your boss might think it is reasonable to tell you to solve the TSP in a particular case using 5 cities, even though the TSDP is NP-complete. Is that a reasonable request? How about solving the TSP in a case using 10 cities? What about 15 cities or 50 cities? Explain your answers.

LINK: ALGORITHMS, ABSTRACTION, AND STRATEGIC PLANNING

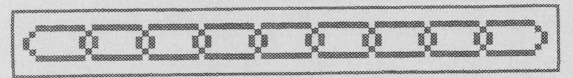

A Wallpaper Pattern

9.9 Suppose you are confronted with a Traveling Salesman Problem in which the 5 cities are Atlanta, Boston, Chicago, Denver, and El Paso. Suppose you also have a second TSP that is identical with the first problem except that the cities are called A, B, C, D, and E. Do you really have two different mathematical problems to solve? Of course not. Clearly, the names of the cities are irrelevant to the mathematical problem of finding a shortest tour through them.

One of the central processes of mathematics is **abstraction**, whereby irrelevant details of problems are removed and only the important mathematical aspects are considered. The abstraction process is demonstrated in most chapters of this book. For example, the subject of geometry, discussed in Chapter 3, is an abstraction originally created by the Egyptians and Greeks to help them investigate measurements of the Earth. The graphs discussed in Chapter 8 are really abstractions of collections of objects, such as cities, with connections, such as roads, between them. A fuller discussion of the abstraction process appears at the beginning of Chapter 6.

By finding the "essence" of a problem and eliminating all superfluous elements, mathematicians can often simplify things to a point where a solution can easily be obtained. In particular, if it turns out that the problem involves the same core idea as a previously solved problem, then the new problem can be quickly solved using the methods for solving the earlier problem. In the same way, the process of abstraction helps mathematicians generalize their results in that their results apply to all problems with the same essential core, regardless of the other details of these problems.

How does one find the essence of a problem? Often it involves the problem-solving tactics discussed in Chapter 1. A mathematician may need to try simpler problems, look for patterns, make diagrams, etc. In this link section we will demonstrate these processes with examples drawn from two commonplace activities—wallpapering and the game of darts. Problems involved in these activities will be reduced to variations of the Traveling Salesman Problem.

The first variation concerns wallpaper patterns. To prepare a bid for a job, a wallpaper contractor must estimate how much wallpaper will be needed. The walls are $7\frac{1}{2}$ feet from the top of the baseboard to the ceiling trim, with relatively few windows and doors. The owner of the building, a person of strong will and peculiar taste, has already chosen the wallpaper. It has a pattern that repeats every 2 feet. The paper must be hung so that the pattern on each vertical strip starts 4 inches higher than it did on the preceding strip, as is shown in *Figure 9.11*.

Now, this wallpaper comes in rolls that are 32 feet long. Since the length of each strip to be hung is $7\frac{1}{2}$ feet and since $4 \times 7\frac{1}{2} = 30$, the contractor hopes to get 4 strips out of each roll, leaving only 2 feet, or 24 inches, of wasted paper per roll. The problem is that the pattern must be matched in exactly the way described. Since the pattern only repeats at 2-foot intervals, after cutting and pasting one strip the contractor might have to waste almost 2 feet of paper when cutting the next strip in order that it fits the required design.

In fact, there are several places where waste can occur in one roll. If the pattern on the first strip cut from the roll doesn't start where the roll starts, some paper will be wasted there. Similarly, if the pattern on the first strip cut from the roll doesn't end where the second strip should begin, paper will be wasted. The same is true for the remaining strips. Finally, if the pattern on the last strip cut from the roll doesn't end where the roll ends, paper will be wasted. Therefore, the contractor's problem can be stated as follows:

> Given these restrictions, can 4 strips of wallpaper be cut out of each roll?

Our solution of this problem will refer to the 4 strips of paper shown in *Figure 9.11*. As a first step, note that we can visualize the roll as being marked

Figure 9.11
A wallpapering problem.

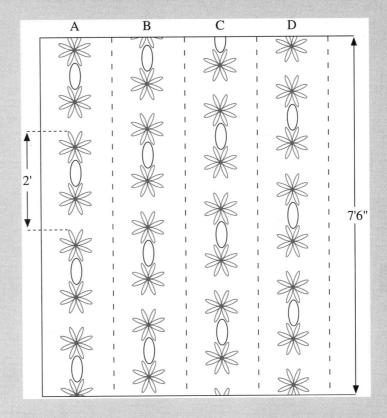

into 2-foot sections because the wallpaper has a repeating pattern every 2 feet. (Do you see that the pattern itself does not matter, only the fact that it repeats every two feet?) We will suppose that each strip has the characteristics shown in *Table 9.8*.

When we say that the starting position is 0 inches, we mean that we have to start the strip 0 inches down from the start of one section of the repeating wallpaper patterns. Similarly, an ending position of 18 inches means that the strip ends 18 inches down from the start of a section. Therefore, if we were to cut strip *A* first and strip *B* next, then between the two cuts we would waste 6 inches of the last section used by *A* and the first 4 inches of the first section used by *B*, for a total waste of 10 inches of paper. On the other hand, if we start with strip *B* and follow it with strip *A*, then, conceivably, we might waste only 2 inches

Strip	Starting Position	Ending Position
A	0 inches	18 inches
B	4 inches	22 inches
C	8 inches	2 inches
D	12 inches	6 inches

Table 9.8 *Characteristics of strips of wallpaper.*

of paper between the two strips, namely the 2 inches not used by strip *B* at the end of a repeating pattern. Therefore, the order in which the 4 strips of wallpaper are cut from the roll can make a big difference in the amount of waste produced. With this in mind, we can rephrase the contractor's problem:

> Is there an order in which the 4 strips can be cut so that they all fit on the 32-foot roll, that is, so that the total amount of waste before the first strip, after the last strip, and between successive strips is at most 24 inches?

To help us answer this question, in *Table 9.9* we list the various amounts that would be wasted by cutting the strips in all possible orders from the roll. For example, the value of 2 in the third row and second column means that if strip *C* were to be cut from the roll followed by strip *B*, the amount of paper wasted between the cuts would be 2 inches. The *S* in the bottom row refers to the start of the roll and the *E* in the last column refers to the end. Thus, the value of 8 in row *S* and column *C* means that 8 inches of paper would be wasted at the beginning of the roll if *C* were the first strip cut. Similarly, the value of 2 in row *B* and column *E* means that 2 inches of paper would be wasted at the end of the roll if B were the last strip cut from the roll.

Figure 9.12 illustrates the information in *Table 9.9*. Each strip of wallpaper is represented by a circle; the arrows from one circle to another are labeled with the amount of paper that would be wasted if the second strip were cut from the roll immediately after the first strip is cut. That is, the arrows between circles are labeled with the numbers from *Table 9.9*. The circle labeled *SE* corresponds to both the start and the end of the roll of wallpaper. The arrows *from* that circle to other circles are labeled by the numbers in row *S* in *Table 9.9*, and the

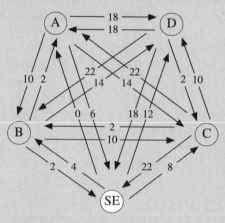

	A	B	C	D	E
A	0	10	14	18	6
B	2	0	10	14	2
C	22	2	0	10	22
D	18	22	2	0	18
S	0	4	8	12	

Table 9.9 *Wastes resulting from cutting the strips in different orderings.*

Figure 9.12 *A variation of the Traveling Salesman Problem as an aid to solving the wallpapering problem.*

arrows pointing *toward* that circle from other circles are labeled by the numbers in column *E*.

Using this illustration, think again about the contractor's problem of finding an order in which the 4 strips of paper can be cut in order to fit all of them in a 32-foot roll. But now consider this in terms of a tour of these 5 "cities" that starts and ends at *SE*. For example, consider the tour *SE–A–B–C–D–SE*. This corresponds to cutting strip *A* from the roll of wallpaper first, followed in order by *B*, *C*, and *D*. The length of this tour, $0 + 10 + 10 + 10 + 18 = 48$, corresponds to the amount of wallpaper (in inches) that would be wasted if the 4 strips were cut from the roll in this order. Clearly, this particular order of cutting the strips from the roll will not work since we only have 24 inches, not 48 inches, of excess. Therefore, the contractor's problem can be rephrased in terms of *Figure 9.12*, as follows:

Is there a tour of the 5 cities with total length at most 24?

In other words, the contractor's problem is essentially the same as the Traveling Salesman Decision Problem, except for one minor difference: The "distance" from one city to another is not necessarily the same as the distance from the second city back to the first. For example, the distance from *A* to *C* is not the same as the distance from *C* to *A*, unlike the TSDP described earlier in this chapter. However, it turns out that this does not significantly change the complexity of the problem. That is, this asymmetric version of the TSDP is still NP-complete, and so it is unlikely anyone will find a polynomial-time algorithm for it.

Luckily, we are only interested in solving the TSDP using 5 cities, which can be done readily by inspection. We see that there is a tour whose length is at most 24, namely *SE–D–C–B–A–SE*. Furthermore, this is the only tour whose length is at most 24. Therefore, the contractor can cut these 4 strips from one roll of wallpaper if and only if he does so in the order *D*, *C*, *B*, *A*.

There is, of course, nothing to stop us from generalizing this idea and applying these techniques in situations where there are more or fewer than 4 strips of wallpaper and when the roll of wallpaper is of length other than 32 feet. The problem of minimizing the wallpaper wasted between cuttings is still essentially the TSP or the TSDP.

In summary, note the techniques from Section 1.2 that were used in attempting to solve this problem. In particular, all details of the wallpapering, such as the particular pattern on the wallpaper, that were irrelevant to the problem were ignored and the remaining information was analyzed in an abstract manner. The result was that the problem turned out to be a variation on the TSDP, and so all known theory and algorithms pertaining to the TSDP apply to this problem. This saves the problem solver a great deal of effort in trying to come up with a solution from scratch.

Our second variation of the Traveling Salesman Problem involves the game of darts. The standard dartboard pattern is shown in *Figure 9.13* on page 514.

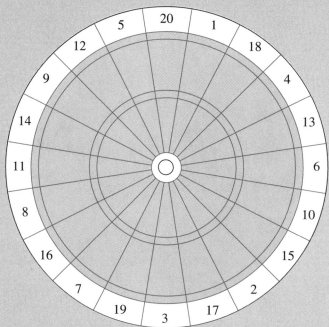

Figure 9.13 A standard dartboard pattern.

Note that the numbers on the board are 1 through 20, arranged in a particular order. The question here is: why in that order? Historically, the dartboard was designed so that if players miss the region they are aiming at, they will hit a region whose value is as different as possible from the desired value. That is, the values have ideally been arranged in such a way as to maximize the penalty players receive for missing their target. This suggests a new problem: Find the optimal arrangement of the numbers on the dartboard so as to maximize this penalty.

Let us see how mathematicians would analyze and solve this problem. As a first step, we simplify the problem a little by assuming that players miss their target by at most one region. Then we can rephrase the problem as follows: How should the numbers be arranged around the board so that the differences between adjacent numbers are as large as possible? Let us look at the example of the standard dartboard pattern. Take the absolute value of each difference and then add these values to get a sum of these **absolute differences**. By going around the dartboard clockwise, starting at the top, we get the sum

$$|20 - 1| + |1 - 18| + |18 - 4| + \ldots + |5 - 20| = 19 + 17 + 14 + \ldots + 15$$
$$= 198$$

Is this the largest possible total, or is there a better dartboard design than the standard pattern for our purpose? This question can be rephrased:

Of all arrangements of the numbers around a dartboard, which pattern will result in the largest sum of absolute differences?

To help us answer this question, we can mentally create a picture that allows us to look at the problem in a different way. In our mental picture, we draw 20 large dots, each one corresponding to a number from 1 to 20, arranged in a big circle. We connect each dot to every other dot with a line labeled by the absolute difference between the numbers on the ends of the line. For example, the dots corresponding to the numbers 4 and 19 would be connected by a line labeled with the number 15. (Such a picture is not actually drawn on paper since it would be almost unreadable due to all the lines—190 of them—and dots in it.) Then our problem can be rephrased as follows: Which *tour* of the dots has the maximum length? This is again a variation on the TSP! This particular variation asks for the maximum-length tour instead of the minimum-length tour, but the complexity turns out to be the same. That is, this variation is still NP-complete, and hence, it is unlikely anyone will create a polynomial-time algorithm to solve it.

Luckily for us, we only want a solution for a problem involving 20 "cities," so the complexity is not quite as important as it might otherwise be. We can use one of the existing algorithms that give exact solutions of the TSP (for example, the Slow TSP given on pages 483–84), and slightly modify the algorithm so that it finds the maximum tour instead of the minimum tour. When we do so, we find that the optimal solutions of the dartboard problem have a sum of absolute differences of 200. One such solution is the ordering

$$20, 1, 19, 3, 17, 5, 15, 7, 13, 9, 11, 10, 12, 8, 14, 6, 16, 4, 18, 2$$

This is not the only ordering with a sum of absolute differences of 200, so there is actually more than one optimal arrangement. However, the arrangement on the standard dartboard is *not* one of these optimal arrangements because, by a prior calculation, the sum of absolute differences for the standard board is only 198.

It might be interesting to let expert throwers use one of these optimal patterns and then compare the results with the results obtained when they use the standard board. However, tradition plays a large role here and so we are unlikely ever to see the standard board changed, regardless of the results of such a comparison.

In summary, we have again started with a problem seemingly different from the TSP; but when we eliminated superfluous elements and reduced it to its essential components, it turned out to be a variation on the TSP. So, we were able to solve the problem using slightly modified TSP algorithms.

The process of abstraction was demonstrated with two examples in this Link section. Even though the examples concerned very different activities—wallpapering and the game of darts—it was shown that each problem could be viewed as a variation of the Traveling Salesman Problem. One of the amazing and wonderful aspects of mathematics is that applications of it appear in unexpected places, which hopefully demonstrates to the reader the value of understanding mathematics and its role in all forms of strategic planning.

TOPICS FOR PAPERS

1. Research one of the following characters in the history of computer science and write a biographical account of the person you choose, concentrating on his/her contribution to computers:

    ■ Charles Babbage
    ■ Ada Lovelace
    ■ John von Neumann
    ■ Alan Turing
    ■ Norbert Wiener

    (Be sure to list the source(s) you used in writing this paper, and be careful that you do not simply copy or paraphrase the material from your sources.)

2. Write a paper of no more than two pages, using your current knowledge and opinions, on the question, "Can a machine think?" Then read Alan Turing's essay of that title. (It originally appeared in *Mind* in 1950; reprinted in *The World of Mathematics*, James R. Newman, ed. New York: Simon & Schuster, Inc., 1956, 1988.) Write a second paper, as a companion to and commentary on your own original paper, covering the following points:

    ■ Which portions of your original paper would you leave unchanged? Which would you change?

    ■ Which issues, if any, did Turing raise that you did not deal with in your paper? Would you characterize these issues as an oversight on your part? Are they points of disagreement with Turing?

3. Read about (deterministic) Turing machines. (See Item 1 in the list *For Further Reading*, which follows, for an elementary introduction to Turing machines.

    (a) Create your own Turing machines as described here:

    i. Create rules for a Turing machine that takes any string of 0s and 1s as input and converts the last 0 (that is, the 0 furthest to the right) in the string to a 1. You may assume that there is at least one 0 in the input string. (*Hint*: Move the read/write head of the Turing machine to the right end of the string and then back up.)

    ii. Create rules for a Turing machine that takes any string of 0s and 1s as input and determines whether there are at least two 0s in the string. If the machine finds two or more 0s, it writes a T at the end of the string. Otherwise, it writes an F.

    (b) Write a paper describing the differences between "step" as used in this chapter and one step in a Turing machine computation. Discuss the significance of these differences.

    (c) Read about nondeterministic Turing machines and summarize the difference between them and deterministic Turing machines.

4. Write a paper on the work that has been done to speed up computers over the last forty years. This should include a discussion of CPUs, main memory (RAM), hard disks, and the connections between them, as well as a discussion of parallel computers (which have more than one CPU).

5. A computer is just a piece of hardware without the innate intelligence that humans possess. How does such a computer actually understand programs? That is, how does a computer know what each step means and which step should be executed next? Do some research and write a paper on how computers and, particularly, computer hardware use electricity to make decisions and to execute programs.

Some variations on the TSP are easier to solve than the TSP itself, and others are harder. Topics 6–9 ask you to explore what happens in some of these variations. In particular, try to determine which of the variations, if any, are easier to solve than the TSP.

6. (a) Suppose that you allow roads to have negative times associated with them. For example, if the road from A to E in *Figure 9.1* on page 461 has a time of -10 instead of 10, will that change the problem at all? Does it ever make sense to allow negative times?

    (b) Assume that you are allowed to visit each city more than once in your tours and that you are allowed to use each road more than once. In *Figure 9.14*, find a shortest route from A that visits all the other cities *at least once* before returning to A. Then use this assumption in *Figures 9.2–9.4* on page 464. Do you get a shorter route than in the case where you are allowed to visit each city only once?

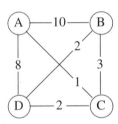

Figure 9.14 Tours for Exercise 6(b).

    (c) What happens if you allow roads with negative lengths of time *and* you allow each city to be visited more than once? (*Hint*: Consider three cities A, B, and C such that the lengths from A to B, from B to C, and from C to A are all -10. One route is A–B–C–A. But A–B–C–A–B–C–A is shorter. Is there a shortest route?)

7. Suppose that you do not care about the return to the starting city. That is, you want to find a shortest path that visits all the cities and stops at the last city visited. Explore the effect of this change.

8. Suppose that you want to visit each city (other than the starting one) twice before returning to the starting city. For example, if you were dropping off a questionnaire or census form and then returning later to pick it up, you would need to visit each city twice. Assume that you do not need to return to the starting city in between your visits to the other cities and that you do not have to visit all the cities the first time before returning to any of them a second time. That is, you could visit a city B for the second time before visiting city D for the first time. What is a shortest route that visits the cities in this manner?

9. Suppose that, unlike our assumptions in the rest of the chapter, the lengths of the roads in one direction are not necessarily the same as the lengths of the roads in the opposite direction. That is, traveling along the road from one city to another may take more or less time than going along the same road in the opposite direction. For example, consider *Figure 9.15*. Here, each road is represented by two arrows pointing in opposite directions and each direction has a different length. Find a shortest tour in this case. Explore the effects of this change with other collections of cities and roads.

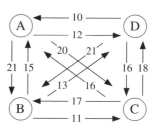

Figure 9.15 *Tours for Exercise 9.*

10. In Exercises 12, 15, and 17 of Section 9.2, you found all the tours of four, five, and six cities. But there are not as many different *lengths* of tours as there are tours because some tours have the same lengths. For example, if two tours use exactly the same set of roads but traverse the roads in different orders, the two tours will have the same length.

    (a) How many different lengths of tours are there among all the tours you found in Exercises 12, 15, and 17 of Section 9.2?

    (b) Suppose that there are four cities and that all the roads between pairs of cities have the same length. How many different-length tours are there?

    (c) Consider *Figure 9.16*. How many different-length tours are there that start at city *A*?

    (d) What can be said about the number of different tours possible among four cities and about the number of different-length tours possible among four cities? Give a formula in each case, if possible.

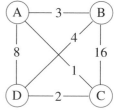

Figure 9.16 *Tours for Exercise 10(c).*

    (e) What can be said about the number of different tours possible among five cities and about the number of different-length tours possible among five cities? Give a formula in each case, if possible.

    (f) What can be said about the number of different tours possible among *n* cities and about the number of different-length tours possible among *n* cities? Give a formula in each case, if possible.

11. Do some research to find out which variations of the TSP can be solved in polynomial time. In your paper explain as much as possible why these variations are solvable in polynomial time even though the TSP itself is most likely not solvable in polynomial time.

12. Read about various programming languages and write about the differences and similarities. Why are there so many languages? Isn't one enough?

FOR FURTHER READING

1. Decker, R., and S. Hirshfield. *The Analytical Engine*. Belmont, California: Wadsworth, 1990.

2. Devlin, Keith J. *Mathematics: The New Golden Age*. New York: Penguin Books, 1988.

3. Garey, Michael R., and David S. Johnson. *Computers and Intractability: A Guide to the Theory of NP-Completeness*. San Francisco: W. H. Freeman and Company, 1979.

4. Gersting, Judith L., and Michael C. Gemignani. *The Computer: History, Workings, Uses & Limitations*. New York: Ardsley House, Publishers, Inc., 1988.

5. Gleick, James. *Chaos, Making a New Science*. New York: Penguin Books, 1987.

6. Goldstine, Herman. *The Computer from Pascal to Von Neumann*. Princeton: Princeton University Press, 1972.

7. Goodman, S. E., L. I. Press, S. R. Ruth, and A. M. Rutkowski. "The Global Diffusion of the Internet: Patterns and Problems," *Communications of the ACM*, Vol. 37, No. 8, Aug. 1994, pp. 27–31.

8. Kirkpatrick, S., C. D. Gelatt, Jr., and M. P. Vecchi. "Optimization by Simulated Annealing," *Science*, Vol. 220, No. 4598, 13 May 1983, pp. 671–80.

9. Knuth, D. E. *The Art of Computer Programming*. 2nd Ed. Vols. 1–3. Reading, MA: Addison-Wesley, 1973.

10. Kolata, Gina Bari. "Trial-and-Error Games That Puzzles Fast Computers," *Smithsonian*, October, 1979, pp. 90–96.

11. Laporte, G. "The Traveling Salesman Problem: An overview of exact and approximate algorithms," *European Journal of Operational Research*, Vol. 59, No. 2, June 10, 1992, pp. 231–47.

12. Lawler, E. L., J. K. Lenstra, A. H. G. Rinnooy Kan, D. B. Shmoys, ed. *The Traveling Salesman Problem: A Guided Tour of Combinatorial Optimization*. New York: John Wiley & Sons, 1985.

13. Lenstra, J. K., and A. H. G. Rinnooy Kan. "Some Simple Applications of the Travelling Salesman Problem," *Operations Research Quarterly*, Vol. 26, No. 4, i, 1975, pp. 717–33.

14. McCorduck, Pamela. *Machines Who Think*. San Francisco: W. H. Freeman and Company, 1979.

15. Miller, D. L. and J. F. Pekny. "Exact Solution of Large Asymmetric Traveling Salesman Problems," *Science*, Vol. 251, February 15, 1991, pp. 754–61.

16. Peterson, Ivars. *The Mathematical Tourist*. San Fransisco: W. H. Freeman and Company, 1988.

17. Poirot, James, and David Groves. *Computers and Mathematics*. Manchaca, TX: Sterling Swift Publishing Company, 1979.

A

BASIC LOGIC

STATEMENTS AND THEIR NEGATIONS

A.1

We present here a brief summary of elementary mathematical logic. In order to find a simple starting point, we leave to the philosophers and scientists many relevant, but difficult, questions, such as "What is truth?" and examine, instead, just the *form* of the reasoning process. Consequently, our study of logic focuses on how the known truth or falsity of some statements can be used to guarantee the truth or falsity of others. Thus, we begin the study of formal logic by regarding the words **true** and **false** just as labels applied (somehow) to some sentences. These two words are called **truth values**; it is a basic principle of our logical system that there are only two of them. (There are other systems of logic that use more than two truth values, but they are not as simple or as widely used as the system presented here.)

All reasoning is based on "statements" of one sort or another; so we consider first what makes a statement meaningful in the context of logic. A **statement** is a sentence that has a truth value, that is, a sentence that can be labeled either *true* or *false*. This truth value may come from various sources. Often we rely on our common-sense view of reality to tell us whether a statement is true or false; in some cases the truth value is just assigned (as in formal axiom systems). Sometimes the truth value of a compound or complex sentence can be derived from knowing the truth values of its simpler parts. In any case, a sentence *must* have a truth value in order for it to be called a statement.

| EXAMPLE A.1 | "There is printing on this page" is a true statement. "No one passes mathematics" is a false statement. |

| EXAMPLE A.2 | "Today is Tuesday" is a statement. It is either true or false, depending on when you are reading this. |

| EXAMPLE A.3 | The sentences "What's your name?" and "Close the book!" are not statements because there is no reasonable way to assign either of them a truth value. |

Standard logic is based on two fundamental assumptions about truth values:

- **The Law of the Excluded Middle:** There are only two truth values, *true* and *false*.
- **The Law of Contradiction:** No statement can be both true and false at the same time (in the same context).

These laws are crucial to the proof process. Specific examples of their use appear later in this Appendix and throughout the book.

| EXAMPLE A.4 | In the first paragraph of Chapter 1, every sentence but the last is a statement; each one is either true or false. (The fact that you might consider them all true stems, no doubt, from your complete faith in the accuracy of the authors. Such faith is flattering, but not always wise.) |

| EXAMPLE A.5 | The sentence "This statement is false" is *not* a statement in standard logic because it cannot have either truth value. If we consider it true, then we must accept what it says, which makes it also false, violating the Law of Contradiction. If we consider it false, then the Law of the Excluded Middle implies that the only alternative is that it is true, again violating the Law of Contradiction. |

Two types of statements deserve special consideration because of their importance in the reasoning process; they are the two basic kinds of "quantified" statements:

- A **universal statement** asserts that all things of a certain kind satisfy some condition.

■ An **existential statement** asserts the existence of at least one thing that satisfies some condition.

Words such as *every, all, some, at least one,* etc., which are used in these types of statements, are called **quantifiers**.

EXAMPLE A.6	"All mice are animals" and "Every building has a flat roof" are universal statements.

EXAMPLE A.7	"There is a fly in my soup," "Wizards exist," and "There really are unicorns" are existential statements.

Note The word *some* is used in mathematics as a shorter form of "there is at least one." It does not necessarily refer to more than one thing, nor does it rule out the possibility that *all* things might satisfy the condition being discussed. Both "Some trout are fish" and "All trout are fish" are true statements.

EXAMPLE A.8	"Today is Thursday" and "The moon is made of green cheese" are neither universal nor existential statements. They are called **particular statements** because they refer to properties of specifically designated things.

Universal, existential, and particular statements can be either true or false.

EXAMPLE A.9	"$x + 3 = 7$" is *not* a statement because not enough is known about x to decide whether the sentence is true or false. However,

$$\text{"For all numbers } x, x + 3 = 7\text{"}$$

is a false universal statement;

$$\text{"There exists a number } x \text{ such that } x + 3 = 7\text{"}$$

is a true existential statement;

$$\text{"There exists a number } x \text{ such that } x + 3 = x\text{"}$$

is a false existential statement;

$$\text{"}4 + 3 = 7\text{"}$$

is a true particular statement.

Every statement has a *logical opposite*. Having said this, let us quickly caution you that "opposite" must be interpreted very carefully. For instance, "This car is going north" is *not* the logical opposite of "This car is going south." To reinforce that caution, we label this idea with a less ambiguous word defined in terms of truth values, considering the values *true* and *false* as logical opposites.

DEFINITION

The **negation** of a statement is a statement whose truth value is always opposite to that of the original statement.

Note

The word "always," as used in the definition of negation, means "in every circumstance." Thus, *whenever* a statement is true, its negation *must* be false; and *whenever* the statement is false, its negation *must* be true.

EXAMPLE A.10

"The wall is completely white" and "The wall is not completely white" are negations of each other. "The wall is completely white" and "The wall is completely green" are *not* negations of each other because, although they cannot both be true at the same time, they can both be false at the same time. (Such statements are called **contraries**.)

NOTATION

In algebra, the task of writing numerical expressions is often simplified by using letters to represent numbers. Similarly, we simplify (and shorten) the writing of logical expressions by using letters, such as s, p, and q, to represent statements. If s represents a statement, then its negation is denoted by $\sim s$.

EXAMPLE A.11

If s represents "Jack owns a car," then $\sim s$ represents "Jack does not own a car." Suppose s is true; that is, suppose that Jack actually owns a car. Then the statement $\sim s$ is false. On the other hand, suppose s is false. This means that Jack does not own a car, which means that $\sim s$ is true.

The truth-value relationships among several statements may be represented conveniently by a diagram called a **truth table**. A very simple truth table can be used to show the relationship between a statement s and its negation $\sim s$. For instance, suppose s is the statement "Jack owns a car," as in Example A.11. Then $\sim s$ can be stated as "Jack does not own a car." We use a column of the table for each of the two statements (*Table A.1(a)*). Because s may be either true or false,

s	$\sim s$		s	$\sim s$		s	$\sim s$
			T			T	F
			F			F	T
(a)			(b)			(c)	

Table A.1 Negation.

s	$\sim s$	$\sim(\sim s)$
T	F	T
F	T	F

Table A.2 Double negation.

we enter both possibilities (symbolized by T and F) in the first column (*Table A.1*(b)). The entries in the second column indicate that the truth value of $\sim s$ must always be the opposite of the truth value of s (*Table A.1*(c)).

As another example, the relationships among a statement s, its negation $\sim s$, and the negation of that, $\sim(\sim s)$, are shown by *Table A.2*. Notice that the first and third columns have identical truth-value entries. This means that s and $\sim(\sim s)$ are the same logically, even though they might be grammatically different. If s is "Jack has a car," then $\sim(\sim s)$ can be phrased "It is not the case that Jack doesn't have a car."

DEFINITION

Two statements are **logically equivalent** (or simply **equivalent**) if they always have the same truth values.

Thus, *Table A.2* shows that, for every statement s,

s and $\sim(\sim s)$ are logically equivalent.

Because they have identical truth values, equivalent statements are interchangeable at any time in any logical argument. They are just different ways to say the same thing. The choice of which form to use is merely a matter of convenience or taste.

EXAMPLE A.12

The negation of the universal statement "All animals have fur" is

"Not all animals have fur."

It is logically equivalent to the existential statement

"Some animals do not have fur."

EXAMPLE A.13

The negation of "All mice are gray" may be phrased as "There is a mouse that is not gray," or "Some mice are not gray," or "Not all mice are gray." The statement "No mice are gray" is *not* a negation of the first statement; it and the first statement are contraries.

Examples A.12 and A.13 illustrate why we refer to *the* negation of a statement, despite previous examples of several apparently different negations of the same statement. The forms of the negation may be *grammatically* different, but they are *logically* equivalent; that is, *from the viewpoint of truth values, every statement has exactly one negation*. These two examples also illustrate a basic relationship between universal and existential statements:

- The negation of a universal statement is (equivalent to) an existential statement.

- The negation of an existential statement is (equivalent to) a universal statement.

EXAMPLE **A.14**	"There are red fire engines" and "No fire engines are red" are negations of each other. "All birds fly" and "Some birds do not fly" are negations of each other.

EXAMPLE **A.15**	"It is false that the wall is not green" is equivalent to "The wall is green." (Why?)

A.1   EXERCISES

In Exercises 1–10, which of these sentences are statements? Why? Label each statement *true* or *false*.

1. The sun is shining.

2. This book is written in French.

3. Nobody loves me.

4. Where are you going?

5. Join the Pepsi generation.

6. Meet me in St. Louis.

7. $3 + 5 = 8$

8. $3 + x = 8$

9. $3 + 5 = 7$

10. For some number x, $3 + x = 8$

In Exercises 11–20, classify each statement as *universal*, *existential*, or *particular*.

11. Some flowers are yellow.

12. All dogs have fleas.

13. There is a unicorn on the lawn.

14. The sun is shining.

15. Some elephants fly.

16. These trees are green.

17. Not all students are sophomores.

18. No textbooks are useful.

19. Everything I need to know I learned in kindergarten.

20. Some important ideas don't fit on a bumper sticker.

In Exercises 21–30, express (an equivalent form of) the negation of each statement.

21. This is a mathematics book.

22. Today is Sunday.

23. All roses are red.

24. Some violets are blue.

25. $8 + 3 = 11$

26. No rabbits chase mice.

27. Some cars do not get good gas mileage.

28. There are at least three trees in the field.

29. The square of every number is positive.

30. Thanksgiving Day is always on a Thursday.

31–40. Express (an equivalent form of) the negation of each statement in Exercises 11–20.

CONJUNCTIONS AND DISJUNCTIONS

Next, we consider truth values for combinations of statements. We focus on pairs of statements because more complex combinations can be analyzed two at a time. The truth value of a combination can be deduced from the truth value of its component parts, according to several simple rules that are generally in agreement with common sense. These rules are actually the *definitions* of the connective words used. In this section we consider the formal logical usage of *and* and *or*.

DEFINITION

The **conjunction** of two statements p and q is a statement of the form "p and q"; it is true if p and q are both true, and is false otherwise. "p and q" is often written "$p \wedge q$."

Note

The word "but," which signals a contrast between two statements it connects, is usually the logical equivalent of "and." That is, "p but q" usually has the same truth-value cases as "p and q." For instance, "Jo was sick, but she came to class, anyway" is logically equivalent to "Jo was sick and she came to class." When you analyze the truth value of a sentence that contains "but," replacing it with "and" will usually simplify your task.

DEFINITION

The **disjunction** of two statements p and q is a statement of the form "p or q"; it is false if p and q are both false, and is true otherwise. "p or q" is often written "$p \vee q$."

Note

This usage is sometimes called the *inclusive or* because it allows both statements to be true. If we were to define "p or q" to be false when both statements are true (as in "That's either a dog or a cat"), it would be called the *exclusive or*. This latter usage is discussed in Exercise 30.

The truth values for these two compound statement forms are summarized in *Table A.3*. Notice that we must consider four different cases, corresponding to the four different truth-value possibilities for the *pair* of statements p, q. Those four pairs of possible truth values are listed in the first two columns of the table.

p	q	p and q	p or q
T	T	T	T
T	F	F	T
F	T	F	T
F	F	F	F

Table A.3 *Conjunction and disjunction.*

EXAMPLE A.16

Consider the statements

p: Today is Friday.

q: The sun is shining.

Then

"p and q" is "Today is Friday and the sun is shining."

"p or q" is "Today is Friday or the sun is shining."

To see how the truth tables behave in this situation, observe from *Table A.3* that "p and q" is true only on sunny Fridays, but "p or q" is true every Friday, regardless of the weather, and is also true on every sunny day.

At this point it is natural to ask how negation affects these compound statement forms. In particular, how are the negations of the conjunction and disjunction of two statements related to the negations of the separate statements? Notice that everyday English usage is based on a precise logical structure, but we seldom pay specific attention to the structure itself. For instance, if you tell someone who wants you to take a message that you do not have paper or pen, you are saying that you do not have paper *and* you do not have a pen. Often this appears as a prepositional phrase—such as "without paper or pen," meaning "without paper and without pen"—in which the preposition itself conveys the negative sense. This shift from *or* to *and* when the compound statement is negated typifies an important general principle governing the negation of conjunctions and disjunctions.

Consider the four combination forms involving negation:

$$\sim(p \text{ and } q) \qquad \sim(p \text{ or } q) \qquad (\sim p) \text{ and } (\sim q) \qquad (\sim p) \text{ or } (\sim q)$$

As usual in mathematical symbolism, we first perform the operations within the parentheses. Thus, in "$\sim(p \text{ and } q)$" we first consider "p and q" and then negate it; on the other hand, in "$(\sim p)$ and $(\sim q)$" we first form the negations $\sim p$, $\sim q$—then we put them together by conjunction.

p	q	p and q	$\sim(p$ and $q)$	p or q	$\sim(p$ or $q)$	$\sim p$	$\sim q$	$(\sim p)$ or $(\sim q)$	$(\sim p)$ and $(\sim q)$
T	T	T	F	T	F	F	F	F	F
T	F	F	T	T	F	F	T	T	F
F	T	F	T	T	F	T	F	T	F
F	F	F	T	F	T	T	T	T	T

Table A.4 *Negation with conjunction and disjunction.*

Now examine *Table A.4*, which gives us the truth values for these four combination forms. The tables show that, as negation is distributed over the separate statements in these combinations, conjunction is changed to disjunction and vice versa.

By examining all the truth-value possibilities, *Table A.4* actually proves

xxxxxxxxxxxxxxxxxxxxxxxxxxxxxxxxxxxxxxxxxxxxxxxxxxxxxxxxxx

De Morgan's Laws:

(A.1)   "$\sim(p$ and $q)$" is logically equivalent to "$(\sim p)$ or $(\sim q)$";

(A.2) $\sim(p$ or $q)$" is logically equivalent to "$(\sim p)$ and $(\sim q)$."

xxxxxxxxxxxxxxxxxxxxxxxxxxxxxxxxxxxxxxxxxxxxxxxxxxxxxxxxxx

These two statements are named after the nineteenth-century British mathematician Augustus De Morgan.

EXAMPLE A.17

A simple, but cumbersome, example of De Morgan's Law (A.1) might be phrased:

"It is false that (both) today is Friday and the sun is shining"

is equivalent to

"Either today is not Friday or the sun is not shining."

An example of De Morgan's Law (A.2) is:

"It is false that either today is Friday or the sun is shining"

is equivalent to

"Today is not Friday and the sun is not shining."

s	$\sim s$	s and $(\sim s)$	s or $(\sim s)$
T	F	F	T
F	T	F	T

Table A.5 *A statement combined with its own negation.*

One more question about the relationships among *and*, *or*, and *not* ($\sim$) is worth pursuing at this point, if only because it is an obvious one to ask: For a single statement s, what can be said about "s and $(\sim s)$" and about "s or $(\sim s)$"? *Table A.5* provides the answer.

Notice that, regardless of the truth value of *s*,

$$s \text{ or } (\sim s)$$

is always true, and

$$s \text{ and } (\sim s)$$

is always false. These are examples of two important types of logical statements:

DEFINITION   A statement that is always true, regardless of the truth values of its component parts, is called a **tautology**; a statement that is always false is called a **contradiction**.

Using the form "*s* or (∼*s*)," we can write many obvious tautologies, such as "Either today is Friday or it's not Friday" and "Either I'll go out or I won't." Similarly, we can construct obvious contradictions, such as "Today is Monday and it's not Monday" and "I'm reading this page and I'm not reading it." Some far more useful instances of tautologies and contradictions occur in formal logic. They form the basis for establishing the validity of logical arguments. Examination of that part of the theory is beyond the scope of this appendix; however, important examples of the use of contradictions in proofs occur in Chapters 3 and 5 and in Appendix B.

A.2   EXERCISES

In Exercises 1–10, consider the statements

    p: **This book is interesting.**

    q: **I am falling asleep.**

Write each of the following as a grammatically correct sentence.

1. *p* and *q*
2. *p* or *q*
3. (∼*p*) and *q*
4. *p* and (∼*q*)
5. (∼*p*) or *q*
6. *p* or (∼*q*)
7. (∼*p*) and (∼*q*)
8. (∼*p*) or (∼*q*)
9. ∼(*p* and *q*)
10. ∼(*p* or *q*)

In Exercises 11–16, write the given statement in symbolic form, using *p* and *q* to represent two simpler, positive statements, along with "and," "or," and "∼," as needed. Specify what the statements *p* and *q* are.

11. Either there are 29 days in February or this is not a leap year.

12. Either I play shortstop or I'm going home!

13. English words are more familiar, but logical symbols are clearer.

14. The President's party does not control the Senate, but it controls the House.

15. It's not true that I went to the movies and ate too much popcorn!

16. The sky was not cloudy, nor was the wind chilly.

In Exercises 17–20, use De Morgan's laws, rules for quantifiers, and common sense to rewrite the given statement as a combination of two simpler, negative statements.

17. It's not true that I went to the movies and ate too much popcorn!

18. It's false that Jack is at least 25 years old but hasn't got a driver's license.

19. It's false that every state governor is either a Democrat or a Republican.

20. It's not true that either every tree will have to be cut down or some people will be out of work.

In Exercises 21–29, let *p*, *q*, and *r* represent statements. Use truth tables to prove:

21. "*p* and *q*" is logically equivalent to "*q* and *p*."

22. De Morgan's Law (A.2).

23. "~(*p* and (~*q*))" is logically equivalent to "(~*p*) or *q*."

24. "~(*p* or (~*q*))" is logically equivalent to "(~*p*) and *q*."

25. "*p* or ~(*p* and *q*)" is a tautology.

26. "*p* and ~(*p* or *q*)" is a contradiction.

(**Warning:** In the next three exercises, your truth table will require eight rows. Why?)

27. "(*p* and *q*) and *r*" is logically equivalent to "*p* and (*q* and *r*)."

28. "(*p* and *q*) or *r*" is *not* logically equivalent "*p* and (*q* or *r*)."

29. "*p* or (*q* and *r*)" is logically equivalent to "(*p* or *q*) and (*p* or *r*)."

30. Although in mathematics *or* is usually interpreted in the inclusive sense unless specifically indicated otherwise, this is not always the case in everyday usage. For each of the following sentences, determine from the context whether *or* is being used in the *inclusive* sense or the *exclusive* sense.

(a) The first prize for the contest is a trip to Hawaii or the cash equivalent.

(b) A student will qualify for financial aid if he/she can demonstrate financial need or high scholastic ability.

(c) Decide whether each statement is true or false.

(d) That tree is either an oak or a maple.

(e) To receive a passing grade you either must have a class average of B or better before the final, or you must earn an A on the final exam. (Interpret both *or*'s.)

(f) To be exempt from the tax, you must be a senior citizen or have an annual income of less than $10,000.

(g) He intends to join the Army or the Navy next week.

In Exercises 31–34, let *s* represent any statement, let *t* represent a tautology, and let *c* represent a contradiction. Use truth tables to prove:

31. "*s* and *t*" is logically equivalent to *s*.

32. "*s* or *c*" is logically equivalent to *s*.

33. "*s* and *c*" is logically equivalent to *c*.

34. "*s* or *t*" is logically equivalent to *t*.

35. Exercise 31 says that any statement formed by conjunction with a tautology can be simplified by eliminating the tautology. Give the analogous interpretations of Exercises 32, 33, and 34.

CONDITIONALS AND DEDUCTION

A.3

One of the most important ways to combine two statements is by the condition-consequence linkage, sometimes called the "if-then" form. As with the other compound logical statements, the formal truth-value structure coincides with normal usage, provided we think of such statements as contracts or agreements between two parties. For instance, suppose you and I have an agreement:

> If you get a perfect score on the final, then I will give you an A for the course.

What does it mean to say that this statement is true (the agreement has been kept) or false (the agreement has been broken)? First of all, observe that this is a compound statement made up of two simple statements:

> p: You get a perfect score on the final.
>
> q: I give you an A for the course.

Each of these statements might be either true or false, so there are four cases to consider:

1. p and q might both be true.
    (You get a perfect score on the final and I give you an A.)
2. p might be true and q false.
    (You get a perfect score on the final, but I don't give you an A.)
3. p might be false and q true.
    (You don't get a perfect score on the final, but I still give you an A.)
4. p and q might both be false.
    (You don't get a perfect score on the final and I don't give you an A.)

Clearly, the agreement is upheld in Case 1 and is broken in Case 2. But what about the other two cases? Since the condition of getting a perfect score on the final is not fulfilled in those cases, I can give you an A or not, as I see fit, without violating our agreement. Thus, the agreement is unbroken (true) *in all cases except* 2. In general:

DEFINITION

A **conditional statement**, or just a **conditional**, is a statement that can be put in the form "if p, then q," where p and q are themselves statements; it is false if p is true and q is false, and it is true otherwise. The statement p is called the **hypothesis** of the conditional and q is called the **conclusion**.

NOTATION

The conditional "if p, then q" is symbolized by "$p \rightarrow q$," which can be read "p implies q."

| EXAMPLE A.18 | "If it is spring, then the grass is green" is a conditional statement. Its hypothesis is "It is spring"; its conclusion is "The grass is green." |

The statement form "if p, then q" is logically equivalent to "q if p." The word "if" *always* labels the hypothesis.

| EXAMPLE A.19 | "I'll be happy if I get an A" is a conditional statement. Its hypothesis is "I get an A"; its conclusion is "I'll be happy." |

| EXAMPLE A.20 | "All dogs are animals" is a universal statement. It is easily rephrased as a conditional: |

"If something is a dog, then it is an animal."

| EXAMPLE A.21 | "If today is Saturday, then the moon is made of green cheese" is a conditional statement. Note that the hypothesis and conclusion need not have any real causal connection. The *form* of the statement determines whether or not it is a conditional. |

p	q	$p \to q$
T	T	T
T	F	F
F	T	T
F	F	T

Table A.6 The conditional.

The truth-value possibilities for the conditional are summarized in *Table A.6*. The four rows correspond exactly to the four cases in the example at the beginning of this section. In every conditional statement, a condition (hypothesis) is asserted to guarantee a consequence (conclusion); this assertion is false *only* if the condition is satisfied but the consequence does not follow.

If the condition is not satisfied, then the conditional statement itself is considered true "by default." In other words, *a conditional statement is true whenever its hypothesis is false.* In such cases we say the conditional is **vacuously true.** (Recall that in our logical system we *must* assign either *true* or *false* to each case.) Thus, to test whether a conditional $p \to q$ is true or false, *it suffices to assume that the hypothesis p is true.* If it is possible for q to be false at the same time, then the conditional $p \to q$ is false; otherwise, $p \to q$ is true.

The claim that a conditional statement is *false*, then, says that there are instances when the hypothesis is true and the conclusion is false. *It does not say that this is the only possible circumstance that can exist.* For example, to say that the statement "If a number ends in 3, then it is divisible by 3" is false means that *there exists* a number that satisfies the hypothesis (ends in 3) but not the conclusion (is not divisible by 3). Thus, to prove the statement false, all we have to do is find one such number.

DEFINITION   An example that proves a statement false is called a **counterexample**.

EXAMPLE A.22   In the example of the preceding paragraph, 13 is a counterexample for the given statement. The fact that there are other numbers, such as 63, that satisfy both the hypothesis and the conclusion does not matter; if the hypothesis does not *guarantee* the conclusion, the conditional statement is false.

EXAMPLE A.23   [*Example A.20 again*] "All dogs are animals" can be regarded as a conditional disguised as a universal statement. We consider this to be a general truth in the sense that, if Phydeau is a dog, we are confident that it is also an animal. On the other hand, if Phydeau turns out to be a cat or a boat, the fact that Phydeau may or may not also be an animal does not alter the truth of the statement, "All dogs are animals."

EXAMPLE A.24   [*Example A.21 again*] "If today is Saturday, then the moon is made of green cheese" is a conditional statement that is (vacuously) true six days out of every week because from Sunday through Friday both the hypothesis and the conclusion are false. Only on Saturdays is the hypothesis true, and then the falsity of the conclusion makes the entire conditional statement false.

Unlike conjunctions and disjunctions, the order of the two simple statements in a conditional makes a difference. "If a figure is a square, then it is a rectangle" says something quite different from "If a figure is a rectangle, then it is a square." Conditionals involving the negations of the hypothesis and the conclusion can complicate matters further. Since these forms and the relationships among them are often useful, it is helpful to have special terms to distinguish one form from another.

As you read through the following definitions, it might help you to think of a simple example of a conditional. For instance:

$$p: \text{Nancy wins.} \qquad q: \text{Tom loses.}$$

Then the original conditional $p \rightarrow q$ is

If Nancy wins, then Tom loses.

DEFINITION   The **converse** of a conditional statement $p \rightarrow q$ is the statement

$$q \rightarrow p$$

formed by interchanging the hypothesis and the conclusion. (If Tom loses, then Nancy wins.)

The **inverse** of $p \rightarrow q$ is the statement

$$(\sim p) \rightarrow (\sim q)$$

formed by negating both the hypothesis and the conclusion. (If Nancy doesn't win, then Tom doesn't lose.)

The **contrapositive** of $p \rightarrow q$ is the statement

$$(\sim q) \rightarrow (\sim p)$$

formed by interchanging the hypothesis and the conclusion and also negating them. (If Tom doesn't lose, then Nancy doesn't win.)

We can summarize these definitions as follows:

a conditional	if p, then q	$p \rightarrow q$
its converse	if q, then p	$q \rightarrow p$
its inverse	if not p, then not q	$(\sim p) \rightarrow (\sim q)$
its contrapositive	if not q, then not p	$(\sim q) \rightarrow (\sim p)$

EXAMPLE A.25

Conditional: "If it is spring, then the grass is green."

Converse: "If the grass is green, then it is spring."

Inverse: "If it is not spring, then the grass is not green."

Contrapositive: "If the grass is not green, then it is not spring."

Universal statements can be put into conditional form. Thus, we can write the converse, inverse, and contrapositive of a universal statement. These related forms can, in turn, be translated back into a universal form, if desired.

EXAMPLE A.26

Universal statement: "All dogs are animals."

Conditional: "If something is a dog, then it is an animal."

Converse: "If something is an animal, then it is a dog."
       "All animals are dogs."

Inverse: "If something is not a dog, then it is not an animal."
       "Anything that is not a dog is not an animal."

Contrapositive: "If something is not an animal, then it is not a dog."
       "Anything that is not an animal is not a dog."

---

p	q	$p \to q$	$q \to p$	$\sim p$	$\sim q$	$(\sim p) \to (\sim q)$	$(\sim q) \to (\sim p)$
T	T	T	T	F	F	T	T
T	F	F	T	F	T	T	F
F	T	T	F	T	F	F	T
F	F	T	T	T	T	T	T

Table A.7
The conditional together with its converse, inverse, and contrapositive.

The foregoing examples suggest interesting logical connections among a conditional and its converse, inverse, and contrapositive. Since these statements are themselves conditionals, their truth values are easily determined, as shown in *Table A.7*. This table shows that

> A conditional statement and its contrapositive are logically equivalent

(because the truth values in those two columns are exactly the same). Thus, "If Nancy wins, then Tom loses" is equivalent to "If Tom doesn't lose, then Nancy doesn't win."

> The converse and the inverse are also logically equivalent statements

(for the same reason). "If Tom loses, then Nancy wins" is equivalent to "If Nancy doesn't win, then Tom doesn't lose." These interrelationships give us convenient alternative forms of conditional statements because logically equivalent statements are interchangeable.

A true conditional may or may not have a true converse. For example, "If something is a rabbit, then it eats lettuce" is true (normally), but its converse, "If something eats lettuce, then it is a rabbit," is not. Thus, a special situation exists when a conditional $p \to q$ and its converse $q \to p$ are simultaneously true. We might describe this by saying that the truth of either of the conditions p or q guarantees the truth of the other. This is sometimes abbreviated by saying, "p is true if and only if q is true."

DEFINITION   A **biconditional statement**, or simply a **biconditional**, is a statement that can be put in the form "$(p \to q)$ and $(q \to p)$" (that is, "p if and only if q"), where p and q are themselves statements. It is true precisely when p and q have the same truth values.

NOTATION   The biconditional "p if and only if q" is denoted by "$p \leftrightarrow q$" or "p iff q."

Notice from *Table A.7* that $(p \to q)$ and $(q \to p)$ are both true precisely when p and q have the same truth values. Thus,

> Saying that $p \leftrightarrow q$ is true is the same as asserting the logical equivalence of p and q.

EXAMPLE A.27	No odd whole number is divisible by 2, and every whole number that is not divisible by 2 is odd. Thus, "A whole number is odd" and "A whole number is not divisible by 2" are equivalent statements. That is, a whole number is odd *if and only if* it is not divisible by 2.

Note   When a word or phrase is *defined* by a condition, that definition is actually a biconditional statement because a defining condition is a characteristic property.[1] It is customary to write such definitions as simple conditionals, but to understand them as biconditionals. Thus,

A triangle is **isosceles** if two if its sides have equal length

is understood as

a triangle is isosceles if and only if two of its sides have equal length.

EXAMPLE A.28	The statements "If the sun is shining, then it is daytime" and "Either it is daytime or the sun is not shining" are logically equivalent (even though they seem to be quite different). We can analyze them with a truth table; but first they should be written in terms of simpler statements. Let p and q be

p: The sun is shining. q: It is daytime.

Then "If the sun is shining, then it is daytime" is the conditional $p \to q$, and "Either it is daytime or the sun is not shining" is the disjunction "q or ($\sim p$)." Now the truth-table analysis is easy; it is shown in *Table A.8*, which establishes a general form of equivalent statements.

p	q	$\sim p$	$p \to q$	q or ($\sim p$)
T	T	F	T	T
T	F	F	F	F
F	T	T	T	T
F	F	T	T	T

Table A.8 Equivalent statements.

Because the last two columns of *Table A.8* agree in all cases, the statements "$p \to q$" and "q or ($\sim p$)" are logically equivalent. In other words, "($p \to q$) iff (q or ($\sim p$))" is a tautology. This gives us a simple way to form the negation of $p \to q$ in terms of p, q, and their negations. We can just form the negation of "q or ($\sim p$)," which, by De Morgan's laws, is equivalent to "p and ($\sim q$)."

Notice that this approach agrees with what we should expect from our earlier discussion about the truth values for $p \to q$. That is, recall that $p \to q$ is false if and only if p is true and q is false. Thus, the truth of "p and ($\sim q$)" guarantees the falsity of $p \to q$, making "p and ($\sim q$)" a likely candidate for the negation. To prove that it is, we construct the appropriate truth table (*Table A.9* on page 538).

---

1. See Section 3.3 for an explanation of *characteristic property*.

p	q	$\sim q$	$p \to q$	$\sim(p \to q)$	p and $(\sim q)$
T	T	F	T	F	F
T	F	T	F	T	T
F	T	F	T	F	F
F	F	T	T	F	F

Table A.9 Negation of the conditional.

In somewhat oversimplified terms, the connection between formal logic and deductive reasoning (including mathematical proofs) comes by way of conditional statements that are tautologies. The hypothesis of an argument is often the conjunction of several (or perhaps many) statements (referred to in the plural as "hypotheses"), and these hypotheses may themselves be complex statements derived from prior arguments. An argument is called **valid** if its conclusion is true whenever all of its hypotheses are true; in other words, it is valid, if

(conjunction of hypotheses) → (conclusion)

is a tautology.

Thus, in order for the conclusion of a valid deductive argument to be guaranteed true, the hypotheses of that argument *must* all be true. This principle is useful in two ways. The more obvious one is a straightforward application: To prove a statement true, we construct an argument starting with hypotheses that are known to be true and ending with the statement in question. This type of argument is called a **direct proof** of the statement. There are many examples of direct proofs throughout the book, from simple number-theoretic statements in Chapter 2 to the fairly complex proof of Lagrange's Theorem in Chapter 6.

The second use of the principle stated at the beginning of the preceding paragraph is based on the definition of the negation of a statement and the Law of the Excluded Middle. Because true hypotheses and a valid argument together must yield a true conclusion,

a valid argument that yields a false conclusion must proceed from a false hypothesis.

Therefore, we may also prove the truth of a statement s by forming its negation, $\sim s$, and using $\sim s$ as the hypothesis of a valid argument whose conclusion is known to be false. This implies that $\sim s$ must itself be false, and hence s must be true. This procedure is known as **proof by contradiction**, or **indirect proof**. A simple example of this type of argument is the proof that $\sqrt{2}$ is not a rational number, which appears in Section B.3 of the Appendix. Other examples of indirect proof appear in several chapters. Perhaps the most important instances occur in the discussion of axiomatic independence and the non-Euclidean geometries (in Chapter 3) and in the proofs of Cantor's diagonalization process and Cantor's Theorem (in Chapter 5).

A.3   EXERCISES

In Exercises 1–10, state the hypothesis, the conclusion, and the converse of each statement.

1. If there is snow on the ground, then this is winter.

2. If birds fly, then fish swim.

3. Rabbits must have good eyesight if they eat carrots.

4. If all men are mortal, then experience is the best teacher.

5. All turkeys are birds. (*Convert to conditional form first.*)

6. All roads lead to Rome. (*Convert to conditional form first.*)

7. No airplanes fly like birds. (*Convert to conditional form first.*)

8. If there are sufficient funds and the carpenters do not strike, the entire building will be renovated by September.

9. A conditional is true if its hypothesis is false.

10. I will be able to sleep through this lecture if nobody asks a question.

11–20. State the inverse and the contrapositive of each statement in Exercises 1–10.

Two political slogans from George Orwell's novel, *1984*, are

"War is peace"   and   "Freedom is slavery."

Let us assume that both of these statements are false. What is the truth value of each statement in Exercises 21–32? Justify your answers.

21. War is not peace.

22. War is not peace and freedom is slavery.

23. Either freedom is slavery or war is peace.

24. If war is peace, then this is 1984.

25. If this is 1984, then war is not peace.

26. If war is not peace, then freedom is not slavery.

27. If war is not peace, then freedom is slavery.

28. If war is peace, then freedom is not slavery.

29. If freedom is slavery and war is peace, then this is 1984.

30. If freedom is slavery and war is peace, then this is not 1984.

31. If war is not peace and freedom is slavery, then war is peace.

32. If war is not peace or freedom is slavery, then war is peace.

33. Is the statement "If ignorance is strength, then ignorance is not strength" a tautology? Is it a contradiction? Justify your answer. (*Note:* "Ignorance is strength" is the third of Big Brother's political slogans in Orwell's *1984*.)

34. Is the statement "Ignorance is strength if and only if ignorance is not strength" a tautology? Is it a contradiction? Justify your answer.

35. Construct a truth table to show that
$$\text{"}\sim(p \to q)\text{" is not equivalent to}$$
$$\text{"}(\sim p) \text{ and } q.\text{"}$$

36. Write the negation of "If birds have wings, then they can fly" in the form "p and $(\sim q)$."

37. Do the same as in Exercise 36 for (the conditional form of) "All elephants have trunks."

38. Use De Morgan's Laws and the fact that any statement s is equivalent to its double negation $\sim(\sim s)$ to write "$p \to q$" in the form of a disjunction. Check your answer by making a truth table.

39. Using Exercise 38, write the original conditionals of Exercises 36 and 37 as disjunctions.

TOPICS FOR PAPERS

We have seen that in mathematical logic there are only two truth values, *true* and *false*, and that every legitimate statement must be assigned one of these truth values. However, in everyday discourse we sometimes run across statements that are not *decidably* true or false. (A simple example: "It will snow in Omaha, Nebraska on December 25, 2050." Must we wait until the middle of the twenty-first century before we decide whether this is true or false?) Suppose that we altered the Law of the Excluded Middle so that our logic system had three truth values—*true, false,* and *maybe.* (The name of the third value is irrelevant.) Write a paper tracing the effects of such a change on the rules of logic presented in this Appendix. When you find a rule that breaks down, try to alter it so that it comes as close as possible to capturing the spirit of the original while allowing for the third truth value. (For example, a new Law of Contradiction might be: "No statement may have more than one truth value at the same time (in the same context).") Each time you adjust a rule, use the new form from there on in your discussion.

FOR FURTHER READING

1. Copi, Irving M. *An Introduction to Logic*, 7th ed. New York: Macmillan Publishing Co., 1986.

2. Exner, Robert M., and Myron F. Rosskopf. *Logic in Elementary Mathematics*. New York: McGraw-Hill Book Co., 1959.

3. Hodges, Wilfred. *Logic*. London: Penguin Books, 1980.

4. Lucas, John F. *Introduction to Abstract Mathematics*, 2nd ed. New York: Ardsley House, Publishers, Inc., 1990, Chapters 1 and 2.

5. Morash, Ronald P. *Bridge to Abstract Mathematics*, 2nd ed. New York: McGraw-Hill Book Co., 1990, Chapters 2–6.

6. Smith, Douglas, Maurice Eggen, and Richard St. Andre. *A Transition to Advanced Mathematics*, 3rd ed. Monterey, CA: Brooks-Cole Publishing Co., 1990, Chapter 1.

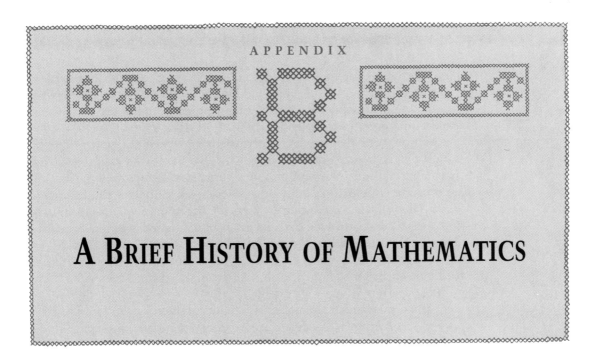

APPENDIX

A BRIEF HISTORY OF MATHEMATICS

INTRODUCTION

B.1

The investigation of any field of human achievement may be placed in perspective by relating individual accomplishments to the overall development of that field. For this reason it is appropriate in an introductory mathematics course to present a brief outline of the history of mathematics. No attempt is made here to detail mathematical theories; we merely offer a historical skeleton of names, places, discoveries, and inventions to provide a context for the material found elsewhere in the book. Our process of selection is unavoidably subjective, but choices and opinions advocated by recognized mathematical historians are followed wherever possible.

We confine our attention predominantly to the mathematics of Western civilization because mathematical development in the Far East was largely independent of the West until modern times and because the genealogy of mathematics as we know it today can be traced almost entirely through Europe and the Near East. This emphasis is not intended to minimize the mathematical achievement of the Orient, but rather, to indicate the relatively small influence it had on present-day Western mathematics.

The chronological sequence is divided into eight major parts, determined mainly by the characteristics of the developments during each period. In more recent times this corresponds roughly to a division by centuries, but the lengths of the intervals prior to the Renaissance vary greatly. Of course, the partitioning is by no means absolute. Ideas and patterns of thought tend to overlap each other, intermingling with the passage of time; consequently, this separation of history into pieces is only an approximation to aid in the analysis of specific events.

As you read through this retrospective overview, you might be struck by the relatively few women whose names appear among the mathematical luminaries of previous centuries. Before drawing any inappropriate conclusions from this observation, consider the following facts. The "place" of women in most of the societies and cultures of Western civilization has, for the most part, denied women access to significant formal education, particularly in the sciences. Moreover, even when a woman succeeded in learning enough about mathematics to make an original contribution to the field, often her results were published anonymously or not at all (for fear of social criticism), or were subsumed under the name of a male mathematician (who had access to the learned societies, the university faculties, and other traditional academic outlets for research). Only in very recent years have historians begun to uncover the full extent of these obscured contributions of female mathematicians.[1] Although most of the formal barriers to women in the sciences had been dissolved by the middle of the twentieth century, some effects of this "uneven playing field" in mathematics and science persist even today. Thus, the perception of mathematics as a male domain has been a remarkably resilient self-fulfilling prophecy that only now is succumbing to the twin realities of careful historical research and outstanding current achievement by twentieth-century women in mathematics.

From the Beginning to 600 B.C.

B.2

Somewhere in prehistoric times, perhaps during the Middle Stone Age, two general concepts began to emerge from the countless diverse phenomena of the physical world. They were *quantity* and *form*, the ancestors of two great families of thought, algebra and geometry, whose true kinship would remain obscure until the seventeenth century A.D. Although accounts of this period are based largely on conjecture, it is believed that ideas about quantity developed from attempts to compare collections of objects by counting, and they slowly evolved into a

---

1. See, for example, the descriptions of the work of Ada Lovelace and Grace Chisholm Young in Item 8 in the list *For Further Reading* at the end of this appendix.

variety of primitive number systems. The earliest of these systems were quite simple, often based on the idea of two or three things, with collections containing more than five or six things classified by "much" or "heap" or some other equally imprecise expression, until the need for exchange and barter gave rise to more extensive number systems. The emergence of form began as primitive art with plaited rushes, woven patterns in cloth, and simple ornamentation on pottery and buildings. The mathematical aspects of form did not become apparent for a while; what we now call geometric figures began simply as decorative designs.

At the beginning of the historic period, about 5000 B.C., mathematics was well into its second stage of development. The quantitative needs of early societies had become so widespread and constant that it was desirable to develop general methods for calculation and to record these rules and results for future use. Fortunately, the Babylonians wrote on almost indestructible clay tablets and the Egyptian papyrus stayed well preserved in the dry climate of northern Africa, so there are enough relics to provide a fairly detailed picture of this work in the early civilizations of the Near East.

The earliest evidence of organized mathematical knowledge seems to indicate the existence of an Egyptian calendar in 4241 B.C., and possibly a Babylonian one before that. By 3000 B.C., Sumerian merchants had a workable arithmetic, and texts from the Third Dynasty, during the reign of King Hammurabi, show that by 1950 B.C., the Babylonians had an algebra capable of handling linear and quadratic equations in two unknowns and some equations of higher degree. Their geometry consisted of "recipes" for finding simple areas and volumes.

Egyptian progress was not far behind that of Babylonia. One of the earliest sources still in existence is the Ahmes Papyrus, written in 1650 B.C.[2] It is a practical handbook containing methods of solving types of linear equations, material on fractions with numerator 1 (a unique feature of Egyptian mathematics),

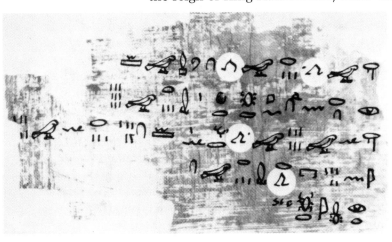

A Calculation from the Ahmes Papyrus. In modern algebraic symbolism:

$$x + \frac{2}{3}x - \frac{1}{3}\left(x + \frac{2}{3}x\right) = 10$$

2. This is also known as the Rhind Papyrus, named after A. Henry Rhind, the nineteenth-century archaeologist who brought the manuscript to England.

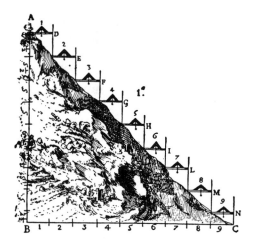

A Method of Leveling Used by the Egyptians and Later by the Greeks and Romans. From Pompodoro's *La Geometria Prattica*, Rome, 1624.

measuring techniques, and problems in elementary series. The scribe Ahmes stated that he was copying an earlier work that had been written about 1800 B.C., so it may well be a compilation of the mathematical knowledge of that time. Thus, the Babylonians and the Egyptians may be credited with producing the first great period of mathematics.

Our knowledge of early Chinese and Indian mathematics is comparatively poor. These peoples wrote on bark or bamboo, so all their manuscripts were highly susceptible to decay. These natural inconveniences sometimes were compounded by man's perversity. For example, in 213 B.C., the Chinese Emperor Shi Huang-ti of the Ch'in Dynasty ordered all existing books burned and had protesting scholars buried alive in order that he might be considered the creator of a new era of learning. Nevertheless, transcriptions of several ancient treatises were preserved, among them the *Chou-pei*, a dialogue concerning astronomy and mathematics. It was written shortly before 1100 B.C., and some of the material in it is attributed to a much earlier period. The *Chou-pei* contains material on the geometry of measurements, the computational principle of the Pythagorean Theorem, some elementary trigonometry and a discussion of instruments for astronomical measurements.

We know even less about Indian mathematics of this period. All that can be said with certainty is that there is evidence of a workable number system, used for astronomical and other calculations, and of a practical interest in elementary geometry.

The outstanding feature of all pre-Hellenic mathematics is the absence of deductive reasoning. In general, attempts to justify statements were not made; the rules were given because they had worked in the past and they were expected to work in the future. Trial-and-error methods were the sources of knowledge, and successful results were noted and passed on to succeeding generations as formulas. Not until the blossoming of Greek civilization did mathematics come of age.

600 B.C. TO A.D. 400

B.3

Early in the first millennium before Christ, sweeping changes occurred in the lands around the Mediterranean Sea. The Age of Iron brought with it increased travel and trade; new towns sprang up along the coasts of Asia Minor and Greece, and the supremacy of the landlord gave way to the rising stature of wealthy merchants. The exchange of goods was accompanied by the exchange of ideas, and wealth begot leisure, so that by the sixth century B.C. there was a class of men whose affluence allowed the luxury of intellectual speculation. With the first "Why?" mathematics entered a new phase of development, a science studied for its own sake.

Credit for that first speculative impulse is given to Thales of Miletus (c. 640 to c. 546 B.C.), a merchant whose travels to Babylon and Egypt acquainted him with Near Eastern mathematics. Until this time geometry had been confined to its literal meaning, measure of the earth, and its propositions were simply rules for accomplishing this task. Thales, however, chose six statements, including:

- A circle is bisected by any of its diameters.

  When two lines intersect, the vertical angles are equal.

  The sides of similar triangles are proportional.

and he demonstrated that they followed logically from previous ones. The statements themselves were well known, but his approach to them was a radical departure from that of traditional mathematics.

One of the most interesting early Greek scholars was Pythagoras, whose life was as mysterious as his work was brilliant. Both the date and the place of his birth are uncertain, but it is generally held that he lived from about 570 to 500 B.C. Much of his life is obscured by myths, for he was a legendary figure in Greece. His followers were banded together in a secret society that worshipped the idea of *Number* and hoarded knowledge as if it were gold. It is alleged that Pythagoras founded his own school at Crotona, a town of Magna Graecia on the southeast coast of the Italian peninsula. There he taught a philosophy based on the immutable elements of nature, embodied in and represented by whole numbers and their ratios. Despite their tendency to indulge in number mysticism, the Pythagoreans contributed greatly to number theory, the theory of music, astronomy, and geometry. They insisted upon deductive proof; the famous theorem about right triangles (shown in *Figure B.1* on page 546 and proved in Section 3.2) still bears their name. The theory of perfect numbers, discussed in Chapter 2, was begun by the Greeks of this era.

Strangely enough, one of the most significant ideas to come from the Pythagorean School was the verification of a concept that destroyed

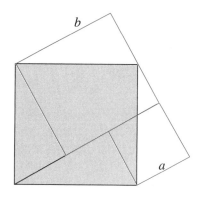

The square of the
hypotenuse of a right
triangle

c^2

equals

$=$

the sum of the squares of the
other two sides.

$a^2 + b^2$

Figure B.1 *The Pythagorean Theorem.*

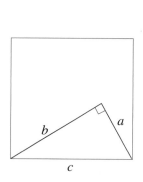

Figure B.2

$d^2 = 1^2 + 1^2 = 2$

the Pythagorean philosophy. The Pythagoreans discovered the existence of incommensurable line segments; that is, they proved the existence of irrational quantities, lengths that cannot be expressed as ratios of whole numbers. Specifically, one of the Pythagoreans proved that the diagonal of a square one unit on a side cannot be expressed as a ratio of whole numbers (in current terminology, as a fraction). The proof, outlined here and illustrated in *Figure B.2*, is interesting for its logical form as well as for its content. It is a clear, brief example of *proof by contradiction*, a type of argument described at the end of Appendix A.

Suppose that a square one unit on a side has a diagonal d that can be expressed as a fraction. Then d can be written in lowest terms, say as $\frac{a}{b}$. By the Pythagorean Theorem,

$$\left(\frac{a}{b}\right)^2 = 1^2 + 1^2$$

so

$$\frac{a^2}{b^2} = 2$$

implying

$$a^2 = 2b^2$$

This means that a^2 is even, so a must be even. Write a as $2s$, where s is some whole number. Then, substituting $2s$ for a in the last equation, we have

$$(2s)^2 = 2b^2$$

or

$$4s^2 = 2b^2$$

so

$$2s^2 = b^2$$

which implies that b^2 is even. But this means b must be even. The fact that a and b must both be even contradicts the hypothesis that $\frac{a}{b}$ is in lowest terms. The only logical escape from this absurdity is to recognize the original hypothesis to be false. That is, the diagonal d *cannot* be expressed as the ratio of whole numbers!

The unfortunate Pythagorean who uncovered this numerical heresy was drowned in a shipwreck. His colleagues tried to suppress the offensive idea he had found, but eventually word leaked out and people began shopping for a new philosophy of nature.

Another disturbing figure in the turbulent world of Greek thought was Zeno of Elea, a philosopher of the fifth century B.C., who speculated on the nature of motion of change of any kind. His contribution to mathematics consists of four paradoxes, which he posed for the thinkers of his day. They involved two diametrically opposite viewpoints of infinity and motion, with two paradoxes for each viewpoint. We give here one example of each type:

- ■ **Achilles** — Achilles is running to catch a plodding tortoise. He must first reach the place where the tortoise started. However, by the time he does this, the tortoise has already left, so it is ahead of him. This process must recur again and again. Hence, Achilles never overtakes the tortoise.

- ■ **The Arrow** — Each moment is an indivisible point in time. A moving arrow, at any given moment, is either moving or at rest. The arrow cannot move within a moment, or else the moment would be divisible. But time is made up of moments. If the arrow cannot move in any one moment, then it cannot move at any time, and so it is always at rest.[3]

Any attempt to resolve these paradoxes involves a consideration of *limits*. Zeno's contemporaries were unsuccessful in their attempts to unravel his verbal tangles, but a seed had been planted that would eventually blossom into calculus. It would, however, require a growing season of 2000 years.

The works of Plato and Aristotle exemplify the type of thought that was the Greeks' greatest contribution to mathematics. Their development of logical principles and axiomatic methods of demonstration put mathematics on a foundation that was considered unshakable until our own century. Under their guidance mathematics partook of the glory of the Golden Age as a philosophical science second to none. In this period arose the "problems of antiquity," perhaps the most famous mathematical problems of all time. They were geometric-construction problems, which allowed only an unmarked straightedge

---

3. A discussion of all four paradoxes can be found on page 24 of Item 1 in the list *For Further Reading* at the end of this appendix.

and a compass as tools for their solution. (See Section 3.2.) With these tools one was asked to:

- Divide a given angle into three equal parts (*Trisection of the Angle*);
- Find the side of a cube whose volume is twice that of a given cube (*Duplication of the Cube*);
- Find a square whose area equals that of a given circle (*Quadrature of the Circle*).

All three problems remained unsolved until modern times, when the constructions were shown to be impossible. Nevertheless, the very fact of their existence bred a tantalizing kind of annoyance which led scholars to many new discoveries.

One of these scholars was Eudoxus (c. 408 to c. 355 B.C.), a pupil of Plato, a physician, a legislator, and a mathematician. He is credited with developing a theory of proportions that overcame the difficulties of dealing with incommensurable quantities. He introduced the "method of exhaustion," a way of dealing with areas and volumes that is remarkably similar to some of the basic concepts of integral calculus.

As Alexander the Great set out to conquer the world in 334 B.C., Western civilization began to change. Greek culture and thought mingled with that of the Orient, and the mathematical center of the Western world shifted to Alexandria. That city's period of ascendancy began with Euclid (c. 300 B.C.), "the most successful textbook writer the world has ever known" and "the only man to whom there ever came or can ever come again the glory of having successfully incorporated in his own writings all the essential parts of the accumulated mathematical knowledge of his time."[4]

Euclid's works were so comprehensive that they superseded all previous writings, and for this reason very few pre-Euclidean Greek manuscripts were preserved. Most of the information regarding work prior to the third century B.C. has had to be reconstructed from second-hand source material. Euclid's greatest work, the *Elements*, is a 13-book treatise that covers plane geometry, Eudoxus' theory of proportions, number theory, a geometric treatment of square roots, and solid geometry. Some of this material is undoubtedly Euclid's own work and he freely acknowledges his indebtedness for other parts, but the exact division is difficult to ascertain. However, even if he had contributed nothing original, the *Elements* would be significant; for it was a remarkable attempt to organize all of mathematics into a system logically deduced from an axiomatic foundation.

---

4. See Item 15, Vol. 1, pp. 103–104 in the list *For Further Reading* at the end of this appendix.

This second great period of mathematics reached its peak with the flourishing of the School of Alexandria. Shortly after Euclid came Archimedes (287–212 B.C.), an astronomer and pioneer in physics and applied mathematics as well as a speculative mathematician. He studied at Alexandria for a short time and then returned to Syracuse, where he had been born. Archimedes masterminded an ingenious scientific defense of that city against the hordes of Marcellus; he was killed by an impetuous Roman soldier when the city was finally taken. His works include major contributions to number theory and algebra, and an enormous amount of geometrical work. Among his greatest achievements were his derivations of the formulas for the volumes and surface areas of spheres and his use of the "method of exhaustion" to solve geometric problems now handled by calculus.[5]

Apollonius of Perga (c. 260 to c. 210 B.C.), "the great geometer," was a contemporary of Archimedes. Seven of his masterful eight books on conic sections have survived intact. In these he systematically developed many basic properties of the ellipse, the parabola, and the hyperbola, investigating such topics as congruence and similarity criteria for these curves, tangent figures, and inscribed and circumscribed polygons. The world was not to see another pure geometer of his stature until Jacob Steiner in the nineteenth century.

With the passing of Apollonius, the tide of Greek mathematics crested and ebbed with the rest of Greek civilization. Only two great waves appeared above the ripples of minor writers of this period. The astronomer Ptolemy (c. A.D. 85 to c. 165) wrote a comprehensive treatise on astronomy known as the *Almagest*, in which the frontiers of computational mathematics were extended to plane trigonometry and the beginnings of spherical trigonometry. About a century later, Diophantus of Alexandria (c. 275) wrote the *Arithmetica*, an unusual blend of Greek and Oriental mathematics, of which six books have survived. It is a milestone in the development of number theory, containing a treatment of indeterminate equations and problems requiring fractional solutions. It is the first treatise in which a type of algebraic symbolism is used systematically. In this regard Diophantus was several centuries ahead of his contemporaries.

One of the last great mathematical figures of this era was Hypatia (c. 370 to 415), daughter of the Alexandrian mathematician Theon. Hypatia wrote commentaries on the works of both Apollonius and Diophantus; she is the first woman known to have written on mathematical subjects. A celebrity in Alexandria at the beginning of the fifth century, she became entangled in the power struggle between

---

5. For a clear, elegant description of Archimedes' work on spherical volume and surface area, see Chapter 4 of Item 6 in the list *For Further Reading* at the end of this appendix.

the Prefect Orestes and the Archbishop Cyril. Her torture and murder in the cathedral of Alexandria by followers of the archbishop brutally signaled the end of a period of prolific scholarship in the Western world and the beginning of what some historians have called the Dark Ages.

A.D. 400 TO 1400

Despite its productivity in many other areas, the Roman Empire was mathematically barren, and its conquest of the Mediterranean world did little for speculative science. Whatever activity there was resulted from surviving Greek and Oriental influences. With the fall of Rome and the dissolution of Latin political domination in 476, Western civilization entered a period of intellectual stagnation. The progress of ideas

Arithmetic and Boethius; Music and Pythagoras. From an illuminated manuscript, Bavaria, 1240.

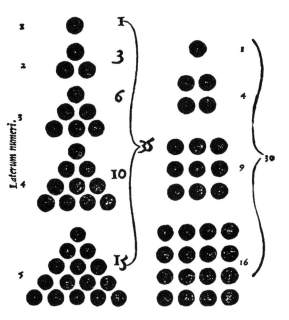

Pyramidum numeri hoc paſto digeruntur.

Laterum numeri.

1

2

3

4

5

1

3

6

10

15

1

4

9

16

Pyramidal Numbers as Sums of Successive Triangular or Square Numbers. From *Ringelbergius*, Opera, 1531.

in Western Europe reached its nadir in the sixth century. The only person worthy of note at this time is Anicius Boethius (c. 475–524), a Roman citizen, statesman, philosopher, and mathematician, whose works include books on arithmetic, geometry, and astronomy, as well as a treatise on music (which was considered to be part of mathematics then). Although his texts lacked originality and were not especially rich in content, they were considered authoritative in schools and universities for many centuries.

Meanwhile, Hindu mathematics was beginning to bear fruit. The influences of Babylonian and Greek science blended with native Indian ideas and there emerged significant results in algebra and arithmetic. Hindu mathematicians of the fifth century worked with *plane* and *solid numbers*—squares, cubes, triangle numbers (as defined in the Exercises for Section 2.1), etc.—obtaining both theoretical and computational results, including an approximation of π as $\frac{62{,}832}{20{,}000}$, or 3.1416. In the next century this work was extended to a treatment of algebraic equations, following the methods of Diophantus.

By far, the most important achievement of Hindu mathematics was the development of the numeration system we use today, a place system based on ten and including a symbol for zero. Decimal systems and place systems had been developed previously by other peoples, but this was the first combination of the two concepts. Evidence of its use dates from A.D. 595; but the zero symbol in Indian arithmetic cannot be traced back beyond the ninth century with any certainty, although the Babylonians had a somewhat similar notational device long before the Christian era.

Starting with Mohammed's Hegira in 622, the Moslems became the dominating influence in Western mathematics. The soldiers of Islam swept across North Africa, into western Asia, and through parts of Europe. Both Greek and Hindu mathematics were absorbed and merged by Arab scholars, who translated most of the important manuscripts into Arabic. They contributed little that was original, but they refined existing material and preserved the continuity of thought. The Mohammedan caliphs of the eighth and ninth centuries were great patrons of the exact sciences, especially astronomy and mathematics, so scientific learning spread throughout the Arab world.

Among the Moslem scholars of this period, one deserves special mention. He is Mohammed ibn Musa al-Khowarizmi (c. 825), who wrote two significant books, one on arithmetic and the other on algebra. Only the second still exists in the original Arabic, but both were translated into Latin by twelfth-century European scholars. The title of the first in translation was *Algorithmi de numero Indorum* (literally, "Al-Khowarizmi on Indian Numbers"). This book was one of the means by which Europe was introduced to the Hindu number system, and its title is the source of the word *algorithm*. The second book was entitled *Al-jabr w'al muqabalah* (literally, "Restoration and Opposition"); it was devoted entirely to the study of linear and quadratic equations. By Latinization, the first word of the title became *algebra*, and, because of the widespread popularity of this text in Europe, the term soon became synonymous with the science of equations.

It was not until the latter part of the eleventh century that the Greek mathematical classics started to filter into Europe, and with its slow emergence from feudalism, the West began to stir from intellectual somnolence. During the twelfth century, the great mathematical works that had been translated into Arabic a few centuries earlier were translated from Arabic into Latin, especially in Spain after the

Arab Mathematician. Although they contributed little that was original, Arab scholars refined and preserved existing material.

Seated Monk Writing. From the *Mer des Hystoires*, Paris 1488–89. Scholastic philosophers speculated on the nature of infinity.

defeat of the Moors in 1085, when many Jewish scholars were employed for this purpose. As trade between East and West expanded, powerful commercial centers were established on the coasts of Italy. European merchants began to visit the Orient to seek and put to practical use whatever scientific information they could find. The first of these to do significant mathematical work was Leonardo of Pisa (c. 1170–1250), more commonly known as Fibonacci. His main work, the *Liber Abaci*, was instrumental in acquainting Western Europe with the Hindu-Arabic numeration system. He also wrote on algebra and geometry, and investigated the infinite sequence that still bears his name.[6]

St. John's College, Oxford. The first universities were chartered in the thirteenth century.

Not all mathematics of this period was done solely for its practical value. Scholastic philosophers speculated on the infinite, and their ideas influenced mathematicians of the seventeenth and nineteenth centuries. Clerical scholars also worked in the fields of geometry and algebra. Outstanding among these men was the Bishop of Lisieux, Nicole Oresme (c. 1323–82), who was also an excellent economist. His mathematical work includes the first known use of fractional exponents and the location of points by numerical coordinates.

As cities sprang up throughout Europe, the church schools began to assume a new role. They became universities, empowered to grant degrees recognized by both Church and State. The first universities were at Paris, Oxford, Cambridge, Padua, and Naples, all chartered in the thirteenth century. This innovation held out great promise for the progress of learning during the fourteenth century. That promise was only partially fulfilled, for although other universities were established, the Hundred Years' War (1337–1453) and the Black Death (1347–51) staggered the emerging European culture and, in particular, postponed the revival of mathematical creativity.

6. The Fibonacci sequence begins: 0, 1, 1, 2, 3, 5, 8, 13, ... Each number (after the second) is the sum of the two preceding terms.

THE FIFTEENTH AND SIXTEENTH CENTURIES

B.5

The renaissance of European art and learning began in earnest during the fifteenth century. When Constantinople fell in 1453, many Greek scholars moved to the cities and new universities of the West. Their migration coincided with the invention of movable-type printing, an invaluable aid in the dissemination of information. The ever-increasing demands of trade, navigation, astronomy, and surveying spurred mathematical study, but at the same time confined it somewhat. The dominating theme of fifteenth- and sixteenth-century mathematics was computation, and great strides were made in achieving accuracy and efficiency.

The leading mathematician of the fifteenth century was Johannes Müller (1436–76), also known as Regiomontanus. Besides translating many classical Greek works and studying the stars, he wrote *De triangulis omnimodis*, the first treatise to be devoted solely to trigonometry. This book differs very little from the trigonometry of today except in notation; it marked the beginning of trigonometry as a study independent of astronomy.

The first printed mathematics books were a commercial arithmetic that appeared in 1478, and a Latin edition of Euclid's *Elements*, which appeared in 1482. Mathematics derived its first real benefits from printing with the appearance in 1494 of *Summa de Arithmetica* by Luca Pacioli, a Franciscan monk. The book was written in Italian (rather than Latin) and was a complete summary of all arithmetic, algebra, and trigonometry known at that time, ending with the remark that the general cubic equation was insoluble with existing mathematical techniques. As if in direct response to this challenge, mathematicians at the University of Bologna attacked the problem and disposed of it within the first quarter of the next century.

The foremost scientist of this age was Leonardo da Vinci (1452–1519), a man of universal brilliance, whose fields of activity included painting, sculpture, biology, architecture, mechanics, and optics. His mathematical work was centered about geometry and its applications to art and the physical sciences. The application of mathematics to art was by no means confined to da Vinci's work. Several famous artists of the sixteenth century were excellent geometers. Notable among them was Albrecht Dürer. He wrote the first printed work dealing with higher plane curves, and his investigation of perspective and proportion is reflected both in his paintings and in the artistic work of his contemporaries.

The father of English mathematics was Robert Recorde (c. 1510–58), a medical doctor, educator, and public servant. He published four books on mathematics, written in English dialogue with clarity, precision, and originality: *The Ground of Artes*, an arithmetic that went through twenty-nine editions; *The Castle of Knowledge*, the first English exposition of the Copernican theory in astronomy; *The Pathwaie to Knowledge*, an abridged version of Euclid's *Elements*; and *The Whetstone of Witte*, an algebra, in which the equality sign "=" appears for the first time.

In the second half of the sixteenth century, a French lawyer by the name of François Viète (1540–1603) began to devote his leisure time to mathematics, and he advanced algebraic technique considerably. He was the first person known to use alphabetical coefficients in solving equations, and his consistent symbolism, including the signs + and −, helped him to pioneer the development of general methods for solving equations. The last name of importance in the sixteenth century is that of Simon Stevin of the Netherlands, developer of the theory of decimal fractions in 1585. Thus ended 200 years of preparation for a scientific explosion that signaled the dawn of the third great mathematical era.

THE SEVENTEENTH CENTURY

Two previously destructive trends began to bear constructive fruit in the seventeenth century. The first and more obvious of these was the frequent occurrence of political and religious wars, both large and small. Ever since the Crusades, Europe had been in turmoil; hardly a year had passed without open conflict somewhere. As has been the case throughout history, war and its ever-insistent need for newer and better weapons pressed the best minds of the age to compete in devising machines for battle. Once the weapons were made, however, the truly great minds that had devised them turned to the investigation of machines in general; Leonardo da Vinci's scientific work is a prime example of this. Slowly, Europe embarked upon the first stages of mechanization, and in an effort to develop the necessary engineering skill, its scientists and scholars turned to the study of motion and change.

The second trend was just as powerful, but it began more subtly. The religious reformation championed by Martin Luther on the Continent and by Henry VIII and his daughter Elizabeth I in England unleashed an intellectual revolt against authority and tradition that had been brewing for many years. The general skepticism rejuvenated philosophy, and the explorers of ideas pushed ahead in all directions. This composite age of science and rationalism produced some of the greatest men of history—astronomers Galileo and Kepler; philosophers Hobbes, Spinoza, and Locke; authors Shakespeare, Milton, and Dryden. Mathematics experienced a period of growth unparalleled until modern times. The number of people who contributed significantly to the advance of science becomes so large from this time on that henceforth we shall be compelled to confine our attention primarily to creative mathematicians of the first rank.

The Tudors had just passed the British crown to the Stuarts and William Shakespeare was in his prime when John Napier (1550–1617) reached the peak of his scientific career. He was a Scottish contemporary of Galileo and Kepler, and a true product of his time. He spent much of his life alternating between attacks on the Catholic Church and the contemplation and design of military devices, envisioning such futuristic fantasies as submarines and self-propelled cannon. His claim to immortality was established just a few years before his death with the publication of *Mirifici Logarithmorum Canonis Descriptio*, in which he set forth the theory of logarithms. This work was followed by another on logarithms, one on computing rods, and one on algebra. Napier's system of logarithms was a bit unwieldy, but several years after his death, it was perfected by a friend and colleague, Henry Briggs, who had originally suggested the use of 10 as the base.

Not to be outdone by its neighbors across the Channel, France produced four brilliant mathematicians within half a century. The first of these was the philosopher-scientist René Descartes (1596–1650). His many years of study and reflection convinced him that all scientific investigations are related and the key to that relation is mathematics. In his famous *Discourse on Method*, published in 1637, Descartes stated:

> The long chains of simple and easy reasonings by means of which geometers are accustomed to reach the conclusions of their most difficult demonstrations, had led me to imagine that all things, to the knowledge of which man is competent, are mutually connected in the same way, and that there is nothing so far removed from us as to be beyond our reach, or so hidden that we cannot discover it, provided only we abstain from accepting the false for the true, and always preserve in our thoughts the order necessary for the deduction of one truth from another.[7]

He explained that his method was a fusion of logic, the "Analysis [geometry] of the Ancients" and the "Algebra of the Moderns." He set forth four basic rules of procedure, saying,

---

7. This and the subsequent quote are from *Discourse on Method* (trans. John Veitch; London and Washington: L. Walter Dunne, 1901).

> In this way I believed that I could borrow all that was best in both Geometrical Analysis and in Algebra, and correct all the defects of one by the help of the other.

To say that Descartes succeeded in unifying all of science would be an exaggeration; but he did achieve a marriage between quantity and form in an appendix to the *Discourse on Method*, simply titled *La Geometrie*. This was the first publication of analytic geometry. In it Descartes applied the methods of algebra to geometry, expressing and classifying various curves and other geometric figures by means of algebraic equations relative to a coordinate system. He used this algebraic approach to settle a number of geometric questions, including some classical problems that hitherto had been insoluble. (Some modern consequences of Descartes' work appear in Chapter 7.)

A countryman and acquaintance of Descartes was Pierre de Fermat (1601–65), called by some the greatest pure mathematician of the seventeenth century. He was certainly one of the foremost scientific amateurs in history. Fermat was a quiet, unobtrusive lawyer and civil servant who indulged in mathematics sheerly for the fun of it, publishing little, but exhibiting his creativity in exchanges of correspondence with Descartes, Pascal, Mersenne, and others. He invented analytic geometry independently of Descartes, conceived the tangential approach to differential calculus before either Newton or Leibniz was born, and was one of the creators of the mathematical theory of probability. (See Chapter 4.)

Despite these monumental achievements, Fermat is best known for his work in number theory on the properties of primes. Ironically, his name has been permanently associated with a statement he neither originated nor (presumably) proved. In his copy of a translation of Diophantus, one of Fermat's many marginal jottings was next to a problem asking for values x, y, and a to satisfy the equation

$$x^2 + y^2 = a^2$$

In this note he states that for any power greater than 2 there are no integer solutions, claiming,

> I have discovered a truly marvelous demonstration [of this] which this margin is too narrow to contain.[8]

Unfortunately, he apparently never wrote it down anywhere else, and for more than 300 years "Fermat's Last Theorem" ranked among the most puzzling unsolved problems of mathematics. Finally, in 1994,

---

8. *Œuvres de Fermat*, ed. Tannery, P., and C. Henry. (Paris, 1891–1912), Vol. 3, p. 241.

Andrew Wiles of Princeton University presented a "truly marvelous demonstration" of this theorem, but one that Fermat could never have envisioned—a tightly reasoned, 200-page proof that draws on some of the most sophisticated results of twentieth-century mathematics.

The third person in this group was Gerard Desargues (1593–1662), soldier, engineer, and geometer. During his lifetime much of his work was eclipsed by the general interest in Descartes's writings; but two centuries later his treatise on conics was republished and was immediately hailed as a classic in pure geometry. Desargues introduced the geometric treatment of points at infinity and worked extensively with perspective, thus becoming the founder of modern projective geometry.

Completing this quartet of renowned French mathematicians is Blaise Pascal (1623–62). From the age of twelve he regarded geometry as recreation, and by sixteen he had proved one of the most beautiful and far-reaching theorems in all of geometry, applying it then to consolidate and extend much of the previous work in this field. The theorem states:

> If a hexagon is inscribed in a conic section, then the points of intersection of the three pairs of opposite sides are collinear. [An example is shown in *Figure B.3*.]

Pascal invented a computing machine when he was nineteen, and in his twenties was recognized as a competent physicist. He was cocreator with Fermat of probability theory, a study to which both were led in a joint attempt to answer questions from members of the gambling nobility. However, at the age of twenty-five, Pascal became a nearly fanatical proponent of Jansenism (a religious movement devoted to personal holiness and mysticism). As a consequence, he considered mathematics as a trifle to be toyed with only occasionally. Most of his later life was devoted to a study of philosophy and religion, from which emerged two literary classics, *Pensees* and *Provincial Letters*.

During the first part of the seventeenth century, the final building block for the foundation of calculus was fashioned in Italy. Bonaventura Cavalieri, a Jesuat professor at the University of Bologna,[9] proposed the "principle of

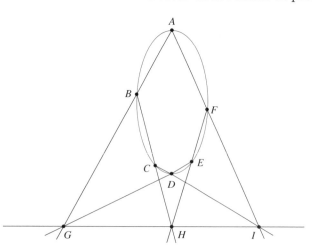

Figure B.3 *Pascal's Theorem for an ellipse. Hexagon ABCDEF is inscribed in the ellipse. Points G, H, and I are collinear.*

9. The Jesuati, not to be confused with the Jesuits, were a congregation within the Catholic Church from 1366 to 1668.

indivisibles" in 1629. This principle sets forth a criterion for comparing areas and volumes of certain geometric figures, based on the assertion that a planar region can be considered as an infinite set of parallel line segments and a solid figure can be considered as an infinite set of parallel planar regions. His work circulated throughout the European scientific community, and in the latter part of the century, two men combined it with Cartesian geometry to erect a towering mathematical structure with a very shaky ground floor.

One of these was Isaac Newton (1642–1727), whose achievements as a physicist are well known. When he was a boy, Newton showed little formal mathematical ability, and his early scholastic record was less than outstanding; but after a brief absence from school, he returned to his studies with renewed enthusiasm. He entered Trinity College at Cambridge as a student at the age of nineteen, and in eight years became a professor of mathematics. In the decade between 1666 and 1676, Newton developed his "theory of fluxions" in three treatises. Although they were not published until many years later, these papers provided a foundation for a subsequent work, *Philosophiae Naturalis Principia Mathematica*, an axiomatic development of physics, in which Newton first set forth a complete mathematical formulation of his famous laws of motion. The *Principia* quickly assumed the role of an indispensable prerequisite for future scientific and technical progress; it ranks among those few mathematical works that have radically affected the history of civilization.

Among Newton's contemporaries and successors, however, acceptance and acclaim were far from unanimous. The man of genius had known what he wanted, and he allowed intuition to suppress a few logical scruples, accepting the theory primarily because it worked. But some of his peers were justifiably skeptical. None of these critics was as caustically witty and incisive as George Berkeley, Anglican Bishop of Cloyne. In *The Analyst*, published in 1734, he contended that scientists who criticized faith in mysteries of religion had precisely the same difficulties in their own field:

Sir Isaac Newton. His view of the universe as being described by Euclidean geometry dominated the physical sciences for two hundred years.

> And what are these fluxions? The velocities of evanescent increments. And what are these same evanescent increments? They are neither finite quantities, or quantities infinitely small, nor yet nothing. May we not call them the ghosts of departed quantities? Certainly . . . he who can digest a second or third fluxion . . . need not, methinks, be squeamish about any point in Divinity.

Credit: IBM Archives

Baron Gottfried Wilhelm von Leibniz. He independently developed calculus shortly after Newton did.

Newton's archrival was Gottfried Wilhelm von Leibniz (1646–1716), Germany's universal genius. His extraordinary talents benefited law, diplomacy, religion, philosophy, physical science, and mathematics. He independently developed differential and integral calculus shortly after his British counterpart had done the same. This last fact was hotly disputed for many years, with charges of plagiarism from both camps of admirers flying back and forth across the English Channel and engendering a bitter patriotic partisanship among the scientific community that was to last for many decades. Far less publicized, yet quite important, is Leibniz's work in combinatorial analysis, the study of specialized counting formulas that are basic to probability and statistics. Moreover, in searching for a "universal characteristic" unifying all of mathematical thought, he became one of the founders of symbolic logic, a study that would not be investigated intensively until two centuries later.

THE EIGHTEENTH CENTURY

B.7

Calculus dominated mathematical development in the eighteenth century. Exploration of this powerful new theory proceeded in two directions—application to other parts of mathematics and to physics, and examination of its theoretical foundation. Research during this era was carried on mainly by Royal Academies subsidized by the "enlightened despots" of the age, while the universities played only a minor part in the production of ideas. The most prominent academies were at Berlin, London, Paris, and St. Petersburg. France held the lion's share of mathematical talent, but Switzerland also contributed significantly to the field.

To avoid religious persecution, a family named Bernoulli fled from Belgium in 1583 and ultimately settled in Switzerland. Their descendants provide a strong argument for the inheritance of intellectual ability. Within three generations they had produced eight eminent mathematicians, of whom four achieved international renown! The first and best known of these were two brothers, Jakob (1654–1705) and Johann (1667–1748). Lured away from careers in theology and medicine by the fascination of Leibniz's pioneering work in calculus, Jakob and Johann Bernoulli entered into a fierce mathematical rivalry with each other and with Leibniz himself that produced much of the material now contained in courses on elementary calculus as well as results in the theory of ordinary differential equations. Much of the Bernoulli brothers' work focused on the properties of various special curves, such as the catenary and the logarithmic spiral. Besides his contributions to calculus, Jakob also did outstanding work in geometry and wrote the first book devoted to the theory of probability. Two of Johann's sons also achieved fame—Nicholaus (1695–1726), for his work in geometry, and Daniel (1700–82), for his extensive publications in astronomy, mathematical physics, and hydrodynamics.

The most productive mathematician of the eighteenth century was Leonhard Euler (1707–83), who was born in Basel, Switzerland. He was instructed in mathematics by Johann Bernoulli, and he also studied theology, medicine, Oriental languages, astronomy, and physics. In 1727, he accepted a post at the St. Petersburg Academy. He became head of the Berlin Academy in 1747, then returned to St. Petersburg twenty years later, where he remained until his death in 1783. Euler was an indefatigable worker whose labors were not appreciably diminished by partial blindness at the age of twenty-eight, nor by complete blindness at fifty-nine. He wrote almost 900 important books and papers, including works on arithmetic, algebra, analysis, mechanics, music, and astronomy. He is responsible for much of our modern notation in algebra and calculus, and he put trigonometry into its present form. One of his papers is considered to mark the beginning of graph theory, a branch of mathematics that is finding widespread applications in this century.[10] But perhaps Euler's most

Leonhard Euler. He established calculus as a purely analytic theory independent of geometry.

10. The paper was a mathematical treatment of the "Königsburg Bridge Problem," presented in 1736. See Section 8.1 for a little more about Euler and the Königsburg Bridge Problem, and see Chapter 8 in general for an introduction to graph theory.

consequential achievements resulted from his efforts to establish calculus as a purely analytic theory independent of geometry.

In France, much of the mathematics in the early part of the century focused on the rival viewpoints of Newton and Leibniz about the mathematical basis of "natural philosophy" (physics). Pierre de Maupertuis (1698–1758) was an ardent, outspoken supporter of Newtonian physics. He found an early ally in the writer-philosopher Voltaire (1694–1778), and together they labored to move French thought beyond the philosophy of Descartes. Their work was profoundly influenced by Emilie du Châtelet (1706–49),[11] who was first Maupertuis's lover and then, for the last 16 years of her life, Voltaire's lover, coauthor, and intellectual companion. Châtelet was the first prominent female mathematical scholar since Hypatia (some 1300 years earlier). She and Voltaire wrote the *Éléments de la Philosophie de Newton* in 1738. Two years later she published a de-tailed exposition of physics based on the principles of Leibniz. In 1745, she turned back to Newtonian physics; the remaining four years of her intellectual life were spent writing a two-volume French translation of and commentary on Newton's *Principia* that made Newton's work accessible to French readers. One historical commentator summarizes Châtelet's mathematical achievements by saying:[12]

> . . . her work of translation, commentary, and synthesis was a valuable contribution to the triumph of Newtonian science in the eighteenth century.

Meanwhile, just across the border in northern Italy, another woman was making her mark in European mathematical history. The first child of a mathematics professor, Maria Gaetana Agnesi (1718–99) was born into a post-Renaissance culture somewhat more tolerant of educated women than was eighteenth-century France. The development of her prodigious intellectual skills was encouraged and supported by her parents, and by 1748 (at only thirty years of age) she had completed and published a 1020-page, two-volume exposition of algebra, analytic geometry, calculus, and differential equations. Since this book brought order and clarity to a large body of hitherto fragmented mathematical writings, it was immediately recognized as an important contribution to the mathematical literature and was quickly translated into French and English. It gained immediate recognition

11. Her full name was Gabrielle-Émilie Le Tonnelier de Breteuil, Marquise du Châtelet-Lomont.
12. Garry J. Tee, in Item 8, page 24, in the list *For Further Reading* at the end of this appendix.

for Agnesi as a first-class mathematician. In 1750, Pope Benedict XIV named her honorary professor at the University of Bologna, but she never taught there. Shortly after 1750, Maria Agnesi turned away from mathematics and dedicated the rest of her life to religious and humanitarian activities, spending almost a half century caring for the poor, the sick, and the dying.

The latter part of the eighteenth century was a period of political turmoil. Great Britain was having a bit of difficulty disciplining her rowdy American colonies, and France's nobility had lost their knack of controlling the peasantry. The upheaval in France was so drastic that it is incredible that scientific investigation not only endured but even thrived throughout the entire period. In fact, French mathematics retained its position of superiority. The Continental mathematicians had a strong advantage over the English in that the differential and integral calculus of Leibniz was much easier to understand and apply than Newton's unwieldy theory of fluxions, and their achievements indicate that they put it to good use.

"The lofty pyramid of the mathematical sciences," according to Napoleon Bonaparte, was Joseph Louis Lagrange (1736–1813), a brilliant, but modest, mathematician who was honored extravagantly by Napoleon and two foreign monarchs, and whose career was advanced immeasurably by the selfless support of Euler. Lagrange improved and organized much of Euler's calculus and worked extensively in the theory of equations, number theory, and mechanics. He was also responsible for inaugurating intensive mathematics curricula in France's two newly established schools, the École Normale and the École Polytechnique.

Pierre Simon Laplace (1749–1827) was the eighteenth century's applied mathematician *par excellence*. Born of a peasant couple, he used his exceptional mathematical talent to rise through the social strata until he was made a count by Napoleon. His most famous works are *Théorie analytique des probabilités* and the five-volume *Mécanique céleste*, which reviewed, unified, and extended all previous work in the fields of probability and celestial mechanics, respectively. Despite an annoying tendency to borrow ideas without acknowledgement, Laplace is generally recognized as an outstanding creative scientist. The size of his works belies their succinctness; in the words of one of his translators:[13]

> I never come across one of Laplace's "Thus it plainly appears" without feeling sure that I have hours of hard work before me to fill up the chasm and find out and show how it plainly appears.

13. Nathaniel Bowditch (1773–1838), American astronomer.

Adrien-Marie Legendre. He worked in analysis and applied mathematics, as well as in geometry.

Brief mention must be accorded to Adrien-Marie Legendre (1752–1833) and Gaspard Monge (1746–1818). Both of these men reached the peaks of their careers at the end of the eighteenth century, but their works differ radically from each other. Although he was responsible for a text that provided a thorough reorganization of Euclidean geometry, a major portion of Legendre's work was in analysis and applied mathematics, the dominant fields of that time. Monge, on the other hand, was strictly a geometer, and geometric ideas pervade all his writings. As one of the first modern mathematical specialists, he is a herald of the next age.

THE NINETEENTH CENTURY

The modern era began with Carl Friedrich Gauss (1777–1855), "prince of mathematicians." Just as Archimedes' influence pervaded the science of the Hellenistic age and Newton's achievements overshadowed the post-Elizabethan period, so did the genius of Gauss dominate nineteenth-century mathematics. A child prodigy, he could do arithmetic at the age of three and was familiar with infinite series by the time he was ten. The modern theory of numbers traces its ancestry back to his monumental *Disquisitiones Arithmeticae*, published in 1801. As a result of his writings on celestial mechanics, Gauss became generally recognized as the leading mathematician of Europe. His work was clear and concise, and was characterized by rigorous proof. Landmarks in almost every branch of mathematics bear his name; but Gauss expressed a strong personal inclination toward one particular field, saying,

> Mathematics is the queen of the sciences, and the theory of numbers is the queen of mathematics.

With the dawn of the new century, mathematical creativity began to increase exponentially, and by 1900, it had produced about five times as much original mathematics as had been accomplished in all previous ages. With this phenomenal wealth of material, mathematics had become a subject so vast that its totality defied the comprehension of a single mind. Except for a few individuals at the highest level of brilliance, mathematicians were forced to confine their efforts to one major branch of the subject, such as algebra, geometry, or analysis (the study of concepts generalized from calculus). At the same time, many other changes were occurring in the mathematical landscape. The dominance of royally supported scientific academies declined rapidly, and research became an important function of the universities. Within mathematics itself, practitioners became increasingly critical of previously accepted results and techniques. Their demands for a new level of rigor in proofs and their distrust of intuition led to the study of symbolic logic and axiomatics.

From this point on, a strictly chronological succession of biographies is no longer sufficient to outline mathematical progress. We provide, instead, a series of narratives, each following some topical subdivision. There is an unavoidable overlapping of names, times, and ideas; but every effort has been made to preserve the perspective and continuity of the overall picture.

As physical science and engineering reaped the rewards of calculus, the mathematicians of France continued to play an important role. Among them was Sophie Germain (1776–1831), a Parisian contemporary of Gauss, who overcame her society's prejudice against female intellectuals to become one of France's finest mathematical scholars.[14] Germain did considerable research in number theory, much of it detailed in correspondence with Gauss and Legendre, including a partial proof of Fermat's Last Theorem. However, her most famous work—for which she won first prize in the 1816 competition held by the Institut de France—was on the mathematical theory of elastic surfaces, a calculus-based solution to an important physical problem. Her work provided the basis and direction necessary for other researchers to

---

14. Molière's satire, *La Femme Savante*, provides a painfully clear illustration of the dominant French attitude in Sophie Germain's time.

perfect the theory of elasticity, which was to be applied so successfully in constructing the Eiffel Tower later in the century.[15]

In Scotland, too, a woman overcame the obstacles of her society to become one of the century's foremost mathematical scientists. Denied access to formal mathematical education and faced with the active disapproval of relatives and acquaintances, Mary Fairfax Somerville (1780–1872) nevertheless pursued for almost 80 years an interest in mathematics that began by a chance encounter with an algebra puzzle in a women's fashion magazine. During that time, aided and encouraged by some of England's foremost scientific scholars, she taught herself enough mathematics and science to do important experimental

Evariste Galois. The night before he was killed in a duel, he dashed off a letter in which he outlined much of the basics of group theory.

work on the physical effects of the sun's rays and to write four major books on science. The first of these books, *The Mechanism of the Heavens* (1831), an annotated and expanded translation of Laplace's *Mécanique céleste*, gave the English scientific community access to French mathematical methods, which were more advanced than the English methods of the day. Her second and perhaps most influential book, *On the Connexion of the Physical Sciences* (1834), viewed physical science as a single entity and, in fact, was instrumental in shaping the scope of the term "physical science."[16]

Meanwhile, theoretical mathematics had begun to benefit from the spirit of revolution that was sweeping the Western world. In France, the overthrow of the monarchy and the subsequent Napoleonic period provided an ideal climate for the cultivation of new ideas. Into that climate was born Évariste Galois (1811–32), a brilliant, temperamental boy, who spent a large part of his short life being thrown out of schools and into jails. Despite these frequent embroilments, he devoted much

---

15. Nevertheless, the name of Sophie Germain is *not* inscribed on the base of the Eiffel Tower among the 72 scholars whose work made possible its construction!

16. John Couch Adams, who, in 1845, predicted the discovery of the planet Neptune (a year later) by calculating its orbit, mass, and size from otherwise unexplained irregularities in the movements of Uranus, claimed to have gotten his idea of looking for Neptune from reading a passage in this book.

of his time to algebra, which at that time was little more than generalized arithmetic, but his writings went unnoticed. Shortly before his
twenty-first birthday, Galois became involved in a duel that cost him
his life. The night before the "affair of honor," he dashed off a hurried
letter to a friend, sketching some of his recent mathematical ideas. He
wrote:

> I have made some new discoveries in analysis. . . . Later there will be,
> I hope, some people who will find it to their advantage to decipher all
> this mess.

"This mess" was the theory of groups, the foundation of modern algebra and geometry. (Group theory is explored in Chapter 6.)

The liberation of algebra from its dependence on arithmetic took a
giant stride forward with the work of William Rowan Hamilton (1805–
65), an Irish astronomer and mathematician. A child prodigy, who at
the age of twelve had a working knowledge of twelve languages,[17]
Hamilton spent much of his early scientific career applying mathematics to physical theories, especially mechanics and optics. In 1835,
he turned his attention to algebra and, eight years later, discovered
"quaternions." Roughly speaking, the system of quaternions is a generalization of the complex-number system. Quaternion multiplication
was the first worthwhile example of a noncommutative operation.
General classes of algebras were soon forthcoming, and the subject
was well on its way to abstraction. England was the center of this
nineteenth-century algebra with its geometric applications, which
flourished under the active guidance of Arthur Cayley (1821–95), the
originator of matrix theory, and James Joseph Sylvester (1814–97).
Sylvester was also a prime mover in the early development of American
mathematics, as a result of a lengthy visit to Johns Hopkins University.

The work of the French mathematician Augustin Louis Cauchy
(1789–1857) and his contemporaries made analysts more aware of
the need for strictly logical demonstration. Cauchy gave a workable
definition of the limit concept, and proceeded to establish a firm
foundation for calculus. He also developed the theory of functions of a
complex variable at about the same time that Gauss published his
complex arithmetic. Bernhard Riemann (1826–66) of Germany also
pioneered in the theory of complex variables. Riemann was a person
whose varied mathematical interests led to profound accomplishments in many other areas, as well, including mathematical physics,
the theory of functions, and the foundations of geometry. The most important single contribution of Gauss, Cauchy, Riemann, Norway's
Neils Henrik Abel (1802–29), and the other mathematicians of the

17. English, Latin, Greek, Hebrew, French, Italian, Arabic, Sanskrit, Syriac, Persian,
Hindustani, and Malay.

early nineteenth century was their meticulous insistence on rigorous proof. Their work paved the way for Karl Weierstrass (1815–97), a mathematician famous for his deliberate, painstaking reasoning. He clarified the notions of *function* and *derivative*, and eliminated all remaining obscurity from calculus.

A protégé of Weierstrass was Sofia Kovalevskaia (1850–91), a brilliant Russian woman, who moved to Berlin to study with him at the prestigious University of Göttingen. Denied access to formal classes because she was female, she studied under Weierstrass's private tutelage for three years and became the first woman to receive a doctorate in mathematics from Göttingen, in 1874. Kovalevskaia's effective use of Weierstrass's function-theoretic techniques in solving problems of mathematical physics brought her recognition as one of the world's leading mathematicians. Tragically, she died of pneumonia at the age of 41, at the peak of her professional career.

Insistence on rigor and logical simplicity was typified by Leopold Kronecker (1823–91), who asserted,

> All results of the profoundest mathematical investigation must ultimately be expressible in the simple form of properties of the integers.

He was a number theorist, but is best known for his prolonged ideological feuds with Weierstrass, Dedekind, and Cantor, whose theories were based on the concept of infinite progressions. Kronecker, on the other hand, would not admit the mathematical existence of anything not actually constructible in a finite number of steps. (See Section 5.9 for a discussion of these ideas.)

Diametrically opposed to this view were Richard Dedekind (1831–1916) and Georg Cantor (1845–1918). Dedekind rigorously developed the concept of irrational number, thus enabling the real-number system to become the basis for all of analysis. Cantor, in his *Mengenlehre* (Set Theory), based the concept of number on that of set, and proceeded to develop different types of infinity, called transfinite numbers, that have an arithmetic of their own. This, in Kronecker's opinion, was a dangerous travesty of mathematics; he attacked both theory and person with such vehemence that Cantor suffered a series of breakdowns and ultimately died in a mental hospital. Set theory, however, has remained as a prominent, though controversial, part of mathematical thought. (Cantor's work is explored in Chapter 5.)

The revolution in geometry was foreshadowed as early as 1733 by the work of Girolamo Saccheri (1667–1733). Ever since Euclid set forth his *Elements*, the fifth postulate (often called the "Parallel Postulate") had been questioned by those who thought it might be provable from the other four. Knowing that all previous attempts to establish this had failed, Saccheri proposed a radically different approach to the problem. He replaced the questionable postulate with its negation, hoping to arrive at two contradictory statements within the new system. If he had done this, it would have meant that the original fifth postulate must be a necessary consequence of the other four; but the new system yielded no contradictions. Disappointed, Saccheri turned back—just when one more step would have yielded one of the great discoveries of the century—and his work was promptly forgotten.

Early in the nineteenth century, three men in three different countries used Saccheri's approach, and they had the insight to realize the meaning of their "failures" as well as the courage to publish their findings. Nicolai Lobachevsky in 1829 (in Russia), Janos Bolyai in 1832 (in Hungary), and Bernhard Riemann in 1854 (in Germany) all published consistent non-Euclidean systems of geometry, each independently of the other two. Gauss had found some of these ideas several decades before, but had refrained from publishing them for fear of criticism. These ideas conflicted with the prevailing Kantian philosophy, which contended that space conception is Euclidean *a priori*, so they remained obscure for several decades more. But the logical floodgates had been opened. No longer were axioms regarded as statements that were intuitively evident; they were viewed as the initial assumptions whose choice was logically arbitrary, subject to no preconditions. This was the start of formal axiomatics. (Chapter 3 discusses these ideas in greater depth.)

Now that geometry was no longer confined to visual images, it expanded at a fantastic rate. In his book *Ausdehnungslehre* (Theory of Extension), Hermann Grassmann (1809–77) gave the world a fully developed geometry of n dimensions for metric spaces.[18] This work established him as one of the founders of vector analysis (along with Hamilton). Jacob Steiner (1796–1863), a pure geometer who detested algebra and analysis, developed much of projective geometry, which is discussed briefly in Section 3.4. On the other hand, Felix Klein (1849–1925) unified all geometries by means of modern algebra, declaring that every geometry is the study of properties of a set which remain unchanged with respect to some group of functions.

The trend toward unification of mathematics was personified by Henri Poincaré (1854–1912). His almost superhuman memory and powers of logical apprehension enabled him to produce valuable contributions to arithmetic, algebra, geometry, analysis, astronomy, and mathematical physics. He also wrote popularizations of mathematics and was actively interested in the psychology of creativity. Poincaré

---

18. *Metric space* is a generalization of the usual concept of physical space. It is a set of points on which some idea of distance is defined.

profoundly influenced the development of topology, a relatively new branch of mathematics, sometimes referred to as "rubber sheet geometry."

The formalistic treatment of algebra in England and the abstract axiomatic approach to geometry on the Continent triggered a sudden interest in logic and the foundations of mathematics, an interest that was redoubled after the appearance of Cantor's controversial set theory. The first significant mathematical studies of logic were *The Mathematical Analysis of Logic* (1847) and *The Laws of Thought* (1854), both by George Boole (1815–64). In these works, Boole exhibited a completely symbolic approach to logic and laid the foundation for future extensions of the field. In 1884, Gottlob Frege (1848–1925) published *Die Grundlagen der Arithmetik*, which offered a derivation of arithmetical concepts from formal logic and thus greatly stimulated efforts to unify logic and mathematics. (See Appendix A for a brief treatment of elementary logic.)

THE TWENTIETH CENTURY

B.9

In 1900, members of the International Congress of Mathematicians, assembled in Paris, listened as one of their foremost colleagues lectured on mathematics in the newborn century. David Hilbert (1862–1943) had just completed his now-famous *Grundlagen der Geometrie* (Foundations of Geometry), a complete renovation of Euclid's *Elements* using modern axiomatic methods. Hilbert outlined twenty-three unsolved problems, a challenge for the new century. His insight was so accurate that every one of these problems has led to important new results, and the solution of even part of one of Hilbert's problems carries with it international recognition for the solver. Almost all of the problems have now been solved and many of them have led the way to major advances in mathematical theory.

Even Hilbert, however, could not foresee how mathematics would mushroom in the twentieth century. The phenomenal growth rate that began in the 1800s has continued, with mathematical knowledge doubling every twenty years or so. Lest the implications of the last statement slip by unnoticed, consider that it means this: About as much original mathematics has been produced *after* astronauts first walked on the moon as there had been in all previous history! In fact, it is estimated that 95% of the mathematics known today has been produced in this century! More than 300 periodicals published in various parts of the world devote a major share of their attention to mathematics. Each year, the abstracting journal *Mathematical Reviews* publishes more than 8000 synopses of recent articles containing new results. The present century is justifiably called the "golden age of mathematics."

Quantity alone, however, is not the key to the unique position the twentieth century occupies in mathematical history. It is essential to understand that beneath this astounding proliferation of knowledge, there is a fundamental trend toward unity, a unity even more profound and productive than that envisioned by Descartes and Leibniz. The basis for this unity is abstraction. Although the non-Euclidean geometries of the nineteenth century paved the way for an abstract axiomatic treatment of mathematics as a whole, many of the fundamental interrelationships among the major branches of the subject did not begin to appear until the 1940s. The recent recognition of these unifying theories and of the vast unexplored fields they unlocked has led some prominent mathematicians to regard the end of World War II as the beginning of a new era in mathematics.[19]

As we approach the present, accurate evaluation of the relative significance of new mathematical results becomes almost impossible; so we have made little effort to compare individual contributions with one another, leaving that task to a later generation. Hence, although all mathematicians mentioned here have achieved prominence through acclaim from the modern scientific world, no claim is made for the completeness of the listings, nor is the topical coverage intended to represent a comprehensive survey of twentieth-century mathematics. The primary purpose of this section is to indicate briefly the scope and power of contemporary mathematical activity.

As a result of Boole's work and the recognition of formal axiomatics following the birth of the non-Euclidean geometries, interest in the logical foundations of mathematics began to spread rapidly. The most notable successor to Boole's efforts in mathematical logic is the *Principia Mathematica*, a monumental two-volume work that appeared during the years 1910–13, in which the philosopher-mathematicians Bertrand Russell (1872–1970) and Alfred North Whitehead (1861–1947) attempted to fulfill Leibniz's dream by expressing all of mathematics in a universal logical symbolism. Hilbert, too, dreamed of unifying mathematics, and he labored for many years to find a

---

19. See, for example, "Recent Developments in Mathematics," by Jean Dieudonné in the *American Mathematical Monthly*, **71** (3): 1964, p. 240.

single provably consistent set of axioms upon which all of mathematics could be based. Thus, mathematics entered a fourth phase. Starting only with an animal skin, it had first fashioned some serviceable garments and then acquired an elaborate wardrobe. Now, elegantly attired, mathematics began looking into its own pockets for unseen holes.

The holes started to appear as Russell found inconsistencies in Cantor's theory of sets, a theory upon which all of mathematics could be based. This made Hilbert's goal all the more desirable, and hence the mathematical community was profoundly shocked when Kurt Gödel (1906–78) proved in 1931 that this goal was unattainable. Gödel also laid the groundwork for one of the century's most spectacular mathematical discoveries. In 1964, Paul J. Cohen, basing his work on Gödel's, proved that both the Continuum Hypothesis and the Axiom of Choice are independent of the currently accepted axioms of set theory.[20] Gödel and Cohen showed that these two statements are unprovable from the other axioms, and that the inclusion of their negations in set theory can lead to entirely new theories of mathematics as a whole!

Once they recovered from their initial surprise, mathematicians accepted the fact that mathematics was not provably consistent from within, and continued to explore their own branches of the subject. In fact, the development of most parts of mathematics was virtually unaffected by the sudden tremor that had rocked its foundation. Algebra became far more general than it had ever been before, and similar tendencies toward abstraction in geometry led to extensive advances in both fields.

Analysis is undergoing an extraordinary metamorphosis during this century. When Henri Lebesgue (1875–1941) revolutionized the theory of integration in 1902, he opened the way for a much more abstract and unified treatment of analysis, as exemplified by the theories of abstract spaces developed by E. H. Moore in 1906 and Maurice Fréchet in 1928. This new generality has linked analysis with both algebra and topology.

According to at least one expert commentator on contemporary mathematics,

> the main fact about our time which will be emphasized by future historians of mathematics is the extraordinary upheaval which has taken place in and around what was earlier called algebraic topology.[21]

---

20. The Continuum Hypothesis is discussed in Section 5.8. Roughly speaking, the Axiom of Choice states that, given any collection of pairwise disjoint sets, we can choose exactly one element from each set in the collection. Although not recognized explicitly as an axiom until 1904, this assumption is basic to much of topology, modern analysis, and abstract algebra.

21. Dieudonné, "Recent Developments . . . ," p. 243.

Algebraic topology, a mixture of the ideas of "rubber sheet geometry" and group theory, began to be a major field of investigation with the work of Henri Poincaré at the end of the last century. During the first half of this century it became the spawning ground for some of the most powerful tools in all of mathematics. Besides extending the frontiers of mathematical knowledge in traditional areas of research, the methods of algebraic topology became the basis for an even newer field, homological algebra, that has forged powerful bonds of unity among previously separate theories in algebra, analysis, and geometry. Homological algebra and a close relative called category theory (which treats entire classes of abstract structures as individual objects) did not begin to appear in print until the 1940s and 1950s, but their techniques have already invaded much of mathematics. Other areas of topology, such as the study of knots and manifolds (generalized surfaces), have also become very active, particularly in the latter part of the century.

The use of mathematics in the physical and social sciences has been so widespread and penetrating in this century that it would be futile to attempt any summary of modern advances in applied mathematics. Nevertheless, there are several men whose works must be mentioned because of their profound effect on the world we live in. At the age of twenty-six, Albert Einstein (1879–1955) revolutionized physical science with his theory of relativity, a radically different analysis of change based in part on non-Euclidean geometry. (See Section 7.1.) Einstein's work made him the best known scientist of his day. John von Neumann (1903–57), a Hungarian-born American, directed the development of some of the first electronic computers at Princeton's Institute for Advanced Study, helped to expand quantum theory in physics, and is considered to be the founder of game theory, a mathematical analysis of strategies that is widely applied in fields such as economics and psychology. Von Neumann also made important contributions to the development of the atomic and hydrogen bombs and to long-range weather forecasting. Finally, we come to the father of automation, Norbert Wiener (1894–1964), whose work on information processes resulted in a field named by the title of his book, *Cybernetics, or Control and Communication in the Man and the Machine.*

The outstanding event of twentieth-century applied mathematics has been the invention of electronic computers. Their ability to perform routine calculations at speeds of many millions of operations per second has radically altered problem-solving methods in mathematics, the physical sciences, the social sciences, business, and many areas of daily life. The social sciences are currently undergoing a fundamental shift of emphasis from qualitative to quantitative considerations, a transition made extremely difficult by the many variables that are usually present in social situations. The proper handling of these variables would be virtually impossible by traditional techniques, but computers can simulate and analyze complex situations involving hundreds, or even thousands, of items, including people, keeping track of random effects and other pertinent data. Thus, electronic computers are fast becoming indispensable research and instructional tools in economics, sociology, psychology, education, and other fields. (See Chapter 9 for a discussion of computer algorithms.)

The "computer age" is affecting pure mathematics as well. It has created the need for a new branch of mathematical logic concerned with problems of machine design and coding (discussed in Section 2.8), and it has aided and encouraged the development of numerical analysis. Rapid advances in machine logic, program efficiency, and hardware technology during the second half of the twentieth century have provided computers with astounding capabilities. These increasingly powerful machines have been used to make major breakthroughs in settling some of the longstanding open questions of pure mathematics. Here are two of the famous problems that fell to the power of the computer in this century.

■ **The Four-Color Problem.** In 1852, a young British mathematician by the name of Francis Guthrie posed a seemingly straightforward question. Simply put, he asked whether four colors were enough to permit every possible (planar) map to be colored so that no two "countries" with a common border would be the same color. It is easy to show that three colors are not enough, and a proof that five colors always suffice had emerged before the end of the nineteenth century. However, the critical case of four colors resisted the efforts of the world's best mathematicians for more than a century and thereby spawned some major advances in graph theory. (See Chapter 8 for a discussion of graph theory.) It was finally resolved in 1976 by Kenneth Appel and Wolfgang Haken, who used powerful, intricate computer techniques to show that four colors were indeed sufficient. Their computer-dependent argument generated quite a bit of controversy about what is acceptable as a mathematical proof, and many mathematicians are still somewhat uncomfortable with procedures whose accuracy cannot be checked "by hand," as it were.[22]

■ **Classification of Simple Groups.** The problem here is a little harder to state, except by analogy. A type of algebraic structure

---

22. For a detailed discussion of the Four-Color Problem, see Chapter 7 of Item 5 in the list *For Further Reading* at the end of this appendix.

called a group, which entered the mathematical scene in the nineteenth century, has become very important in various areas of pure and applied mathematics. (See Chapter 6 for a detailed description of groups.) Among all the possible finite group structures, there are some—called *simple groups*—that can be used as "building blocks" for all the rest, like the prime numbers are used in number theory. As in the case of prime numbers, the natural question of finding some comprehensive classification that described all simple groups turned out to be a very hard problem. Its solution was finally completed in 1980, after some 40 years of computer-assisted work by more than one hundred mathematicians around the world. The published proof stretches over nearly 15,000 pages of journal articles![23]

While these and other previously intractable mathematical problems have fallen to the power of the computer in this century, a few famous questions have so far resisted even the most sophisticated combined efforts of mathematician and machine. Among them is a problem in number theory discussed elsewhere in this book—the characterization of perfect numbers (see Chapter 2). In this case, (as in other problems like it), computers have been able to test the relevant conjectures up to astronomically large numbers. However, since such testing can never exhaust the infinite set of integers, it appears that disposition of such questions will have to await ingenious logical breakthroughs in the theory.

Besides settling old questions, the computer age has begun to generate new areas of mathematical investigation. Perhaps the most striking of these, in conceptual originality and in sheer visual impact, is the theory of fractals. The term *fractal* refers to the notion that, rather than being 1-, 2-, 3-, or 4-dimensional (as traditional geometry would lead us to expect),[24] the geometric objects of this theory are considered to be of fractional dimension, such as $\frac{4}{3}$ or 1.2618. Fractal geometry was created almost singlehandedly in the decade 1967–77 by the French mathematician Benoit Mandelbrot, a researcher working for IBM in the United States. Perhaps the most striking feature of this theory is that it reveals an underlying mathematical order in much seemingly chaotic behavior that occurs in nature. Since 1980, the theory and its applications have been extended by increasing numbers of mathematicans, scientists, and even artists; but we are still much too early in the development of this beautiful concept to assess its power or its limitations accurately. Suffice it to say that fractal geometry

23. For more details about this problem and its solution, see Chapter 5 of Item 5 in the list *For Further Reading* at the end of this appendix.
24. See Chapter 7 for a discussion of the usual notion of dimension.

typifies the modern symbiotic relationship between human logical theory and electronic computational power.[25]

The growth of mathematics in the United States during this century has indeed been phenomenal. American mathematical scholarship is usually considered to have truly begun with the extended visit of British mathematician James Joseph Sylvester at Johns Hopkins University late in the nineteenth century; his presence brought a new level of respectability to mathematical research in this country. Sylvester was also instrumental in providing women equal access to higher education in mathematics. Through his efforts, Johns Hopkins grudgingly admitted Christine Ladd-Franklin (1847–1930) to graduate study in 1878; but it did not award her a doctorate for the work until 1926, 44 years after she had finished her dissertation! Throughout her career as a lecturer at Johns Hopkins and then at Columbia, Ladd-Franklin, in turn, devoted much of her time, energy, and money to giving women equal access to graduate education.

Similar efforts in late nineteenth-century England were also to affect the American mathematical scene. The algebraist Arthur Cayley at Cambridge actively encouraged promising young mathematics students at nearby Girton College, England's first college for women.[26] Among the Girton mathematics students helped by Cayley were Grace Chisholm Young (1868–1944), who went on to work with Felix Klein and David Hilbert and who got her doctorate at Göttingen, and Charlotte Angas Scott (1858–1931), who came to America in 1885 as the (only) mathematician among the eight founding faculty members of Bryn Mawr College in Pennsylvania. Bryn Mawr was a college dedicated to providing undergraduate and graduate education for women. Scott's long, productive career there had a profound impact on American mathematical education. She spent 40 years building Bryn Mawr's mathematics program into one with a well-deserved reputation of excellence at all levels, including the granting of doctorates; she personally supervised the dissertations of seven doctoral students. She was instrumental in attracting to the Bryn Mawr faculty Emmy Noether (1882–1935), one of the truly great algebraists of modern times. Noether, who earned her doctorate at Erlangen, had been invited by David Hilbert and Felix Klein to Göttingen in 1915, where she stayed until the rise of the Nazis in 1933. She accepted a guest professorship

25. A beautiful visual display of this geometry appears in *The Beauty of Fractals*, by H. O. Peitgen and P. H. Richter (Springer-Verlag, 1986). A description of the theory and its application can be found in *The Fractal Geometry of Nature*, by Benoit B. Mandelbrot (W. H. Freeman & Co., 1982).

26. Girton College opened in 1869 with five students. In 1873, it relocated to within three miles of Cambridge, where Girton students were allowed to attend lectures by some (but not all) of the professors. See page 194 of Item 8 in the list *For Further Reading* at the end of this appendix.

at Bryn Mawr that year, and lectured both there and at Princeton's Institute for Advanced Study until her sudden, premature death two years later.

American mathematics became truly self-supporting with the emergence in 1913 of George David Birkhoff (1884–1944) as an internationally renowned mathematician. Birkhoff was the first American-educated mathematician to achieve such stature, and he, in turn, directed and guided many of the most productive American mathematicians of this century through his position as a professor, first at Princeton and then, from 1912 until his death, at Harvard.

As the prestige of the United States has grown, more and more Asian and European mathematicians have migrated to the major American research centers, such as Princeton's Institute for Advanced Study (whose original faculty included Einstein and von Neumann). Today, the combined membership of the three major American mathematical societies is approximately 30,000, as compared with a mere handful in 1913.[27] Moreover, the employment focus of this mathematical talent has shifted and broadened over the years. In the early part of this century, almost all mathematicians worked in academic settings. Now, however, a variety of industrial, commercial, and government needs, often occasioned by the ever-widening use of computers, has drawn more and more mathematicians away from higher education and into applied-mathematical endeavors. In this way, mathematics is riding the historical pendulum, swinging back from the abstractions of the past hundred years toward applications of these theoretical results to "real" problems in other areas. Perhaps this remarriage of mathematics and reality will carry with it an appropriate social attitude change toward the subject, and mathematics once again will be viewed not as an area of irrelevant esoterica, but as a vital subject whose pursuit is essential to the progress of the human race.

TOPICS FOR PAPERS

1. Choose a mathematical event or a mathematician mentioned in this Appendix and write a paper that describes what was going on in the world at that time in history. Do not confine yourself to the region of the person or event you choose, but consider all parts of the world. Also consider all major historical features of the time—political changes, social trends, scientific breakthroughs, religious movements, literature, art, etc. Then

27. These professional societies are the American Mathematical Society, the Mathematical Association of America, and the Society for Industrial and Applied Mathematics.

discuss how (if at all) the mathematics of your person or event influenced and/or was influenced by the historical situation you described. It is legitimate to present your own opinions about this, but they should be accompanied by supporting arguments based on the historical evidence.

2. Do most women think about mathematics differently than most men do? If so, how? (Your *opinion* is solicited here; there are no definitive psychological research findings on this question yet.) If your answer to the first question is *no*, then this paper topic is not for you. If your answer is *yes*, read on. Assuming your opinion is correct, in what ways might the mathematics that you studied in this course have developed differently if it had been originated primarily by women? Provide specific illustrations from the course; then try to generalize your comments to encompass the development of mathematics as a whole.

3. The first African-American to be awarded a Ph.D. in mathematics by an American university was Elbert F. Cox, who received his doctorate from Cornell University in 1925. Do some library research to find out more about him and about at least three other African-Americans who have earned Ph.D.'s in mathematics. Write up the results of your research as a series of brief (1-page) biographical sketches. If you wish, end your paper with some general comments about the contributions of African-American mathematicians to twentieth-century mathematics. Be sure to identify the sources you use in gathering your information.

FOR FURTHER READING

1. Bell, E. T. *Men of Mathematics*. New York: Simon & Schuster, Inc., 1937.

2. Boyer, Carl B. *History of Mathermatics*, 2nd ed. New York: John Wiley and Sons, Inc., 1989.

3. Burton, David M. *The History of Mathematics*. Boston: Allyn and Bacon, Inc., 1985.

4. Dantzig, Tobias. *Number, The Language of Science*, 4th ed. New York: Macmillan Publishing Co., Inc., 1954.

5. Devlin, Keith. *Mathematics: The New Golden Age*. London: Penguin Books, 1988.

6. Dunham, William. *Journey through Genius: The Great Theorems of Mathematics*. New York: John Wiley and Sons, Inc., 1990.

7. Eves, Howard. *An Introduction to the History of Mathematics*, 4th ed. New York: Holt, Rinehart and Winston, 1976.

8. Grinstein, Louise S., and Paul J. Campbell, eds. *Women of Mathematics*. New York: Greenwood Press, 1987.

9. Keyser, C. J. *Mathematics as a Culture Clue*. New York: Scripta Mathematica, Yeshiva University, 1947.

10. Kline, Morris. *Mathematical Thought from Ancient to Modern Times*. New York: Oxford University Press, 1972.

11. ———. *Mathematics in Western Culture*. New York: Oxford University Press, 1964.

12. Newman, James R., ed. *The World of Mathematics*, Vols. 1–4. New York: Simon & Schuster, Inc., 1956.

13. Osen, Lynn M. *Women in Mathematics*. Cambridge, MA: The MIT Press, 1974.

14. Perl, Teri. *Math Equals*. Reading, MA: Addison-Wesley Publishing Co., 1978.

15. Rossiter, Margaret. *Women Scientists in America*. Baltimore: Johns Hopkins University Press, 1982.

16. Smith, David Eugene. *History of Mathematics*, Vols. 1 and 2. Boston: Ginn and Company, 1923.

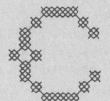

LITERACY IN THE
LANGUAGE OF MATHEMATICS

INTRODUCTION

Every now and then, serendipity brings together two independent works that seem to have been made for each other. Such is the case with this book and Dr. James O. Bullock's article, "Literacy in the Language of Mathematics." This article originally appeared in the October 1994 issue of the *American Mathematical Monthly*. However, its intended audience reaches far beyond the mathematical community. Dr. Bullock, a senior scientist at the Cancer Research Center in Columbia, Missouri, is a research physiologist who describes mathematics as a liberal art. In this essay, he argues that mathematics is a powerful tool in science largely *because* it is independent of any specific piece of reality. He views mathematics as a sophisticated language with a literature rich in powerful metaphors. In Bullock's words:

> Mathematics is not a way of hanging numbers on things so that quantitative answers to ordinary questions can be obtained. It is a language that allows one to think about extraordinary questions. . . . Getting the picture does not mean writing the formula or crunching the numbers, it means grasping the metaphor.

To be sure, there are mathematicians and scientists who find some of Dr. Bullock's ideas controversial. Even the authors of this book disagree with some details here and there. That's as it should be. Commenting on the purpose, Bullock says,

> In writing this essay, it was my intention to be provocative, not to compile a canon. . . . If my exposition truly changes your thinking, I expect you to be inclined to argue some points, or take a different direction in elaborating on them. Every flaw found in a canon weakens it, but each new thought brought into an intellectual discourse enriches it. . . . [I]t is the latter activity that is the very essence of liberal education.

580

This is a deeply thoughtful essay that richly repays several readings. Read it once, think about it, then read it again. Ask yourself: Does this fit my idea of the mathematics I have seen so far? In what ways do I agree with it? In what ways do I disagree? Expect to read it a third or fourth time. The effort you invest can have a profound, positive impact on *your* liberal education as it shapes and refines your understanding of the complex, productive, artistic enterprise called "mathematics."

Literacy in the Language of Mathematics
James O. Bullock

The American educational system has treated mathematics as skill in numerical manipulations, and has used the term "quantitative reasoning" to describe the application of mathematics to other areas of study. This serious misconception has severely hampered the ability of our students to comprehend important developments in scientific and philosophic thought. Mathematics can be more properly regarded as a form of language, developed by humankind in order to converse about the abstract concepts of numbers and space. In addition to the intellectual appeal of its intrinsic linguistic structure, the language should be appreciated for the richness of the literature composed with it.

The immediate cause for the recent concern about mathematics education is the widespread difficulty students apparently experience with this subject. While some educators have expressed an interest in improving pedagogy, students have tended to opt for a more practical approach whenever it has been open to them: avoiding the subject altogether. Predictably, the reaction of the educational system has been an attempt to force students to learn mathematics anyway. The mandate to hold students' feet to the fire has been carried out with varying degrees of resolve, the debate over how much pain to inflict being dominated by two conflicting lines of reasoning. According to the first, there must be some minimum standards which can be applied to all students. In addition, since it builds character, everyone deserves their fair share of suffering. Those determined to adopt a more merciful attitude, on the other hand, often argue that the dose of unpleasantness should be limited to what is absolutely necessary. By this logic, mathematics courses need be required only when they are listed as prerequisites for courses that are required for some other reason.

In the course of such arguments, the question of mathematics itself has been relegated to a position of secondary importance. Implicit in this view is the premise that mathematics is a highly esoteric subject which, despite having some practical applications, possesses only tenuous connections to other areas of scholarship. Any consideration of the intellectual and aesthetic appeal possessed by the discipline has been dismissed as irrelevant for all, save the mathematicians themselves. Within most curricula, mathematics has been so thoroughly dissociated from all other subjects that students generally encounter little evidence that these assumptions might be mistaken. Physical science, for example, is nearly always taught as if students know little or nothing of mathematics. Similarly, students are expected to learn biology without any reliance upon the concepts and formalism of the physical sciences. Many statisticians now insist that their subject is something quite apart from mathematics, so that statistics courses do not require any preparation in mathematics. Some computer scientists have gone so far as to suggest that if you understand the mathematics of their machines, it is not necessary to understand the mathematics of natural phenomena. To the students, nothing seems to be amiss. They observe that the factors which determine whether a course is classified as hard or soft, general or specialized, are the complexity of the problem assignments and the thickness of the textbook. As the values of these parameters increase, the demand for *proficiency* with specific mathematical techniques also increases. Across the entire spectrum of course offerings, however, students are exposed only to the vaguest notions of what we mean by *rigor*.

A careful consideration of the nature of mathematics as an intellectual activity reveals the utter incongruity of this situation. Not only have our students failed to appreciate the beauty of mathematics, they have little grasp of the profound insights about the natural world which mathematics has made possible. Rather than a mere intellectual curiosity or useful skill, mathematics is an important facet of the most distinctive capability of the human species.

Mathematics Is a Language

As a general definition, mathematics could be called the study of numbers and space. In order to talk about such things, mathematicians first had to devise an appropriate vocabulary and alphabet. It is apparent that the objects defined by mathematicians are entirely *abstract*, and can never actually be observed in any way except by the human imagination. Although many students seem to be disturbed by this fact, mathematics is comprehensible precisely *because* it is abstract. We are not dependent upon observation to know that lines are completely straight and parallel lines never meet.[1] It is possible to make these assertions with confidence because mathematics was not discovered, it was invented. It must also be remembered that imagination and abstraction are universal parts of the human experience, and not the exclusive domain of the mathematician. The idea of a line or a point is no more abstract than the idea of loyalty or freedom. Mathematics differs from other languages not because of its inclusion of abstraction, but because of its complete detachment from the complications of what we experience by direct observation.

1. See p. 70.

If mathematics consisted only of new words and symbols, it could properly be considered as an extension of existing language. The reason mathematics is a new and separate language is that it also has its own syntax and grammar. Having devised both vocabulary and rules for the language, mathematicians seek to discover what things are possible or not possible to say about numbers and space. In this ideal, ordered world of the imagination, it is possible to apply relentless logic to any questions that arise. Euclid's geometry should be appreciated for the beauty which can result from this process. His system of axioms, definitions, theorems, and constructions clearly lays out both the vocabulary and the rules for conversing about his chosen topic. By limiting his considerations to fanciful but specific abstractions of his imagination, he was able to produce an indisputably logical and thorough exposition. If one understands the language, one will not only know what a triangle is, but exactly what can be said about triangles. The lexicon and grammar of language are quite precise; the conclusions are quite inescapable.

From the Linguistic to the Literary

The natural sciences seek to explain the behavior of the things we encounter in nature. Experimental observation is the court of highest appeal for all such explanations. What then is the role of the abstract, detached language of mathematics in experimental science? The answer is not different than that for the roles of language in other forms of scholarship. The scientist uses the language of mathematics to construct *metaphors* which represent insights into the workings of nature. The use of metaphors is as commonplace in science as it is in poetry. We speak of radio "waves," subatomic "particles," and celestial "bodies." What distinguishes the mathematical metaphor is the extraordinary power of this language to uncover implications of the underlying idea. Calling the earth a sphere is an extremely powerful statement, for we know a great deal about the things we can say about spheres. The fact that most scientists prefer to call such constructions "models" or "laws" does not change their essential character. It is just as metaphorical to call the world a sphere as it is to call it a stage.

Getting the Picture

It is generally agreed that the object of studying science is to understand its fundamental concepts, but how does one come to understand a mathematical metaphor? It is widely believed that the first step is to find a way to remove the mathematical details so that the essential picture can be seen unobscured. Mathematics, then, need be introduced only if one is confronted with a specific problem that requires a numerical answer. Many educators are fond of using the term "quantitative reasoning" as a substitute for "mathematics." The idea that mathematical questions can be answered qualitatively without actually using any mathematics represents one of the most serious misconceptions about the nature of this discipline. Mathematics is not a way of hanging numbers on things so that quantitative answers to ordinary questions can be obtained. It is a language that allows one to think about extraordinary questions. Saying that the earth has a round shape means only that it has no edges. This nonmathematical picture is not simply "qualitative," "verbal," or "intuitive"; it is primitive and empty. If we wish to construct a meaningful metaphor about the shape of the earth, we must use the language of shapes, which is mathematics. To those prepared to examine its implications, even the mathematically simple picture of the spherical shape has a great richness when used as a metaphorical device. Getting the pictures does not mean writing the formula or crunching the numbers, it means grasping the metaphor. Without mathematics, one cannot even read the words.

Reading Critically

Students of the physical sciences are often left with the impression that facility in solving mathematical problems is all that is necessary to understand scientific ideas. After all, the entire subject of classical physics can be reduced to seven compact, simple-looking equations, and most study time in this subject is spent working problems. This view, however, has the same basic flaw as the belief that mathematics can be replaced by less abstract language. Just as excluding numerical calculations from an argument does not eliminate the need for mathematics, one cannot bring mathematics to bear on a question by simply assigning numbers and symbols to its elements. Metaphors do not have answers, they have implications. When we struggle to find solutions to mathematical problems, we are exploring the specific implications of a writer's metaphors in particular physical circumstances.

In physics, as in all other intellectual disciplines, understanding derives from critical thought, not just hard work. The equations of electrodynamics, for example, are important not simply because they give rise to interesting and complicated mathematical problems. Maxwell's picture of flowing force fields which diverge and swirl according to precise mathematical relationships is an incredibly rich metaphor. In order to understand this literature, it must not only be read, it must be pondered carefully. By thoughtful consideration of wisely chosen examples, you can begin to discover the things about different physical situations that have important influences on the outcomes one observes. You begin to understand the metaphor. As Paul Dirac put it: "I understand what an equation means if I have a way of figuring out the characteristics of its solution without actually solving it." This is what we mean by physical insight; it is an intuition based upon critical analysis of key questions. Einstein's surmisal that magnetism is fundamentally a relativistic effect of electricity, one of the truly remarkable insights of human history, came directly from his contemplation of Maxwell's equations. Building such intuition may require considerable effort. Electricity and magnetism are more difficult to understand than the shape of the earth, and Maxwell's metaphor is complex in language and construction. On the other hand, one could spend a lifetime working out solutions to very complex and difficult problems in electrodynamics and never discover the possibility of electromagnetic radiation, even though Maxwell's equations directly imply its existence. A mason may indeed build great strength in his arm muscles, but his goal is to build the house. Likewise, the goal of the scholar is not technical dexterity, but insight.

Analyzing Mathematical Literature

While it is important to choose interesting examples to think about, as a practical matter we are still left with mathematical problems to solve. In asserting that mathematics is an idealized abstraction which lends itself to elegant, logical treatments, we have overstated our case in several particulars. First, not all mathematical problems have solutions. In fact, the only way a mathematical system can be made to give uniformly consistent answers is for some questions to be left completely unanswered. Even if solutions are known to exist, it is sometimes true that there can be no general method for finding them. (The most famous example is the problem of trisecting an angle.[2]) Such philosophical difficulties with the positivist viewpoint pale in comparison to the practical difficulties in finding solutions, even when you know that they exist and that you should be able to find them. Specific mathematical problems, no matter how well posed, often lead one into an impenetrable thicket from which it can be exceedingly difficult to extract solutions by any means. In attempting to understand science, one should not let these kinds of mathematical difficulties take over. What the scientist wants to know is whether the original metaphor can reveal anything about what will be expected to happen in a particular situation. Often, a good approximate solution can lead out of the quagmire toward great insight. The concepts of field lines, resistance, and capacitance are examples of insightful approximations of Maxwell's equations which allow one to picture and analyze the behavior of certain physical systems by avoiding purely mathematical entanglements. Hence, the use of these heuristic devices is another way to achieve the kind of understanding to which Dirac referred. One must never lose sight of the goal of this process. By itself, an individual solution or the particular technique used for finding it is important only insofar as it provides clues about the nature of the metaphor. It is in the original equations, not in a catalog of solutions, that the complete law, the entire metaphor, is to be found. The goal is to grasp the metaphor, not just to compile a catalog.

---

2. See p. 548.

The Nature of Knowledge

Finally, we have to consider that the metaphors themselves may have limited usefulness. Newton's laws of motion have turned out to need corrections in order to be consistent with the theory of relativity. The problems with Maxwell's laws are more serious: they are internally inconsistent at some points. Even the most imaginative metaphor cannot bring every facet of truth into plain view. In nonmathematical literature, we are used to the fact that meanings become distorted when ideas are stretched too far. (For example, the cartoonist Johnny Hart observed some years ago that the trouble with a melting pot is that the bottom gets burnt and the scum comes to the top.) In interpreting mathematical language in the scientific literature, we tend to become confused by the Platonistic view that mathematical objects themselves actually exist. Regardless of whether this assertion about mathematics is philosophically correct, the use of this language does not imbue one's ideas about nature with an independent existence. The use of mathematical objects is still a metaphorical device. Mathematical metaphors are not different from other forms of human understanding. They may provide some significant insights, but are not a holy grail of ultimate truth. This should come as no surprise. We know that the earth is not *exactly* a sphere. Furthermore, the holy grail is a heretical idea, not just to Christian theology, but to Western philosophy as well. Despite our desires to the contrary, nature is not obliged to limit its complexity to the confines of the human imagination. Despair is not an appropriate reaction to this revelation. The importance of the insights provided by any single idea is not diminished because it fails to explain everything else in nature. In addition, our capabilities are not forever fixed. Mathematicians have continued to explore new territories in the abstract landscape they have created, and certainly do not believe that this is a task which will one day be completed. If all of human understanding is based on metaphors, then extending the reach of the languages we use to construct them can greatly expand the boundaries of our imagination. We have cited but a scant few of the important advances in the natural sciences which have been made possible by the development of mathematics. In a fuller accounting, we would need to discuss its influences in philosophy and in the social sciences, as well as the fresh insights provided by new mathematics into old conundrums. Mathematics, therefore, does not simply represent a narrow but important class of intellectual achievement. Its distinct and prominent place in the realm of human language and literature make it an essential foundational element of education.

How Does One Become Literate?

As educators, we seek to engage students in the exploration of literature, in the understanding of other people's ideas, and in the expression of their own views. If the starting place of this process is *language*, its goal must be *literacy*. This goal can be attained only through thoughtful reading and critical analysis. Having seen that all forms of language share the same basic literary device, we can draw rather extensive parallels between the meaning of literacy in the language of mathematics and that which applies to other languages. The following list of assertions about the nature of literacy hardly requires explanation or justification. When applied to mathematics education, however, they imply that our current approaches are grossly ill conceived, and require wholesale revision.

Individuals can learn more than one language. The assertion that mathematics requires a very special innate talent is often used as an excuse for not learning this subject and a reason for not trying to teach it. This sort of condescension must not be permitted in the educational community. Mathematics is not so different from other expressions of the human intellectual capacities for communication and imagination. The fact one can learn to say things in mathematical language that cannot be said in other languages is beside the point, as is the possibility that different regions of the brain may be used. Mathematics is not the exclusive provence of either the gifted or the deranged; it is for all who would seek to be truly educated.

One must learn to read before trying to study literature. This point seems so obvious that it is surprising that it is violated so systematically, especially in engineering and science curricula. Unsuspecting students are routinely confronted by passages of unintelligible mathematics. The accompanying explanation, if one is to be found, is typically cursory. Usually, it is unclear whether the mathematical interlude is given as an aside, whether it is a point of central importance, or whether it is something the students were supposed to have known already. Often, such narratives become more obscure as they progress. In reading any form of literature, one may occasionally find it necessary to consult a dictionary, and footnotes can be very helpful in clarifying points of grammar usage. When one has to look up every other word and the footnotes begin to swallow up the text, however, a reader has very little chance of putting together the meaning of even a single complete sentence. If students are given the chance to study enough mathematics *first*, it can be startling how facilely they are able to grasp the central ideas of other disciplines.

Vocabulary is not enough. We all remember vocabulary tests from grade school: look up the word in the dictionary, memorize its definition, learn how to spell it, and go on to the next word. We compile the same sorts of lists for mathematics students: quadratic equations, logarithms, antiderivatives, Bessel functions, etc. Furthermore, we forbid our students to use a dictionary to look up the unfamiliar mathematics they may encounter in their studies. We insist that they commit the entire dictionary to memory. The difficulty with this approach is that knowing the formal definitions of a large number of words is no guarantee that one can actually say anything comprehensible. One must obtain practice in expressing complete thoughts with language. In mathematics, this means constructing examples which deal with specific objects and events; that is, applications. It does not matter so much whether the objects are observable or abstract, just as it does not matter whether the words are in the form of prose or verse. What matters is that words must be given a context if we are to be enriched by understanding their meanings.

Conversational fluency is of limited value. We spend a great deal of time telling our students "this is how you solve this kind of problem." To divide by a fraction, for example, you must invert and multiply. This is just like saying "when you answer the telephone, pick up the receiver and say 'hello'." This might seem to be a step forward from the vocabulary lists, but most often it actually represents a retreat. You do not even have to know what the individual words mean in order to be able to fire back the correct response. Once you have learned the drill, you will never have to think about it again. In short, this is *training* rather than *education*. Whether or not any sort of training has a proper role in university curricula is a matter of longstanding controversy, but in this case, the point is moot. We have trained our students to become highly proficient in answering questions they will never be asked again. Outside the classroom, it is extremely unlikely that you will be asked to "solve for x." The harm we inflict by insisting on this sort of raw skill development goes far beyond simply wasting our students' time, however. Since they never really understood the meaning of the answers they were trained to give, they will be unable to answer the questions that do come up. Because they come to perceive mathematics as a collection of specialized skills, rather than a way of thinking and talking, our students are lead to conclude that mathematics is either worthless or impossible to master. Imagine what we would think if they came to this conclusion about, say, Spanish. There was a line in one of our high school dialogues that went: "they–always–serve–meatballs–in–the–cafeteria–on–Wednesday." Since I have never had the opportunity to use Spanish to tell someone this, am I to conclude that Spanish is worthless for communicating with my fellow humans, or that it is too much to expect that I will ever be able to learn how to say anything really useful? By offering training rather than education in mathematics, we have produced a legion of mathematical sophomores, possessed of an extensive but superficial knowledge. Breaking this cycle is one of the most important pedagogical challenges in mathematics education.

Writing is not the same as penmanship. Whereas penmanship used to be considered a highly valuable skill, most of us would now refuse to accept a handwritten document, no matter how legible or graceful the hand. We would not say that a manuscript had been "written" by a machine just because it was typed. In the case of mathematics, our attitude is exactly the opposite. We insist on work which has been done by hand, and disallow all forms of mechanical or electronic assistance because it amounts to cheating. Whether the students are in college learning calculus or in the third grade learning multiplication, we insist that they concentrate on developing their skill at the mechanical manipulation of symbols required to arrive at the desired product. The time for us to abandon this antiquated and misguided attitude is long past. We are wasting students' time by teaching them how to write with a quill pen when they should be using a typewriter. The use of modern technology is beneficial to mathematical undertakings and literary tasks alike. One's attention can be focused on the conceptual content of what is being written, rather than the manipulative processes required to produce the writing itself. The skills of using the typewriter, the calculator, and the computer are useful ones. The value of being able to add large columns of numbers in one's head or remember the Fourier transforms of complicated functions is dubious at best. More to the point, in the outside world, it will be considered inappropriate to waste time performing such tasks by hand.

Complicated is not the same thing as sophisticated. Nearly everyone has had experience with long, tedious problems involving only elementary arithmetic. There is no trick to constructing similar problems at any level of mathematics. When a problem appears to be difficult, it is important to distinguish whether obtaining the solution requires new insights, a bit of cleverness, or just plain drudgery. The simple problems are not the only ones that are worth solving, but it is appropriate to offer some justification for tackling difficult ones.

The study of grammar can be overdone. While systematic consideration of mathematical grammar is generally ignored in lower level courses in mathematics, it often overwhelms everything else in advanced courses. For someone who is studying mathematics in order to do a bit of advanced reading, it is not always necessary to grasp all of the subtleties of mathematical arguments in order to appreciate the usefulness of their conclusions. Some such problems, like proving the existence of the derivative, turn out to be quite formidable. It is sometimes enough to understand the kinds of things mathematicians worry about, but to leave the actual worrying to them.

It is possible to be fluently ignorant. The ability to read the English language does not imply any knowledge of Shakespeare's plays, nor is an understanding of the playwright's ideas derived simply from hearing his words. Further, because a piece of writing is grammatically correct and draws from a large vocabulary is no indication that it contains any worthwhile ideas. Similarly, the ability to perform mathematical gymnastics is not necessarily indicative of any underlying understanding of either the literature of mathematics or the nature of the language. Mathematical exercises should always be chosen in such a way as to provide the student with a keener insight into great concepts, not just practice in manipulations. Education is not just learning to read and write. The questions of what is read and what is written about are central to any practical definition of literacy.

Literature must be studied from original sources. The application of this principle to science and mathematics is often misunderstood. We are not suggesting that the only proper way to learn Newtonian physics is to study the *Principia*. Ideas and arguments presented in the language of mathematics can be reproduced very concisely. When that part of the exposition written in other languages is paraphrased or even condensed, little or no loss of meaning need occur as long as the mathematics is preserved. On the other hand, major distortions of perception will inevitably result if mathematics is replaced by non-mathematical pictures or lists of specific examples. It is essential to present the original formulation, the equations, which define the author's metaphor. This is true even if the students' command of the language is limited, and they are able to grasp only some of the simpler implications of the metaphor. Having seen the original idea in its entirety, they will understand the underlying basis of however much they are ready to learn. Furthermore, they can approach the subject again in the future without having anything to unlearn.

The accomplishments of the past cannot be dismissed. We have become accustomed to considering pre-modern science as so much primitive nonsense: domed sky, flat earth, flies erupting spontaneously from dead meat, et cetera. In the humanities, such an attitude would appear silly. The works of Shakespeare did not relegate those of Sophocles to the dust bin. Because mathematics is generally, and quite incorrectly, categorized with science rather than with philosophy, most people, including our students, tend to think that the mathematics being taught in the present day represents modern developments. In fact, most college students are exposed to very few mathematical ideas which originated since the fifteenth century. The accomplishments of the early Greek, Hindu, and Islamic scholars are of more than historical interest. They were not made obsolete by subsequent developments, but rather formed the basis for those very advances. What is distressing about the present state of education is that so few students have even an inkling of the progress made in mathematics during the last half millennium.

There is no substitute for learning how to read. The reason mathematics was devised in the first place was the inability of existing language to deal with this subject matter. Newton, for example, found it necessary to invent the calculus in order to develop and express his ideas. Trying to understand Newton without calculus is *not* like trying to understand Sophocles without Greek, it is like trying to understand Sophocles without *words*. One could represent *Antigone* as a series of pictures. If done with sufficient skill, the major points of the story line could be made evident; but the play is not just a story about some people who end up dead under tragic circumstances. To pretend that one can understand Newton without using his mathematics would be equally delusional. At best, one might hope to be trained in solving certain types of problems involving throwing things up in the air or sliding them down inclined planes.

With few exceptions, such training will prove useless in the long run. Visual images, no matter how important they may be to human communication, do not constitute a form of language. Creating a picture book is not an act of translation; and, despite the aphorism to the contrary, no number of pictures can match the power of the written or spoken word. Mathematics, on the other hand, is precisely a form of language. If we truly wish to understand the things that have been said with this language, we must eventually take up some books with words in them.

Illiteracy has unfortunate consequences. In demanding picture books, one chooses illiteracy and is cut off from some of the most important ideas in the history of western civilization. This is a decision that should not be taken lightly by anyone who truly desires an education. That Americans seem unaware of the consequences of such a decision is hardly surprising, since they do not make it for themselves. If it makes us feel better about our own ignorance to say that we know that the earth is spherical, it is only because we were never told that the educated people of the world had known that for at least 1800 years prior to the time Columbus set sail. Americans may not think that the earth is flat, but most do believe that Michael Jordan can hang in the air and change direction in midflight. Only the ignorant could be convinced of this. The educated have known better for three hundred years. Our own educational system has given out nothing but picture books, and as a result, our society has been left illiterate and vulnerable to manipulation. We have told the public that they should leave the technical stuff to the experts, that they are not smart enough to understand anyway. They have been convinced that they are stupid. They are not stupid; they have been cheated.

Literacy is the goal of education. The study of mathematics not only allows one to read and understand the work of others, it also increases one's own powers of thought, imagination, and expression. That so many regard this subject as frightening, boring, or otherwise unpleasant represents a disappointing failure of our educational efforts. Literacy in mathematics is not simply a question of how much or for whom. To improve our present condition will require substantive reform across the spectrum of curricula. It will not be enough for our departments of mathematics to teach our students how to read. All the rest of us must see to it that they do read, that what they read is worthwhile and important, and that they are able to comprehend what they read. Becoming literate requires a *real* education, the kind that does not become out of date, but prepares one for a lifetime of learning. Just as there is no aspect of the human experience unworthy of serious study, there is no student unworthy of a genuine education. For once, a popular buzz word has captured the true essence of a national problem. Our difficulty is not that education has failed to keep pace with rapidly changing knowledge, it is that we have misplaced the goal of all true education: literacy.

References

1. Adler, M. J. and C. Van Doren. 1972. *How to Read a Book*, Simon and Schuster. New York.

2. Feynman, R. P., R. B. Leighton, and M. Sands. 1964. *The Feynman Lectures on Physics*. Addison-Wesley Publishing Co. Reading, Massachusetts.

3. Whorf, B. L. 1956. *Language, Thought, and Reality*. Technology Press and John Wiley & Sons, Inc. New York.

Department of Physiology
University of Missouri—Columbia
Columbia, MO 65212

Answers to Most Odd-numbered Exercises

Note: In a few cases, answers have been omitted because the processes of seeking and/or constructing those solutions are far more instructive than the answers themselves.

CHAPTER 1

1.2 page 8

1. *Angle*: A figure formed by two line segments with a common endpoint. *Angle measure* (in degrees): A multiple of $\frac{1}{90}$th of a right angle. *Decagon*: A plane figure bounded by ten line segments. *Sum*: The result of addition. **7.** $8 \times 180° = 1440°$ **11.** If n is an integer, then 6 is a factor of $n(n+1)(n+2)$. **13.** Prove that the product has both 2 and 3 as factors. **15.** Either n or $n+1$ is a multiple of 2, so the product is, as well; either n or $n+1$ or $n+2$ is a multiple of 3, so the product is, as well. Thus, the product is a multiple of 6. **19.** A spiral starts 5 cm from a point P and ends 30 cm from that point after 750 cycles around P. What is its length?

1.3 page 13

1. If two copies of *Figure 1.1* are fit together along their "saw-tooth" edges, the result is a rectangle 999,999 by 1,000,000 units. The area is 999,999,000,000 square units; so each half has area 499,999,500,000 square units. **3.** The sum of all positive integers from p to q, inclusive, is $\frac{q(q+1)}{2} - \frac{(p-1)p}{2}$. **5.** How many squares are determined by the lines of an n-by-n chessboard? What is the sum of $1^2 + 2^2 + 3^2 + ... + n^2$? **7.** 1296 **9.** 1225 **11.** $2^{64} - 1$ moves; about 58.5 billion years.

A1

CHAPTER 2

2.1   page 24

1. 200   **3.** 1988   **5.** 1987   **7.** 289   **9.** 498
11. $100^2 = 10,000$ **13.** $498^2 = 248,004$
15. $2161^2 = 4,669,921$
17. $1988 \times 1167 + 1167^2 = 3,681,885$
19. 21   **21.** 600   **23.** 19   **25.** 598   **27.** 52

29. 1403   **31.** $(10 + 1403) \times 200 \div 2 = 141,300$
33. $a + 4d$ **35.** $a + (n-1)d$ **37.** They record
the number of points in a triangular array.
39. 55   **41.** 5050   **43.** 90

2.2   page 28

1. True   **3.** False   **5.** False   **7.** True
9. True   **11.** True   **13.** True   **15.** True
17. True   **19.** True   **21.** False   **23.** True
25. 1, 2, 3, or 6   **27.** b and c are even; so $2 \mid b$

and $2 \mid c$. Thus, $2 \mid (b + c)$; that is, $b + c$ is even.
29. One of the two must be even, so the product
is even.   **31.** See *Table 2A*.

2.3   page 40

1. 7×13 **3.** 13×23 **5.** 347
7. $2^2 \times 11 \times 19$ **9.** 11×347
11. $2^4 \times 3^3 \times 7^2$ **13.** $2^2 \times 7 \times 11^2 \times 19$
15. $7 \times 11 \times 13^2 \times 23$ **17.** 4   **19.** 4   **21.** 2
23. 12   **25.** 4   **27.** 60   **29.** 36   **31.** 24
33. 4; 8; 2^k **35.** For each divisor a of n, there is
a "partner" divisor b, determined by $a \times b = n$.
Each divisor can be paired with its partner; so
$D(n)$ is even unless n is the square of c, in which
case c is its own partner and $D(n)$ is odd.
37. False. *Counterexample*: $D(3) = 2$; $D(13) = 2$;

$D(16) = 5$ **39.** p^9 or $p^4 \times q$ **41.** p^{11} or $p^5 \times q$
or $p^3 \times q^2$ or $p^2 \times q \times r$ **43.** $2^2 \times 3^2 \times 5 = 180$
45. $7 \times 11 \times 13$. Let pqr be a three-digit number.
Then $1001 \times pqr = pqrpqr$. Thus, division by 7,
11, and 13 necessarily yields the quotient pqr.
47. (a) $3 \times 7 \times 13 \times 37$ (b) Pick any two-digit
number; form a six-digit number by repeating the
two-digit number twice; divide this number by
13; divide the quotient by 21; divide this quotient
by 37; the final quotient will equal the original
number.

2.4   page 45

1. 1, 7, 13, 7×13
3. 1, 2, 2^2, 2^3, 2^4, 2^5, 3, 2×3, $2^2 \times 3$, $2^3 \times 3$,
$2^4 \times 3$, $2^5 \times 3$, 3^2, 2×3^2, $2^2 \times 3^2$, $2^3 \times 3^2$,
$2^4 \times 3^2$, $2^5 \times 3^2$ **5.** 1, 2, 2^2, 2^3, 2^4, 31, 2×31,
$2^2 \times 31$, $2^3 \times 31$, $2^4 \times 31$ **7.** 112   **9.** 336
11. 348   **13.** 1680   **15.** 4176   **17.** 70,680
19. 148,960   **21.** 421,632   **23.** 63   **25.** 22,932

27. If $z = 1 + n + n^2 + \ldots + n^{k-1}$, then
$n \times z = n + n^2 + n^3 + \ldots + n^k$. Subtracting corresponding sides of the first equation from the
second, we see that $n \times z - n = n^k - 1$; dividing
both sides by $n - 1$, we obtain the formula.
29. $(2^{31} - 1)$ cents = \$10,737,418.21

2.5   page 47

1. See *Table 2A*.   **3.** 21   **5.** 37   **7.** 1
9. 844   **11.** 359   **13.** 49,512   **15.** 84,588
17. 122,333   **19.** Deficient   **21.** Deficient

23. Deficient   **25.** Abundant   **27.** Deficient
29. Abundant   **31.** Abundant   **33.** Deficient
35. If n is prime, $P(n) = 1$. Since $1 < n$, n is deficient.

n	Divisors	$D(n)$	$S(n)$	$P(n)$	Type
1	1	1	1	0	Deficient
2	1, 2	2	3	1	Deficient
3	1, 3	2	4	1	Deficient
4	1, 2, 4	3	7	3	Deficient
5	1, 5	2	6	1	Deficient
6	1, 2, 3, 6	4	12	6	Perfect
7	1, 7	2	8	1	Deficient
8	1, 2, 4, 8	4	15	7	Deficient
9	1, 3, 9	3	13	4	Deficient
10	1, 2, 5, 10	4	18	8	Deficient
11	1, 11	2	12	1	Deficient
12	1, 2, 3, 4, 6, 12	6	28	16	Abundant
13	1, 13	2	14	1	Deficient
14	1, 2, 7, 14	4	24	10	Deficient
15	1, 3, 5, 15	4	24	9	Deficient
16	1, 2, 4, 8, 16	5	31	15	Deficient
17	1, 17	2	18	1	Deficient
18	1, 2, 3, 6, 9, 18	6	39	21	Abundant
19	1, 19	2	20	1	Deficient
20	1, 2, 4, 5, 10, 20	6	42	22	Abundant
21	1, 3, 7, 21	4	32	11	Deficient
22	1, 2, 11, 22	4	36	14	Deficient
23	1, 23	2	24	1	Deficient
24	1, 2, 3, 4, 6, 8, 12, 24	8	60	36	Abundant
25	1, 5, 25	3	31	6	Deficient
26	1, 2, 13, 26	4	42	16	Deficient
27	1, 3, 9, 27	4	40	13	Deficient
28	1, 2, 4, 7, 14, 28	6	56	28	Perfect
29	1, 29	2	30	1	Deficient
30	1, 2, 3, 5, 6, 10, 15, 30	8	72	42	Abundant

Table 2A

37. Label the proper divisors of an abundant number n as follows: 1, d_1, d_2, d_3, ... , d_k. Then x, d_1x, d_2x, d_3x, ... , d_kx are distinct proper divisors of nx, which implies that $P(nx) \geq x + d_1x + d_2x + d_3x + ... + d_kx = (1 + d_1 + d_2 + d_3 + ... + d_k)x > nx$. Thus, nx is abundant.   **39.** 945

2.6   page 54

1. $2^5 \times 63 = 2016$ **3.** $P(2016) = 4536 > 2016$
5. False. *Counterexample*: $2 \times 5 \times 11$ is deficient.
7. True: $E_k = (2^k - 1)2^{k-1} = \dfrac{(2^k - 1)2^k}{2}$; that is,

the kth Euclidean number is the $(2^k - 1)$th triangle number.   **9.** Because $2 \times 2 = 4$, $2 \times 4 = 8$, $2 \times 8 = 16$ and $2 \times 16 = 32$; the final digit of 2^k cycles through 2, 4, 8, 6 as k increases.

2.7   page 60

1. Composite   **3.** Prime   **5.** Composite
7. Prime   **9.** Not perfect   **11.** Perfect
13. Not perfect   **15.** Perfect   **17.** 2^{88}
19. $M_{13} = 2^{13} - 1 = 8191$ **21.** False   **23.** True
25. False   **27.** True   **29.** False   **31.** True
33. True   **35.** 23 or 89   **37.** 524,287

39. 7 or 8191   **41.** 23, 89   **43.** 3, 5, 11, 31, 41
45. As shown in Exercise 10 of Section 2.6, the final digit of E_k, starting with $k = 2$, cycles through 6, 8, 0, 6. Furthermore, E_k is perfect only if k is prime, and hence, for $k > 2$, k is odd. Therefore, the final digit of E_k must be 6 or 8.

CHAPTER 3

3.2   page 82

Notation: $\cong$: "is congruent to," $\perp$: "is perpendicular to," $<$: "is less than," $>$: "is greater than."

1. The figure is different; the argument still works.
3. (a) Yes   (b) The given segment suffices as the one desired.   **5.** Let A be the vertex of the given angle; let B be any point on one side of the angle.

Construct C on the other side of the angle so that $\overline{AC} = \overline{AB}$; draw $\overline{BC}$. Construct equilateral $\triangle BCD$; draw $\overline{AD}$. $\triangle ABD \cong \triangle ACD$, by SSS; so $\angle BAD = \angle CAD$, and $\overline{AD}$ is the angle bisector. (See *Figure 3A*.)   **7.** Let A be the given point; let B be any other point on the segment. Construct C on the segment (or its extension) on the opposite side of A from B such that $\overline{AC} = \overline{BA}$. Construct equilateral $\triangle BCD$. $\triangle DBC \cong \triangle DCB$, by SSS; so $\angle DBC = \angle DCB$. Draw $\overline{DA}$. $\triangle DAB \cong \triangle DAC$, by SAS; so $\angle DAB = \angle DAC$. These angles are supplementary, so each is a right angle. (See *Figure 3B*.)   **9.** On l, copy $\overline{BA}$ as $\overline{QP}$; with one end at P, copy $\overline{AC}$ as $\overline{PT}$ and $\overline{BC}$ as $\overline{QS}$. Construct a circle with center P and radius $\overline{PT}$, and a second circle with center Q and radius $\overline{QS}$. Let R be one of the points where these circles intersect. Then $\triangle PQR \cong \triangle ABC$, by SSS. (See *Figure 3C*.)   **11.** Given $\overline{AB}$, construct $\overline{AC} \perp \overline{AB}$ at A and $\overline{BD} \perp \overline{AB}$ at B, with C and D on the same side of $\overline{AB}$. Construct E on $\overline{AC}$ (or an extension of $\overline{AC}$) such that $\overline{AE} = \overline{AB}$. Construct F on $\overline{BD}$ (or an extension of $\overline{BD}$) such that $\overline{BF} = \overline{AB}$; draw $\overline{EF}$. To verify that $ABFE$ is a square, we must show that $\angle AEF = \angle BFE = $ a right angle and that $\overline{EF} = \overline{AB}$. Construct M as midpoint of $\overline{AB}$; draw $\overline{ME}$ and $\overline{MF}$. Now, $\triangle AME \cong \triangle BMF$, by SAS; so $\overline{ME} = \overline{MF}$ and $\angle AEM = \angle BFM$. $\triangle MEF \cong \triangle MFE$, by SSS; so $\angle MEF = \angle MFE$. Then, by adding angles,

$\angle AEF = \angle BFE = $ a right angle. Draw $\overline{AF}$. $\triangle ABF \cong \triangle FEA$ by hypotenuse-leg; so $\overline{EF} = \overline{AB}$, implying that $ABFE$ is a square. (See *Figure 3D*.)   **13.** Let $\triangle ABC$ be isosceles with $\overline{AB} = \overline{AC}$; extend $\overline{BC}$ through C to D, and draw $\overline{AD}$. Then $\overline{AB} = \overline{AC}$, $\overline{AD} = \overline{AD}$, and $\angle ADB = \angle ADC$. Thus, two sides of $\triangle ABD$ are equal, respectively, to two sides of $\triangle ACD$, and angles opposite one pair of corresponding sides of each triangle are also equal. But clearly, these two triangles are not congruent. (See *Figure 3E*.)   **15.** Let $\triangle PQR$ be the given triangle, with $\angle PQR = \angle PRQ$. Suppose, contrary to the desired conclusion, that $\overline{PQ} \ne \overline{PR}$; then one of these segments, say $\overline{PR}$, is longer than the other. On $\overline{PR}$, copy $\overline{PQ}$, the shorter segment, as $\overline{PS}$; draw $\overline{QS}$. Then $\triangle PQS$ is isosceles, and hence, $\angle PQS = \angle PSQ$. But $\angle PQR > \angle PQS$ (the whole is greater than any of its parts) and $\angle PSQ > \angle PRQ$ (Exterior Angle Theorem); so $\angle PQR > \angle PRQ$, contrary to the hypothesis. (See *Figure 3F*.) **17.** Let $\triangle PQR$ be the given triangle, with $\overline{PQ} < \overline{PR}$. On $\overline{PR}$, the longer segment, copy $\overline{PQ}$, the shorter segment, as $\overline{PS}$; draw $\overline{QS}$. Then $\triangle PQS$ is isosceles, and hence $\angle PQS = \angle PSQ$. But $\angle PQR > \angle PQS$ (the whole is greater than any of its parts) and $\angle PSQ > \angle PRQ$ (Exterior Angle Theorem); so $\angle PQR > \angle PRQ$. (See *Figure 3F*.)

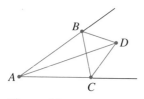

Figure 3A

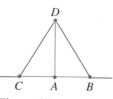

Figure 3B

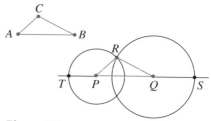

Figure 3C

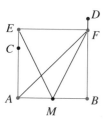

Figure 3D

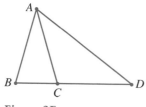

Figure 3E

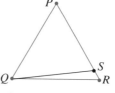

Figure 3F

19. (a) Exercise 11   (b) Axiom E2   (c) Exercise 4
(d) The sum of the angles of a triangle is 180°.
(e) The definition of *complement*   (f) Axiom 1
(g) The definition of *square*   (h) SAS   (i) Exercise 9   (j) Corresponding parts of congruent triangles   (k) Steps (d) and (j)   (l) $\angle EDB$ is a right angle; so $\angle EDG + \angle EDB + \angle BDH = 180°$.
(m) $BCJH$ has a right angle at B (because $\angle DBH$ = $\angle ABC$, so $\angle CBH = \angle ABD$), at C (because its supplement is a right angle), at H (because $\angle DHA$ = $\angle ACB$), and at J (because the sum of all the angles must be 360°), and $\overline{BH} = \overline{CB}$.   (n) is similar to Step (m).   (o) Corresponding parts of congruent triangles   (p) Step (i) and Common Notion 2.

3.3   page 88

1. Circular (b)   **3.** Noncharacteristic (b). The ray exterior to $\angle ABC$ that forms equal angles (c)
5. Noncharacteristic (b). An acute, nonequilateral triangle (c)   **7.** Noncharacteristic (b). Any non-right triangle (c)   **9.** Noncharacteristic (b). A right isosceles triangle (c)   **11.** Circular (b)
13. Noncharacteristic (b). A scalene triangle (c)
15. Accurate (a)   **17.** Noncharacteristic (b). A regular hexagon (c)   **19.** Accurate, though non-standard (a)   **21.** Circular (b)   **23.** Accurate, though nonstandard (a)   **25.** Noncharacteristic (b). An equilateral triangle (c)   **27.** By Exercise 26, there are at least three trees, X, Y, and Z, and, by supposition, only these three. By Axiom 1, there are exactly three squirrels, A, B, and C. Suppose some squirrel, say A, climbs all three trees. By Axiom 2, squirrel B climbs two trees, say X and Y; by Axiom 3, no other squirrel can climb these two trees. That leaves only one tree that is climbed by squirrel C, contradicting Axiom 2. Hence, the supposition must be false.   **29.** By Axiom 1, there are at least three squirrels. By Axiom 2, each squirrel climbs at least two trees. By hypothesis, no tree is climbed by more than one squirrel; so all the trees that are climbed must be distinct from each other. Hence, there must be at least six trees.

3.4   page 95

1. P1: There is at least one brass ring (for example, D) and one wire (for example, 2). P2: $A\,\&\,B$—1, $A\,\&\,C$—1, $A\,\&\,D$—2, $A\,\&\,E$—2, $A\,\&\,F$—3, $A\,\&\,G$—3, $B\,\&\,C$—1, $B\,\&\,D$—4, $B\,\&\,E$—5, $B\,\&\,F$—4, $B\,\&\,G$—5, $C\,\&\,D$—6, $C\,\&\,E$—7, $C\,\&\,F$—7, $C\,\&\,G$—6, $D\,\&\,E$—2, $D\,\&\,F$—4, $D\,\&\,G$—6, $E\,\&\,F$—7, $E\,\&\,G$—5, $F\,\&\,G$—3.
P3: $1\,\&\,2$—A, $1\,\&\,3$—A, $1\,\&\,4$—B, $1\,\&\,5$—B, $1\,\&\,6$—C, $1\,\&\,7$—C, $2\,\&\,3$—A, $2\,\&\,4$—D, $2\,\&\,5$—E, $2\,\&\,6$—D, $2\,\&\,7$—E, $3\,\&\,4$—F, $3\,\&\,5$—G, $3\,\&\,6$—G, $3\,\&\,7$—F, $4\,\&\,5$—B, $4\,\&\,6$—D, $4\,\&\,7$—F, $5\,\&\,6$—G, $5\,\&\,7$—E, $6\,\&\,7$—G. P4: 1—ABC, 2—ADE, 3—AFG, 4—BDF, 5—BEG, 6—CDG, 7—CEF.
P5: Obvious.   **3.** A1: There is at least one wire (for example, 3). A2: Every wire is attached to at least two brass rings (for example, 1—$A\,\&\,B$, 2—$A\,\&\,D$, 3—$A\,\&\,F$, 4—$B\,\&\,D$, 5—$B\,\&\,E$, 6—$C\,\&\,D$, 7 = $C\,\&\,E$). A3: Every brass ring is attached to at least two wires (for example, A—1 & 2, B—1 & 4, C—1 & 6, D—2 & 4, E—2 & 5, F—3 & 4, G—3 & 5).

A4: Given any two wires, there is exactly one brass ring attached to both of them (for example, $1\,\&\,2$—A, $1\,\&\,3$—A, $1\,\&\,4$—B, $1\,\&\,5$—B, $1\,\&\,6$—C, $1\,\&\,7$—C, $2\,\&\,3$—A, $2\,\&\,4$—D, $2\,\&\,5$—E, $2\,\&\,6$—D, $2\,\&\,7$—E, $3\,\&\,4$—F, $3\,\&\,5$—G, $3\,\&\,6$—G, $3\,\&\,7$—F, $4\,\&\,5$—B, $4\,\&\,6$—D, $4\,\&\,7$—F, $5\,\&\,6$—G, $5\,\&\,7$—E, $6\,\&\,7$—G).   **5.** Let *squirrel* mean "one of the numbers 1, 2, 3, 4, 5, or 6"; let *tree* mean one of the sets $\{1, 2\}$, $\{3, 4\}$, or $\{5, 6\}$"; let *climb* mean "is an element of." This model differs from the preceding one in that it has six "squirrels" rather than three.   **7.** Here is one example. Undefined terms: *giraffe, taller*. Axioms: (a) If p and q are distinct giraffes, then either p is taller than q or q is taller than p. (b) Given any giraffe, there is a taller giraffe. (c) There is a giraffe that is not taller than any giraffe. Model 1: *giraffe*—counting number; *taller*—greater. Model 2: *giraffe*—A, B (the letters themselves); A *taller* than A, A *taller* than B (by definition). **9.** *Cats*: 1, 3; *mice*: 2, 5; *x* catches *y* provided *x* + *y* is even. **11.** The model of Example 3.4 with *heart* as number, *spade* as set, and *trumps* as "is contained in." **13.** *gumps*: *a*, *b*, *c*; *lumps*: {*a*, *b*}, {*a*, *c*}, {*a*, *b*, *c*}, {*d*}; *bumps*: "is an element of," or "contains as an element." **15.** By **A1**, there is a *point*, *p*, and, by **A2** there are *lines*, *L* and *M* on *p*; by **A3**, *L* is on another *point* *q*, distinct from *p*, and *M* is on another *point*, *r*, distinct from *p*. The points *q* and *r* must be distinct, for otherwise, by **A4**, *L* and *M* would be the

same. **17.** Given the *point p*, by **A2**, there are *lines L* and *M* on *p*; by **A3**, *L* is on another point, *q*, distinct from *p*, and *M* is on another *point*, *r*, distinct from *p*. The *points q* and *r* must be distinct; for otherwise, by **A4**, *L* and *M* would be the same. By **A4**, there is a *line N* on *q* and *r*. Line *N* cannot be on *p*; for otherwise, *N* would be equal to *L* and *M* (by **A4**), and they would then equal each other. **19.** Construct two models of the axiom system so that the given statement is verified in one, but not in the other.

3.5   page 102

1. *Envelopes*: *E*, *F*; *letters*: *a*, *b*, *c*, *d*, *e*, *f*. *E* contains *a*, *b*, *c*; *F* contains *d*, *e*, *f*. **3.** *Envelopes*: *E*, *F*; *letters*: none. **5.** *Envelopes*: *E*, *F*; *letters*: *a*, *b*, *c*, *d*. *E* contains *a*, *b*, *c*; *F* contains *a*, *b*, *d*. **7.** Dependent. *Proof*: by Axiom (**a**), there is some envelope, which, by Axiom (**b**), contains exactly three letters, say *p*, *q*, and *r*. By Axiom (**c**), *p* is not in all envelopes, so there is some envelope that does not contain *p*. But by Axiom (**b**), this envelope contains three letters; so there must be at least four letters. **9.** Label the four points of Axiom (**b**) as *p*, *q*, *r*, and *s*; by Axiom (**a**), there must be lines on each of the pairs of points *p*–*q*, *p*–*r*, *p*–*s*, *q*–*r*, *q*–*s*, *r*–*s*. By Axiom (**c**), no line is on three points; so each of these six pairs must be on distinct lines, which contradicts Axiom (**d**). **11.** (a) Either there is at most one *X* or there is at most one *Y*. (b) There is (at least) one pair of *X*s that is either related to no *Y* or to at least two *Y*s. (c) There is (at least) one *Y* that is related to no *X*. **13.** Yes; *X*s: *p*, *q*; *Y*s: *A*, *B*; *related to* (denoted by ∗): *p* ∗ *A*, *q* ∗ *A*

15. *Boxes*: 1, 2, 3, 4; *crates*: {1, 2, 3}, {1, 4}, {2, 4}, {3, 4}; *in*: an element of **17.** Axiom (**a**)—*boxes*: 1, 2, 3; *crates*: {1, 2}, {1, 3}, {2, 3}. Axiom (**c**)—*boxes*: 1, 2, 3, 4; *crate*: {1, 2}. Axiom (**d**)—*boxes*: 1, 2, 3, 4; *crates*: {1, 2}, {3, 4}, {1, 2, 3, 4} **19.** Axiom (**e**): Every box is in exactly three crates. The interpretation—*boxes*: 1, 2, 3, 4; *crates*: {1, 2, 3}, {1, 2, 4}, {1, 3, 4}, {2, 3, 4}—shows consistency. The model for Exercise 15 shows independence. **21.** *Points*: 1, 2, 3, 4, 5; *lines*: {1, 2}, {3, 4, 5}; *contains*: contains **23.** Axiom (**e**) is dependent. By Axiom (**b**), there are two lines with no points in common, and each, by Axiom (**c**), contains at least two points. **25.** Independent; model: *points*: 1, 2, 3, 4; *lines*: {1, 2}, {3, 4}, {1, 2, 3}; *contains*: contains **27.** Dependent; Axiom (**b**) (or Axiom (**c**)) implies Axiom (**a**). **29.** Dependent. By Axiom (**b**), there are books. By Axiom (**d**), there are two students, *A* and *B*, who read books. By Axiom (**c**), there is a student, *C*, distinct from *A* and *B*, who does not read books. **31.** Vacuously true

3.6   page 112

1. Congruent triangles have all corresponding angles equal; thus, equal angle sum and equal excess. **3.** exc (*ABC*) = ∠1 + ∠2 + ∠4 + ∠5 − 180° = ∠3 + ∠6 + ∠4 + ∠5 − 180° = ∠3 + ∠6 + ∠4 + ∠5 + 90° + 90° − 360° = exc (*BCHG*) **5.** Using the labeling and results of Exercise 4, triangles *ABD* and *ACD* cannot both have excess more than half that of triangle *ABC*. **7.** 360°; very close to 540°.

9–13. The arguments and conclusions for 9, 11, and 13 are analogous to those for Exercises 1, 3, and 5, respectively. **15.** 180°; very close to 0°. **17.** (a) A great circle; (b) Any other circle; (c) The arc between two points on any circle other than a great circle is not the shortest path between those two points. **19.** (a) Greater than; (b) No; (c) Very close to π; (d) No limit

CHAPTER 4

4.2   Page 136

1. $\{1, 3, 5, 7, 9\}$; U is the set of all digits.
3. $\{0, 1, 2, 3, 4, 5, 6, 7, 8, 9, 10, 11, 12, 13, 14, 15,$
$16, 17, 18, 19, 20\}$; U is the set of all whole numbers   **5.** False   **7.** False   **9.** False   **11.** False
13. False   **15.** $A' = \{0, 6, 7, 8, 9\}$ **17.** $\{1, 2, 3, 4,$
$5, 6\}$ **19.** $\{1, 2, 3, 5, 6\}$ **21.** $\{3, 4, 5, 6\}$
23. $\{(1, 5), (2, 5), (3, 5), (1, 6), (2, 6), (3, 6)\}$
25. $\{0, 7, 8, 9\}$ **27.** $\{0, 1, 2, 4, 5, 6, 7, 8, 9\}$

29. 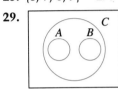   **31.**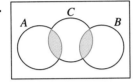

Figure 4A   Figure 4B

33. By the definition of *union*, A is a subset of $A \cup$ any set. In particular, $A \subseteq A \cup \varnothing$. The reverse inclusion also follows from the definition of *union*. Anything in $A \cup \varnothing$ is either in A or in $\varnothing$; but $\varnothing$ contains no elements; so every element of $A \cup \varnothing$ must be in A. Therefore, $A \cup \varnothing = A$, by the definition of *equality*.

35. By the definition of *union*, $U \subseteq A \cup U$. Conversely, any element of $A \cup U$ is either in A or in U. But every element of A is also in U, by the definition of universal set; so $A \cup U \subseteq U$, implying equality.   **37.** *Counterexample*: In the universal set $\{1, 2, 3, 4, 5\}$, let $A = \{1, 2\}$ and $B = \{2, 3, 4\}$. Then $A' \cap B = \{3, 4\}$, but $A \cup B' = \{1, 2, 5\}$.
39. Always true. By De Morgan's Laws, $(A' \cap B)'$ $= (A')' \cup B'$, and the definition of *complement* (applied twice) says that $(A')' = A$.   **41.** Always true. Any element in $A \cap (B \cup C)$ is in A and is in either B or C. If it is in B, then it is in $A \cap B$; if it is in C, then it is in $A \cap C$. Either way, it is in $(A \cap B) \cup (A \cap C)$; so $A \cap (B \cup C) \subseteq (A \cap B) \cup (A \cap C)$. To show that inclusion holds the other way, too, consider any element of $(A \cap B) \cup (A \cap C)$. Either it is in $A \cap B$ or it is in $A \cap C$ (or in both). Either way, it is in A and it must also be in either B or C. Thus, it is in $A \cap (B \cup C)$; so $(A \cap B) \cup (A \cap C) \subseteq A \cap (B \cup C)$. Since each set is a subset of the other, they must be equal.

4.3   page 142

1. $\{hhh, hht, hth, thh, htt, tht, tth, ttt\}$ **3.** $P(E)$ $= \frac{1}{2}$, $P(F) = \frac{1}{2}$, $P(G) = \frac{1}{4}$ **5.** $E \cup F =$ $\{hhh, hht, hth, tht, tth, ttt\}$; $P(E \cup F) = \frac{3}{4}$
7. $F \cup G = \{hhh, hth, tht, ttt, htt\}$; $P(F \cup G) = \frac{5}{8}$
9. $E = \{2h, 4h, 6h, 2t, 4t, 6t\}$, $F = \{5h, 5t\}$, $G = \{1h, 2h, 3h, 4h, 5h, 6h\}$, $H = \{1t, 2t, 3t\}$
11. $E \cap G = \{2h, 4h, 6h\}$; $P(E \cap G) = \frac{1}{4}$
13. There are 36 of them.   **15.** $E = \{(1, 1)\}$, $H = \{(1, 2), (1, 3), (2, 2), (3, 1), (2, 1)\}$ **17.** None; $\{(1, 1), (1, 6), (6, 1), (2, 5), (5, 2), (3, 4), (4, 3)\}$
19. $\{(1, 1), (2, 2), (1, 3), (3, 1)\}$; $\{(1, 1), (1, 2), (2, 1),$ $(2, 2), (1, 3), (3, 1), (3, 3), (2, 4), (4, 2), (4, 4), (1, 5),$ $(5, 1), (3, 5), (5, 3), (5, 5), (2, 6), (6, 2), (4, 6), (6, 4),$ $(6, 6)\}$ **21.** There are 64.   **23.** $P(E) = 0$, $P(F) = \frac{12}{64} = \frac{3}{16}$, $P(G) = \frac{1}{2}$, $P(H) = \frac{1}{16}$ **25.** 0;

$\frac{12}{64} = \frac{3}{16}$ **27.** $\frac{3}{64}$; $\frac{33}{64}$ **29.** It has 42 outcomes.
31. $\frac{19}{21}$ **33. (a)** About 25 times, on the average, but not exactly 25 *heads* every time.   **(b)** Results will vary widely. No; it is even possible (but unlikely) to get no *heads* or all *heads* in 50 tosses of a fair coin. A single experiment is not definitive. **(c)** If the results of both 50-toss experiments are far from the 25-*heads* result in the same direction, you might suspect that the coin should be checked further; but even then, you can't be sure. "Fairness" is a matter of equal likelihood *in the long run*, and 100 tosses do not constitute a very long run. **(d)** Roughly speaking, the more tosses you check, the closer your ratio of *heads* to tosses should approach $\frac{1}{2}$. However, you can't expect a uniform rate of improvement because even fair

coins will generate long runs of one outcome from time to time. **(e)** This answer will depend on the results you get. In most cases, the results will support assigning a probability of $\frac{1}{2}$ to each outcome. **35. (a)** One would think so. **(b)** This will depend on the page chosen. There might be wide variability in the frequencies of the numbers. This could indicate that not all the digits are equally likely, or it could just be the variation one would expect in a relatively small sampling of numbers. **(c)** It might be interesting to see if the distribution pattern is the same for both pages. **37. (a)** No. The first three digits of a phone number indicate its local calling area. There are relatively few of these calling-area codes in any particular phone directory, and they usually are not evenly distributed. **(b)** and **(c)** (Answers will vary.)

4.4   page 153

1. $4 \cdot 4 = 16$ **3.** $4 \cdot 2 = 8$ **5.** $3 \cdot 2 \cdot 5 = 30$
7. The answer is the same for both: $3^7 = 2187$
9. 360   **11.** 720   **13.** 15   **15.** 120   **17.** 780
19. 11,881,376   **21.** 3125   **23.** 30,240
25. $3! = 6$ **27.** $6! = 720$ **29.** 2520
31. $2 \cdot 3 = 6$ **33.** $2 \cdot 3 \cdot 10 \cdot 9 = 540$ **35.** 26^4
$= 456,976$ **37.** $36^4 = 1,679,616$ **39.** Seven

letters. 8,031,810,176 passwords will be available.
41. $26 \cdot 36^5 = 1,572,120,576$ **43.** Beginning with a letter, there are $26^4 \cdot 10^3 = 456,976,000$ available passwords. Beginning with a digit, there are $26^3 \cdot 10^4 = 175,760,000$ available passwords.
45. $\frac{1}{1,048,576}$

4.6   page 164

1. 0   **3.** .55   **5.** 0   **7.** .30   **9.** 75%   **11.** 15%
13. 85%   **15.** They do not add up to 1.
17. $P(E \cup F)$ must be greater than or equal to both $P(E)$ and $P(F)$.   **19.** $E \subseteq F$ implies $P(F \mid E) = 1$; so

they are not independent.   **21.** $\frac{3}{20}$ **23.** $\frac{4}{5}$
25. $\frac{11}{40}$ **27.** $\frac{3}{13}$ **29.** $\frac{3}{52}$ **31.** $\frac{1}{2}$ **33.** $\frac{1}{2}$
35. 40   **37.** 50   **39.** $\frac{2}{5}$

4.7   page 173

1. $\frac{1}{4}$ **3.** $\frac{4}{17}$ **5.** $\frac{1}{17}$ **7.** $\frac{1}{2}$ **9.** $\frac{33}{100}$ **11.** $\frac{4}{25}$
13. $\frac{1}{4}$ **15.** $\frac{3}{4}$ **17.** $\frac{4}{33}$ **19.** $\frac{49}{100}$ **21.** $\frac{2401}{10,000}$
23. $\frac{1}{216}$ **25. (a)** They all contain 30 days. **(b)** $\frac{1}{4}$
(c) $\frac{1}{30}$ **(d)** $\frac{1}{120}$ **(e)** Yes, if the occurrence of either event does not affect the probability of the other.
(f) The product of the answers to **(b)** and **(c)** should equal the answer to **(d)**.   **27. (a)** $\frac{31}{365}$ **(b)** $\frac{28}{365}$
(c) $\frac{1}{365}$ **(d)** $\frac{1}{31}$ **(e)** The product of the answers to

(a) and **(d)** should be the answer to **(c)**. (If the approximate, but inexact, answer $\frac{1}{12}$ was given for **(a)**, this relationship will not work out properly.)
(f) No. If they were, the answers to **(c)** and **(d)** would be the same. **(g)** No. If either occurs, the probability of the other becomes 0. They are mutually exclusive.   **29. (a)** *Hint*: Apply Property (7) twice. **(b)** $P(E_1 \cap \ldots \cap E_n) =$
$P(E_1) \cdot P(E_2 \mid E_1) \cdot \ldots \cdot P(E_n \mid E_1 \cap E_2 \cap \ldots \cap E_{n-1})$

4.9   page 186

1. The numerical information is easier to read from the bar graph (*Figure 4C*), but the size of each piece relative to the entire federal budget amount is more apparent from the pie chart (*Figure 4D*).

The pie chart is more informative if you indicate the total amount of the "whole pie"—$1475 billion.   **3.** See *Figures 4E* and *4F*. A pie chart can be made (as shown), but it's somewhat misleading because there is no meaningful "whole pie." Each country's tax is an amount that is independent of all the others. **5.** See *Figures 4G* and *4H*. Again, a pie chart can be made (as shown), but it's somewhat misleading because there is no meaningful "whole pie." **7.** Mean = $295 billion, the average amount of spending per category. **9.** $2.245, the average cigarette tax per pack in the countries shown. **11.** Mean = 19.5 million BTU (approximately), the average heating value of the types of wood shown. **13. (a)** 6.5 **(b)** 7.0 **15.** See *Figure 4I*. Data items: 23, 21, 25, 20, 37, 37, 32, 31, 36, 53, 59, 60, 64, 64, 62, 64, 68, 72, 71, 72; mean: 48.55

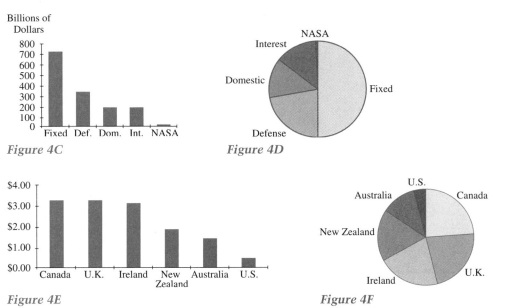

Figure 4C

Figure 4D

Figure 4E

Figure 4F

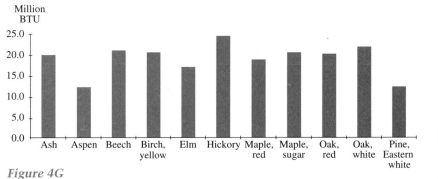

Figure 4G

Figure 4H

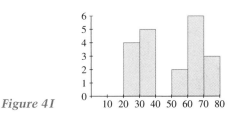

Figure 4I

17. (a) 22 **(b)** 5 **(c)** 2 **(d)** 8 **(e)** 4 **(f)** No
(g) One way is to multiply the mean of each score
interval (on the horizontal axis) by the frequency
of scores in that interval; add; then divide by the
total number of students: 83.4 (approximately).

4.10 page 197

1. (a) Mean: 4; median: 3; mode: 1 **(b)** 1, 1, 3,
6.5, 10 **(c)** Average deviation: 2.7; variance: 10.8;
standard deviation: 3.3 **3. (a)** Mean: 5.0;
median: 4.6; no useful mode **(b)** .11, 2, 4.6, 8.05,
10.6 **(c)** Average deviation: 3.0; variance: 15.0;
standard deviation: 3.9 **5. (a)** Mean: 693.3;
median: 665; no useful mode **(b)** 450, 520, 665,
810, 1050 **(c)** Average deviation: 160; variance:
46,826.7; standard deviation: 216.4 **7. (a)** Mean:
$27.43; median: $18.76; no useful mode **(b)** $1.96,
$9.17, $18.76, $39.87, $82.27 **(c)** Average deviation: $19.27; variance: 612.6; standard deviation:
$24.75 **9. (a)** Mean: 82.4; median: 85 **(b)** 23,
80, 85, 92, 99 **(c)** New mean: 86.6; new median:
85.5; 5-number summary for new boxplot: 73, 82,
85.5, 92, 99 **11. (a)** Mean: $56,153.85; median:
$100,000; mode: $100,000 **(b)** Mean: $53,846.15;
median: $10,000; mode: $10,000 **13. (b)** In the
third quarter **(c)** At least 20 **(d)** 10 **(e)** 30
15. (a) First boat: mean = 3, median = 3; second
boat: mean = 3, median = 3 **(b)** Five-number
summaries for boxplots: First boat: 3, 3, 3, 3, 3;
second boat: 0, 0, 3, 6, 6 **(c)** First boat: range = 0,
standard deviation = 0; second boat: range = 6,
standard deviation = 3 **17. (a)** All three data
sets have the same mean, median, and mode:
$20,000 **(b)** Set 1: $10,000, $20,000, $20,000,
$20,000, $30,000; set 2: $20,000, $20,000,
$20,000, $20,000, $20,000; set 3: $1000, $20,000,
$20,000, $20,000, $39,000 **(c)** Standard devia-
tions—set 1: $6454.97; set 2: $0; set 3: $5484.83

4.11 page 209

1. See *Figure 4J*. **3.** See *Figure 4K*. **5.** See
Figure 4L. **7.** See *Figure 4M*. **9.** Since all six
possible outcomes are equally likely, the Law of
Large Numbers implies that the histograms for
larger numbers of trials should become more like
a horizontal line at a frequency level of $\frac{1}{6}$ the number of trials. **11. (a)** $P(0h) = P(4h) = \frac{1}{16}$,
$P(1h) = P(3h) = \frac{1}{4}$, $P(2h) = \frac{3}{8}$ **(b)** See *Figure 4N*.

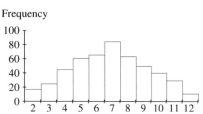

Figure 4J

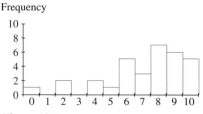

Figure 4K

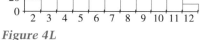

Figure 4L

Figure 4M

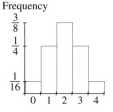

Figure 4N

4.12   page 220

1. $(-1, 1)$ **3.** $(-3, 3)$ **5.** $(80, 120)$
7. $(6.85, 9.35)$ **9.** $(4.35, 11.85)$ **11.** $(-.08, .72)$
13. 95%   **15.** 99.7%   **17.** 95%   **19.** 197
21. 75   **23.** 10.4   **25.** 4.65   **27.** 1.90   **29.** .46
31. .50   **33.** (a) 1.5% (b) 197
35. (a) $\sigma \approx \sqrt{(.58)(.42)} \approx .49$, so $\bar{\sigma} \approx \frac{.49}{\sqrt{50}} \approx .07$.
Thus, a 95% confidence interval about this sample
mean of $\frac{29}{50} = 58\%$ is (44%, 72%). Since the break-
even vote is well within this interval, the manager
could not be 95% confident that the candidate
would win. (b) The two phone polls together
constitute a 150-voter sample; so this time, $\bar{\sigma} \approx$
$\frac{.49}{\sqrt{150}} \approx .4$. The corresponding 95% confidence

interval is (50%, 64%). Although the 50% break-
even vote is at one end of this interval, the likelihood that the actual vote for the candidate is *less*
than 50% is only 2.5%, well within the manager's
desired confidence level.   **37.** (a) $\bar{\sigma} = \frac{.06}{3} = .02$.
Thus, a 99.7% confidence interval for this sample
is (14.98, 15.10)—completely within the allowable variation. The company can be 99.7% confi-
dent. (b) $\bar{\sigma} = \frac{.12}{3} = .04$. Thus, a 95% confidence
interval for this sample is (14.87, 15.03) and a
68% confidence interval is (14.91, 14.99). In this
case the company cannot be 95% confident, but it
can be 68% confident.

CHAPTER 5

5.1   page 231

1. True   **3.** False   **5.** True   **7.** True   **9.** False
11. True   **13.** False   **15.** False   **17.** True
19. False   **21.** False   **23.** False   **25.** True
27. True   **29.** False   **31.** False   **33.** True
35. False   **37.** False   **39.** (a) {5, 6, 7, 8}
(b) Answers will vary; for example, $\frac{11}{2}, \frac{13}{2}, \frac{29}{4}, \frac{799}{100}$,
etc.   **41.** There are infinitely many correct

possibilities. Any ordered pair of integers is correct.
43. There are infinitely many correct possibilities.
Any ordered pair of nonintegral rationals is correct.
45. [4, 9]   **47.** {−2, −1, 0, 1, 2}. (The order is not
important.)   **49.** {−2, −3, −4, −5, −6}. (The order
is not important.)   **51.** $\mathbf{E} \times \mathbf{E}$ **53.** $\mathbf{R} \times [1, 8]$

5.2   page 237

1. There are 24 in all.   **3.** Any 5-element set can
be put in 1-1 correspondence with {0, 1, 2, 3, 4}; no
other set can be.   **5.** Yes; yes; both represent **N**.
7. Yes; match each element with itself.
9. {2, 4, 6, ..., 2n, ...}
      ↕ ↕ ↕   ↕
   {5, 10, 15, ..., 5n, ...}
(Other correct answers to Exercises 11, 13, 15, and
17 are possible.)   **11.** The days of the week
13. All members of the U.S. House of Represen-
tatives   **15.** {f, g, h, i, j}   **17.** {a, b, c}
19. {2, 4, 6, ..., 2n, ...}   **21.** $\{2, \frac{5}{2}, \frac{8}{3}, \frac{11}{4}, ..., 3 - \frac{1}{n}, ...\}$

23. {2, 4, 6, ..., 2n, ...}
      ↕ ↕ ↕   ↕
   {4, 8, 12, ..., 4n, ...}
25. {6, 12, 18, ..., 6n, ...}; {3, 6, 9}
27. {6, 7, 8, ..., n + 5, ...}; {5, 6, 7, ..., 20}
29. {4, 16, 36, ..., (2n)², ...}; {9}
(There are many ways to do Exercises 31 and 33.
We give typical, but not unique, solutions.)
31. In this case, a natural extension of the even–
odd breakdown implicit in this section is to let
$A = \{3, 6, 9, 12, ..., 3n, ...\}$, $B = \{2, 5, 8, 11, ...,$
$3n - 1, ...\}$, and $C = \{1, 4, 7, 10, ..., 3n - 2 ...\}$.

That is, separate the natural numbers into three sets based on their remainders when divided by 3. **33.** Generalizing the answer to Exercise 31, you get the k subsets: $\{k, 2k, 3k, \ldots, nk, \ldots\}$, $\{k - 1, 2k - 1, 3k - 1, \ldots, nk - 1, \ldots\}$, $\{k - 2, 2k - 2, 3k - 2, \ldots, nk - 2, \ldots\}, \ldots$, $\{1, k + 1, 2k + 1, \ldots, nk - (k - 1), \ldots\}$ **35.** This can de done with a little ingenuity. One way is to subdivide the set of multiples of 3 into the even multiples and the odd multiples, and subdivide the set of nonmultiples of 3 into those that leave a remainder of 1 when divided by 3 and those that

leave a remainder of 2. Match the even multiples of 3 with one of the remainder sets, and match the odd multiples of 3 with the other. For instance, match $3n$ with $\frac{3n - 1}{2}$ if n is odd, and with $\frac{3n - 2}{2}$ if n is even. That correspondence looks like this:

$$3, \quad 6, \quad 9, \quad 12, \quad 15, \quad 18, \quad 21, \quad 24, \quad \ldots$$
$$\updownarrow \ \updownarrow \ \updownarrow \ \ \updownarrow \ \ \ \updownarrow \ \ \ \updownarrow \ \ \ \updownarrow \ \ \ \updownarrow$$
$$1, \quad 2, \quad 4, \quad 5, \quad 7, \quad 8, \quad 10, \quad 11, \quad \ldots$$

For this correspondence, the number that corresponds to 300 is 149.

5.3   page 243

(Other correct answers to Exercises 1, 3, 5, and 13 are possible.)
1. $\{5, 4, 3, 2, 1\}$, $\{2, 1, 3, 4, 5\}$, $\{1, 2, 4, 3, 5\}$
3. $\{2, 1, 3, 4, 5, 6, \ldots\}$, $\{2, 1, 4, 3, 6, 5, \ldots\}$, $\{3, 2, 1, 6, 5, 4, 9, 8, 7, \ldots\}$ **5.** Adjust the ordering determined by *Figure 6.3* in ways similar to the answers to Exercise 3. **7.** 50 **9.** −37 **11.** $2p$
13. $\{1, \quad 2, \quad 3, \quad \ldots, \quad 17, \quad \ldots, \quad 50, \quad \ldots, \quad n, \quad \ldots\}$
$\quad\ \ \updownarrow \ \ \updownarrow \ \ \updownarrow \qquad\ \updownarrow \qquad\quad \updownarrow \qquad\quad \updownarrow$
$\quad\{30, \ 60, \ 90, \ \ldots, \ 510, \ \ldots, \ 1500, \ \ldots, \ 30n, \ \ldots\}$
15. $\frac{4}{3}$ **17.** $-\frac{4}{3}$ **19.** 16 **21.** −19 **23.** Those same numbers (in lowest-terms form) are already matched with integers. **25.** $n \leftrightarrow \frac{2}{2n - 1}$; $20 \leftrightarrow \frac{2}{39}$; $21 \leftrightarrow \frac{2}{41}$ **27.** This exercise and the next can be done by mimicking the $\mathbf{N} \leftrightarrow \mathbf{I}$ correspondence of this section, with slight adjustments. If n is even, $n \leftrightarrow \frac{5}{\frac{n}{2}}$; $1 \leftrightarrow 0$; if $n \ (\neq 1)$ is odd, $n \leftrightarrow \frac{-5}{\frac{n - 1}{2}}$. Thus, $20 \leftrightarrow \frac{5}{10} \left(= \frac{1}{2}\right)$; $21 \leftrightarrow \frac{-5}{10} \left(= \frac{-1}{2}\right)$
29. The "even–odd" device runs into a little trouble here because of the "lowest terms" issue. The easiest way is by a simple zigzag pattern, skipping all fractions that are not in lowest terms,

as shown in *Figure 5A*. Using this pattern, $20 \leftrightarrow \frac{2}{13}$ and $21 \leftrightarrow \frac{1}{14}$. The regularity of this pattern suggests that there probably is a not-too-complex algebraic formula for it. However, finding the formula is not necessary; the clear pattern is enough. Here is another possible answer: Pair n with $\frac{2}{n}$ and reduce to lowest terms. (A little checking is required to verify that this is a 1-1 correspondence.) In this way, $20 \leftrightarrow \frac{1}{10}$ and $21 \leftrightarrow \frac{1}{21}$.
31. This is a generalization of Exercise 29. If you arrange these rational numbers as in *Figure 5.3*, each column contains only finitely many (nine) numbers. Thus, you can form a 1-1 correspondence with $\mathbf{N}$ by counting down each successive column, skipping the numbers that are not in lowest terms. If you do this, $20 \leftrightarrow \frac{8}{3}$ and $21 \leftrightarrow \frac{1}{4}$.
33. Think of the fractions in *Figure 5.3* as ordered pairs of natural numbers. (For instance, $\frac{4}{5}$ in the 4th row and 5th column is just $(4, 5)$.) Then you can use the zigzag pattern of *Figure 5.3* for your correspondence. The only difference in this case is that, since all these ordered pairs are different elements of $\mathbf{N} \times \mathbf{N}$, you don't skip any.

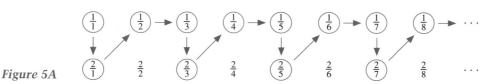

Figure 5A

5.4   page 249

1. 4; 1   **3.** 36; 3; yes   **5.** 6; 6; yes   **7.** 13; 0; no (if you consider repeated 0s)   **9.** 9989; 4; no   **11.** $.4 = .4\overline{0}$ **13.** $.68\overline{1}$ **15.** $.\overline{153846}$ **17.** $.\overline{428571}$ **19.** $3.\overline{142857}$ **21.** $.0\overline{45}$ **23.** $\frac{2}{90}$

25. $\frac{2}{99}$ **27.** $\frac{14}{90}$ **29.** $\frac{61{,}554}{990}$ **31.** $\frac{40}{99}$ **33.** $\frac{68{,}446}{9990}$ **35.** $.03 < .\overline{03} < .0\overline{3} < .3 < .33 < .\overline{3}$ **37.** $.567 < .567566756667... < .\overline{567} < .5\overline{67} < .56\overline{7} < .\overline{6}$

5.5   page 255

(There are many correct answers for Exercises 1 and 3; typical examples are given here.)

1. $\frac{1}{2}, \frac{1}{3}, \frac{1}{4}, \frac{1}{5}, \frac{1}{6}$ **3.** .51, .52, .515, .525, .525555 ...   **5. (a)** .121112 ...   **(b)** .775777 ..., .888848..., .555555 ...   The answers for Exercises 7, 9, 11, and 13 depend on the choice of P; the answers given here are for P chosen 3 inches beyond the 2-inch end of T.   **7.** $1\frac{1}{4}$ in.   **9.** $2\frac{5}{8}$ in.   **11.** $1\frac{5}{16}$ in.   **13.** $1\frac{13}{16}$ in.   **15.** 0 in.   **17.** No

19. Yes   **21.** Yes   **23.** No   **25.** .515151 ...   **27.** .73530303 ... $= .7353\overline{03}$ **29.** (.777 ..., .777 ...)   **31.** (.5, 0) $= \left(\frac{1}{2}, 0\right)$ **33. (a)** $\frac{3}{4}, \frac{3}{2}, \frac{9}{4}, 1, 1.2, 0, 3$ **(b)** $0, \frac{1}{3}, \frac{2}{3}, 1, \frac{1}{6}, \frac{1}{9}, .081, \frac{5}{6}, .9$ **35.** The formula is $y = 18x + 5$. **(a)** $9\frac{1}{2}, 14, 18\frac{1}{2}, 11, 12.2, 5, 23$ **(b)** $0, \frac{5}{18}, \frac{10}{18} = \frac{5}{9}, \frac{15}{18} = \frac{5}{6}, \frac{3}{36} = \frac{1}{12}, \frac{42}{54} = \frac{7}{9}, \frac{12.357}{18} = .6865$ **37.** $y = (b - a)x + a$

5.6   page 259

(There are many correct answers for Exercises 1 and 5; typical examples are given here.)
1. {a, b, c, d, e}, {p, q, r, s, t}, {a, e, i, o, u}   **3.** 26
5. {♠, ♡, ◇, ♣}

$\updownarrow$ $\updownarrow$ $\updownarrow$ $\updownarrow$

{a, b, c, d, e, f, g}

7. In general, a set is not equal to the number of elements it contains. The cardinality of {a, b, c, d} is 4. (The set contains 4 elements.)   **9.** The "<" relation is defined for numbers, not sets. The cardinality of **N** is less than the cardinality of **R**. ($\aleph_0 < c$)   **11.** In general, a set is not equal to the number of elements it contains. The cardinality of [0, 1] is c.   **13.** The relation "$\subset$" is not defined

between numbers. $\aleph_0$ contains *as elements* all the sets that are equivalent to **N**, and none of these sets are elements of c. $\aleph_0 < c$ **15.** They are equal because they have equivalent reference sets.   **17.** No; yes   **19.** **N** would have to be equivalent to a proper subset of X, but not to all of X. Moreover, X would have to be equivalent to a proper subset of **R**, but not to all of **R**.   **21.** A reference set for 3, {a, b, c}, is equivalent to {1, 2, 3}, which is a proper subset of **N**; so, **N** represents the larger number, $\aleph_0$.   **23.** Any reference set for n is equivalent to {1, 2, ..., n}, which is a proper subset of **N**; so, **N** represents the larger number, $\aleph_0$.   **25.** For any natural number n, $n < \aleph_0$ and $\aleph_0 < c$; so $n < c$.

5.7   page 263

(There are many correct answers for Exercises 3, 5, 7, 9, and 13; typical examples are given here.)
1. {∅, {a}, {b}, {c}, {a, b}, {a, c}, {b, c}, {a, b, c}}
3. {2, 4, 6}   **5.** {1, 2, 3, ..., 500}   **7.** $\left\{1, \frac{1}{2}, \frac{1}{3}\right\}$
9. $\left\{1, \frac{1}{2}, \frac{1}{3}, ..., \frac{1}{n}, ...\right\}$ **11.** $S = \{∅, \{a\}, \{b\}, A\}$.

The set T of all subsets of S is {∅, {∅}, {{a}}, {{b}}, {A}, {∅, {a}}, {∅, {b}}, {∅, A}, {{a}, {b}}, {{a}, A}, {{b}, A}, {∅, {a}, A}, {∅, {b}, A}, {{a}, {b}, A}, {∅, {a}, {b}, A}. 4; 16

13. (a) { u,   v,   w,   x,   y,   z }
  $\updownarrow$ $\updownarrow$ $\updownarrow$ $\updownarrow$ $\updownarrow$ $\updownarrow$
  {{u, v}, {w, x, y}, {z}, {x, z}, {v, y}, $\varnothing$}
(b) {v, w, z}   **15. (a)** $1 \leftrightarrow \{1\}$, $2 \leftrightarrow \{1, 2\}$,
$3 \leftrightarrow \{1, 2, 3\}$, $4 \leftrightarrow \{1, 2, 3, 4\}$, $5 \leftrightarrow \{1, 2, 3, 4, 5\}$
(b) $W = \varnothing$ **(c)** 32 is not an element of W.
(d) The set that corresponds to 32 contains 32 numbers; none of them are in $\varnothing$.   **17. (a)** $1 \leftrightarrow \{1\}$,
$2 \leftrightarrow \{1, 4\}$, $3 \leftrightarrow \{1, 4, 9\}$, $4 \leftrightarrow \{1, 4, 9, 16\}$,

$5 \leftrightarrow \{1, 4, 9, 16, 25\}$ **(b)** $W =$
$\{n \mid n$ is not a perfect square$\}$ **(c)** 32 is an element of W. **(d)** 2 is in W, but not in the set that corresponds to 32. (In fact, the two sets are disjoint.)
19. (a) $1 \leftrightarrow \{2\}$, $2 \leftrightarrow \{3, 4\}$, $3 \leftrightarrow \{4, 5, 6\}$,
$4 \leftrightarrow \{5, 6, 7, 8\}$, $5 \leftrightarrow \{6, 7, 8, 9, 10\}$ **(b)** $W = \mathbf{N}$
(c) 32 is an element of W. **(d)** 1 is in W, but not in the set that corresponds to 32.

5.8   page 267

(There are many correct answers for Exercises 1, 7, and part of 9; typical examples are given here.)

1. (a) $[0, 1]$ **(b)** $1, \frac{1}{2}, \frac{1}{3}$ **(c)** $\left\{1, \frac{1}{2}, \frac{1}{3}, \frac{1}{4}, \frac{1}{5}\right\}$

(d) $\left\{1, \frac{1}{2}, \frac{1}{3}, \ldots, \frac{1}{n}, \ldots\right\}$ **3.** $\aleph_6$ **5.** Subset; 1
7. $\{\{x\} \mid x \in K\}$ **9.** Three: $\aleph_0$, $\aleph_1$, $\aleph_2$;
$\mathbf{N}, \mathbf{R}, \{S \mid S \subseteq \mathbf{R}\}$

CHAPTER 6

6.1   page 285

(There are many correct answers for Exercises 1 and 3; typical examples are given here.)
1. $3 \cdot 1 + 3 \cdot 2 = 3 \cdot 3$; $3 \cdot 2 + 3 \cdot 5 = 3 \cdot 7$;
$3 \cdot 20 + 3 \cdot 5 = 3 \cdot 25$ **3.** $(1 + 2) + 3 = 1 + (2 + 3)$;
$7 + 5 = 3 + 9$; $25 + 10 = 20 + 15$ **5.** $2x = x + x$
7. $xy + xz = x(y + z)$ **9.** $2 \cdot 2^x = 2^{x+1}$

11. $a \cdot a^x = a^{x+1}$ **13.** The operation is still "followed by," but the objects of the set are different; they are (clockwise) rotations of 0°, 120°, and 240°. If we call these objects 0, 1, and 2, respectively, then the operation is described by *Tables 6A and 6B*.   **15.** See *Tables 6C and 6D*.

0	1	2	3	4
1	2	3	4	0
2	3	4	0	1
3	4	0	1	2
4	0	1	2	3
0	1	2	3	4

$0f0 = 0$	$0f1 = 1$	$0f2 = 2$
$1f0 = 1$	$1f1 = 2$	$1f2 = 0$
$2f0 = 2$	$2f1 = 0$	$2f2 = 1$

Table 6A

0	1	2
1	2	0
2	0	1

Table 6B

0	1	2		
1	2	0		
2	0	1		
3	4	0	1	2
4	0	1	2	3
0	1	2	3	4

Table 6C

$0f0 = 0$	$0f1 = 1$	$0f2 = 2$	$0f3 = 3$	$0f4 = 4$
$1f0 = 1$	$1f1 = 2$	$1f2 = 3$	$1f3 = 4$	$1f4 = 0$
$2f0 = 2$	$2f1 = 3$	$2f2 = 4$	$2f3 = 0$	$2f4 = 1$
$3f0 = 3$	$3f1 = 4$	$3f2 = 0$	$3f3 = 1$	$3f4 = 2$
$4f0 = 4$	$4f1 = 0$	$4f2 = 1$	$4f3 = 2$	$4f4 = 3$

Table 6D

6.2   page 290

1. b, d, a, c   **3.** 5   **5.** 5   **7.** 6   **9.** 4   **11.** 4
13. There are sixteen such tables in all; six of them, *Tables 6E–6J*, are shown.   **15.** See *Table 6K*.   **17. (a)** two ⋆ words = two; qwerty ⋆ uiop = qwerty **(b)** It is an operation on S and also on E. The (alphabetically) first of two

strings is a string, and the first of two words is a word.   **19. (a)** two ⋆ words = 5; artificial ⋆ example = 10 **(b)** This is not an operation on S or on E because the results are numbers, not strings of letters. For counterexamples, see part **(a)**.   **21.** $1 + 3 = 4$; $3 + 7 = 10$

	1	2
1	1	1
2	1	1

Table 6E

	1	2
1	1	2
2	1	1

Table 6F

	1	2
1	2	1
2	1	2

Table 6G

	1	2
1	2	2
2	2	2

Table 6H

	1	2
1	1	1
2	2	1

Table 6I

	1	2
1	1	2
2	2	1

Table 6J

o	1	2	3	4	5
1	1	2	3	4	5
2	2	1	3	4	5
3	3	3	1	4	5
4	4	4	4	1	5
5	5	5	5	5	1

Table 6K

6.3   page 294

(There are many ways to do Exercises 1, 9, and 11; one correct answer for each is given here.)
1. $(12 \div 6) \div 2 = 1$; $12 \div (6 \div 2) = 4$ **3.** Commutative, associative   **5.** Not commutative, associative   **7.** Not commutative, not associative. (Check the groupings of Exercises 2, 4, and 6, in that order, for example.)

9.

	p	q	r
p	p	q	r
q	q	r	p
r	r	p	q

Table 6L

11.

	p	q	r
p	p	q	r
q	p	q	r
r	p	q	r

Table 6M

13. Associative, but not commutative
15. Commutative and associative   **17.** The *followed by* process is both associative and commutative *on these sets of rotations*. It is *not* commutative on some other, more general sets, as we shall see in Section 6.9.   **19.** Commutative, but not associative. (Try the groupings of Exercises 1, 2, and 3, in that order.)   **21.** Neither commutative nor associative   **23.** Associative, but not commutative   **25.** Neither commutative nor associative

6.4   page 299

1. 2   **3.** 6   **5.** 2   **7.** 2   **9.** 1   **11.** 0   **13.** 4
15. 1   **17.** *Table 6.17* is the one that is not associative.   **19.** Possible   **21.** Not possible   **23.** Not possible (in this case)   **25.** *Table 6N* shows the entries that are determined by the given information.

6.5   page 304

7. See *Table 6O*.   **9.** Not possible   **11.** Not possible   **13.** Not possible   **15.** *Table 6P* shows the entries that are determined by the given information.   **17.** See *Table 6Q*. The identity is 0°, which is its own inverse. 180° is also its own inverse; 90° and 270° are inverses of each other. This group is commutative; the table is symmetric with respect to its main upper-left to lower-right diagonal.

*	1	2	3	4	5
1	1			2	
2	1	2	3	4	5
3		3	2		
4		4	1		
5	2	5			

Table 6N

	r	s
r	s	r
s	r	s

Table 6O

*	a	b	c	d	e
a			e		a
b				e	b
c	e				c
d		e			d
e	a	b	c	d	e

Table 6P

•	0°	90°	180°	270°
0°	0°	90°	180°	270°
90°	90°	180°	270°	0°
180°	180°	270°	0°	90°
270°	270°	0°	90°	180°

Table 6Q

19. See *Table 6R*. The identity is 0°, which is its own inverse. 72° and 288° are inverses of each other; 144° and 216° are inverses of each other. This group is commutative; the table is symmetric with respect to its main diagonal.   **23.** Definition of

group; definition of *operation*; commutativity of $\star$; associativity of $\star$; definition of *inverse*; definition of *identity*; definition of *inverse*; Property (6.3) (Why?)

6.6   page 309

3. You should find four subgroups in all, of sizes 1, 2, 3, and 6 elements.   **5. (a)** See *Table 6S*. The smallest nonzero rotation in this group is 60°. **(b)** Besides the group itself, there are three others: {0°}, {0°, 180°}, and {0°, 120°, 240°}. The smallest nonzero rotations of these subgroups are divisors of 360°.   **7. (a)** See *Table 6T*.   **(b)** (Verify by checking.)   **(c)** {1}, {1, −1}, {1, i, −1, −i}   **9.** 1 **13.** 1, 2, 3, 4, 6, 12; all factors of 12   **15.** Closure: For any multiples of 3, say $3a$ and $3b$, the sum,

$3a + 3b = 3(a + b)$, is a multiple of 3. Identity: $0 = 3 \cdot 0$ is a multiple of 3. Inverses: The additive inverse of $3a$ is $-3a = 3(-a)$, which is also a multiple of 3.   **17.** Closure: For any multiples of 6, say $6a$ and $6b$, the sum, $6a + 6b = 6(a + b)$, is a multiple of 3. Identity: $0 = 6 \cdot 0$ is a multiple of 6. Inverses: The additive inverse of $6a$ is $-6a = 6(-a)$, which is also a multiple of 6. This is a subgroup of the group of all multiples of 3 because every multiple of 6 is a multiple of 3.

6.7   page 315

1. a   **3.** No; not closed—$b \star b = c$ **5.** Three, because three elements are their own inverses. **7.** No; no identity   **9.** No; not closed   **11.** {0}, {0, 3, 6}, S **13. (a)** Y has the most; X has the fewest.   **(b)** 1, 2, 11, 22 in W; 1, 5, 25 in Z **(c)** W and Y must; X and Z cannot.   **15. (a)** {0}, {0, 5, 10}, {0, 3, 6, 9, 12 }, and $\mathbf{I}_{15}$ itself.   **(b)** Yes; there are two such chains: {0} ⊂ {0, 5, 10} ⊂ $\mathbf{I}_{15}$ and {0} ⊂ {0, 3, 6, 9, 12} ⊂ $\mathbf{I}_{15}$.   **(c)** No. If there were, then, by Lagrange's Theorem, the number of elements in each subgroup would have to be a proper divisor of the number of elements in the next subgroup in the chain. But there are no four divisors of 15 that are related in this way.   **17. (a)** 16;

{0} ⊂ {0, 8} ⊂ {0, 4, 8, 12} ⊂ {0, 2, 4, 6, 8, 10, 12, 14} ⊂ $\mathbf{I}_{16}$ **(b)** 81   **(c)** 16. By Lagrange's Theorem, the number of elements in each subgroup must be a (larger) multiple of the number of elements in the preceding subgroup in the chain. The smallest multiple of a number, other than the number itself, is 2 times the number. The first number is 1, the number of elements in {0}. Thus, the smallest numbers of elements in the successive subgroups are 1, 2, 4, 8, 16 (the size of the whole group). **(d)** No; n can be any number of the form p^4, where p is prime. Since there is no largest prime number, there is no largest number of this form.

$\cdot$	0°	72°	144°	216°	288°
0°	0°	72°	144°	216°	288°
72°	72°	144°	216°	288°	0°
144°	144°	216°	288°	0°	72°
216°	216°	288°	0°	72°	144°
288°	288°	0°	72°	144°	216°

Table 6R

$\cdot$	0°	60°	120°	180°	240°	300°
0°	0°	60°	120°	180°	240°	300°
60°	60°	120°	180°	240°	300°	0°
120°	120°	180°	240°	300°	0°	60°
180°	180°	240°	300°	0°	60°	120°
240°	240°	300°	0°	60°	120°	180°
300°	300°	0°	60°	120°	180°	240°

Table 6S

$\cdot$	1	i	−1	−i
1	1	i	−1	−i
i	i	−1	−i	1
−1	−1	−i	1	i
−i	−i	1	i	−1

Table 6T

6.8   page 323

1. {2, 4, 6}   **3.** {2, 4, 6}   **5.** {1, 3, 5}   **7.** {2, 5, 8, 11}
9. {2, 5, 8, 11}   **11.** {1, 5, 9}   **13.** {3, 7, 11}
15. (a) {a, c} and {b, d}   **(b)** {a, b} and {c, d}   **(c)** {a, d}
and {b, c}   **17.** See *Table 6U*.   **19. (a)** 5
(b) $H = \{0, 6, 12, 18, 24\}$. The given coset must be
of the form $a \oplus H$ for some positive integer a. Now,
4 is the smallest number in this coset, and 0 must
be in the subgroup H; so, one form of this coset is
$4 \oplus H$. By subtracting 4 from each element of the
coset, then, we can find the subgroup itself.
(c) $1 \oplus H = \{1, 7, 13, 18, 25\}$ **(d)** 6. Only the numbers 0 through 5, when used for a, form distinct
cosets $a \oplus H$; all the others are repetitions of one
of these.   **(e)** 30, because the total number of

elements in the group is the number of elements
in a subgroup times the number of its distinct
cosets.   **21. (a)** See *Table 6V*.   **(b)** Yes   **(c)** {1, 4},
{2, 3}, {3, 6}, {4, 1}, {5, 2}, {6, 5}   **(d)** Not a group;
the cosets of {1, 4} are not disjoint.

Subgroup Size	Number of Cosets
1	55
5	11
11	5
55	1

Table 6U

⋆	1	4
1	1	4
4	4	1

Table 6V

6.9   page 328

3. Denote the eight rigid motions as: 0—rotation
of 0°; 1—rotation of 90°; 2—rotation of 180°;
3—rotation of 270°; v—reflection on vertical
median; h—reflection on horizontal median;
d—reflection on left-right "down" diagonal;
u—reflection on left-right "up" diagonal. This
group is not commutative. See *Table 6W*.
5. (a) There are only two possible rigid motions of
this figure: 0—rotation of 0°; f—flip (reflection)
about the bisector of the right angle. See *Table 6X*.
(b) Yes   **(d)** It matches any of the subgroups {j, f},
{j, k}, or {j, g}.   **7.** A regular pentagon has 5 lines
of symmetry, one through each vertex to the
midpoint of the opposite side.   **9.** A regular
n-sided polygon with an odd number of sides has
n lines of symmetry, one through each vertex to
the midpoint of the opposite side.   **11.** Partial
answer: There are exactly n rotations of a regular
n-sided polygon that take it back into its own
"imprint," corresponding to the multiples from 0
to $n - 1$ of $\frac{360°}{n}$.   **13.** No (except for the case
$n = 2$), even if the identity is included. An easy
way to see why not is to think of the two sides of
the polygon as being different colors, say red and
green. Each reflection reverses the color that is
"face up." Suppose you start with the red side up,

•	0	1	2	3	v	h	d	u
0	0	1	2	3	v	h	d	u
1	1	2	3	0	d	u	h	v
2	2	3	0	1	h	v	u	d
3	3	0	1	2	u	d	v	h
v	v	u	h	d	0	2	3	1
h	h	d	v	u	2	0	1	3
d	d	v	u	h	1	3	0	2
u	u	h	d	v	3	1	2	0

Table 6W

∘	0	f
0	0	f
f	f	0

Table 6X

reflect around one line of symmetry and then reflect around a different line of symmetry. The
polygon will again have its red side facing up. But
a single reflection changes red to green, so this
cannot be the result of a single reflection.
15. (a) Yes, you can make them all, regardless of
which reflection you have. If the set of rotations
and reflections form a group of size 40, then (by
Lagrange's Theorem) any subgroup containing
more than 20 elements must be the whole group.
There are 20 rotations; if you also include a single
reflection, you have a 21-element subset. This
cannot be a proper subgroup, even though it contains the identity and inverses for all its elements.

This means that it is not closed; so, another reflection can be made from what you have. But this gives you a 22-element subset with identity and all inverses, which still cannot be a proper subgroup. Repeat the argument until the total number of elements obtained is 40.

CHAPTER 7

7.2   page 343

1. 4   **3.** 17   **5.** 0   **7.** 4   **9–15.** See *Figure 7A*.
17. 4   **19.** 7   **21.** 0   **23.** $2\sqrt{2}$ **25.** $|6 - x|$
27. $\{x \mid 2 \le x \le 3\}$; length 1   **29.** $\{x \mid -2 \le x \le 1\}$; length 3   **31.** $\left\{x \mid -5 \le x \le -\frac{1}{2}\right\}$; length $\frac{9}{2}$

33–37. See *Figure 7B*.   **39.** A circle in 1-space is just a pair of points, one on each side of the center, the distance of each from the center being the radius.   **41.** $\{x \mid |x - 8| = 5\} = \{3, 13\}$

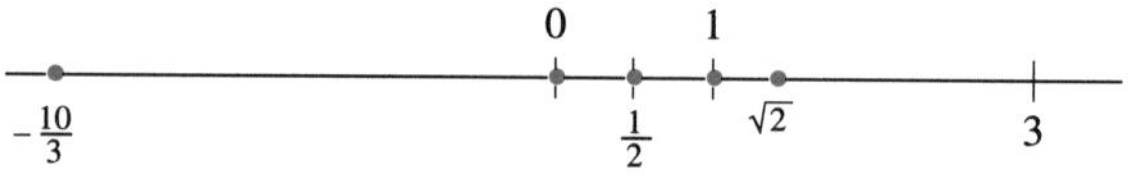

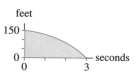

Figure 7A

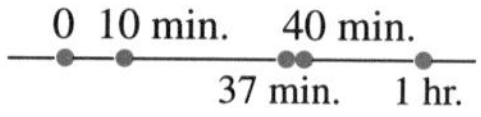

Figure 7B

7.3   page 352

1–7. See *Figure 7C*   **9.** 10   **11.** $\sqrt{2}$ **13.** 5
15. Inside   **17.** On   **19.** On
21. $\{(x, y) \mid x^2 + y^2 = 144\}$. Any y such that $y^2 < 95$ yields an inside point; any x such that $x^2 > 137.75$ yields an outside point. The two points on the circle with x-coordinate 6 are $(6, \sqrt{108})$ and $(6, -\sqrt{108})$.   **23–31.** See *Figure 7D*.
33. $\{(x, 0) \mid 0 \le x \le 3\} \cup \{(3, y) \mid 0 \le y \le 4\}$; $(3, 0)$. $\{(0, y) \mid 0 \le y \le 4\} \cup \{(x, 4) \mid 0 \le x \le 3\}$; $(0, 4)$
35. $\{(x, 2) \mid 1 \le x \le 5\} \cup \{(5, y) \mid -1 \le y \le 2\}$; $(5, 2)$. $\{(1, y) \mid -1 \le y \le 2\} \cup \{(x, 2) \mid 1 \le x \le 5\}$; $(1, -1)$
37. $\{(x, 0) \mid -1 \le x \le 2\} \cup \{(-1, y) \mid -1 \le y \le 0\}$; $(-1, 0)$. $\{(2, y) \mid -1 \le y \le 0\} \cup \{(x, -1) \mid -1 \le x \le 2\}$; $(2, -1)$ **39.** $\left\{(2, y) \mid \frac{-5}{2} \le y \le \frac{7}{3}\right\}$; no intersection point. $\left\{\left(x, \frac{-5}{2}\right) \mid 1 \le x \le 2\right\} \cup \left\{(2, y) \mid \frac{-5}{2} \le y \le \frac{7}{3}\right\}$

$\cup \left\{\left(x, \frac{7}{3}\right) \mid 1 \le x \le 2\right\}$; $\left(2, \frac{-5}{2}\right), \left(2, \frac{7}{3}\right)$ **41.** See *Figure 7E*.   **43.** See *Figure 7F*.
45. $\{(3, y) \mid 0 \le y \le 1\} \cup \{(x, 1) \mid \frac{1}{2} \le x \le 3\}$ $\cup \left\{\left(\frac{1}{2}, y\right) \mid 0 \le y \le 1\right\}$ **47. (a)** $\{(x, y) \mid x^2 + y^2 \le 25\}$
(b) $\{(x, y) \mid x^2 + y^2 \le r^2\}$ **49. (a)** The distance between these two points is
$$\sqrt{(6 - (-1))^2 + (8 - 4)^2} = \sqrt{49 + 16} = \sqrt{65}.$$
Therefore, the circle is given by $\{(x, y) \mid (x + 1)^2 + (y - 4)^2 = 65\}$.   **(b)** $(8, 0)$ is outside; $(0, 8)$ is inside.   **51.** The straight path is $\{(x, 2) \mid -5 \le x \le 1\}$. Thus, the path crosses the circle at a point with y-coordinate 2. Substitute into the equation for the circle; then solve for x: $x^2 + 2^2 = 9$. Thus, $x = \pm\sqrt{5}$. The negative root fits the description of the interval; so, $(-\sqrt{5}, 2)$ is the point of intersection.

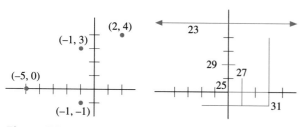

Figure 7C   *Figure 7D*

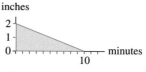

Figure 7E   *Figure 7F*

7.4   page 363

1. $\sqrt{3}$ **3.** $\sqrt{66}$ $(= 8.124)$ **5.** 5   **7.** 3
9. Inside   **11.** Inside   **13.** Outside
15. $\{(x, y, z) \mid x^2 + y^2 + z^2 = 144\}$. Any y such that $y^2 < 70$ yields an inside point; any x such that $x^2 > 127.75$ yields an outside point. There are two points of the form $(-8, 8, z)$ on the sphere: $(-8, 8, 4)$ and $(-8, 8, -4)$.   **17.** $\{(x, 6, 1) \mid x \in \mathbf{R}\}$ and $\{(x, 6, 1) \mid 4 \leq x \leq 7\}$ **19.** $\{(-1, 2, z) \mid z \in \mathbf{R}\}$ and $\{(-1, 2, z) \mid 0 \leq z \leq 2\}$ **21.** $\{(x, 0, 0) \mid 0 \leq x \leq 2\}$ $\cup \{(2, y, 0) \mid -5 \leq y \leq 0\} \cup \{(2, -5, z) \mid -1 \leq z \leq 0\}$; $(2, 0, 0)$, $(2, -5, 0)$ **23.** $\{(x, 3, 5) \mid -4 \leq x \leq 7\}$ $\cup \{(7, y, 5) \mid -1 \leq y \leq 3\} \cup \{(7, -1, z) \mid -9 \leq z \leq 5\}$; $(7, 3, 5)$, $(7, -1, 5)$ **27. (a)** See *Figure 7G*.
(b) $(0, 0, 0)$, $(30, 0, 0)$, $(0, 30, 0)$, $(30, 30, 0)$, $(0, 30, 5)$, $(30, 30, 5)$
(Many correct answers are possible for Exercises 29 and 31; these are typical examples.)
29. $\{(1, 3, z) \mid 0 \leq z \leq 1\} \cup \{(1, y, 1) \mid 1 \leq y \leq 3\}$ $\cup \{(1, 1, z) \mid 0 \leq z \leq 1\}$ **31.** $\{(0, 6, z) \mid 0 \leq z \leq 1\}$ $\cup \{(0, y, 1) \mid 1 \leq y \leq 6\} \cup \{(x, 1, 1) \mid 0 \leq x \leq 1\}$ $\cup \{(1, 1, z) \mid 0 \leq z \leq 1\}$ **33. (a)** $\{(x, y, z) \mid x^2 + y^2 + z^2 \leq 64\}$ **(b)** $\{(x, y, z) \mid x^2 + y^2 + z^2 \leq r^2\}$
35. (a) The distance between these two points is

$\sqrt{(5 - 2)^2 + (3 - (-1))^2 + (6 - 4)^2} =$ $\sqrt{9 + 16 + 4} = \sqrt{29}$. Therefore, the sphere is $\{(x, y, z) \mid (x - 2)^2 + (y - (-1))^2 + (z - 4)^2 = 29\}$.
(b) $(-1, -1, 0)$ is inside the sphere; $(0, -1, -1)$ is on it; $(-1, 0, -1)$ is outside it.   **37.** The straight path is $\{(1, 2, z) \mid 3 \leq z \leq 7\}$. Thus, the path crosses the sphere at a point with x-coordinate 1 and y-coordinate 2. Substitute into the equation for the sphere, then solve for z: $1^2 + 2^2 + z^2 = 25$; so, $z = \pm\sqrt{20}$. The positive root fits the description of the interval; so, $(1, 2, \sqrt{20})$ is the point of intersection.

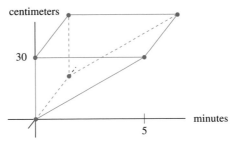

Figure 7G

7.5   page 371

1. 2   **3.** 4   **5.** 17.0024 (approximately)   **7.** 4
9. Inside   **11.** On   **13.** Inside
15. $\{(x, y, z, t) \mid x^2 + y^2 + z^2 + t^2 = 144\}$. Any y such that $y^2 < 34$ yields an inside point; any z such that $z^2 > 118$ yields an outside point. There are two points of the form $(2, -9, 6, t)$ on the sphere: $(2, -9, 6, \sqrt{23})$ and $(2, -9, 6, -\sqrt{23})$.
17. $\{(x, 6, 4, 2) \mid x \in \mathbf{R}\}$; $\{(x, 6, 4, 2) \mid -1 \leq x \leq 5\}$
19. $\{(-3, 7, z, 2) \mid z \in \mathbf{R}\}$; $\{(-3, 7, z, 2) \mid -8 \leq z \leq -4\}$
21. $\{(x, 0, 0, 0) \mid 0 \leq x \leq 1\} \cup \{(1, y, 0, 0) \mid 0 \leq y \leq 1\}$ $\cup \{(1, 1, z, 0) \mid 0 \leq z \leq 1\} \cup \{(1, 1, 1, t) \mid 0 \leq t \leq 1\}$; $(1, 0, 0, 0)$, $(1, 1, 0, 0)$, $(1, 1, 1, 0)$
23. $\{(x, -3, 4, -1) \mid 2 \leq x \leq 7\} \cup$ $\{(2, y, 4, -1) \mid -3 \leq y \leq 5\} \cup \{(2, 5, 4, t) \mid -1 \leq t \leq 0\}$; $(2, -3, 4, -1)$, $(2, 5, 4, -1)$
25. $\{(x, 7, 6, -1) \mid 2 \leq x \leq 5\} \cup$ $\{(5, 7, 6, t) \mid -1 \leq t \leq 3\}$; $(5, 7, 6, -1)$

(*Note*: The answers to Exercises 27, 29, and 31 are given in space-time units of meters and hours. They assume that $t = 0$ is the present time.)
27. From $(0, 0, 0, 0)$, the path is 1 meter along the positive x-axis, followed by 1 meter parallel to the y-axis in the positive direction, followed by 1 meter (positive) parallel to the z-axis, followed by a 1-hour wait.   **29.** Start at $(7, -3, 4)$ in the 3-space of an hour ago. In that 3-space, go 5 meters in the positive x-direction and 8 meters in the positive y-direction. Then wait 1 hour, up to the present.   **31.** Starting at $(2, 7, 6)$ in the 3-space of an hour ago, go 3 meters (positive) parallel to the x-axis to $(5, 7, 6)$, then a 4-hour wait, until 3 hours from now.   **35. (a)** $(0, 0, 0, 0)$, $(2, 0, 0, 0)$, $(0, 2, 0, 0)$, $(0, 0, 2, 0)$, $(2, 2, 0, 0)$, $(2, 0, 2, 0)$, $(0, 2, 2, 0)$, $(2, 2, 2, 0)$ **(b)** $\{(x, 1, 1, 0) \mid 1 \leq x \leq 3\}$

intersects the cube wall at (2, 1, 1, 0). **(c)** (Many different correct answers are possible.)
$\{(1, 1, 1, t) \mid 0 \le t \le 1\} \cup \{(x, 1, 1, 1) \mid 1 \le x \le 3\}$
$\cup \{(3, y, 1, 1) \mid 1 \le y \le 3\} \cup \{(3, 3, z, 1) \mid 1 \le z \le 3\}$
$\cup \{(3, 3, 3, t) \mid 0 \le t \le 1\}$
37. (a) $\{(x, y, z, t) \mid x^2 + y^2 + z^2 + t^2 \le 81\}$
(b) $\{(x, y, z, t) \mid x^2 + y^2 + z^2 + t^2 \le r^2\}$
39. (a) The distance between these two points is
$\sqrt{(5-2)^2 + (8-(-1))^2 + (6-4)^2 + (7-3)^2}$
$= \sqrt{9 + 81 + 4 + 16} = \sqrt{110}$. Therefore, the
hypersphere is $\{(x, y, z, t) \mid (x-2)^2 + (y-(-1))^2$

$+ (z-4)^2 + (t-3)^2 = 110\}$. **(b)** (6, 1, −5, 0) is
on the hypersphere; (−5, 6, 0, 1) is outside it;
(0, −5, 6, 1) is inside it.   **41.** The straight path is
$\{(1, 2, 3, t) \mid -10 \le t \le 4\}$. Thus, the path crosses
the hypersphere at a point with x-coordinate 1,
y-coordinate 2, and z-coordinate 3. Substitute into
the equation for the hypersphere, then solve for t:
$1^2 + 2^2 + 3^2 + t^2 = 36$. Thus, $t = \pm\sqrt{22}$. The
negative root fits the description of the interval;
so, $(1, 2, 3, -\sqrt{22})$ is the point of intersection.

7.6   page 379

1. $\{(4, y, z) \mid 0 \le y \le 2, 0 \le z \le 3\}$, a 2-by-3-inch rectangular region 1 inch in from the far right face
3. $\{(x, y, 0) \mid 0 \le x \le 5, 0 \le y \le 2\}$, a 5-by-2-inch rectangular region that is the bottom face of the block
5. $\{(x, 2, z) \mid 0 \le x \le 5, 0 \le z \le 3\}$, a 5-by-3-inch rectangular region that is the back face of the block
7. The line segment $\{(4, 1, z) \mid 0 \le z \le 3\}$
9. $\{(2, y, z, t) \mid 0 \le y \le 6, 0 \le z \le 8, 0 \le t \le 10\}$,
a 6-by-8-by-10-unit solid rectangular block
11. $\{(x, y, 8, t) \mid 0 \le x \le 4, 0 \le y \le 6, 0 \le t \le 10\}$, a
4-by-6-by-10-unit solid rectangular block that is
the "top" face of the figure in the z direction
13. $\{(1.2, y, z, t) \mid 0 \le y \le 6, 0 \le z \le 8, 0 \le t \le 10\}$,
a 6-by-8-by-10-unit solid rectangular block
15. $\{(x, y, 3\sqrt{2}, t) \mid 0 \le x \le 4, 0 \le y \le 6, 0 \le t \le 10\}$,
a 4-by-6-by-10-unit solid rectangular block
17. The rectangular region $\{(2, 5, z, t) \mid 0 \le z \le 8,$
$0 \le t \le 10\}$ **19.** Using inches and minutes as the
units, this is a 4-by-6-by-8-inch rectangular block
that exists for 10 minutes.   **21.** $t = 0$: (0, 0, 0, 0),
(2, 0, 0, 0), (0, 2, 0, 0), (0, 0, 2, 0), (2, 2, 0, 0),
(2, 0, 2, 0), (0, 2, 2, 0), (2, 2, 2, 0). $t = 1$: (0, 0, 0, 1),
(1, 0, 0, 1), (0, 1, 0, 1), (0, 0, 1, 1), (1, 1, 0, 1),
(1, 0, 1, 1), (0, 1, 1, 1), (1, 1, 1, 1). $t = \frac{3}{2}$: $\left(0, 0, 0, \frac{3}{2}\right)$,
$\left(\frac{1}{2}, 0, 0, \frac{3}{2}\right)$, $\left(0, \frac{1}{2}, 0, \frac{3}{2}\right)$, $\left(0, 0, \frac{1}{2}, \frac{3}{2}\right)$, $\left(\frac{1}{2}, \frac{1}{2}, 0, \frac{3}{2}\right)$,
$\left(\frac{1}{2}, 0, \frac{1}{2}, \frac{3}{2}\right)$, $\left(0, \frac{1}{2}, \frac{1}{2}, \frac{3}{2}\right)$, $\left(\frac{1}{2}, \frac{1}{2}, \frac{1}{2}, \frac{3}{2}\right)$
23. A $\frac{5}{4}$-meter cube with corners at $\left(0, 0, 0, \frac{3}{4}\right)$,
$\left(\frac{5}{4}, 0, 0, \frac{3}{4}\right)$, $\left(0, \frac{5}{4}, 0, \frac{3}{4}\right)$, $\left(0, 0, \frac{5}{4}, \frac{3}{4}\right)$, $\left(\frac{5}{4}, \frac{5}{4}, 0, \frac{3}{4}\right)$,
$\left(\frac{5}{4}, 0, \frac{5}{4}, \frac{3}{4}\right)$, $\left(0, \frac{5}{4}, \frac{5}{4}, \frac{3}{4}\right)$, $\left(\frac{5}{4}, \frac{5}{4}, \frac{5}{4}, \frac{3}{4}\right)$

25. A pyramid-type structure like *Figure 7.34(a)*,
with base corners (0, 0, 0, 0), (2, 0, 0, 0), (0, 2, 0, 0),
and (2, 2, 0, 0) and peak at (0, 0, 0, 2)   **27.** As in
Figure 7.34(b), with base corners at (0, 1, 0, 0),
(2, 1, 0, 0), (0, 1, 2, 0), and (2, 1, 2, 0) and "top"
corners at (0, 1, 0, 1), (1, 1, 0, 1), (0, 1, 1, 1), and
(1, 1, 1, 1)   **29.** A single square "sheet" with corners at (0, 2, 0, 0), (2, 2, 0, 0), (0, 2, 2, 0), and
(2, 2, 2, 0)   **31.** $x = 0$: $\{(0, y, z) \mid y^2 + z^2 = 25\}$,
a circle of radius 5 in the yz-plane. $y = 0$:
$\{(x, 0, z) \mid x^2 + z^2 = 25\}$, a circle of radius 5 in the
xz-plane   **33. (a)** They are circles of the same
radius, 4, on parallel planes, each of which is 3
units away from the yz-plane, on opposite sides.
(b) If $|a| < 5$, they are circles of the same radius,
$\sqrt{5^2 - a^2}$, on parallel planes, each of which is a
units away from the yz-plane, on opposite sides.
If $|a| = 5$, they are the two points (0, 0, 5) and
(0, 0, −5). If $|a| > 5$, both cross sections are empty.
35. $\{(x, y, z) \mid x^2 + y^2 + z^2 \le 25\}$ **37.** $x = -1$:
$\{(-1, y, z) \mid y^2 + z^2 \le 24\}$, a disk of radius $\sqrt{24}$
parallel to the yz-plane (1 unit to the left of it).
$x = -2$: $\{(-2, y, z) \mid y^2 + z^2 \le 20\}$, a disk of radius
$\sqrt{20}$ parallel to the yz-plane (2 units to the left of
it). $x = -3$: $\{(-3, y, z) \mid y^2 + z^2 \le 16\}$, a disk of
radius 4 parallel to the yz-plane (3 units to the left
of it). $x = -4$: $\{(-4, y, z) \mid y^2 + z^2 \le 9\}$, a disk of
radius 3 parallel to the yz-plane (4 units to the left
of it). $x = -5$: $\{(-5, y, z) \mid x^2 + y^2 = 0\}$, the point
(−5, 0, 0).   **39.** $\{(0, y, z) \mid y^2 + z^2 \le 25$ and $z > 0\}$
(or $z \ge 0$). It is a half-disk of radius 5, lying above
the y-axis on the yz-plane.   **41.** The sphere

$\{(2, y, z, t) \mid y^2 + z^2 + t^2 = 21\}$; $(2, 1, 2, 4)$
43. The sphere $\{(x, y, 0, t) \mid x^2 + y^2 + t^2 = 25\}$;
$(2, 3, 0, \sqrt{12}\,)$ **45.** $y = 0$: $\{(x, 0, z, t) \mid x^2 + z^2$
$+ t^2 \le 49\}$, a solid ball of radius 7 in xzt-space,
centered at the origin. $z = 0$: $\{(x, y, 0, t) \mid x^2 + y^2$
$+ t^2 \le 49\}$, a solid ball of radius 7 in xyt-space,
centered at the origin. These two solid balls intersect in the xt-plane. Their intersection is
$\{(x, 0, 0, t) \mid x^2 + t^2 \le 49\}$, a disk of radius 7 centered at the origin.   **47.** This intersection can be
found by inserting the two given values as con-
stants in the algebraic description of the hyperball:
$\{(3, y, z, 2) \mid 3^2 + y^2 + z^2 + 2^2 \le 49\}$, so
$\{(3, y, z, 2) \mid y^2 + z^2 \le 36\}$. This is a disk of radius
6 centered at $(3, 0, 0, 2)$.   **49.** This intersection
is empty. Inserting the two given values as con-
stants in the algebraic description of the hyperball,
we get $\{(x, y, 5, 5) \mid x^2 + y^2 + 5^2 + 5^2 \le 49\}$
which reduces to $\{(x, y, 5, 5) \mid x^2 + y^2 \le -1\}$. But
$x^2 + y^2$ must be positive; so no real-number
values for x and y can satisfy this condition.

7.7   page 391

(Many correct answers are possible for Exercises 1
and 3; typical examples are given here.)
1. $(3, 4, 6)$, $(-2, \sqrt{21}, 1)$ **3.** $(1, \sqrt{3}, 6)$,
$(-2, \sqrt{12}, 2)$ **5.** $\{(x, y, z) \mid x^2 + y^2 = 9$ and
$0 \le z \le 7\}$ **7.** $\{(x, y, z) \mid x^2 + y^2 = \left(4 - \frac{1}{5}z\right)^2$ and
$0 \le z \le 20\}$ **9.** 4   **11.** 3   **13.** 0   **15.** 0
17. 10   **19.** 2   **21.** $\{(x, y, 1) \mid x^2 + y^2 = 25\}$, a
circle of radius 5 inches   **23.** $\{(-4, y, z) \mid y^2 = 9$
and $0 \le z \le 12\}$, two parallel 12-inch line segments
that are 6 inches apart   **25.** $\{(x, y, z, 3) \mid x^2 + y^2$
$+ z^2 = 25\}$, a sphere of radius 5 inches
27. $\{(x, 0, z, t) \mid x^2 + z^2 = 25$ and $0 \le t \le 12\}$, a
cylinder of radius 5 inches and height 12 inches
(in xzt-space)   **29.** $\{(1, y, z, t) \mid y^2 + z^2 = 24$ and
$0 \le t \le 12\}$, a cylinder of radius $\sqrt{24}$ inches and
height 12 inches   **31.** $\{(5, 0, 0, t) \mid 0 \le t \le 12\}$,
a 12-inch line segment   **33.** $x^2 + y^2 = 25$
35. $x^2 + y^2 = 1$ (Several different correct answers are possible for Exercises 37 and 39; we give
two typical 2-dimensional cross sections for each
exercise.)   **37.** $\{(x, y, 9) \mid x^2 + y^2 = 25\}$;
$\{(-3, y, z) \mid y^2 = 16$ and $0 \le z \le 12\}$
39. $\{(x, \sqrt{7}, z) \mid x^2 + 7 = \left(5 - \frac{1}{2}z\right)^2$ and $0 \le z \le 10\}$;
$\{(x, y, 2) \mid x^2 + y^2 = 16\}$ **41.** (Many different cor-
rect answers are possible for this exercise; here is
a typical example.) $\{(7, 8, z) \mid 9 \le z \le 13\}$
$\cup \{(x, 8, 13) \mid 0 \le x \le 7\} \cup \{(0, y, 13) \mid 1 \le y \le 8\}$
$\cup \{(0, 1, z) \mid -1 \le z \le 13\} \cup \{(x, 1, -1) \mid 0 \le x \le 7\}$
$\cup \{(7, y, -1) \mid 1 \le y \le 8\} \cup \{(7, 8, z) \mid -1 \le z \le 9\}$

CHAPTER 8

8.2   page 410

1. All the drawings except (d) and (h) are graphs.
3. (b), (c), (g), and (i)   **5. (a)** Z_1WX
(b) $W_1ZYXW_2Z_1WX$ **(c)** W_2ZYXW_1Z **(d)** Six
(e) None   **(f)** Infinitely many   **7. (a)** True
(b) False   **(c)** Cannot be determined   **(d)** True
(e) Cannot be determined   **(f)** True   **(g)** Cannot
be determined   **(h)** True   **(i)** Cannot be determined   **(j)** Cannot be determined   **(k)** Cannot be
determined   **(l)** Cannot be determined

9.

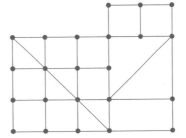

Figure 8A

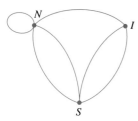

Figure 8B

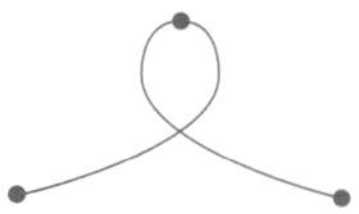

Figure 8C

Figure 8D

11. See *Figure 8B*; N—north bank, S—south bank, I—island.   **13. (a)** See *Figure 8C*. **(b)** See *Figure 8D*. **(c)** Impossible; a vertex path guarantees a path from any vertex to any other; hence, the graph must be connected. **(d)** Impossible; if the graph is connected, then there must be edges joining at least two (of the three possible) pairs of vertices. Thus, some vertex (say P) must be joined to each of the others (say Q and R). Then there is a vertex path QPR. **(e)** See the answer to Part **(b)**. **(f)** Impossible; by the reasoning of Part **(c)**, the existence of a vertex path guarantees that the graph is connected. With only three vertices, there must be an edge path, regardless of loops or multiple edges. **(g)** See the answer to Part **(a)**.

8.3   page 418

1. (a) Even: *A, C, D,* and *F*; odd: *B* and *E*
(b) Even: *G* and *L*; odd: *H, J, K,* and *M* **(c)** All the vertices are even. **(d)** All the vertices are even. **(e)** Even: *N* and *P*; odd: *K, L, M, Q, R,* and *S* **(f)** All the vertices are even. **(g)** Even: *F, G, H, J, K,* and *L*; odd: *D* and *E* **3.** *D*, 5; *E*, 5; *F*, 4; *G*, 4; *H*, 4; *J*, 6; *K*, 4; *L*, 4 **5. (a)** Graphs (b), (d), and (e) **(b)** Graphs (a) and (g) **(c)** Graphs (c) and (f) **(d)** Graph (f) **7.** Step 1 is modified so that a path is randomly constructed starting at any vertex, *P*. The path must eventually end at *P* because all the vertices are even. The rest of the proof is the same. **9.** Start a path at any odd vertex, proceeding as described in Step 1. The path must end at some other odd vertex. If there is a third odd vertex, start another path there; it will end at a fourth odd vertex. Proceeding in this fashion, odd vertices can be paired, thus showing that there must be an even number of them. **11.** See *Figure 8.12*. All four vertices are odd, so no edge path exists; the Königsburg Bridge Problem has no solution. **13.** Adding one edge that joins distinct vertices in *Figure 8.12* changes two odd vertices to even ones, leaving only two odd vertices. **15.** Two. Several locations are possible, as long as each one of the four land masses is connected to another one by one of the new bridges. **17.** $\frac{n}{2}$ **19.** The closed edge path must pass through every vertex because the graph is connected. If there are *n* vertices, this path must contain at least *n* edges in order to pass through every vertex at least once and return to its starting point. **21.**  •———•

Figure 8E

23. Two vertices, each with one loop

8.4   page 426

1. Graph (a) **3.** Graph (a) **5.** Graph (a)
7. Graph (f) **9.** Yes. For example, if the vertices (intersections) are labeled alphabetically from left to right (top row: *A, B, C*; next row: *D* through *I*; and so on). One possible path is
ABCIRWVUTSNOPQMLKJDEFGH
11. (a) All **(b)** *Figure 8.22: RPNQSTXYWUVZ*;
Figures 8.23 and *8.24: RSPQT*; *Figure 8.25:*

ABCDEFGH; Figure 8.26: RQVSPUT; Figure 8.27: AFGNPQJKCBDEHLM   **13.** If a vertex path omits an *H*-edge at a vertex, it must use the (only) other edge at that vertex; thus, this vertex must be one end of the path. Since there can be only two ends to the path, there can be at most

two H-edges omitted.   **15.** Suppose that *A* is a terminal vertex, connected by its one edge to *B*, which is, in turn, connected to other vertices in the graph. The deletion of *B* and its edges disconnects *A* from the rest of the graph; so *B* is critical.

8.5   page 432

1. See *Figure 8F*.   **3. (a)** 4 **(b)** 4 **(c)** 2 **(d)** 2 **(e)** 0 **(f)** 4 **(g)** 2 **(h)** 4   **5.** Graphs (c), (d), (e), and (g)   **7.** See *Figure 8G*.   **9.** See *Table 8A*. **11.** See *Figure 8H*.   **13. (a)** Loop **(b)** Face of

degree one (face inside a loop) **(c)** Face of degree zero (that is, the entire plane). Note that the only polygonal graph with an isolated vertex is the null graph on one vertex, a self-dual graph.

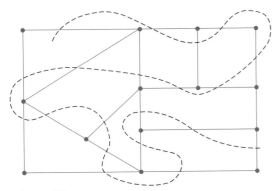

Figure 8F

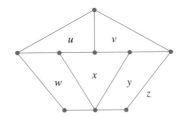

Figure 8G

Face	Vertex (Dual)	Edge	Edge (Dual)
ABC	A	AB	AC
BCE	B	BE	BC
CED	E	ED	EC
ACD	D	DA	DC
ABDE	C	AC	DA
		BC	AB
		EC	BE
		DC	ED

Table 8A

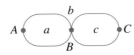

Figure 8H

8.6   page 439

1. (a) $V = 6, E = 9, F = 5$ **(c)** $V = 11, E = 16, F = 7$ **(d)** $V = 8, E = 11, F = 5$ **(e)** $V = 9, E = 12, F = 5$ **(f)** $V = 9, E = 14, F = 7$ **(g)** $V = 12, E = 16, F = 6$ **(h)** $V = 16, E = 24, F = 10$ **(i)** $V = 16, E = 24, F = 10$

3. $V = 7, E = 7, F = 3; V - E + F = 3$ **5.** $V - E + F = 4$ **7.** Not necessarily; a loop at a vertex increases the degree by two, but adds only one face; a terminal edge increases the degree by

one, but adds no face.   **9. (a)** $V = 5, E = 8, F = 5$
(b) $V = 24, E = 36, F = 14$ **(c)** $V = 8, E = 16,$
$F = 10$ **11.** $V - E + F + 2H = 2$ **13.** All are;

in the figures, the thick lines identify the paths.
See *Figure 8I*.

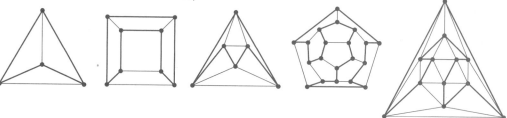

Figure 8I

CHAPTER 9

9.2   page 463

1. $A–D–C–B–E–A$ **3.** Every tour is a shortest
tour.   **5.** A tour is shortest if and only if it does
not include the road from A to B.   **7.** A tour is
shortest if and only if it includes neither the road
from A to E nor the road from B to C.   **9.** Every
tour is a shortest tour.   **11.** One possibility: See
Figure 9A.   **13.** $A–C–B–D–A$, for a distance of
19.   **15.** There are the 12 tours: $A–B–C–D–E–A$,
$A–B–C–E–D–A$, $A–B–D–C–E–A$, $A–B–D–E–C–A$,
$A–B–E–D–C–A$, $A–B–E–C–D–A$, $A–C–B–D–E–A$,
$A–C–D–B–E–A$, $A–C–B–E–D–A$, $A–C–E–B–D–A$,
$A–D–C–B–E–A$, and $A–D–B–C–E–A$; there are
also their reversals, for a total of 24 tours.
17. There are 120 tours.   **19.** Yes, a new shortest
tour is $A–B–D–C–A$, of length 18.   **21.** Using
cities A, B, C, D, and E, let the roads $A–B, B–C,$
$C–D, D–E$, and $E–A$ have length 2, and let all
other roads have length 20. Then a shortest tour
will be $A–B–C–D–E–A$.   **23.** Add a new road
from A to B and give it a length greater than the

sum of the lengths of all other roads. Due to its
large length, that road will never appear in a
shortest tour; thus, the result is the same as if the
road were never there.   **25.** In the TSP, you are
looking for the shortest Hamilton cycle in a graph
in which every pair of vertices is connected by an
edge. A TSP graph always is a Hamilton graph because every city is connected to every other city,
whereas a general graph, in which not all vertices
are connected by edges, may not be a Hamilton
graph. The graphs in Chapter 8 do not have numbers or lengths associated with them.

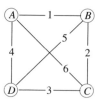

Figure 9A

9.3   page 468

3. (a) .004 seconds   **(b)** .128 seconds   **(c)** 400
seconds = 6.67 minutes   **5 (a)** .145 seconds
(b) 9.7×10^{10} seconds = 3086 years (not practical)
(c) 4×10^{150} seconds = 10^{143} years (not practical)
7. Approximately 15 feet   **9.** Approximately 12
billion instructions per second   **11.** Good for

parallel computers: Computing students' final
grades when each student's grade does not depend
on other students' grades. Bad for parallel computers: Computing something where each computation depends on the results of the preceding one.

9.4   page 475

1. Change the second step of the algorithm from "repeat the following steps *n* times" to "repeat the following steps $n - 1$ times."   **3.** Selection Sort will not run faster on an already sorted list. Slow Sort will check the initial ordering, see that it is sorted, and halt. Thus, it will be very fast.
5. Let $k = 5$.
    Find the largest number from the first 5 numbers.
        (46 is the largest of the 5 numbers.)
    Swap this number with the number in position 5.
        (After swapping, our list becomes 18, 2, 23, 1, 46.)
    Let $k = 4$.
    Find the largest number from the first 4 numbers.
        (23 is the largest of the 4 numbers.)
    Swap this number with the number in position 4.
        (After swapping, our list becomes 18, 2, 1, 23, 46.)
    Let $k = 3$.
    Find the largest number from the first 3 numbers.
        (18 is the largest of the 3 numbers.)
    Swap this number with the number in position 3.
        (After swapping, our list becomes 1, 2, 18, 23, 46.)
    Let $k = 2$.
    Find the largest number from the first 2 numbers.
        (2 is the largest of the 2 numbers.)
    Swap this number with the number in position 2.
        (After swapping, our list becomes 1, 2, 18, 23, 46.)
    Let $k = 1$.
    Find the largest number from "the first 1 number."
        (1 is the only remaining number, and so is the "largest.")
    Swap this number with the number in position 1.
        (After swapping, our list becomes 1, 2, 18, 23, 46.)
7. 24: ABCD, ABDC, ACBD, ACDB, ADBC, ADCB, BACD, BADC, BCAD, BCDA, BDAC, BDCA, CABD, CADB, CBAD, CBDA, CDAB, CDBA, DABC, DACB, DBAC, DBCA, DCAB, DCBA   **9.** 720   **11.** The sorts still work. It doesn't speed up Selection Sort. It speeds up Slow Sort on average, since it is more likely to encounter a sorted list sooner than with a list in which no two numbers are equal.
13. Let $k = n$.
    Repeat the following instructions *n* times.
        Find the smallest number from the last *k* numbers.
        Swap this number with the number in position $n - k + 1$.
        Let $k = k - 1$. ·
    The speed of both versions is the same.
15. The letters of the alphabet are encoded as numbers.

9.5   page 482

1. 6, 46, 4320   **3.** 96 millionths of a second
5. 36.3 seconds   **7.** 96 billionths of a second
9. .0363 second
11. Step 1: Let $k = 2$.
    Steps 2–3: Find the largest of the first two numbers.
    Steps 4–6: Swap the largest number into the second position.
    Step 7: Decrement *k*.
    Step 8: Find the largest of the first number.
    Steps 9–11: Swap the largest number into the first position.
    Step 12: Decrement *k*.
15. See the first few paragraphs of Section 9.5.
17. Step 1: Let $k = 1$.
    Step 2: Extract the first number, 9, from the list.
    Step 3: Increment *k* to 2.
    Step 4: Extract the second number, 13, from the list.

Step 5: Insert it in its proper place relative to the first number. (List: 9, 13, 7, 6.)

Step 6: Increment k to 3.

Step 7: Extract the third number, 7, from the list.

Steps 8–9: Insert it in its proper place relative to the first two numbers. (List: 7, 9, 13, 6.)

Step 10: Increment k to 4.

Step 11: Extract the fourth number, 6, from the list.

Steps 12–14: Insert it in its proper place relative to the first number. (List: 6, 7, 9, 13.)

Step 15: Increment k to 5.

19. $1 + \frac{1}{2}n^2 + \frac{3}{2}n$ **21.** Selection Sort has the same best-case and worst-case complexities. Insertion Sort will have a best-case complexity of $1 + 3n$, if the algorithm starts the insertion process by checking the element immediately preceding the current element. In this case, there are only 3 steps each time through the loop. **23.** .5; 2; 9; 48; 300; 18,144,000; 7.6×10^{65}. Slow Sort is still worse.   **25.** The algorithm in Exercise 10 of Section 9.4 has two instructions outside the loop, and two instructions inside the loop. Since the loop is executed n-1 times, the total number of steps is $2 + 2(n - 1)$, which is $2n$.

9.6   page 492

1. 5   **3.** 4, 16, 16, 4   **5.** 100; 1.27×10^{30}; 1.27×10^{30}; 100   **7.** 15; 3.9; 3.9; 59; 225; 3375; 759,375; 32,768; 1.3×10^{12} **9.** 75; 6.2; 8.7; 467; 5625; 421,875; 2.4×10^9; 3.8×10^{22}; 2.5×10^{109} **11.** 15, 3.9 μsec, 3.9 μsec, 59 μsec, 225 μsec, 3.4 msec, 759 msec, 32.8 msec, 15 days   **13.** 75, 6.2 μsec, 8.7 μsec, 467 μsec, 5.6 msec, 422 msec, 40 min, 10^9 years, 10^{96} years   **15.** $n!$ **17.** n^3 **19.** n **21.** n^2 **23.** Polynomial: Exercises 14, 16, 17, 19, and 21. Exponential: Exercises 15, 18, and 20.   **25.** Selection Sort: 2.5 msec.

Slow Sort: 10^{50} years   **27.** .0000072 μsec, .0000123 μsec, .000326 μsec, .0225 μsec, 3 μsec, 76 msec, 10^{25} years, 10^{242} years   **29.** .0072 μsec, .0123 μsec, .326 μsec, 22.5 μsec, 3 msec, 76 sec, 10^{28} years, 10^{245} years   **31.** Polynomial: $\sqrt{\sqrt{n}}$, $n^2(\log n)$. Exponential: $n(n!)$, n^n. Ordering: $\log n$, $\sqrt{\sqrt{n}}$, $\sqrt{n}$, n, $n(\log n)$, n^2, $n^2(\log n)$, n^3, n^5, 2^n, $n!$, $n(n!)$, n^n **33.** 12   **35.** (a) A (b) $n > 10^{11}$ (c) $n < 10^{11}$ (d) $n = 10^{11}$ **37.** $2^{100} \cong 10^{30}$; so, 100 years of doubling results in computers that are 10^{30} times faster.

9.7   page 500

1. No such tour   **3.** A–C–B–D–A, with length 19   **5.** No such tour   **7.** The TSDP is a *decision* problem, which means that the output should be "yes" or "no." The TSP is a problem for which the output is a minimum-length tour. The TSDP is easier to work with theoretically, and so is useful when proving theorems about problems.   **9.** Verification algorithm:

Input: a tour T of n cities and a positive number M

Output: Yes or no, indicating whether T has length less than M

Method: (We will denote the kth city in the tour by "city k.")

Let $k = 1$.

Let $total = 0$.

Repeat the following steps n–1 times.

Add the length of the road from city k to city $k + 1$ to $total$.

Let $k = k + 1$.

Add the length of the road from city n to city 1 to $total$.

If $total < M$, then output "Yes."

Otherwise, output "No."

11. 3 bins

13. Bin Packing Decision Problem (BPDP):

Input: A list of n positive numbers (corresponding to the weights of n objects), a positive number C (corresponding to the capacity of the bins), and a positive number M

Problem: Does there exist a way to place the n objects with the given weights into at most M bins so that the total weight in each bin is at most C?

15. Given a graph, mark each edge with a 0. That is, treat each existing edge as if it were a road of length 0. Now add new edges (roads) everywhere they don't already exist and mark them all with 1s. Determining whether the original graph is a Hamilton graph is the same as determining whether the new graph with the markings has a tour that skips all roads of length 1 or more, that is, whether there exists a tour of length 0.

17. 16   **19.** $a = b = c = d = e = f = $ true
21. Not possible   **23.** $a = $ false, $b = $ false, $c = $ false, $d = $ true   **25.** Let $a = $ true, $b = $ true, and let the remaining variables have any value.

9.8   page 507

1. If the circles overlapped then, by the definition of NP-Completeness, all problems in NP would be solvable by polynomial-time algorithms. This would imply that P = NP.   **3.** No. A–C–E has length 18, but A–E has length 999.   **5.** Possible   **7.** Not possible   **9 (a)** A longest tour is A–B–D–E–C–A. It isn't any easier to find than a shortest tour.   **(b)** A shortest tour is A–B–D–E–C–A. It is the same tour.   **(c)** Subtract the length of each road from a large number. A shortest (respectively, longest) tour in the new map is a longest (respectively, shortest) tour in the original map.

APPENDIX A

A.1   page 526

1. Statement   **3.** Statement   **5.** Not a statement   **7.** Statement   **9.** Statement
11. Existential   **13.** Existential   **15.** Existential
17. Existential   **19.** Universal   **21.** This is not a mathematics book.   **23.** Not all roses are red. (Some roses are not red.)   **25.** $8 + 3 \neq 11$
27. All cars get good gas mileage.   **29.** The square of some number is not positive.   **31.** No flowers are yellow.   **33.** There is not a unicorn on the lawn. (No unicorns are on the lawn.)
35. No elephants fly.   **37.** All students are sophomores.   **39.** Some thing(s) I need to know I did not learn in kindergarten.

A.2   page 530

1. This book is interesting and I am falling asleep.   **3.** This book is not interesting and I am falling asleep.   **5.** (Either) this book is not interesting or I am falling asleep.   **7.** This book is not interesting and I am not falling asleep.   **9.** It is not true that this book is interesting and I am falling asleep.   **11.** p or ($\sim q$) p: There are 29 days in February. q: This is a leap year.   **13.** p and q p: English words are more familiar. q: Logical symbols are clearer.   **15.** $\sim(p$ and $q)$ p: I went to the movies. q: I ate too much popcorn.   (For Exercises 17 and 19, the answers are not unique; there are many grammatically different, but logically equivalent correct answers.)   **17.** Either I didn't go to the movies or I didn't eat too much popcorn.   **19.** Some state governor is not a Democrat and *that same governor* is also not a Republican. [Some state governor is neither a Democrat nor a Republican.]

21.

p	q	p and q	q and p
T	T	**T**	**T**
T	F	**F**	**F**
F	T	**F**	**F**
F	F	**F**	**F**

23.

p	q	$(\sim p)$ or q	$\sim(p$ and $(\sim q))$
T	T	**T**	**T**
T	F	**F**	**F**
F	T	**T**	**T**
F	F	**T**	**T**

25.

p	q	$\sim(p$ and $q)$	p or $\sim(p$ and $q)$
T	T	F	**T**
T	F	T	**T**
F	T	T	**T**
F	F	T	**T**

27.

p	q	r	p and q	$(p$ and $q)$ and r	q and r	p and $(q$ and $r)$
T	T	T	**T**	**T**	T	**T**
T	T	F	**T**	**F**	F	**F**
T	F	T	**F**	**F**	F	**F**
T	F	F	**F**	**F**	F	**F**
F	T	T	**F**	**F**	T	**F**
F	T	F	**F**	**F**	F	**F**
F	F	T	**F**	**F**	F	**F**
F	F	F	**F**	**F**	F	**F**

29.

p	q	r	p or $(q$ and $r)$		$(p$ or $q)$ and $(p$ or $r)$		
T	T	T	**T**	T	**T**	**T**	**T**
T	T	F	**T**	F	**T**	**T**	**T**
T	F	T	**T**	F	**T**	**T**	**T**
T	F	F	**T**	F	**T**	**T**	**T**
F	T	T	**T**	T	**T**	**T**	**T**
F	T	F	**F**	F	**T**	**F**	**F**
F	F	T	**F**	F	**F**	**F**	**T**
F	F	F	**F**	F	**F**	**F**	**F**

31.

s	t	s and t
T	**T**	**T**
F	**T**	**F**

33.

s	c	s and c
T	**F**	**F**
F	**F**	**F**

A.3   page 539

1. *Hypothesis:* There is snow on the ground. *Conclusion:* This is winter. *Converse:* If this is winter, then there is snow on the ground.
3. *Hypothesis:* Rabbits eat carrots. *Conclusion:* Rabbits have good eyesight. *Converse:* Rabbits must eat carrots if they have good eyesight.
5. *Hypothesis:* Something is a turkey. *Conclusion:* Something is a bird. *Converse:* All birds are turkeys.   **7.** *Hypothesis:* Something is an airplane. *Conclusion:* Something does not fly like a bird. *Converse:* Anything that does not fly like a bird is an airplane.   **9.** *Hypothesis:* The hypothesis of a conditional is false. *Conclusion:* A

conditional is true. *Converse:* If a conditional is true, then its hypothesis is false.   **11.** *Inverse:* If there is no snow on the ground, then this is not winter. *Contrapositive:* If this is not winter, then there is no snow on the ground.   **13.** *Inverse:* Rabbits must not have good eyesight if they don't eat carrots. *Contrapositive:* Rabbits must not eat carrots if they don't have good eyesight.
15. *Inverse:* Anything that is not a turkey is not a bird. *Contrapositive:* Anything that is not a bird is not a turkey.   **17.** *Inverse:* Anything that is not an airplane flies like a bird. *Contrapositive:* Anything that flies like a bird is not an airplane.

19. *Inverse*: A conditional is false if its hypothesis is true. *Contrapositive*: If a conditional is false, then its hypothesis is true.   **21.** True; it is the negation of a false statement.   **23.** False; the disjunction of two false statements is false.
25. True (whether or not it is 1984); the conclusion is true, so that the conditional is true, regardless of the truth value of the hypothesis.
27. False; the hypothesis is true, but the conclusion is false.   **29.** True; a false hypothesis makes the conditional true.   **31.** True; a false hypothesis makes the conditional true. In this case, the hypothesis is false because it is the conjunction of a true statement and a false statement.
33. Neither; since the hypothesis and the conclusion are negations of each other, they always have opposite truth values. This means that both "T → F" and "F → T" occur as cases. The first of these cases is *false*, and the second is *true*. Therefore, the statement is neither always true nor always false.

35.

p	q	$\sim(p \to q)$		$(\sim p)$ and q	
T	T	F	T	F	F
T	F	T	F	F	F
F	T	F	T	T	T
F	F	F	T	T	F

37. Something is an elephant and it does not have a trunk.   **39.** (Exercise 37) Either something is not an elephant or it has a trunk.

Index